Nuclear Science

Utilisation and Reliability of High Power Proton Accelerators (HPPA5)

Workshop Proceedings
Mol, Belgium
6-9 May 2007

© OECD 2008
NEA No. 6259

NUCLEAR ENERGY AGENCY
ORGANISATION FOR ECONOMIC CO-OPERATION AND DEVELOPMENT

ORGANISATION FOR ECONOMIC CO-OPERATION AND DEVELOPMENT

The OECD is a unique forum where the governments of 30 democracies work together to address the economic, social and environmental challenges of globalisation. The OECD is also at the forefront of efforts to understand and to help governments respond to new developments and concerns, such as corporate governance, the information economy and the challenges of an ageing population. The Organisation provides a setting where governments can compare policy experiences, seek answers to common problems, identify good practice and work to co-ordinate domestic and international policies.

The OECD member countries are: Australia, Austria, Belgium, Canada, the Czech Republic, Denmark, Finland, France, Germany, Greece, Hungary, Iceland, Ireland, Italy, Japan, Korea, Luxembourg, Mexico, the Netherlands, New Zealand, Norway, Poland, Portugal, the Slovak Republic, Spain, Sweden, Switzerland, Turkey, the United Kingdom and the United States. The Commission of the European Communities takes part in the work of the OECD.

OECD Publishing disseminates widely the results of the Organisation's statistics gathering and research on economic, social and environmental issues, as well as the conventions, guidelines and standards agreed by its members.

* * *

This work is published on the responsibility of the Secretary-General of the OECD. The opinions expressed and arguments employed herein do not necessarily reflect the official views of the Organisation or of the governments of its member countries.

NUCLEAR ENERGY AGENCY

The OECD Nuclear Energy Agency (NEA) was established on 1st February 1958 under the name of the OEEC European Nuclear Energy Agency. It received its present designation on 20th April 1972, when Japan became its first non-European full member. NEA membership today consists of 28 OECD member countries: Australia, Austria, Belgium, Canada, the Czech Republic, Denmark, Finland, France, Germany, Greece, Hungary, Iceland, Ireland, Italy, Japan, Luxembourg, Mexico, the Netherlands, Norway, Portugal, Republic of Korea, the Slovak Republic, Spain, Sweden, Switzerland, Turkey, the United Kingdom and the United States. The Commission of the European Communities also takes part in the work of the Agency.

The mission of the NEA is:

− to assist its member countries in maintaining and further developing, through international co-operation, the scientific, technological and legal bases required for a safe, environmentally friendly and economical use of nuclear energy for peaceful purposes, as well as
− to provide authoritative assessments and to forge common understandings on key issues, as input to government decisions on nuclear energy policy and to broader OECD policy analyses in areas such as energy and sustainable development.

Specific areas of competence of the NEA include safety and regulation of nuclear activities, radioactive waste management, radiological protection, nuclear science, economic and technical analyses of the nuclear fuel cycle, nuclear law and liability, and public information. The NEA Data Bank provides nuclear data and computer program services for participating countries.

In these and related tasks, the NEA works in close collaboration with the International Atomic Energy Agency in Vienna, with which it has a Co-operation Agreement, as well as with other international organisations in the nuclear field.

© OECD 2008

No reproduction, copy, transmission or translation of this publication may be made without written permission. Applications should be sent to OECD Publishing: *rights@oecd.org* or by fax (+33-1) 45 24 99 30. Permission to photocopy a portion of this work should be addressed to the Centre Français d'exploitation du droit de Copie (CFC), 20 rue des Grands-Augustins, 75006 Paris, France, fax (+33-1) 46 34 67 19, (*contact@cfcopies.com*) or (for US only) to Copyright Clearance Center (CCC), 222 Rosewood Drive Danvers, MA 01923, USA, fax +1 978 646 8600, *info@copyright.com*.

Cover credits: SCK•CEN, Belgium; Institut für Angewandte Physik (IAP) of Frankfurt University and ACCEL Instruments GmbH, Germany; Oak Ridge National Laboratory, USA.

FOREWORD

The accelerator-driven system (ADS) has been receiving considerable interest due to its possible use in radioactive waste transmutation. The performance of such hybrid systems largely depends on the reliability of the high power proton accelerators (HPPA), as well as the integration of the spallation targets and subcritical systems.

Since 1998, the OECD/NEA Nuclear Science Committee has been organising workshops on the Utilisation and Reliability of High Power Proton Accelerators to discuss technical issues, to provide state-of-the-art reports and to identify any specific needs for developing ADS. Previous workshops were held in Mito (Japan, 1998), in Cadarache (France, 1999), in Santa Fe, NM (USA, 2002) and in Daejon (Korea, 2004). The fifth workshop was held in Mol, Belgium on 6-9 May 2007 and organised in co-operation with the SCK•CEN.

A total of 41 papers were presented. The workshop included a special session on the MEGAPIE programme and five technical sessions: accelerator programmes and applications; accelerator reliability; spallation target development and coolant technology; subcritical system design and ADS simulations; and finally, ADS experiments and test facilities.

These proceedings include all the papers presented at the fifth workshop. The opinions expressed are those of the authors only and do not necessarily reflect the views of the NEA or its member countries.

Acknowledgements

The OECD Nuclear Energy Agency (NEA) gratefully acknowledges the SCK•CEN for hosting the 5th Workshop on Utilisation and Reliability of High Power Proton Accelerators.

TABLE OF CONTENTS

Foreword 3

Executive summary 9

Opening 17

    Chairs: P. D'hondt, A. Mueller

    F. Groeschel, S. Dementjev, H. Heyck,
    W. Leung, K. Thomson, W. Wagner, L. Zanini
    MEGAPIE – Irradiation Experience of the First Megawatt Liquid Metal
    Spallation Target 19

MEGAPIE Programme 27

    Chairs: J. Knebel, F. Groeschel

    L. Podofillini, V.N. Dang
    Safety Evaluation of the MEGAPIE Experimental Facility: Results
    and Insights from the Application of Probabilistic Safety Assessment 29

    K. Thomsen, P.A. Schmelzbach
    A Dedicated Beam Interrupt System for the Safe Operation of the
    MEGAPIE Liquid Metal Target 41

    M. Dierckx
    The Role of SCK•CEN in the MEGAPIE Project 51

Session I   **ACCELERATOR PROGRAMMES AND APPLICATIONS** 57

    Chairs: H. Klein, P. Pierini

    J-L. Biarrotte
    Status of the EUROTRANS R&D Activities for ADS Accelerator
    Development 59

    S. Barbanotti, N. Panzeri, P. Pierini, J-L. Biarrotte,
    S. Bousson, E. Rampnoux, H. Saugnac
    Design of a Test Cryomodule for the High-energy Section of the
    EUROTRANS Linac 69

    Y.S. Cho, J.Y. Kim, B.Y. Choi
    Proton Engineering Frontier Project 81

    A. Ponton, J-L. Biarrotte, S. Bousson, C. Joly, N. Gandolfo,
    J. Lesrel, L. Lukovac, F. Lutton, A.C. Mueller, G. Olry, E. Rampnoux
    Development of Superconducting Spoke Cavities for an ADS Linac 93

N. Panzeri, S. Barbanotti, A. Bosotti, P. Pierini
Status of the Preparation of the Elliptical Cavity System for the
EUROTRANS Cryomodule 103

H. Podlech, A. Bechtold, M. Busch, G. Clemente, H. Klein,
H. Liebermann, R. Tiede, U. Ratzinger, C. Zhang
Development of Room Temperature and Superconducting CH Structures
for High-power Applications 115

P.A. Schmelzbach
Upgrade of the PSI Proton Accelerator Facility to 1.8 MW 125

D. Schumann, J. Neuhausen
ERAWAST – A New Production Route for Exotic Long-lived Radionuclides 135

M. Tanigaki, Y. Mori, T. Uesugi, K. Okabe,
M. Aiba, Y. Ishi, K. Mishima, S. Shiroya, M. Inoue
150-MeV FFAG Accelerator Complex as a Neutron Production Driver
for ADS Study 141

V.G. Vaccaro, S. Albanese, A. Andreone, E. Di Gennaro,
M.R. Masullo, M. Panniello
Hybrid PBG Structure: A Challenging Cavity System for High-power,
High-intensity Accelerators 149

Session II ACCELERATOR RELIABILITY 159

Chairs: B.H. Choi and J-L. Biarrotte

J. Galambos, S. Henderson, A. Shishlo, Y. Zhang
Operational Experience of a Superconducting Cavity Fault Recovery
System at the Spallation Neutron Source 161

P. Pierini, L. Burgazzi
Reliability Studies for a Superconducting Driver for an ADS Linac 171

H. Takei, K. Tsujimoto, N. Ouchi, H. Oigawa,
M. Mizumoto, K. Furukawa, Y. Ogawa, Y. Yano
Comparison of Beam Trip Frequencies Between Estimation from
Current Experimental Data of Accelerators and Requirement from
ADS Transient Analyses 181

V.G. Vaccaro, A. D'Elia, C. Serpico, M. Fraldi, L. Nunziante
Cavity Detuning Due to Power Dissipation: A New Numerical Approach
Combining the Thermo-mechanical and the E-M Codes 195

Session III SPALLATION TARGET DEVELOPMENT AND
COOLANT TECHNOLOGY 203

Chairs: H. Oigawa, Y. Gohar

J. Heyse, H. Aït Abderrahim, T. Aoust, M. Dierckx, K. Rosseel,
P. Schuurmans, K. Van Tichelen, A. Guertin, J-M. Buhour,
A. Cadiou, M. Abs, B. Nactergal, D. Vandeplassche
WEBEXPIR: Windowless Target Electron Beam Experimental Irradiation 205

S. Heusdains
RELAP Model of the MYRRHA draft 2 Spallation Loop 213

J. Neuhausen, F. von Rohr, S. Horn, S. Lüthi, D. Schumann
Polonium Behaviour in Eutectic Lead-bismuth Alloy 227

V. Sobolev
A Brief Review of Thermodynamic Properties and Equation of State
of Heavy Liquid Metals 233

*P. Schuurmans, A. Guertin, J.M. Buhour, A. Cadiou, M. Dierckx,
J. Heyse, K. Rosseel, K. Van Tichelen, R. Stieglitz, D. Coors,
L. Mansani, F. Roelofs, H. Aït Abderrahim*
Design and R&D Support of the XT-ADS Spallation Target 245

*K. Abe, T. Iwasaki, S. Gunji, C-H. Pyeon,
H. Shiga, M. Aiba, H. Yashima, T. Sanai*
Study of Measurement Method of High-energy Neutrons for ADS 253

Th. Aoust, J. Wagemans, J. Cugnon, A. Boudard, S. Leray, Y. Yariv
Application of the INCL+ABLA Reaction Model to the Study of the
Evolution of Spallation Targets 261

*A. Herrera-Martínez, Y. Kadi, M. Ashrafi-Nik,
K. Samec, J. Freibergs, E. Platacis*
Engineering Design of the EURISOL Multi-MW Spallation Target 271

*D. Ridikas, A. Barzakh, V. Blideanu, J-C. David,
D. Doré, X. Ledoux, F. Moroz, V. Panteleev, A. Plukis,
R. Plukiene, A. Prévost, O. Shcherbakov, A. Vorobyev*
Delayed Neutron Yields and Time Spectra from 1-GeV Protons
Interacting with ^{nat}Fe, ^{nat}Pb and ^{209}Bi Targets 283

Session IV **SUBCRITICAL SYSTEM DESIGN AND ADS SIMULATIONS** 293

Chairs: P. Baeten, S. Monti

P. Richard, G. Rimpault, J.F. Pignatel, B. Giraud, M. Schikorr, R. Stainsby
Pre-conceptual Design of a Helium-cooled ADS: He-EFIT 295

C. Artioli
A-BAQUS, a Multi-entry Graph Assisting the Neutronic Design
of an ADS. Case Study: EFIT 309

D. Naberezhnev, Y. Gohar, H. Belch, J. Duo, I. Bolshinsky
Comparative Analysis of Neutron Sources Produced by Low-energy
Electrons and Deuterons for Driving Subcritical Assemblies 321

M. Suzuki, T. Iwasaki, T. Sato
Improvement of Dynamics Calculation Code DSE for
Accelerator-driven System 337

A. Khalid Rivai, M. Takahashi
Corrosion Resistance of Refractory Metals and Ceramics in
Lead-bismuth at 700°C 347

*G. Van den Eynde, V. Sobolev, E. Malambu, D. Maes, D. Lamberts,
H. Aït Abderrahim, L. Mansani, B. Giraud, A. Hogenbirk, P. Vaz,
Y. Romanets, M.C. Vincente, P. Coddington, K. Mikityuk, D. Struwe,
M. Schikorr, G. Rimpault, C. Artioli*
Neutronic Design of the XT-ADS Core with In-pile Sections 357

P. Xia, Y. Shi, Z. Zhao, Y. Li, Q. Zhu, J. Li,
W. Zhang, J. Cao, Y. Quan, H. Luo, X. Wu
The Evaluation of Preliminary Extrapolation Experimental Results
of the Chinese ADS Subcritical Experimental Assembly VENUS-1 367

Session V **ADS EXPERIMENTS AND TEST FACILITIES** 375

Chairs: H. Aït Abderrahim, M. Tanigaki

P. Baeten, H. Aït Abderrahim, G. Vittiglio,
F. Vermeersch, G. Bergmans, B. Verboomen, D. Maes
The GUINEVERE Project at the VENUS Facility 377

H. Oigawa, K. Nishihara, T. Sasa, K. Tsujimoto, T. Sugawara,
K. Iwanaga, K. Kikuchi, Y. Kurata, H. Takei, S. Saito, D. Hamaguchi,
H. Obayashi, Y. Tazawa, M. Tezuka, N. Ouchi
Research and Development Programme on ADS in JAEA 387

J.M. De Conto, S. Albrand, M. Baylac, J.L. Belmont, A. Billebaud,
J. Bouvier, A. Fontenille, E. Froidefond, M. Fruneau, A. Garrigue,
M. Guisset, D. Marchand, R. Micoud, J.C. Ravel, C. Vescovi
The GENEPI Neutron Sources at Grenoble: Prospectives for the
GUINEVERE Programme 401

M. Carta, G.R. Imel, C.C. Jammes, S. Monti, R. Rosa
Reactivity Measurements in Subcritical Core: RACE-T Experimental Activities 413

E.J. Pitcher
The Materials Test Station: A Fast Spectrum Irradiation Facility 427

C.H. Pyeon, T. Misawa, H. Unesaki, K. Mishima, S. Shiroya
Benchmark Experiments of Accelerator-driven Systems (ADS) in
Kyoto University Critical Assembly (KUCA) 435

Workshop conclusions 445

List of participants 447

EXECUTIVE SUMMARY

Reducing the radiotoxicity of nuclear wastes is one of most pressing requirements for the future of the nuclear fuel cycle. One viable option is the subcritical accelerator-driven system (ADS), since it is capable of transmuting the long-lived minor actinides to short-lived or stable elements, with an internal neutron source driven by high-power proton accelerators (HPPA). The improvement of beam reliability as well as lead-bismuth technology, windowless spallation target and subcritical core design are identified as challenging technologies.

To promote the development of such hybrid nuclear systems and to provide a forum for discussion, the OECD/NEA has held a series of international workshops on the utilisation and reliability of high-power proton accelerators since 1998. Four workshops have been organised and hosted by the following major national laboratories: JAERI (former JAEA) in 1998 (Mito, Japan), CEA in 1999 (Cadarache, France), LANL in 2002 (Santa Fe, USA) and KAERI in 2004 (Daejon, Korea). The fifth workshop, hosted by SCK•CEN, was held on 5-9 May 2007 in Mol, Belgium.

The workshop opened with welcome addresses from Hamid Aït Abderdahim (Belgium), SCK•CEN, Director of the Advanced Nuclear Systems Institute and Pierre J. D'hondt (Belgium), the vice chair of the NEA Nuclear Science Committee. Two plenary speeches were given, chaired by P. D'hondt and A. Mueller (France):

- *Y. Jongen* (Belgium) presented accelerator applications used in cancer therapy for their ability to produce radioactive isotopes. Medical radioactive isotopes such as ^{99}Tc have been produced in the reactors. Recently, some very important isotopes (^{201}Tl, ^{123}I) have been produced by high-energy cyclotrons, and it is now possible to distribute them world wide with powerful cyclotrons owned by radiopharmaceutical companies. The application of brachytherapy, proton therapy and carbon therapy were also presented.

- *F. Groeschel* (Switzerland) reported on the MEGAPIE irradiation experience of the first megawatt liquid metal spallation target SINQ. MEGAPIE is an experiment that is being carried out in the SINQ target location at the Paul Scherrer Institute (PSI), and aims to demonstrate the safe operation of a liquid metal spallation target at a beam power in the region of 1 MW. Target material is Pb-Bi eutectic mixture. He concluded that the MEGAPIE target operation was successful and the neutronic performance yielded the expected flux increase. The thermal-hydraulic behaviour was stable and beam trips could be well-controlled.

One special session and five technical sessions were organised. Each session was comprised of invited speeches and presentations of contributed papers. The contents of the different sessions are described below, together with a brief explanation of the papers' content.

Special session on MEGAPIE programme (*Chair: F. Groeschel*)

Three contributed papers were presented during this session.

- *L. Podofillini* (Switzerland) gave an overview of the main issues for the safety evaluation of the MEGAPIE experimental facility. The operational characteristics of safety procedures of the MEGAPIE were reported. He demonstrated that probability safety assessment can provide insights into safety and identify measures for informing designers of the safety of experimental installations. He concluded that the lack of data could be covered by studies on uncertainties. Identifying weaknesses by order of priority, rather than by degree of risk, is necessary.

- *K. Thomsen* (Switzerland) introduced MEGAPIE safety systems which help concentrate the beam so that it reaches the SINQ target. Three independent systems are installed: a dedicated current monitoring system, a beam collimating slit and a novel beam diagnostic device named VIMOS. All these systems have to meet basic requirements to ensure the beam's reliability.

- *M. Dierckx* (Belgium) presented the role of the SCK•CEN in the MEGAPIE project. SCK•CEN is planning to build an experimental accelerator-driven subcritical system (XT-ADS) using the liquid metal spallation target. He introduced the successful completion of the MEGAPIE irradiation phase, which will ease future ADS design. He concluded that the PIE of MEGAPIE is important for developing XT-ADS, as it will provide crucial information on the behaviour of XT-ADS structural materials.

Session I: Accelerator programmes and applications (*Chairs: H. Klein, P. Pierini*)

Two invited speeches and five contributed papers were presented at this session.

- *J-L. Biarrotte* (France) was invited to present the status of the EUROTRANS R&D activities for ADS accelerator development. He described the purpose, strategy and systems features and characteristics of EUROTRANS and ADS accelerator reference solution, based on a reliability-oriented linear superconducting accelerator and summarised the ongoing R&D on prototypical accelerator components, which is being carried out within the 6th Framework Programme of the EC's EUROTRANS project.

- *S. Barbanotti* (Italy) introduced the technical features and design characteristics of the test cryomodule for the high-energy section of the EUROTRANS linear accelerator (linac), which is equipped with elliptical superconducting niobium cavities. She also said that the module design would be completed soon and pointed out that the time line for finalising the assembly would be determined in the autumn of 2008.

- *Y.S. Cho* (Korea) gave an overview of the Korean project on proton accelerator utilisation (PEFP). Korea is now developing technology that will enhance the reliability of the linac by installing 20 and 100 MeV machines. The high-power proton beam facility will be constructed in 2009. The beam will be available to users as of 2010, and will be extended to the 100 MeV linac.

- *A. Ponton* (France) reported on the current status of developing the superconducting spoke cavities for ADS linac in France. Redundancy elimination and fault tolerance design are important for linac construction in order to enhance reliability. The result of the development of the vertical cryostat for the digital LLRF and a new horizontal small-scale cryostat was reported. He concluded that the spoke components exist in order to reach good reliability, which in turn fulfils the XADS linac requirements.

- *N. Panzeri* (Italy) presented the status of the preparation of the elliptical cavity system for the EUROTRANS cryomodule. The tuner and helium tank have been designed to fulfil the cavity requirements in terms of stiffness and tuning range. The reliability is based on tests performed on the TTF blade tuner, which share the same design. The tuner and helium tank are ready, and the low power test is scheduled for the beginning of 2008, after cavity preparation and assembly in the HT.

- *H. Podlech* (Germany) addressed the development status of the room temperature and superconducting CH structures for high power linacs. The CH structure is a highly efficient DTL structure that is suitable for room temperature and superconducting operations. Currently, the prototypes have been developed and tested successfully, under the projects of FAIR p-Linac, EUROTRANS and IFMIF.

- The second invited speech in this session was delivered by *P. Schmelzbach* (Switzerland). He analysed the potential for improvements from the ion source to the spallation target and gave an overview of the work in progress on the proton accelerator facility of the PSI including upgrading the 1.8 MW and 3 mA beams. They will also upgrade Al cavities to Cu cavities. New bunchers and additional cavities in the injector cyclotron will be installed. The conclusion indicated that space charge effects will play a larger role.

- *D. Schumann* (Switzerland) presented the ERAWAST project: exotic radionuclides from accelerator waste for science and technology. The main goal of this project is to reduce the burden of radioactive waste from the accelerators by applying exotic long-life isotopes to the scientific or medical industry. Currently, Cu and C samples are available and works on Pb targets are ongoing. More isotopes are studied and separated for use in medical applications, for instance. Plans for the future were also introduced.

- *M. Tanigaki* (Japan) reported on the development status of the Japanese 150 MeV FFAG accelerator complex for utilisation of ADS. Kyoto University launched the KART (Kumatori Accelerator-driven Reactor Test) project in 2002 to demonstrate the feasibility of ADS. The test operation for the FFAG accelerator is in its final stage. The current status of the project and future plans related to it were presented.

- *V. Vaccaro* (Italy) presented the simulations and measurement status of the hybrid PBG cavities based on the triangular lattice. The hybrid structure showed expected mono-modal behaviour and a good frequency agreement between the simulated and measured scattering parameters. TM_{02} hybrid resonator was also introduced, which is expected to be more confined in the central region. Simulations were performed using a free software package or commercial 3-D codes.

Session II: Accelerator reliability (*Chairs: D. Van de Plassche and J-L. Biarrotte*)

Two invited speeches and two contributed papers were presented at this session:

- *J. Galambos* (USA) gave an invited speech that reported on the experimental results of the superconducting cavity fault recovery system of the US spallation neutron source (SNS). He provided an overview of the SNS facility, normal and superconducting linac, RF systems and beam test results. The development and test status of the fault recovery scheme for superconducting cavity failure followed. To date, its primary application has been for quick recovery from events involving multiple cavities that run successfully at high energies.

- *P. Pierini* (Italy) presented an invited speech that focused on reliability studies for the superconducting driver for an ADS linac. After HPPA4, the reliability study and development of the RBD models were extended to the sensitivity study. A large degree of fault tolerance is provided in order to promote fault detection, isolation and correction procedures. The minimal tweaking of the MTBF simple model for an accelerator system can be altered.

- *H. Takei* (Japan) introduced the conceptual design of developing ADS linac in Japan. Beam trip frequency on the experimental data estimation is presented. The acceptable frequency of beam trips ranges from 50 to 25 000 times per year. The beam trip frequency (t < 5 s) is within the acceptable value at the current level of accelerator technology. Under the shorter VSWR assumption, the beam trip frequency (t < 10 s) is within the acceptable level, while that exceeding 10 seconds should be reduced by about 1/30 to satisfy thermal stress conditions.

- *V. Vaccaro* (Italy) presented the cavity detuning method by finite element numerical codes. Temperature rises in the accelerator cavity wall may be hazardous to detuning, and accurate distribution is needed to estimate mechanical deformation of the structure. Using the balance of electric and magnetic energy at resonance allows good results. The application of this new approach to PALME linac cavities and a comparison with other techniques were also reported.

Session III: Spallation target development and coolant technology (Chairs: *H. Oigawa, Y. Gohar*)

One invited speech and eight contributed papers were presented at this session.

- In his invited speech, *J. Hayse* (Belgium) gave an overview of the WEBExpIr: the windowless electron beam experimental irradiation. WEBExpIr was initially developed to study the interaction of a proton beam with a liquid lead-bismuth eutectic (LBE) free surface. After various tests at beam currents up to 10 mA (= 40 × XT-ADS), neither shockwave effects nor significant droplet ejection effects and significant evaporation enhancement were detected. It was concluded that WEBExpIr free surface flow was not disturbed by the interaction with the electron beam and that vacuum conditions stay well within the design specifications.

- *S. Heusdains* (Belgium) presented RELAP code implications on the LBE loop at steady and transient states. RELAP version mod 3.2 β was used. The following challenges, which placed restrictions on the code, have been successfully dealt with: non-condensable pressure control above the free surfaces; flow and free surface target levels and stability; and rapid tower increase from 0 to 1.43 MW.

- The behaviour of polonium in eutectic lead-bismuth alloy was presented by *J. Neuhausen* (Switzerland). For LBE, significant amounts of high radiotoxic polonium isotopes are formed due to spallation, fission and neutron capture. Investigations of polonium in solidified LBE as well as its release from the eutectic under vacuum conditions were reported.

- *V. Sobolev* (Belgium) gave an overview of the thermodynamic properties and the equation of state associated with heavy liquid metals of interest: Pb, Bi, Hg and Pb-Bi eutectic. A brief history including the range of temperature and availabilities was also presented. A modified Redlich-Kwong equation of state was introduced. This can be used for temperature dependence of main thermodynamic properties of Hg, Bi, Pb and Pb-Bi eutectic.

- *P. Schuurmans* (Belgium) presented the conceptual design of XT-ADS spallation target loop. MYRRHA draft 2 provides no target window, compact vertical confluent flow for target formation; off-axis servicing and two pump uncoupled LBE pumping systems. Recently the experimental facility MYRRHA draft 2 has modified its boundary conditions. It has also detached flow spallation target nozzle, active pump-in feeder line and vacuum system.

- *K. Abe* (Japan) presented development status of the measurement method of high-energy neutrons for the ADS. The high-energy neutrons were measured by the bismuth activation method and followed by an experimental verification. A measurement system for the entire range of neutron energy spectrum existing in the ADS was developed. Future work will cover improving accuracy.

- *Th. Aoust* (Belgium) reported the results of a study on the evolution of spallation targets by extending the Liège intra-nuclear cascade model INCL coupled to ABLA evaporation-fission reaction model. This model is found to be suitable for proton-induced spallation reactions on thin targets in the 200 MeV-2 GeV range. The study covered comparison and validation of models.

- *A. Herrera-Martínez* (Switzerland) gave an overview of the European Isotope Separation On-line Radioactive Ion Beam Project (EURISOL) design study devoted to the coupled neutronics of the mercury proton-to-neutron converter and the fission targets. Challenging technical issues as well as future plans were also presented.

- *D. Ridikas* (France) presented the study results on the delayed neutron yields and time spectra from 1 GeV protons interacting with the ^{nat}Pb, ^{209}Bi and ^{nat}Fe targets of variable thicknesses from 5 cm to 55 cm. Both experimental and numerical studies by MCNPX and PHITS codes were performed. Significant dependence on the choice of the intra-nuclear cascade and fission-evaporation models were shown. By determining cross-sections of interesting isotopes, results permitted examination of both fission and fragmentation.

Session IV: Subcritical systems design and ADS simulations (*Chairs: P. Baeten and S. Monti*)

One invited speech and six contributed papers were presented at this session.

- *P. Richard* (France) presented an invited speech that reported on the work status of the pre-conceptual design of helium-cooled ADS: He-EFIT. A summary of the first two years of the project as well as facility features such as proton beam characteristics; core power, inlet/outlet temperatures, etc. were presented. A discussion on the He-EFIT core as well as spallation module design followed.

- A-BAQUS, a multi-entry graph assisting the neutronic design of an ADS case study on EFIT, was presented by *C. Artioli* (Italy). A-BAQUS is developed to facilitate the design of EFIT's reactor core. A-BAQUS shows the performance (MA burning capability and the conjugated Pu breeding), burn-up reactivity swing, power of the core and proton current as well as range during the cycle. Features and characterisations of the A-BAQUS and future plans were presented.

- *Y. Gohar* (USA) presented two different external neutron source options for driving subcritical assembly: by low-energy particles, either electrons (<200 MeV) or deuterons (<100 MeV). The comparative analysis of both options was performed. It was concluded that the energy range of 150 to 200 MeV is preferable in the production of neutron sources. An assessment of uranium targets, which produces the highest neutron yield per electron, was also reported.

- *R. Sheffield* (USA) filled in for an absent speaker. He gave an overview on the US LANL's ADS programme. The LANL is pursuing a step-by-step realisation of a 4 MW class accelerator research complex, clustering existing facilities and new installations in order to obtain 0.8/3/20 GeV protons (the full paper is not provided in the proceedings).

- *M. Suzuki* (Japan) reported on the improvement of a dynamic calculation code for subcritical systems with an external neutron source (DSE) for ADS. DSE consists of two sub-codes: diffusion code DSE-C and the Monte Carlo method based transport code DSE-M, which were only presented at the workshop. Characterisations and features of the DSE-M as well as applications were presented and followed by a summary and plans for the future.

- *M. Takahashi* (Japan) presented corrosion resistance tests of various metals, which were probably used in the ADS, to investigate compatibility in stirred lead-bismuth eutectic under high temperatures (700°C). The results showed that while Al-Fe coated steel, W, Mo, SiC and Ti_3SiC_2 exhibited good corrosion resistance, Nb did not. SiC/SiC composite showed that LBE penetrated into the matrix due to the material's high porosity and a thin crack layer appeared in LBE at 700°C.

- *G. Van den Eynde* (Belgium) reported on the work status of developing the XT-ADS core and neutronic design with in-pile positions. The XT-ADS core is composed of 72 fuel assemblies surrounding three spallation target modules. Eight positions were identified as so-called in-pile sections that can hold experimental facilities while remaining accessible during the operation. A neutron physics analysis has been performed to quantify the core's characteristics.

- *P. Xia* (China) gave an overview of the developing status of Chinese ADS subcritical assembly, VENUS-1. Since 2000, the Chinese government has been funding basic research on ADS physics and related technologies. The structure, design goals and results of calculations and experiments as well as future plans were presented.

Session V: ADS experiments and test facilities (*Chairs: H. Aït Abderdahim, M. Takahashi*)

Two invited speeches and four contributed papers made up this session.

- *P. Baeten* (Belgium) presented an invited speech on the GUINEVERE project at the VENUS facility. The GUINEVERE project is especially devoted to experiments on the accelerator, target and subcritical core coupling to investigate operation characteristics of the VENUS reactor. In conclusion, it is expected that the GUINEVERE project will provide a unique experiment with a continuous beam coupled to a fast (sub) critical assembly, allowing full investigation of the methodology of reactivity monitoring for XT-ADS and EFIT.

- *H. Oigawa* (Japan), the second invited speaker gave an overview of the research and development programme on ADS in JAEA. JAEA is currently pursuing an ADS of lead-bismuth eutectic (LBE) cooled, tank-type subcritical reactor with the thermal power of

800 MWth driven by a 30-MW superconducting linac. Two aspects are being conducted: one on the development of large scale ADS and one on transmutation experiments under J-PARC. He concluded by emphasising international co-operation and the use of a common roadmap.

- *J-M. de Conto* (France) presented the GENEPI neutron sources at Grenoble, France, and the perspectives for the GUINEVERE project at VENUS. GENEPI is a deuteron, high-intensity, electrostatic accelerator producing neutrons on a tritium target located in the reactor core. After installing two GENEPI accelerators, a new machine (GENEPI-3C) is now under preparation. The presentation covered characteristics and features of the GENEPI series.

- *M. Carta* (Italy) presented reactivity measurements in a subcritical core, the RACE-T experiment, a part of experiment on the coupling of an accelerator (ECATS) of EUROTRANS. The RACE-T project comprises fission rate measurement, the study of different subcritical configurations and the development of devoted instruments and acquisition systems. The experiment is also participating in the IAEA benchmark study.

- *E. Pitcher* (USA) gave an overview of the Material Test Station (MTS): a fast spectrum irradiation facility. The MTS provides a fast neutron spectrum that meets programmatic requirements for transmutation fuel testing by using Pb-Bi liquid metal coolant that meets experimenters' needs for elevated irradiation temperatures. MTS will produce irradiation data results for the development of transmutation fuels starting in 2012.

- *C-H. Pyeon* (Japan) reported on the benchmark study of ADS in Kyoto University, Japan. The benchmark programme study on both 14 MeV neutrons generated from pulsed neutron generator and 150 MeV protons generated from the fixed field alternating gradient (FFAG) accelerator. General presentation and results followed. The presentation also covers Japanese collaboration on developing the ADS as well as international co-operation.

Summary session

Each session chair summarised their sessions. H. Aït Abderdahim gave a presentation on the importance of leaving theoretical assessment and entering experimental demonstration. He emphasised achievements and milestones as well as the future of ADS development.

During the discussion, it was concluded that the accelerator's reliability has been significantly improved during the last decade and lead-bismuth eutectic technology has demonstrated its reliability. Moreover, compared to previous editions of the workshop, a great deal of interest has been generated not only in accelerator development but also in ADS applications, in subcritical core design and in spallation target development, etc. From this interest, one can deduce that improvement of accelerator reliability leads to a greater interest in ADS development. It was also agreed that the study of ADS safety and coolant thermal-hydraulics are more needed. There was a consensus among participants that future workshops could be restructured to include many disciplines in the field of ADS.

The workshop was closed by P. D'hondt, the chair of the workshop.

OPENING

Chairs: P. D'hondt, A. Mueller

MEGAPIE – IRRADIATION EXPERIENCE OF THE FIRST MEGAWATT LIQUID METAL SPALLATION TARGET

F. Groeschel, S. Dementjev, H. Heyck, W. Leung, K. Thomson, W. Wagner, L. Zanini
On behalf of the MEGAPIE Initiative
Paul Scherrer Institut, Switzerland

Abstract

A key experiment in the roadmap for accelerator-driven systems [1] is the demonstration of the feasibility, licensibility and reliable operation of a high-power spallation target. The MEGAwatt PIlot Experiment (MEGAPIE) was initiated in 1999 in order to design and build a liquid lead-bismuth spallation target, then to operate it at the Swiss spallation neutron facility SINQ at Paul Scherrer Institute (PSI) [2]. The project is supported by an international group of research institutions.

After completion of the manufacturing in mid-2005, the target and the ancillary systems were first integrated into a test bench to verify the proper functioning and design predictions and to train the handling operations like filling and draining. The test bench was able to deliver about 200 kW heating power, which allowed to simulate the principal thermal-hydraulic processes. The results confirmed the proper functioning and readiness for beam.

The target system was integrated and tested in the SINQ facility at the beginning of 2006. This process, as well as the upgrading of some of its supporting systems, took several months. The license for active operation was obtained in August.

First protons were delivered on 14 August 2006 on the target. During a short start-up procedure of about a week, the proton beam was ramped up in steps to full power and the response of the target and its safety systems was verified. With all systems working satisfactorily and according to predictions, standard (24 hour) operation delivering neutrons for the regular users of the facility was started, first in manned operation, then unmanned. Until its scheduled stop on 21 December 2006 an accumulated proton current of 2.8 Ah has been achieved and more than 6 000 beam trips have been experienced.

Introduction

A key experiment in the roadmap for accelerator-driven systems [1] is the demonstration of the feasibility, licensibility and reliable operation of a high-power spallation target. The MEGAwatt PIlot Experiment (MEGAPIE) was initiated in 1999 in order to design and build a liquid lead-bismuth spallation target, then to operate it into the Swiss spallation neutron facility SINQ at Paul Scherrer Institute (PSI) [2]. The project is supported by an international group of research institutions: PSI (Switzerland), CEA (France), FZK (Germany), CNRS (France), ENEA (Italy), SCK•CEN (Belgium), DOE (USA), JAEA (Japan), KAERI (Korea) and the European Commission. The objective was to reach at least three months of operation at full power.

The design of the target and of the main ancillary systems has previously been described in Ref. [3]. Manufacturing took place in different European countries and the components were tested individually in dedicated tests to prove their compliance with design. Special attention was paid to the pump system [4], which was tested before integration into the target. Whereas the electromagnetic pumps were in full compliance with the specification, the performance of the electromagnetic flow meters was not satisfactory. Improvement of electronics and the algorithm to compensate parasitic, in particular thermal, effects were expected to increase sensitivity and stability.

Integral test and main outcome

In the second half of 2005, the target and the main ancillary systems for heat removal, cover gas handling and LBE filling and draining were first integrated in a test stand at PSI to allow testing of the target functions, the temperature control and training of the main operations. Figure 1 shows the tests stand with the main systems. The operations consisted of preheating, filling and draining of the target with LBE and simulating gas sampling. Dedicated tests [5] were carried out to:

- Assess the performance of the electromagnetic pumps. Both performed according to design.

- Recalibrate the electromagnetic flow meters. The accuracy and stability of the signals was insufficient for the planned flow control. Further improvements to compensate noise effects were necessary. In parallel, the development of a flow analysis technique based on the temperature distribution in the target was initiated.

- Determine the heat removal capacity of the system using a 200-kW electric heater [6]. Measurements based on thermal balance indicated a 25% increase in heat transfer capacity of the target heat exchanger compared to predictions. The reason was found in the inadequate modelling of the secondary (oil) channel of the pins. As a consequence, the predicted oil inlet and outlet temperatures at full beam power operation had to be raised from 138 and 175°C to 182 and 217°C.

- Assess and optimise the temperature control of the target during steady state and transients simulating beam trips and interrupts. Scaling laws were applied to extrapolate to the expected 580 kW power during beam operation. The temperature control proved adequate, although the strong non-linearity of the three-way valve yielded rather large fluctuations at low power.

- Visualise the effectiveness of the bypass jet to cool the beam entrance window using infrared technology [7]. An effective bypass jet washing across the centre of the beam window was observed under nominal flow conditions; i.e. 4 l/s main flow and 0.3 l/s for the bypass flow. The flow pattern was stable and rather insensitive to small changes in the flow ratio.

**Figure 1. Target integrated in MEGAPIE test stand
and connected to heat removal and fill and drain system**

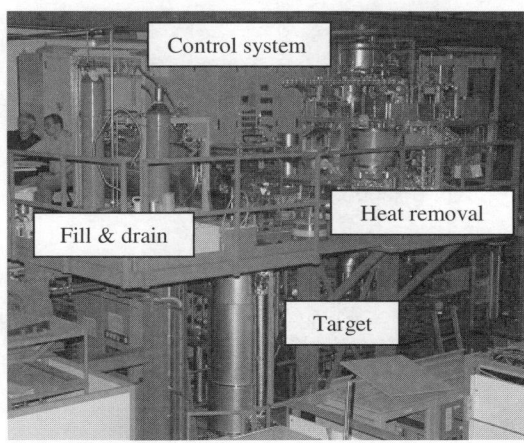

The test phase lasted over four months and the target was operated for 133 hours with LBE. The failure of the three-way valve of the oil loop due to seizure of the valve stem was observed. The thermal expansion of the pipes had not been sufficiently compensated and caused a bending moment on the valve stem. Fortunately, the valve could be removed, repaired and re-installed.

Target assembly and integration in SINQ

The target had been tested without its second enclosure and with only temporary cable connections. It therefore had to be completed. Integration of the target and the ancillary systems in the SINQ facility was done in the first half of 2006. The integration was made without major changes to the existing systems in order to allow a rapid change back to the standard, heavy-water-cooled solid Pb-target in case of premature failure of the liquid metal target.

Some of the support systems had to be upgraded in order to comply with the increased hazard of the liquid metal target under severe accidents; i.e. break of the target:

- The ventilation system had to be strengthened by adding additional active carbon filters to the existing HEPA filters in order to retain gaseous volatiles in case of an accidental release. Earthquake-proof shut-off valves were installed in the in- and outlet ducts to seal the penetrations in case of a heavy earthquake, in which functioning of the ventilation system was no longer expected. In addition, an autonomous, earthquake-proof filter unit was installed in order to assure a controlled evacuation of contaminated atmosphere from the target block. Figure 2 shows this unit, which consists of a filter unit, a ventilator unit, a battery set and a diesel-driven power generator. The start of the unit is triggered by an earthquake switch.

- The use of Diphyl THT coolant required an upgrade in fire prevention. We opted for reduction of the oxygen content in the critical compartments by injection of nitrogen to below 13%, the oil flash point. The nitrogen was generated from compressed air by a membrane separator.

- The control system had to be integrated into the existing system of the SINQ facility and the beam monitoring systems and the beam trip logic had to be adapted to the MEGAPIE specific requirements. In addition to the trip command triggered by the control system, a separate hard-wired set of signals acted directly and independently on the fast beam shut-off system as well as on the slow beam shut-off system.

Figure 2. Autonomous, earthquake-proof filter unit

Readiness for beam and licensing approval

The licensing process of the experiment was undertaken in 2004 by the competent authorities, but four milestones had to be passed to demonstrate the proper functioning of the main systems and 52 additional requirements had to be cleared to obtain permission for irradiation. A review technique was adopted to assess the readiness of the target system for the final steps, comprising the readiness for filling of the target with LBE and the beam readiness. The licensing authority verified the status of the target system with corresponding inspections, in which the proper functioning was demonstrated.

On 28 July 2006 the target was filled with 926 kg of LBE and operated for 17 days off-beam in hot standby mode, i.e. maintaining a LBE temperature of 230°C. During this time, the filling vessel was removed and the facility was closed and prepared for irradiation. After several tests including simulations of loss of pump and loss of heat sink and their effects on the cooling water loops, the

Beam start

The first protons on MEGAPIE were received on 14 August 2006. The beam current was 45 µA, corresponding to a beam power of 25 kW. This phase served to check the beam alignment and the response of the surveillance systems. For about two hours a stable beam was sent to the target. The thermocouples in the target recorded the small heat input and the temperature control activated the heat removal system. The first neutrons were detected at the instruments. The analysis of the data recorded yielded that all systems responded as expected.

Therefore, the second phase of the start-up procedure was carried out, in which the beam power was stepwise increased to 150 kW (250 µA proton current). The goal of this phase was to verify the response of the heat removal system at power conditions comparable to those used when operated out of beam at the test stand in the autumn of 2005. During this phase, a "hot" test of the beam shut-off system was also carried out and gas samples were taken from the target expansion volume.

The third and final phase of the start-up procedure was successfully accomplished on 17 August, when the power was stepwise increased to 700 kW (1 200 µA proton current). Figure 3 shows the beam history during the ramp-up phase. At each power level the beam was interrupted after some 10 minutes of stable proton beam to verify the predicted temperature transients in the target.

Figure 3. Beam history during ramp-up to full power at day three of the start-up of MEGAPIE

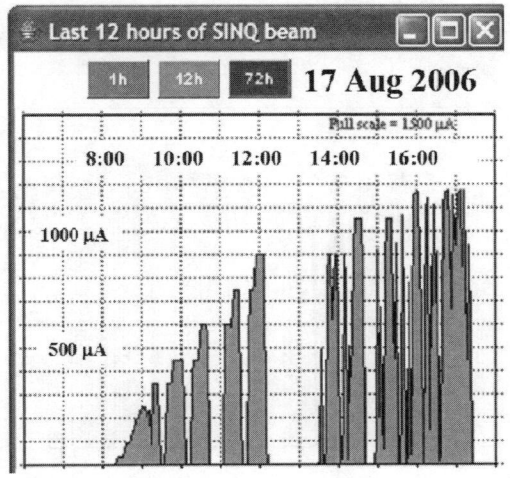

Normal operations

As all systems behaved as predicted, continuous operation was started on 21 August. During the first four days, 24-hour surveillance of the operation was maintained. After 24 August, the target was operated unmanned, which is the standard procedure for the SINQ facility. The operation parameters were only surveyed remotely via internet. Figure 4 shows the start of the normal operation phase.

Figure 4. Beam current on target during first 12 hours of normal operation. Beam current is 1 000 μA

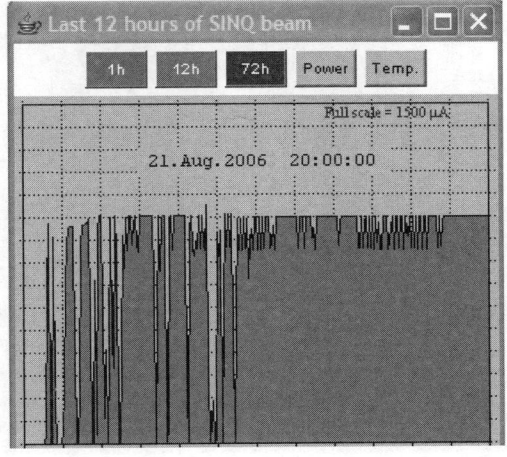

Normal operation continued until 21 December 2006. The accumulated charge was 2.86 Ah. The proton current peaked at 1 400 μA, corresponding to beam power of 805 kW. During the operation period, we registered 5 800 beam trips of less than 1 minute duration and 570 interrupts of longer durations. Among these interrupts were scheduled weekly beam shutdowns for servicing purposes lasting between 8 and 36 hours. Figure 5 shows the evolution of the beam current over the whole irradiation period.

Figure 5. Beam current on target during the irradiation period

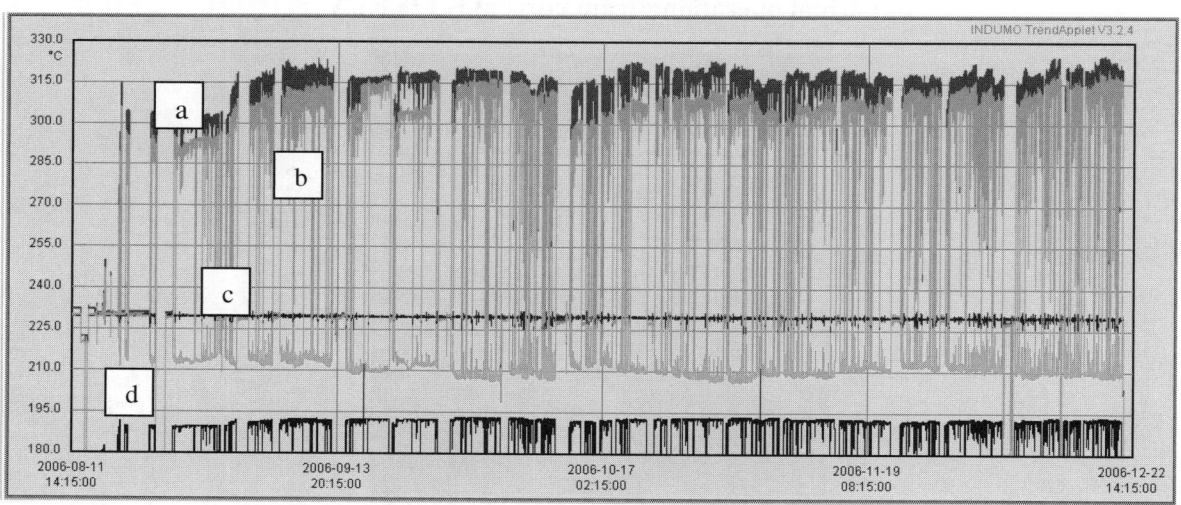

The temperature control and the heat removal from the target functioned as predicted. Figure 6 shows the evolution of the main target temperatures over the irradiation period. The highest LBE temperatures were recorded just above the spallation zone (a). During the upflow, some heat was dissipated mainly to the annular downflow through the guide tube. The target heat exchanger (THX) temperature (b) was therefore slightly lower. The THX outlet temperature (c) was maintained at 230°C. The oil outlet temperature (d) was about 210°C as predicted after the integral test.

Figure 6. Main target temperatures and beam current (black) during the irradiation period

The target temperature control could well manage power transients caused by the beam trips and irradiation interrupts as shown in Figure 7. The drop in temperature above the spallation zone during a trip was between 100 and 130°C. The transient was reduced to about 70°C at the THX inlet due to the thermal inertia of the target. At the THX outlet, the temperature swing was less than ±10°C.

Figure 7. LBE temperature variations in the target during power transients caused by beam trips and interrupts

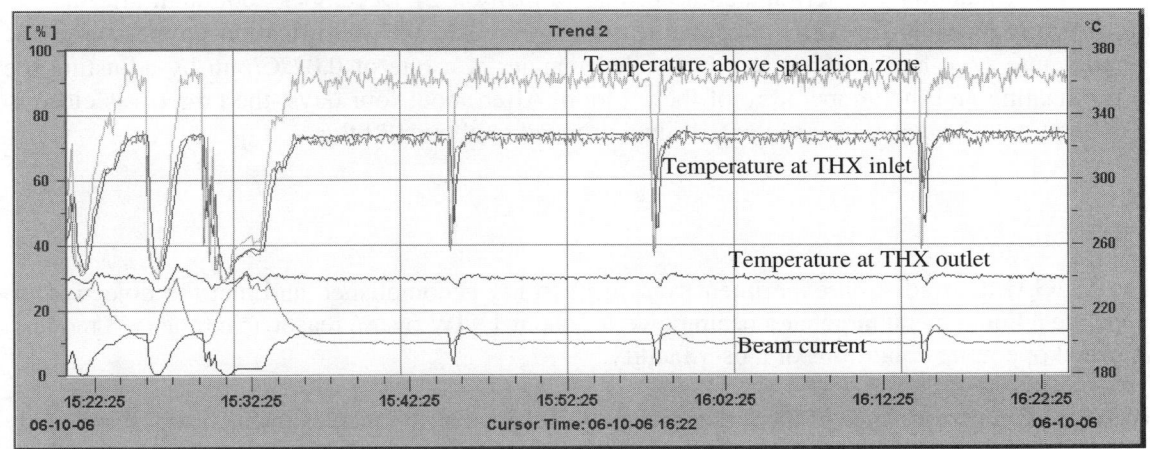

Measurement programme

An extensive measurement programme was carried out to assess the neutronic, nuclear and thermal-hydraulic performance of the target.

Neutron fluxes were measured using micro-fission chambers placed in the central rod and gold foils positioned in the SINQ irradiation position NAA and at several beam lines at the instrument positions. Figure 8 shows the increase of the flux in comparison to that measured with the previous solid lead target. The increase was about 80%, which was more than predicted by pre-calculations due to the overestimated flux of the solid target.

Figure 8. Increase in neutron yield at the SINQ irradiation position NAA and three beam lines in comparison to the previous solid lead target

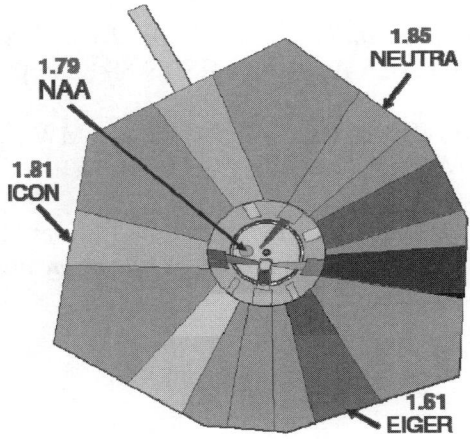

With respect to the thermal assessment, a key figure is the amount of heat deposited in the target and the surrounding systems. The heat deposited in the target itself was evaluated to 415 kW/mA of beam current, which corresponds to about 72% of the beam power. A large fraction of the power difference is dissipated in the moderator (about 13%), whereas the remainder is found in the beam window cooling and the cooled shielding structure of the target block.

End of experiment and target cooldown

The target irradiation was terminated according to plan on 21 December 2006 and, after a final measurement programme, the LBE temperature was lowered to the solidification temperature. The LBE temperature was then lowered in a controlled manner at a rate of 0.02°C/min by adjusting the central rod heating and the temperature of the oil loop. After about four days, the target was close to room temperature. The decay heat of the LBE had dropped to about 100 W.

Conclusions

The MEGAPIE irradiation experiment was successfully accomplished and met the objectives to safely operate a liquid metal target at a beam power of about 1 MW over a reasonable time (> 3 months) under normal operating conditions of a user facility.

The neutronic performance of the target yielded the expected flux increase, which was surprisingly high compared to the standard solid targets used before.

The thermal-hydraulic behaviour of the target system was stable and the beam trips and interrupts could be well managed.

The large amount of data collected and experience gained requires further evaluation and documentation. The experience gained is extremely important for future LM target design and operation.

REFERENCES

[1]   *A European Roadmap for Developing Accelerator Driven Systems for Nuclear Waste Incineration*, European TWG, ENEA April 2001, ISBN 88-8286-008-6.

[2]   Bauer, G.S., M. Salvatores, G. Heusener, "MEGAPIE, a 1 MW Pilot Experiment for a Liquid Metal Spallation Target", *J. Nuclear Mat.*, 296, 17-35 (2001).

[3]   Groeschel, F., A. Cadiou, C. Fazio, Th. Kirchner, G. Laffont, K. Thomsen, "Status of the MEGAPIE Project", *Workshop Proceedings of the Utilisation and Reliability of High-power Proton Accelerators (HPPA-4)*, Daejon, May 2004, 125-135.

MEGAPIE PROGRAMME

Chairs: J. Knebel, F. Groeschel

SAFETY EVALUATION OF THE MEGAPIE EXPERIMENTAL FACILITY: RESULTS AND INSIGHTS FROM THE APPLICATION OF PROBABILISTIC SAFETY ASSESSMENT

Luca Podofillini, Vinh N. Dang
Paul Scherrer Institute, Villigen PSI, Switzerland

Abstract

A scoping-level probabilistic safety assessment (PSA) was applied to selected systems of the experimental facility hosting the Megawatt Pilot Target Experiment (MEGAPIE) at the Paul Scherrer Institut (PSI). MEGAPIE demonstrated the feasibility of a liquid lead-bismuth target for spallation facilities at a proton beam power level of 1 MW (website: http://megapie.web.psi.ch/).

In performing a PSA to this type of facility, a number of challenges arise, mainly due to: 1) the extensive use of electronic and programmable components and of one-of-a-kind components; 2) difficulties in estimating the probabilities of initiating and failure events. A previous work by the same authors emphasised the qualitative results obtainable from the application of PSA. No attempt was made to evaluate probabilities of initiating and failure events. Some level of importance/significance evaluation was nevertheless feasible, and practical and detailed recommendations concerning potential system improvements were derived.

In this paper, a preliminary quantification of the facility risk profile is attempted. This provides more information on risk significance, which allows prioritising the insights and recommendations obtainable from the PSA. At the present stage, the limited knowledge on initiating and failure events leads to uncertainties in their probabilities; the propagation of these uncertainties through the PSA model results in large uncertainties in the quantitative results. Consequently, the reported results should be considered as examples of how quantification can support risk-informed decisions, rather than realistic figures of the facility risk profile.

Introduction

A scoping-level PSA was recently performed for the experimental facility hosting MEGAPIE [1]. The PSA, reported in [3,4], considered scenarios initiated by excessive beam focusing, which may have exposed the target window in the liquid metal container (LMC) to an excessive intensity, eventually causing a breach of the container [2].

The PSA study identified a number of challenges for the application of PSA to experimental facilities, connected with the extensive use of digital components and software systems, for which error modes are difficult to predict and quantify [5]. In addition, experimental facilities often use one-of-a-kind components that are designed especially for the needs of the particular experiment. Past experience with similar components is also difficult to use as it is often the case that the software developed for such components are revised to accommodate new hardware, fix errors or add new functionality. The operating environment and performance requirements may also differ.

As a result of these challenges, in the study of [3,4], emphasis is placed on the qualitative insights obtainable from the PSA. However, quantification of the risk profile with the associated uncertainty is an important output of the PSA, since it allows comparing sequences from different initiating events and quantifying the contribution of the events to the risk. Recommendations for system improvement can also be prioritised based on their impact on risk.

As a first step in this direction, the present paper reports on a preliminary attempt to quantify the PSA in order to obtain the risk profile of the facility. The PSA builds on previous safety evaluations of MEGAPIE. Deterministic studies of fluid dynamics and structural mechanics can be found in [6]. These give the bases for the MEGAPIE safety concepts and requirements presented in the MEGAPIE safety report [2]. Probabilistic approaches were followed by [7] and [8]. In Ref. [7] a feasibility study was performed on the reliability assessment of the MEGAPIE safety systems. The study gives a positive evaluation on the feasibility and highlights difficulties from the lack of data available to estimate the components reliability. In Ref. [8], PSA is used to evaluate the relevant causes of LMC failure and the severity of different scenarios following the failure. The Ref. [9] study confirms that target failure due to an excessively focused beam is the scenario of most concern for the experiment and its safety.

PSA in brief

The PSA methodology, which is widely used in the nuclear power industry as well as in other domains with complex systems, consists of a set of analysis methods. The methodology is based on systematically: 1) postulating potential accident scenarios; 2) identifying the technical and non-technical "defences" against these scenarios; 3) decomposing the systems and defences into components, their failure modes and the probabilities associated with these. Two elements of the PSA methodology typically stand out:

- The event tree (ET), which is used to model the accident scenarios. It represents the main sequences of functional success and failure events (of the system functions that are required in response to an initiating event, i.e. the "barriers" and "defences" against accidents) and the consequences for each sequence. These consequences or end states, as a minimum, include a safe end state and a failure or accident end state.

- The fault tree (FT), which documents the systematic, deductive analysis of the possible causes for the failure or a required function within the accident scenario model (within the ET). A fault tree analysis is performed for each of the system functions required in the response to an IE.

Assigning the safe end state to a sequence means that the scenario has been successfully terminated and undesired consequences have not occurred. In contrast, the failure or accident end state means that the sequence has resulted in undesired consequences. An important feature of PSA is that the end states may be defined in terms of a range of consequences; in other words, small accidents vs. more severe accidents. This represents additional information with which to prioritise safety improvements and recommendations.

The PSA analysis tasks are:

- Initiating event (IE) analysis. This analysis considers how accidents can start. More specifically, it identifies the events that require the response of the system to avoid undesired consequences.

- Accident sequence analysis examines the systems that are needed to respond to an IE. This is the process of building the scenarios. While the ET is the product of accident sequence analysis, it is often helpful to begin with an event sequence diagram, which maps out the sequence of event successes and failures that could occur starting with the IE.

- Both IE analysis and the accident sequence analysis are supported by information in the design documentation. In addition, the experience of the facility or of similar facilities is also analysed to assess the comprehensiveness of the IE and the scenarios (review of experience). This experience may include incidents, accidents or precursor events (near misses).

- In systems analysis, the possible causes for the failure of a function required in the scenarios are systematically identified. This is where the fault trees are developed. In addition to component failures such as mechanical defects, the systems analysis considers how the system is maintained, tested and returned to service, in terms of its impact on system failures.

- Human reliability analysis (HRA). The purpose of the HRA is to identify and analyse the personnel-related failure events that need to be included in event trees or fault trees. The HRA considers the personnel tasks in maintenance and testing as well as those tasks required in the response to an initiating event.

- Data analysis refers to the estimation of the frequencies of initiating events and of the probabilities of the causes of function failures (the "basic events" in the fault trees).

- Model quantification. In this step, the ETs and FTs are quantified using the frequencies and probabilities of the events. Some of the results of model quantification include the probabilities of accidents, the most probable (dominant) scenarios and their contributors. The quantitative contribution of the IEs, of failure events and of system failures are measured with (risk) importance values. These provide a ranking of the contributions that identify what is risk-significant. In addition, when modifications are considered, the importance values provide an indication of how much the risk may be reduced by these modifications.

Description of the facility

The MEGAPIE set-up

MEGAPIE has been carried out at the SINQ neutron source at PSI [1]. Figure 1 shows a rough schematic of the beam line and of the components and systems relevant for this PSA study. A detailed description of the experiment is outside the scope of this paper and can be found in [1] and on the

Figure 1. Simplified schematic of MEGAPIE

Only components relevant for this study are shown

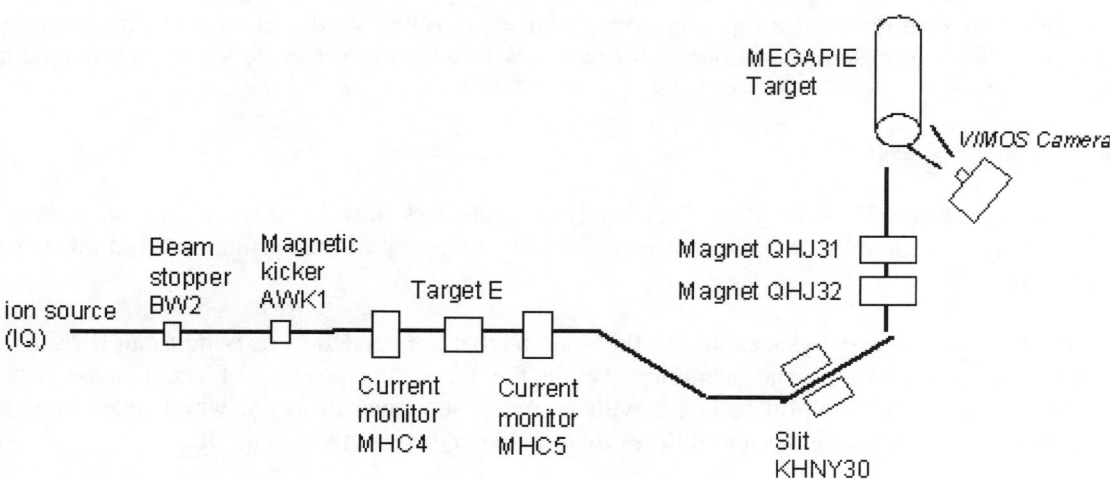

experiment website http://megapie.web.psi.ch/. The target is contained in a liquid metal container (LMC). A helium-filled gap separates the LMC from the lower target enclosure (LTE), a double-walled safety hull that encapsulates the LMC.

MEGAPIE safety concepts

Safety concepts and requirements for MEGAPIE are presented in the safety report [2]. One of the most important concerns for the safe operation was to maintain a correct intensity distribution of the incident beam onto the target. A focused beam may invest with excessive intensity the target window in the LMC, causing overheating of the window and, eventually, a breach of the LMC.

Of special importance for the beam intensity distribution is scattering Target E [9] (Figure 1), which spreads the beam intensity distribution to a full-width at half-maximum (FWHM) of about 7 cm. If for any reason the protons were not scattered, their footprint on the target would shrink to a FWHM of less than 2 cm, leading to a rise in the maximum density of the incident beam by a factor of 25 [6]. At the corresponding current density of 700 $\mu A/cm^2$ it would take 170 ms until the beam causes a breach of the LMC.

Improper beam focusing can also be due to erroneous settings of the bending magnets and quadrupole magnets along the beam line. In particular, the quadrupole magnets QHJ31 or QHJ32 have a critical role due to their position. They are located right underneath the MEGAPIE target (Figure 1, last two components of the beam line) and downstream to the monitors associated to two out of three safety systems. Therefore, as discussed later, if the incorrect beam intensity distribution were due to incorrect settings in these magnets, only one safety system, the Visual Monitor System (VIMOS) would be able to react.

In addition to the MEGAPIE safety system, the accelerator Machine Run Permit System (RPS) constitutes an additional safety barrier. Among other functions, the RPS supervises the beam parameters and the beam line component settings, and if these have out-of-specification values, the RPS does not give the permission of starting the beam, or shuts it down if it is operating. The additional safety barrier from the RPS is not modelled in this study, which focuses only on the safety systems that are dedicated exclusively to the MEGAPIE safety.

The safety systems for MEGAPIE

The defence against an excessively focused beam is based on monitoring the beam and its footprint and on shutting down the beam when the monitored characteristics deviate from acceptable values. Not all of the components and systems are presented here. A more detailed overview of the systems can be found in [9].

There are three monitor systems: the MHC4/5 transmission monitor system, the slit KHNY30 current monitor system and the VIMOS monitor. In particular, the function of VIMOS is to monitor the beam intensity distribution and trigger a beam shutdown signal in case the distribution is out-of-specification [13]. VIMOS visually monitors the light emitted by a glowing mesh, which is heated by the beam. A camera takes pictures of the mesh light emission. A software processes the images and computes the average beam intensity over two regions of interest (ROI1 and ROI2) of the beam section and evaluates a number of parameters. If any of these exceeds predefined threshold values, a beam shutdown signal is raised.

The Schnelles Abschalt System (SAS, German for "fast shutdown system") has the function to process the beam shutdown signals. It is composed by six logic modules (designated: M-SAS A, M-SAS B, S-SAS A, S-SAS B, SAS-AM, AWK1-U) and a sophisticated signal transmission system that handles electrical and optical signals conversions. Simplifying the discussion, SAS delivers an AWK1 actuation signal and a BW2 actuation signal in case it receives a beam shutdown signal from any of the monitoring devices.

The AWK1 magnetic kicker is the main device used to divert the proton beam from the target. The magnetic kicker allows diverting the beam within 100 µs from the time when the actuation signal reaches the magnet power supply.

In parallel to the AWK1 actuation signal, SAS delivers an actuation signal to the BW2 shutter, which is a copper block that, when actuated, slides into the beam trajectory and intercepts it. Its response time is slower than that of AWK1, i.e. about 700 ms. When the BW2 feedback is confirmed, the actuation signal of AWK1 is cancelled and the magnet is switched off. The AWK1 magnet cannot be kept on in the long term, as beam diversion is expected to damage the walls of the magnet against which the beam is diverted. The SAS also supervises that the beam has been actually diverted by the AWK1 magnetic kicker, on the basis of two AWK1 feedback signals. In case at least one AWK1 actuation feedback signal does not reach the SAS back, then fast (300 µs) beam switch-off is performed acting at the level of the ion source (Ionquelle – IQ) of the main PSI accelerator.

Scope of the PSA: Scenarios and initiating events

Three initiating events were selected for the PSA as representative scenarios with challenges that may lead to an excessively focused beam:

- TE100: total (100%) bypass of Target E by the proton beam.

- WSET1: Wrong settings of the quadrupole magnets QHJ31 or QHJ32. Wrong settings loaded into the component control devices.

- WSET2: Wrong settings of the quadrupole magnets QHJ31 or QHJ32. Correct values loaded into the component control devices but failure of the components to set or of the control devices to command the quadrupole current.

The two quadrupoles were selected for their critical location, right underneath the MEGAPIE target, downstream to the MHC4/5 and the KHNY30 systems (Figure 1).

Accident sequence modelling

Total bypass of Target E (TE-BY)

The events sequence starts with the beam totally bypassing Target E (Figure 2). For brevity, the discussion below addresses only some representative blocks of the diagram. The detailed model can be found in [4].

Figure 2. Event sequence diagram for total (100%) Target E bypass

Dotted blocks are not modelled in the PSA

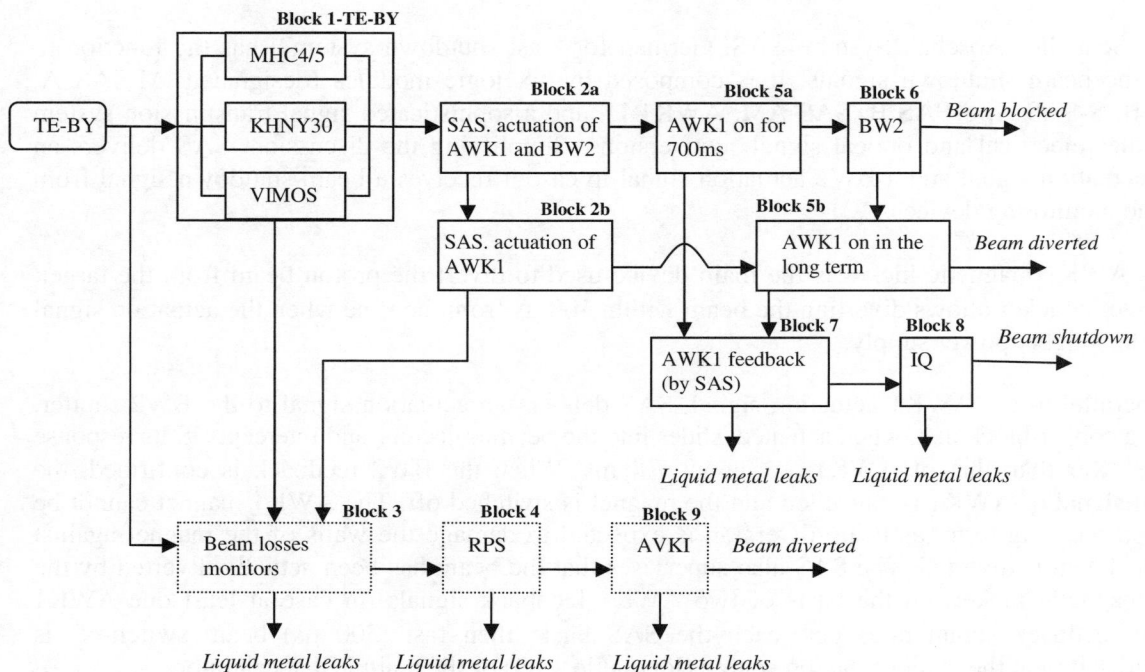

Block 1, TE-BY, models the response to the initiating event of the three monitoring systems, the MHC4/5 system, the KHNY30 system and VIMOS. Success of this block entails that at least one shutdown signal is delivered to the inputs of M-SAS A or of M-SAS B. Each of the three systems can deliver two redundant shutdown signals to SAS, so that the total number of signals to SAS is 6 = 3*2.

Block 2a models the processing by the SAS (inclusive of the M-SAS, the S-SAS, the SAS-AM and AWK1-U) of the shutdown signals from the three monitoring systems. Success of Block 2a entails that the two signals for actuation of the AWK1 kicker magnet and the BW2 beam shutter are available at the outputs of the SAS-AM (a module of the SAS).

Block 2b models the same process as Block 2a, but it has a different success criterion. Success of Block 2b entails that only the AWK1 actuation signal is available at the SAS-AM output. This has the consequence that AWK1 needs to stay on for long-term shutdown (Block 5b), as the BW2 block will not be available.

Block 5a models the actuation of the AWK1 kicker magnet, following the actuation signals from the SAS. The signal at the output of the Abschaltmodul is sent to the AWK1 power supply, which provides the current to the magnet. Success of this block entails that the magnetic field in the AWKI magnet is successfully switched on, and remains on for at least 700 ms.

Wrong settings in the quadrupoles (WSET1 – wrong values loaded and WSET2 – magnets failure)

The WSET1 sequence starts with incorrect settings being loaded into control devices for the quadrupole magnets QHJ31 or QHJ32, producing a highly focused beam distribution that could lead to the breach of the safety hull. The only difference between this sequence and the TE100 scenario is that WSET1 can be detected by VIMOS only (see Figure 1). The WSET2 sequence starts with failures of the quadrupole magnets QHJ31 or QHJ32 that result in fields producing an excessively focused beam distribution that could lead to a safety hull breach. The difference between this sequence and WSET1 is that the IE can also be revealed by the magnet control devices which check the actual magnet parameters (e.g. current in the magnet, resulting magnetic field) against the loaded (demand) settings.

Preliminary risk quantification

As mentioned in the introduction, in a first phase of this work, qualitative results were used to obtain safety insights and to derive safety-enhancing recommendations. These recommendations were oriented to make sure that the VIMOS system is actively working [3]. They addressed concerns that were found with potential failure modes of the system in which VIMOS would continue to evaluate the same frames (either due to a failure to load new images or to a failure of the frame grabber to save new images), thus not recognising a disturbance in the beam intensity distribution.

A first step towards quantification is attempted here. The approach, like the one used in [10], is to obtain qualitative ranks for the IEs or failure probabilities, which express relative expectations of likelihood, i.e. operational, likely, unlikely, extremely unlikely and hypothetical. The next step is to associate a probability or frequency range to these ranks. Given the lack of information, large uncertainties enter here and the ranges are correspondingly broad. It is worth noting that the PSA methodology systematically addresses uncertainties in parameters and probabilities and can propagate these; this application is remarkable and problematic in terms of the ranges of the uncertainties rather than the uncertainties in themselves, which are present in all analyses.

The initiating events WSET1 and WSET2 are both considered unlikely (range 1E-4/y to 1E-2/y). The WSET1 value to some extent represents an upper bound which assumes frequent setting changes as well as neglecting the opportunities for detection provided by the progressive increase in beam current. WSET2 is expected to be less frequent and with a lower upper bound (than WSET1) because the failures have to result in excessive focus (the incorrect field could lead to an unfocused beam as well). In this scoping-level analysis, the TE-BY frequency includes IEs that are expected to be less severe (less than 100% bypass) and is therefore set relatively high. Consequently, all of these distributions (shown in Table 1) are considered very conservative.

For the failure events, reference probabilities from programmable logic components and hardwired connection components are derived from [12], with the process described in the note at Table 1. Then, upper and lower percentiles of the distributions are computed by multiplying and dividing by 10 the values obtained from [12], i.e. this corresponds to an error factor of 10, which is relatively large for hardware. The probability values of the other events have been derived based on the judgment of the authors on how the likelihood of the events compares to the one derived

Table 1. Data of the parameters uncertainty distributions (lognormal) used for the preliminary quantification

Initiating event	Mean	Median	5th perc.	95th perc.
TE-BY	2.67E-3	1E-3	1E-4	1E-2
WSET1	1.33E-2	5E-3	5E-4	5E-2
WSET2	2.02E-3	158E-3	5E-4	5E-3
Failure event	**Mean**	**Median**	**5th perc.**	**95th perc.**
Cables: stuck at failure	2.67E-5	1E-5	1E-6	1E-4
Cables: unplugged or contact failure	2.67E-3	1E-3	1E-4	1E-2
Optical/digital converters: stuck-at failure of one output	2.67E-4	1E-4	1E-5	1E-3
Optical/digital converters: stuck-at failure of all outputs	2.67E-5	1E-5	1E-6	1E-4
Human errors	2.67E-3	1E-3	1E-4	1E-2
Hardwired connections: stuck at failure of one or all outputs	2.67E-5	1E-5[1]	1E-6	1E-4
Optical connection panels: stuck-at failure of one output	2.67E-5	1E-5	1E-6	1E-4
Optical connection panels: stuck-at failure of all outputs	2.67E-6	1E-6	1E-7	1E-5
Programmable logic components: all failures modes involving one output	2.67E-4	1E-4[1]	1E-5	1E-3
Programmable logic components: all failures modes involving multiple output	2.67E-5	1E-5	1E-6	1E-4
All other events	1E-2	Dummy value, no uncertainty distribution		

[1] Derived from Ref. [12] = $\lambda_{DU} * \tau/2$, where λ_{DU} is taken from Table 5.2.1 of [12]) and τ is the interval between functional tests (1 week = 176 hours). Upper and lower percentiles are computed by multiplying and dividing by 10 the median value.

from [12]. Common cause failures are modelled among components of the same type (β-factor model, with β = 0.05). Other events, for which not enough information is available to derive a credible range, are assigned an artificially high dummy probability value of 1E-2, without a distribution.

Figure 3 (left) and Table 2 present the results. The risk profile distribution is obtained by propagating via Monte Carlo simulation (10^4 samples) the uncertainty on failure and initiating events probabilities. Then the profile obtained is interpolated with a lognormal distribution. Figure 3 shows that the risk profile is dominated by the WSET1 scenarios. This results from two factors. First, the WSET1 IE has a higher likelihood compared to the other IEs (Table 1). Second, as discussed earlier in the paper, this scenario can be detected by VIMOS only, while the other two scenarios have additional protective barriers. This confirms the important contribution of VIMOS to the MEGAPIE safety.

Figure 3. Left: Results from the preliminary quantification of the risk profile associated with the considered scenarios. Right: Effect of the recommendations on the risk distribution.

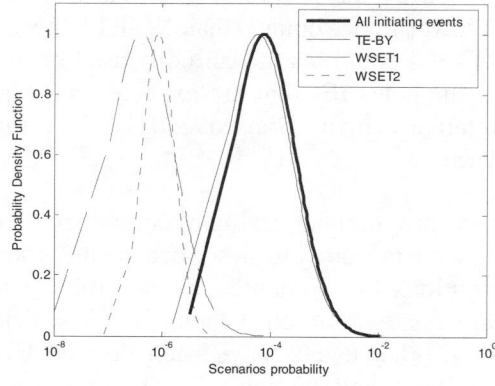
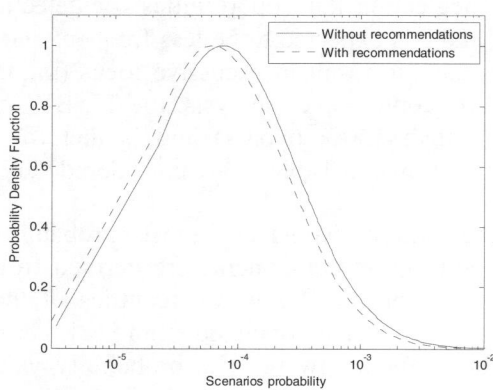

Table 2. Results from the preliminary quantification of the risk profile

Initiating event	Mean (/y)	5% (/y)	95% (/y)
TE-BY	8E-6	3E-7	3E-5
WSET1	1E-3	4E-5	4E-3
WSET2	2E-6	4E-7	4E-6
Total	1E-3	5E-5	4E-3
Total, after recommendations	9E-4	4E-5	3E-3

A relevant outcome of the quantification is the importance of the basic events, indicated by importance measures (IMs) [11]. Table 3 lists the events of top importance with respect to the Fussell-Vesely (FV) IM (i.e. with FV > 5E-2). FV ranks events based on their contribution to the measure of the risk. Table 3 shows that the two events addressed by the two recommendations derived from the qualitative study [3] (VIMOSSW-SA and FRAMEGR-BUFFER-SA) are among the top importance events, thus confirming the appropriateness of the recommendations.

Table 3. Top-importance basic events (with FV > 5E-2)

ID	Description	Mean probability	FV
VBOX-O_SAS_AB-SA	VIMOS box – HW failure causes common cause stuck-at OK of outputs SAS A and B	1E-2	1.14E-1
FRAMEGR-O_DGALL-SA	Frame grabber – HW failure causes common cause stuck-at OK of all digital outputs	1E-2	1.14E-1
VIMOSSW-SA	VIMOS SW – stuck-at while executing due to programming error or operating system failure	1E-2	1.14E-1
VIMOSSW-EF	VIMOS SW – execution failure (failure to set outputs 1 and 2 to 0)	1E-2	1.14E-1
MESH-WB	Mesh – excessive number of mesh wires break	1E-2	1.14E-1
FRAMEGRR-BUFFER-SA	Frame grabber – memory buffer stuck-at due to buffer failure or failure to save new picture	1E-2	1.14E-1
CAMERA-TM	Camera – clock failure, pictures delivered with frequency lower than 50 Hz	1E-2	1.14E-1
VIMOSSW-TM	VIMOS SW – slowed execution due to software failure or operating system failure	1E-2	1.14E-1

As a final analysis it is worthwhile to compare the risk profile before and after the recommendations are implemented, to verify their impact on the safety of the facility. To model this, the probability of the affected events (VIMOSSW-SA and FRAMEGR-BUFFER-SA) is changed from 1E-2 to 1E-5. Figure 3 (right) shows that as an effect of the recommendations the risk distribution is shifted towards lower values; the shift corresponds to a reduction in the mean of about 20%. In reviewing the results shown in Table 2, the conservatisms applied in many parts of the PSA as well as the large uncertainties in the results should be kept in mind.

Work still needs to be done to achieve a risk profile with an acceptable level of realism. In particular, as can be seen in Table 3, the risk is dominated by events quantified with the dummy value of 1E-2. Therefore, information should be collected to assign uncertainty ranges to these events as well. In addition, proper quantification entails improving the PSA model for systems testing. So far, the impact of test is modelled by miscalibration errors that may occur while performing the test. However, the impact of the test on the failure events probabilities should also be represented.

Conclusions

In previous work, a scoping-level PSA was applied to an experimental facility, the MEGAPIE facility at PSI. Emphasis was placed on the qualitative results obtainable from the PSA, given difficulties to quantify the probabilities of initiating and failure events. Recommendations on potential system improvements were also derived.

In this paper, a first step to quantify the facility risk profile is attempted. Failure and initiating events probabilities are assigned values with distributions with significant uncertainties. Other events for which not enough information was available were assigned dummy probabilities. The results from this first quantification attempt confirm the appropriateness of the recommendations derived from the qualitative results obtained in previous work.

Improvement of the quantification should still be the direction to pursue. In particular, information should be collected to reduce the uncertainty in the events' probabilities and to assess the events with probabilities quantified with dummy values. This latter is an important aspect since these events still dominate the risk profile. The PSA model for systems testing should also be improved. Reaching an acceptable level of realism in the quantification of the scenario probabilities, with their uncertainty ranges, is of utmost importance to support decision making. Indeed, quantification reveals the different priorities in protecting against the various failure modes. In addition, the quantification can be used to evaluate the impact on safety of further design and operational changes (e.g. redundancy allocation and/or test prioritisation).

As demonstrated by this study, PSA methodologies can indeed provide a useful tool for informing the designers of the safety of experimental installations and provide insights on the contributions of the different components and systems to the safety. In particular, in this study the PSA results could be used to inform specific and practical measures oriented to enhance the safety of the installation.

Acknowledgements

The authors would like to offer their thanks to Leonid Rivkin, Knud Thomsen, Timo Korhonen, Pierre Schmelzbach, Pierre-Andre Duperrex, Stefan Staudenmann and all the people of the MEGAPIE project for their support and suggestions during the development of the study.

REFERENCES

[1] Bauer, G.S., M. Salvatores, G. Heusener, "MEGAPIE, a 1 MW Pilot Experiment for a Liquid Metal Spallation Target", *J. Nuclear Mat.*, 296, 17-35 (2001).

[2] *MEGAPIE Safety Report. Sicherheitsbericht zum MEGAPIE-Experiment an einem Target mit flüssigem Blei-Bismut-Eutektikum in der Neutronenquelle SINQ des PSIs.* Paul Scherrer Institut (PSI), Villigen, Switzerland (2002).

[3] Podofillini, L., V.N. Dang, "Probabilistic Safety Assessment of Complex Experimental Facilities: Lessons Learned from the Application to a Neutron Source Facility (MEGAPIE)", *Proceedings of the European Safety and Reliability Conference (ESREL 2007)*, Stavanger, Norway, 25-27 June 2007.

[4] Podofillini, L., V.N. Dang, *Probabilistic Safety Assessment (PSA) of the MEGAPIE Safety Systems – Beam Footprint Deformation Scenarios*, PSI Report (forthcoming).

[5] Garret, C.J., G.E. Apostolakis, "Automated Hazard Analysis of Digital Control Systems", *Reliability Engineering and System Safety*, 77 1-17 (2002).

[6] Smith, B.L., *Summary Report For Megapie R&D Task Group X4: Fluid Dynamics And Structure Mechanics*, PSI Report PSI-06-01 (2006).

[7] Mock, R., S. Wagner, *Zuverlässigkeitsanalyse des SINQ-MEGAPIE-Sicherheitssystems. Eine Machbarkeitsstudie, Version 2.1*, Laboratorium für Sicherheitsanalytik, ETH Zürich, draft report (2005).

[8] Bertrand, F., *MEGAPIE Experiment, Reliability and Safety Study on the Failure of the Lower Liquid Metal Container of the Target*, Report of the Commissariat à l'Energie Atomique (CEA), France CEA/DEN/CAD/DER /SESI/LCFR/NT DO 27 19/12/05 (2006).

[9] Thomsen, K., P. Schmelzbach, *Report on Beam Safety Systems for MEGAPIE*, MPR-3-TK34-167/3 (2006).

[10] Burgazzi, L., "Probabilistic Safety Analysis of an Accelerator – Lithium Target Based Experimental Facility", *Nuclear Engineering and Design*, 236, 1264-1274 (2006).

[11] Cheok, M.C., G.W. Parry, R.R. Sherry, "Use of Importance Measures in Risk Informed Applications", *Reliability Engineering and System Safety*, 60, 213-226 (1998).

[12] Hauge, S., P. Hokstad, *Reliability Data for Safety Instrumented Systems*, SINTEF Report STF38 A04423 (2004).

[13] Thomsen, K., *VIMOS, A Novel Optical Safety Device for MEGAPIE*, PSI Scientific and Technical Report 2002, Volume III, 144 (2002).

A DEDICATED BEAM INTERRUPT SYSTEM FOR THE SAFE OPERATION OF THE MEGAPIE LIQUID METAL TARGET

Knud Thomsen, P.A. Schmelzbach
Paul Scherrer Institut, Switzerland

Abstract

MEGAPIE is an experiment which aimed to design, build and demonstrate the safe operation of a liquid metal spallation target at a beam power level of 1 MW in the SINQ target station at the Paul Scherrer Institute. MEGAPIE is an international collaboration joining ten partners with a strong interest in ADS. At PSI the spallation target inside SINQ is located at the far end of the proton beam line. Of special importance for its safe operation is a meson production target located upstream of SINQ. The proton beam is scattered in this target, leading to a wide intensity distribution on the SINQ target. If for any reason the protons were not scattered their footprint on the SINQ target would shrink leading to a rise in the maximum density of the incident beam by a factor of 25. At the resulting high current density it would have taken only 170 ms until a hole was burned through both the liquid metal container inside the target and the lower target enclosure. The liquid metal would have spilled into the beam line and the catcher vertically below the SINQ target. This constituted a most severe accident scenario for MEGAPIE.

In order to prevent an insufficiently scattered beam from reaching the SINQ target three independent safety systems have been installed: a dedicated current monitoring system, a beam collimating slit and a novel beam diagnostic device named VIMOS. The latter monitors the correct glowing of a tungsten mesh closely in front of the liquid metal target. All these systems have to meet the basic requirement to switch off the beam within 100 ms when 10% of the protons bypass Target E (corresponding to an increase in peak intensity by a factor of two to 50 μAcm^{-2}).

General background

A correct intensity distribution in the incident beam was of highest importance for the MEGAPIE liquid metal spallation target at PSI. High peaks in the beam intensity distribution could have damaged the target. MEGAPIE was more sensitive in this respect than a standard solid state target in SINQ and the potential damage was larger.

Situation of the beam at PSI

At PSI the spallation target inside SINQ sits at the far end of the proton beam line. Of special importance for SINQ is Target E. During normal operation the proton beam is scattered in Target E leading to a widening in beam distribution on the SINQ target with a FWHM in the order of 7 cm. If for any reason the protons were not scattered in Target E their footprint on the SINQ target would shrink to a FWHM of less than 2 cm. Detailed simulations yield a rise in the maximum density of the incident beam by a factor of 25 [1]. At the corresponding current density of 700 µA/cm^2 it would have taken only 170 ms until both the liquid metal container inside the target and probably subsequently the lower target enclosure would have been perforated, with the liquid metal spilling into the beam line and the catcher vertically below the SINQ target. This constituted a most severe accident scenario for MEGAPIE [2,3].

Besides incomplete scattering of the beam, improper settings of bending magnets and quadrupoles can also result in higher than nominal proton intensities.

Figure 1. Layout of the proton beam line at PSI, passing Target E, leading to the SINQ target

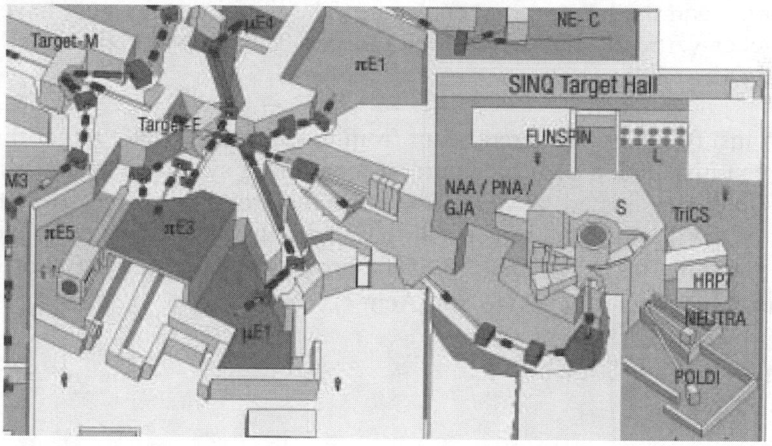

Requirements on beam safety systems

The MEGAPIE target can withstand a non-scattered fraction of 17% of the nominal beam for an extended period of time. Thus the following basic requirement for all beam safety systems had conservatively been derived:

The beam had to be switched off within 100 ms if 10% of the protons bypassed Target E (corresponding to an increase in peak density by a factor of two) [1].

Four safety systems for MEGAPIE (front end)

In order to detect, i.e. prevent, an insufficiently scattered beam from reaching SINQ three independent primary safety systems are in place, i.e. current monitors, a collimating slit and the beam density monitoring system VIMOS. As part of the machine run permit system (RPS) the monitoring of the distribution of the beam losses along the beam line from Target E to SINQ also contributed to the safety of MEGAPIE.

Scattering of the protons in Target E has three direct or "local" effects:

- The beam current is reduced.
- The energy of the protons is reduced.
- The opening angle of the beam is enlarged.

These effects are each monitored with three dedicated and completely independent sensors/systems:

- current monitors (MHC4/MHC5);
- slit (KNHY30);
- VIMOS.

These systems formed a genuine group of sensors to directly trigger switching off the beam if a critical situation was detected. The accelerator operation crew had no access to the settings of VIMOS and to the current monitor outputs dedicated to MEGAPIE safety. The slit position was key-locked and could not be changed by the operator during beam production. Additionally, there is always:

- loss distribution monitoring.

As an addition to the three direct effects listed above comes the specific distribution of the beam losses along the entire beam line which is monitored by the machine crew with a dedicated system in order to ensure the correct low values up to the SINQ target.

Current monitor

Beam current is measured at diverse locations; of greatest interest for MEGAPIE were the current sensors immediately in front of Target E and behind it: MHC4 and MHC5. They consist of a coaxial resonator (the so-called "current transformer") tuned at the 2nd harmonic of the proton beam pulse frequency [4]. After calibration, the total beam current measurement can be measured absolutely to a level of few perc ent [5,6]. The important value for the SINQ target is the ratio between the currents upstream and downstream of Target E.

The current transformers MHC4 and MHC5 have been installed in the beam line for decades. Each monitor has two RF sensors enabling two independent beam current measurements. One of these sensors is being used for the standard beam current measurements. The second sensor provides an independent measurement for MEGAPIE that was handled via the SINQ Schnelles Abschalt System (SAS). Possible improvements were identified with respect to stability and the requirement for regular calibration [7], e.g. the installation of new cables providing similar temperature conditions for both MHC4 and MHC5 measurements.

Performance of the current monitor

Electronics similar to the standard beam measurement equipment were prepared in the beginning of 2005 and adapted to MEGAPIE needs; in particular, the computerised on-line calibration had been disabled to start with. The installation took place during the summer of 2005. Sufficient experience could be gained demonstrating a stability of the MEGAPIE current measurement of a few per cent of the beam intensity [8]. First results indicated that a threshold of 10% could be set, which meet the MEGAPIE requirements and did not interfere with the separate RPS operations (which actually work with a tighter limit, the measurements being calibrated at short intervals) [9].

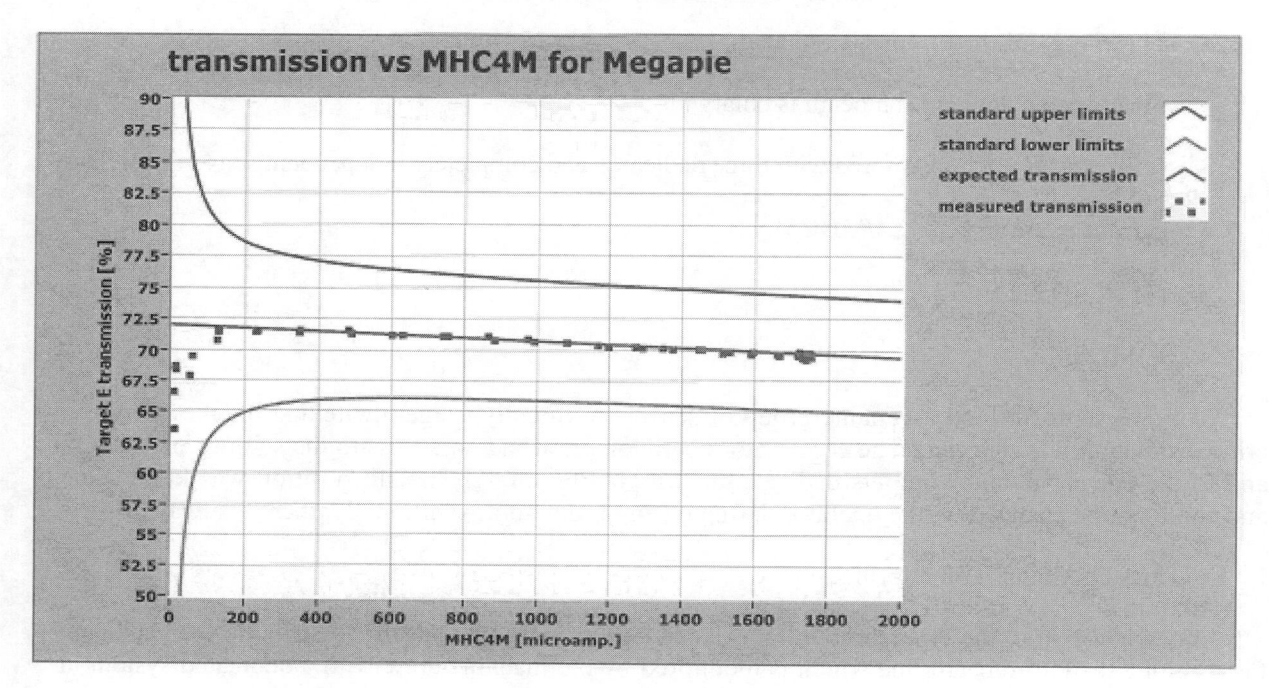

Figure 2. Data from transmission measurement

Slit KHNY30

A slit was installed at a suitable location in the beam line in order to limit the allowed variation of the beam path to the SINQ target, and to detect and to stop protons missing Target E.

As a result of the loss of energy that protons suffer from scattering in Target E, their bending radius in the dipole magnets below the SINQ target is smaller than for non-scattered protons. At a specific location inside the quadrupole QHJ30 the dispersion between the two beam fractions reaches a maximum of about 20 mm [1]. Centering the opening of the slit on the trajectory of the correctly scattered protons it is possible to actually block any protons with higher energy. A predefined current limit on the lower jaw triggers an interrupt signal.

The jaws are built as massive copper bars. In case a large fraction of the beam hits one of them, it would even provide a short-term passive safety: the jaw would not evaporate completely in the time before the beam is switched off. The position of the jaws can be adjusted and tuned remotely [10,11]. For MEGAPIE operation the remote control of the position was disabled (key-locked).

Current alarms from the slit were fed into the SAS. Further redundancy was given by sending an interrupt signal to the Run Permit System of the accelerator. In addition to the direct alarm signal from the slit, it is very likely that the vacuum breakdown would be detected by the gauges included in the SAS and generates this way a further interrupt signal.

Figure 3. Locations of the slit KHNY30 and the main VIMOS components

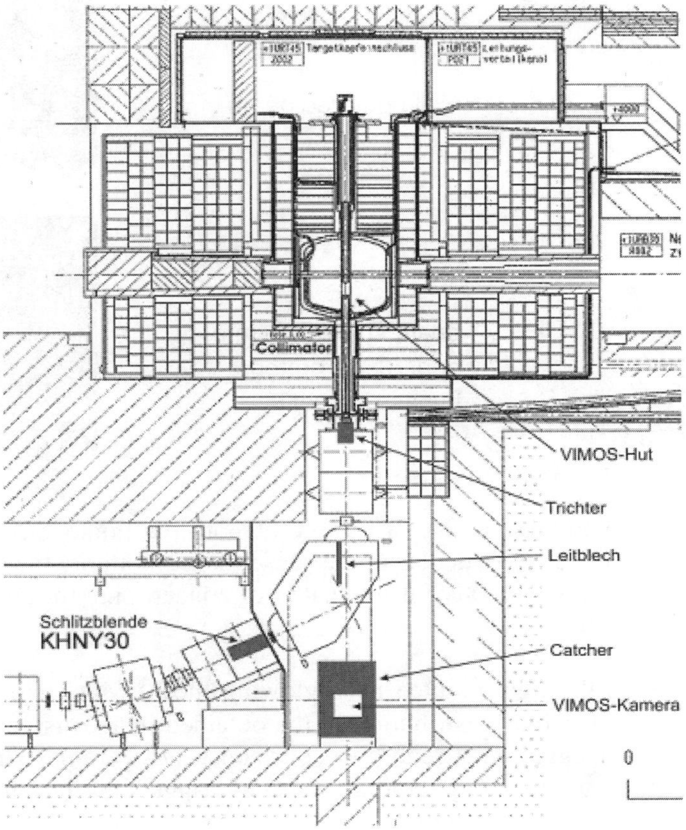

Performance of the slit

The slit KHNY30 was installed inside the quadrupole QHJ30 in the shutdown at the beginning of 2004. Afterwards many tests were performed concerning the optimum positioning of the jaws: they should be close enough to reliably detect any relevant deviation from nominal conditions while not hampering normal operations. Results of the commissioning phase confirmed the expectations and verified the excellent sensitivity of the slit; a fraction of less than one per cent of the protons missing Target E already generates a reliable interlock signal [11]. The slit is now positioned routinely as for MEGAPIE operation. The reproducibility of the beam set-ups was such that readjustments (with their potential source of error) were not needed during MEGAPIE operation.

VIMOS

In contrast to the other beam safety systems, which all employ well known and proven technology at the proton beam at PSI, VIMOS is a newly-developed, rather specific device. The acronym stands for VIsual MOnitor and Sieve, indicating that besides optical monitoring one part of the system should

Figure 4. Comparison of simulation results with actual measurements. On the left side changes in the beam profile are clearly visible leading to the exceeding of thresholds in several monitor signals (lower right).

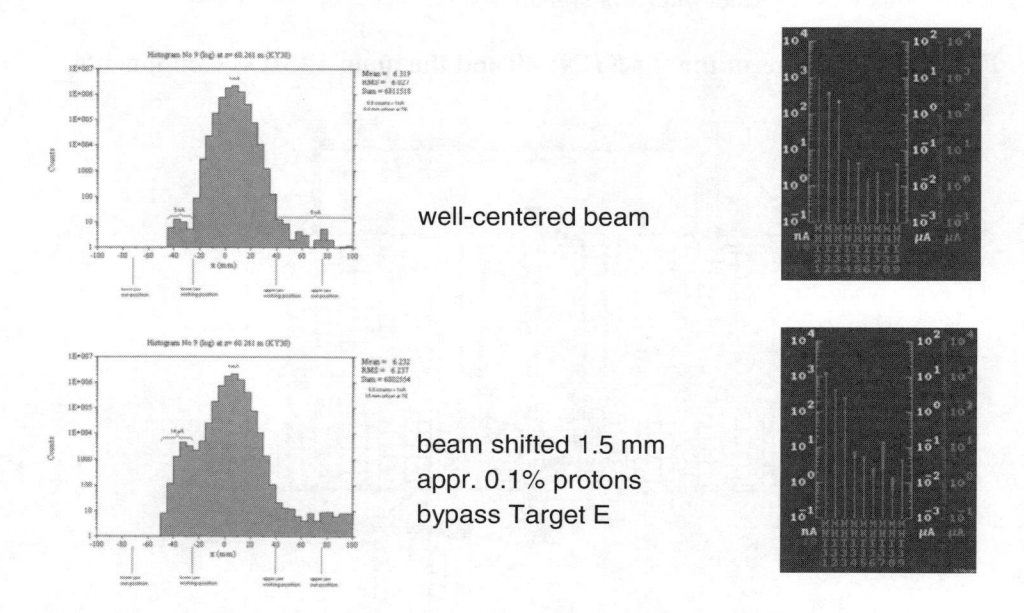

well-centered beam

beam shifted 1.5 mm
appr. 0.1% protons
bypass Target E

also prevent clogging of the beamline in case of rupture of both the liquid metal container and the lower target enclosure. VIMOS is also particular in the sense that it monitors the beam profile closest to the SINQ target and watches most directly for the matter of concern, i.e. for possible "hot spots" in the beam density distribution [12].

VIMOS visually monitors the light emitted by a glowing mesh, which is heated by the beam. Employing fast analysis the proper light distribution in the obtained images is verified at a repetition rate of 50 Hz. Basing the alarm signal on a trend over four consecutive frames guarantees that minor fluctuations do not trigger false alarms. Any large discrepancy triggers an interrupt already based on one frame [13-15].

On top of the last collimator in front of the SINQ target sits a "hat" which bears a glowing foil made of tungsten wires carefully dimensioned to provide just enough glowing signal during normal beam conditions that the correct operational state of the system can be verified. The light from the glowing mesh is collected and focused by a large parabolic first surface mirror at the bottom of the catcher right below the SINQ target. The newly built catcher for MEGAPIE features a window where a special radiation-resistant camera samples the images. The video signal is processed at twice the normal frame rate using a state-of-the-art frame grabber and a powerful PC [13]. An alarm signal generated by VIMOS was handled by the SAS.

Performance of VIMOS

VIMOS was installed for the first time during the 2004 shutdown. In extended commissioning and test phases the proper working of the system has been confirmed [16]. One incident, wherein improper magnet parameters led to a peaked beam intensity distribution, was detected which proved the functioning of VIMOS already during the commissioning phase.

Figure 5. Narrow beam footprint detected by VIMOS

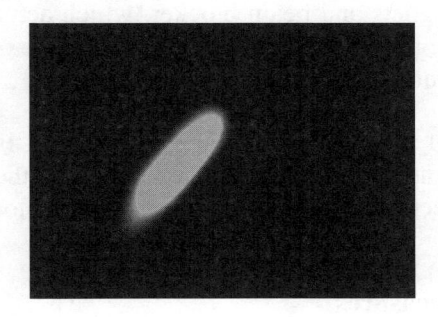

Beam loss distribution monitoring

Loss monitoring is a standard procedure belonging to the operation of the accelerator and the beam lines [17]. The machine protection system is not designed to protect other equipment, but nevertheless provides additional redundancy and could be taken into account as a well proven powerful additional safety system which was already available.

The beam losses are measured by means of halo monitors and ionisation chambers. Based on operational experience and protection requirements of the specific components each individual monitor has been attributed a particular range of allowed values. Special attention has been focused on the distribution of the losses along the beam line. All these monitors in their combination contribute to ensuring nominal beam conditions [17,18].

Fifteen (15) loss monitors are deployed along the beam line between Target E and SINQ. An interrupt triggered by the loss distribution monitoring is processed via the machine Run Permit System.

Performance of the loss distribution monitoring

The loss monitors are the backbone of standard operations and well proven over decades at PSI. The reached sensitivity of the system was much higher than that required by MEGAPIE [17,18].

Effector mechanisms to switch off the beam

Signals from the three specific beam safety systems, i.e. current monitor, slit and VIMOS, were all wired to the SINQ Schnelles Abschalt System (SAS). Parameters reflecting the proper working of two of the systems were also fed into the SINQ Langsames Abschalt System (LAS). This slow channel would not have triggered a proper alarm if an insufficiently scattered beam was impinging on the MEGAPIE target but a loss of the slit current measurement or the VIMOS video signal would have tripped the beam. SAS and LAS were installed along with the SINQ and were cleared for operation at that time [19].

The interrupt signals of the SINQ Schnelles Abschalt System (SAS) and the back-up signals of the Run Permit System of the accelerator were fed to a unit consisting of kicking magnets and a beam stopper to intercept the beam at the Cockcroft-Walton pre-accelerator. This unit worked very reliably, handling over the past 20 years a rate of several 10 thousands interrupts per year without failure. For the MEGAPIE experiment an additional duplication of the fast source switch was activated directly by the SAS [21].

The layout and working of the complete effector chain starting from the signal generation in the diverse sensors up to the kicker magnets and beam blocker, which actually interrupt the proton beam is documented in Refs. [20,21]. The maximum time delay between the occurrence of an alarm and the tripping of the beam had been measured to be less than 5 ms.

Each year before irradiation of a SINQ target starts a check-list is worked through with a well prescribed number of tests of all interrupt conditions generated by the SAS. At the same time the proper functioning of the actual effectors, i.e. SINQ kicker and beam blocker is verified.

Operational and administrative measures

For and during the irradiation of MEGAPIE special precautions had be taken in terms of operations in order to avoid the emergence of improper setting of the beam optics. The probability of human errors was minimised by suppressing access to critical parameters usually free during normal operation:

- Hardware windows set on the power supply settings of elements of the beam line between Target E and SINQ forbade the use of optics differing from the reference one.

- Accelerator operators were especially instructed about the requirements of MEGAPIE. Dedicated rules, e.g. for set-up of the beam, were compiled [22].

- For normal operation with the liquid metal target in SINQ the remote control of the jaws of slit KHNY30 were disabled, i.e. the slit was to be locked in its effective position.

- The beam set-up was supervised by the MEGAPIE operations manager.

- Before the irradiation of the MEGAPIE target was started dedicated tests of the SAS devices were performed.

Operational performance during the MEGAPIE irradiation phase

In summary it could be claimed even before starting with the irradiation of the MEGAPIE target that every reasonable possible effort was undertaken to assure the switching-off of the proton beam if it threatened to damage the MEGAPIE target. The above assessment was affirmed during the actual operation of MEGAPIE. All systems met their respective expectations.

- A few interlocks were triggered by the MEGAPIE transmission monitoring. In several cases where the RPS transmission monitors triggered an interrupt, the installation for MEGAPIE also produced an interlock. No substantial disturbance of the facility operation resulted.

- The collimating slits produced even fewer interlocks than the transmission monitoring, basically for the same events.

- Two interlocks were triggered by VIMOS during the irradiation period of the MEGAPIE target [23].It was quickly confirmed that an alarm criterion had actually been met, due to the change of the settings of some quadrupoles upstream of Target E. As this setting had not yet been used before with the MEGAPIE target, the corresponding VIMOS signal clearly lay outside the expected range.

- No unusual loss configuration was observed during MEGAPIE operation.

- An initial inconvenience with the switching-off of the ion source resulted in the beginning from an over-responsiveness of the system. After correction, the trigger electronics performed as well as the other hardware.

After MEGAPIE irradiation phase

Based on the good performance of VIMOS and the other beam safety systems installed for MEGAPIE these will become part of the standard instrumentation at the beam line leading to SINQ [24].

Acknowledgements

The following persons provided valuable contributions and support to the work on the dedicated MEGAPIE safety systems, in particular on VIMOS, up to now: P. Baron, M. Djiango, M. Dubs, T. Dury, R. Dölling, Y. Foucher, G. Frei, R. Fütterer, K. Geissmann, G. Heidenreich, H. Heyck, A. Irniger, A. Kalt, E. Lehmann, A. McKinnon, R. Mock, M. Moser, R. Mock, the late L. Ni, Ch. Perret, E. Pitcher, M. Reder, U. Rohrer, B. Sigg, B. Smith, A. Strinning, M. Vatré, E. Wagner, S. Wagner and H. Walther, A. Wrulich.

REFERENCES

[1]   Rohrer, U., *A Novel Method to Improve the Safety of the Planned MEGAPIE Target at SINQ*, PSI Scientific and Technical Report, Volume VI, pp. 34-35 (2001).

[2]   Perret, C., *Sicherheitsbericht zum Megapie-Experiment an einem Target mit Blei-Bismut-Eutektikum in der Neutronenquelle SINQ des PSI*, June 2002.

[3]   Smith, B.L., *Summary Report for the MEGAPIE R&D Task Group X4: Fluid Dynamics and Structure Mechanics*, draft 21/06/2005.

[4]   Reimann, R., M. Rüede, "Strommonitor für die Messung eines gepulsten Ionenstrahls", *Nuclear Instruments and Methods*, 129, 53 (1975).

[5]   Duperrex, P-A., U. Frei, U. Müller, L. Rezzonico, Current & Transmission Measurement Systems in the P-canal, PSI Memorandum, 8 March 2001.

[6]   Duperrex, P-A., Transmissionskontrolle und Neues Target E, vs 3.2.1, PSI Memorandum, 13 July 2005.

[7] Müller, U., *Verbesserungsvorschläge für die bestehenden Strommonitore*, TM 84-05-02, 7 August 2002.

[8] Duperrex, P-A., U. Müller, private communication, September 2005.

[9] Duperrex, P-A., U. Müller, Test Report on the MEGAPIE Current Monitoring, document in preparation.

[10] Rohrer, U., *Funktionsbeschreibung der Schlitzblende (KHNY30) für MEGAPIE im P-Strahl zur SINQ*, MPR-1-RU85-005/2, July 2003.

[11] Rohrer, U., *First Beam Tests with the New Slit Collimator in the Proton Beam Line to SINQ*, PSI Scientific and Technical Report, Volume VI, pp. 23-26 (2004).

[12] Thomsen, K., *VIMOS, a Novel Optical Safety Device for MEGAPIE*, PSI Scientific and Technical Report, Volume III, p. 144 (2002).

[13] Thomsen, K., *VIMOS, Design Considerations & First Results*, MPR-3-TK34-003/3, 17 February 2005.

[14] Thomsen, K., *Specifications for the VIMOS s/w*, MPR-3-TK34-029/1, 23 October 2003.

[15] McKinnon, A., M. Moser, K. Thomsen, P. Hengel, *VIMOS s/w, Monitoring Program*, MPR-3-MA34-001/3, 13 December 2005.

[16] Thomsen, K., *VIMOS, Commissioning Report*, MPR-3-TK34-117/0, 20 July 2005.

[17] Rohrer, U., *The Multilevel Protection System for the Vacuum Chambers of the High-intensity 590 MeV Proton Beam Lines*, PSI Scientific and Technical Report, Volume VI, pp. 45-48 (2003).

[18] Dölling, R., R. Rezzonico, P-A. Duperrex, U. Rohrer, K. Thomsen, E. Erne, U. Frei, M. Graf, U. Müller, "Beam Diagnostics at the High Power Proton Beam Lines and Targets at PSI", *Proceedings of DIPAC05*, Lyon, France (2005).

[19] *Gutachten der HSK zum Gesuch des PSI bezüglich Bau und Betrieb der Spallatiuons-Neutronenquelle SINQ*, HSK 2/167, December 1992.

[20] Staudenmann, S., *Megapie-Abschaltsystem: Spezifikation und Lösungs-konzept*, PSI Report TM-93-05-25 a, 2 February 2006.

[21] Dzieglewski, G., *SINQ Schnelles Abschaltsystem für das MEGAPIE-Experiment*, TM-84-05-05.

[22] Mezger, A., U. Rohrer, P. Schmelzbach, Checkliste Strahl Set-Up für MEGAPIE.

[23] Thomsen, K., "Experience with VIMOS During the Irradiation Phase of MEGAPIE", *Proceedings of ICANS-XVIII*, Dongguan Citac, China, 26-29 April 2007.

[24] Blarer, B., Draft Concept for the Implementation of the Beam Safety Systems After MEGAPIE.

THE ROLE OF SCK•CEN IN THE MEGAPIE PROJECT

Marc Dierckx
SCK•CEN, Belgium

Abstract

SCK•CEN has the ambition to design, build and operate an experimental accelerator driven subcritical system (XT-ADS). With the liquid metal spallation target being one of its main components, SCK•CEN decided to join the MEGAPIE project, which aims to design and operate a first-of-a-kind megawatt liquid Pb-Bi spallation target. This paper compares the MEGAPIE and XT-ADS target, highlighting the common use of liquid Pb-Bi technology. An overview is also given of past and future scientific involvement of SCK•CEN in the MEGAPIE project.

Introduction

SCK•CEN is one of the major partners in the European 6th Framework Programme EUROTRANS, which aims to show the feasibility of an accelerator-driven system (ADS) for transmutation of nuclear waste. Part of the project involves the design of XT-ADS, a small-scale ADS demonstrator. In a later stage SCK•CEN is candidate to host this XT-ADS installation on site; this is why SCK•CEN is heavily involved in the design phase of XT-ADS.

One of the main components of an ADS is the spallation target module which converts high-energy protons from an accelerator into neutrons to feed the subcritical core. In XT-ADS a proton beam of 3 mA @ 600 MeV will hit a Pb-Bi liquid metal target. The feasibility and validation of such a high-power spallation target is of crucial importance for the continuing development and licensing of XT-ADS.

With the successful completion of the irradiation phase, MEGAPIE has become the first of a kind Pb-Bi spallation target that has been demonstrated to operate safely during an extended period of time, thus providing valuable input for future ADS systems.

MEGAPIE versus XT-ADS target

Figure 1. MEGAPIE target

Table 1 shows the main operational parameters for the MEGAPIE and XT-ADS targets. Both accept a 600 MeV proton beam to produce neutrons via a spallation reaction in liquid Pb-Bi. The XT-ADS target will need a beam current of at least 3 mA to produce the required amount of neutrons to feed its subcritical core. The current of 1.4 mA for the MEGAPIE target is mainly determined by the maximum current of the accelerator. The MEAGPIE target itself is capable to handle well over a 2 mA beam current.

Table 1. Comparison of MEGAPIE and XT-ADS target specifications

	MEGAPIE	XT-ADS target
Coolant/target	Liquid Pb-Bi	Liquid Pb-Bi
Beam energy	595 MeV	600 MeV
Beam current	1.4 mA max	3 mA
Target lifetime	4 months	9 months
Accumulated charge for total target	2.8 Ah	20 Ah
Target diameter	Ø 20 cm	<Ø 10 cm
Accumulated charge/m^2	90 Ah/m^2	2 500 Ah/m^2
Beam interface	Window	Windowless

The cylindrical hole in the centre of the XT-ADS core contains the spallation target and must be kept as small as possible in order to limit the total power of the reactor. In XT-ADS the diameter of the target is limited to 10 cm. A bigger hole in the centre of the core would mean that fuel has to move from the centre of the core to the edge and extra fuel needed to compensate for the extra neutron leakage of the core due to an increased core diameter. The MEGAPIE target has an outer diameter of 20 cm. This means that MEGAPIE has four times more surface available for beam impact.

The XT-ADS target has a lifetime of at least one operational cycle of the reactor, which is nine months of continuous operation. The total accumulated charge of the target at the end of a cycle is 20 Ah. The total irradiation time of MEGAPIE was four months, but not always at full power and with regular shutdowns, thus reaching a total accumulated charge of 2.8 Ah.

The "window" in MEGAPIE is used to separate the liquid Pb-Bi target material from the beam line vacuum. The window is therefore heavily loaded by high-energy protons and spallation neutrons causing embrittlement of the window material. The amount of material damage in the window at the end of a cycle is proportional to the "accumulated charge per unit target area". For MEGAPIE this is 90 Ah/m^2, while for XT-ADS this is 2 500 Ah/m^2 which is a factor 27 higher. This is one of the main reasons why in XT-ADS it is decided to mitigate the risk of window breakage and make a design that eliminates the window and lets the proton beam directly hit the liquid Pb-Bi.

Importance of the MEGAPIE project for development of XT-ADS

Although specific differences exist between both targets (window versus windowless), there is a general correspondence which is related to the fact that both targets use liquid Pb-Bi as target material. Pb-Bi technology developed for the MEGAPIE target can therefore easily be translated to the XT-ADS application: electromagnetic pumping, heat exchanger design, Pb-Bi conditioning, corrosion protection, etc.

During the MEGAPIE irradiation phase neutron fluxes were measured at different locations outside as well as inside the MEGAPIE target. Linking measured neutron fluxes with measured beam currents provides an excellent opportunity to benchmark neutronic codes and enhance the accuracy and reliability of XT-ADS neutronic calculations. Inventory of the spallation products in both the cover gas as well as the Pb-Bi target material of the MEGAPIE target provides insight in the solubility and evaporation of spallation products, which is of crucial importance for the design of the XT-ADS vacuum/cover gas system.

T91 and 316L steels are foreseen as the main structural materials for the construction of the XT-ADS reactor. The same structural materials are used to build the MEGAPIE target. As such the post-irradiation examination (PIE) of the irradiated MEGAPIE target will provide essential information on the behaviour of these materials under XT-ADS-like conditions (proton and neutron flux, Pb-Bi flow, thermal stress, corrosion, erosion, …).

Beside these technical aspects the MEGAPIE target creates an important precedent that demonstrates to the authorities that a megawatt spallation target is licensable and the XT-ADS concept is built upon known and proven technology.

SCK•CEN contribution in the MEGAPIE project

With the MEGAPIE project liquid Pb-Bi technology would be developed to a stage where a fully licensed megawatt spallation target could be built and operated for an extended time period. MEGAPIE gives unique hands-on experience on liquid Pb-Bi technology and provides a proven concept for future ADS targets. That is why SCK•CEN decided to join the MEGAPIE initiative in 2001 and contributed 11% of the total budget, becoming one of main partners in this consortium.

In addition to this financial investment, SCK•CEN has also contributed to the design, validation and operation of the target and will undertake a large part of the PIE on the structural materials of the irradiated target.

SCK•CEN has developed a target temperature control strategy to keep the temperature inside the target between certain limits to avoid LBE freezing or overheating which would cause excessive thermal stresses on the beam window and other parts of the system. The accelerator of the SINQ facility has on average two to three beam trips per hour, during which the total heat input will instantly drop from 600 kW to 10 kW. If the corresponding 600 kW target cooling is not immediately switched off, the temperature will drop rapidly, with the risk of freezing liquid Pb-Bi in the target. This shows that a fast reaction of the target temperature control system is required A Matlab Simulink model was constructed to deduce the system dynamics and test different control strategies. Although more complex control schemes were evaluated, a simple PI controller was chosen for easy implementation into the MEGAPIE control system.

The temperature control strategy was verified during the MEGAPIE Integral Test phase [1,2]. A 200-kW resistive heater was mounted on the target to simulate the heat input from a proton beam. Tests revealed that the control valve in the oil loop has a highly non-linear characteristic (Figure 2), severely limiting the performance of the temperature control. A digital compensation was implemented in the MEGAPIE control system to solve this problem. The tests also showed that the efficiency of the target heat exchanger was 25% higher than initially calculated. The MEGAPIE Integral Test phase has shown that the target will be able to safely handle the full beam power of the SINQ accelerator.

During the start-up phase of MEGAPIE in SINQ the beam power was gradually increased (Figure 3) while target temperatures and other vital system parameters were closely monitored. No irregularities were detected and within one day the target was at nearly full beam power. The temperature control system performed as expected under steady-state as well as under transient conditions. No additional adjustments to the temperature control parameters were needed.

Figure 2. Highly non-linear control valve characteristic

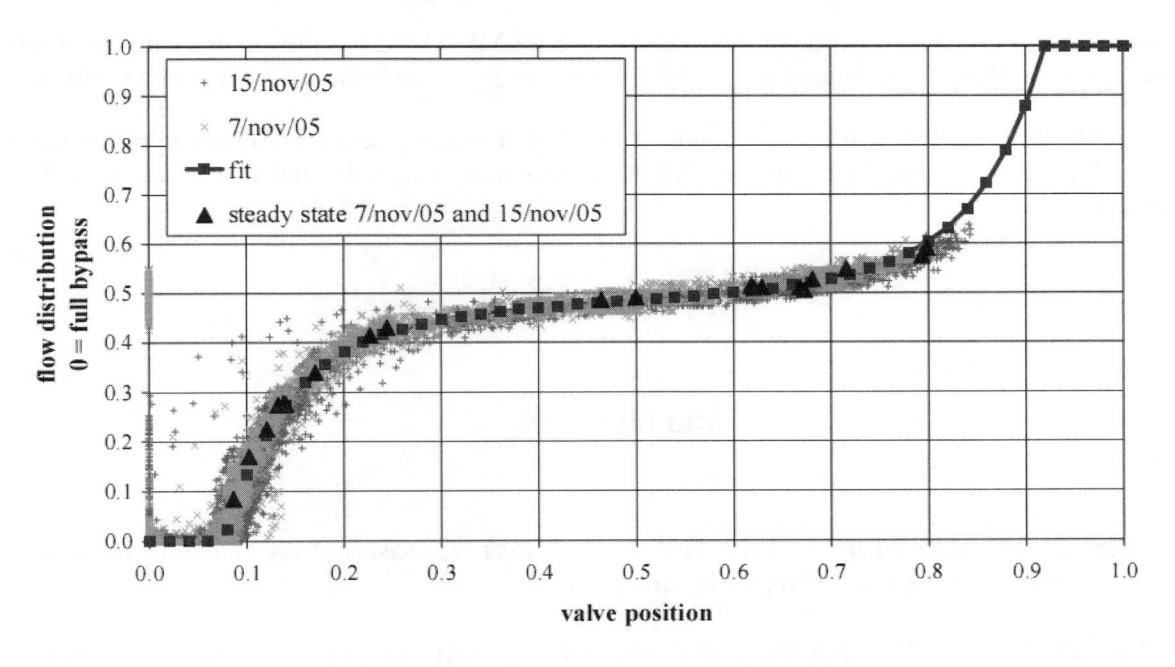

Figure 3. Start-up of MEGAPIE in SINQ

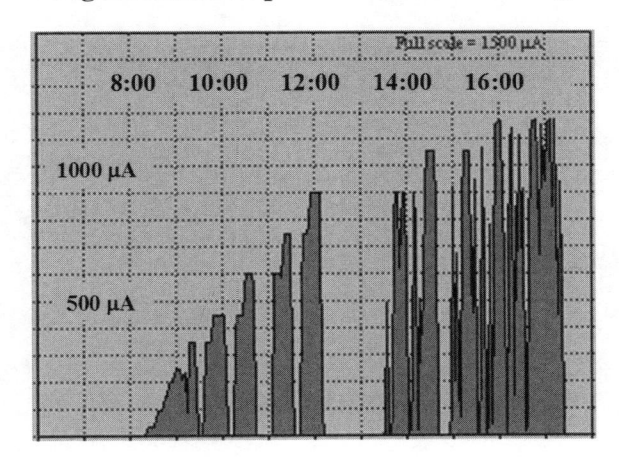

Future contributions to the MEGAPIE project

In 2008 the MEGAPIE target will be dismantled and irradiated structural materials will be extracted for further examination (PIE phase). Part of these irradiated structural materials will be examined at SCK•CEN to quantify the effect of high-energy proton and neutron radiation together with Pb-Bi embrittlement, thermal stress and fatigue.

Based on data gathered during the irradiation phase of MEGAPIE an extensive study on the performance of the heat removal system will be made. SCK•CEN will mainly focus on the overall performance of the heat removal system, looking for signs of possible degradation in the heat removal system. The performance of the target temperature control will be assessed both during stable operation and during beam transients.

Conclusion

With the successful completion of the irradiation phase MEGAPIE has proven that safe operation of a megawatt Pb-Bi spallation target is possible, making the licensing of future ADS systems easier.

The PIE of MEGAPIE will be very important for further development XT-ADS as it will provide crucial information on the behaviour of XT-ADS structural materials under relevant XT-ADS conditions.

REFERENCES

[1] Dierckx, M., *MEGAPIE Integral Test Results Part 1: Steady State Thermo Hydraulic Characterisation*, SCK•CEN report R-4468 (2006).

[2] Dierckx, M., *MEGAPIE Integral Test Results Part 2: Target Temperature Control*, SCK•CEN report R-4467 (2006).

SESSION I

Accelerator Programmes and Applications

Chairs: H. Klein, P. Pierini

STATUS OF THE EUROTRANS R&D ACTIVITIES FOR ADS ACCELERATOR DEVELOPMENT

Jean-Luc Biarrotte
*On behalf of the EUROTRANS WP1.3 collaboration**
CNRS/IN2P3, Institut de Physique Nucléaire d'Orsay, France

Abstract

An accelerator-driven system (ADS) for transmutation of nuclear waste typically requires a 600 MeV-1 GeV accelerator delivering a proton flux of a few mAs for demonstrators, and of a few tens of mAs for large industrial systems. This paper briefly describes the reference solution adopted for such a machine, based on a reliability-oriented linear superconducting accelerator, and focuses on the status of the R&D presently ongoing on some prototypical accelerator components. This work is performed within the 6th Framework Programme EC project "EUROTRANS".

---

* EC Contract N° FI6W 516520, "EUROTRANS".

Introduction

The basic purpose of accelerator-driven systems (ADS) is to reduce the nuclear wastes' radiotoxicity, volume and heat load before their underground storage in deep geological depositories. This issue is particularly significant in Europe where 2 500 tonnes of spent fuel are produced every year by the 145 reactors of the European Union [1].

The proposed solution relies on the partitioning and transmutation strategy: the different elements of the spent fuel are chemically separated, minor actinides, plutonium and long-lived fission products are isolated and recombined to obtain new fuel assemblies to be used and "burnt" in dedicated "transmuter" systems. Such a strategy should significantly – by orders of magnitude – decrease the long-term radiotoxicity of the spent fuel: with a separation efficiency of 99.9% of the long-lived products from the waste, followed by transmutation, the natural uranium ore radioactivity level can be reached in less than 1 000 years, instead of millions of years if no action is performed [1].

An ADS transmuter system is composed of two main parts: a subcritical reactor ($k_{eff} < 1$), in which the chain reaction can not be self-sustained, that greatly relieves the safety problem aspect, and an intense spallation source that provides the "missing" neutrons needed to maintain the reaction. Such a neutron source, composed by a liquid lead target subjected to a high-energy proton flux, also produces the suited broad energy spectrum required to "burn" the minor actinides components, that are otherwise accumulated in conventional thermal spectrum critical reactors.

The EUROTRANS research programme

The European research programme for the transmutation of high-level nuclear waste in accelerator-driven systems (EUROTRANS) is funded by the European Commission within the 6th Framework Programme, and involves 31 partners (research agencies and nuclear industries) with the contribution of 16 universities. EUROTRANS is a four-year programme (2005-2009) extending previous activities (Preliminary Design Study for an Experimental Accelerator-driven System – PDS-XADS) and paving the road towards the construction, during subsequent EC framework programmes, of an experimental facility demonstrating the technical feasibility of transmutation in an accelerator-driven system (XT-ADS).

Within the EUROTRANS programme, activities are split into five main technical areas (called Domains), respectively devoted to: the design of the ADS system and its sub-components; small-scale experiments on the coupling of an accelerator, a spallation target and a subcritical core; studies on advanced fuels for transmuters; investigations on suited structural materials and heavy liquid metal technology; collection of nuclear data for transmutation. The main objective is to work towards a European Transmutation Demonstration (ETD) in a step-wise manner, i.e.:

- provide an advanced design of all the components of an XT-ADS system, in order to allow its realisation in a short-term view (~10 years);

- provide a generic conceptual design of modular European Facility for Industrial Transmutation (EFIT) for the long-term objective of the programme.

Machine design for the European Transmutation Demonstration

The XT-ADS machine, that will be loaded with conventional MOX fuel, is meant to be built and tested in the near future (before 2020) so as to fulfil three main objectives: 1) demonstrate the ADS concept (coupling of proton accelerator, spallation target and subcritical assembly) at significant but reasonable core power levels (50-100 MWth); 2) demonstrate the transmutation using some dedicated positions able to accept minor actinides assemblies; 3) provide a multi-purpose irradiation facility for the neutron community in general, and for the testing of different EFIT components in particular (samples, fuel pins…).

The EFIT facility will be an industrial-scale transmutation demonstrator system, loaded with transmutation dedicated fuel. Its characteristics are meant to maximise the efficiency of transmutation, the ease of operation and maintenance, and the high level of availability in order to achieve an economical transmutation. Despite these sometimes contradictory objectives, the XT-ADS and the EFIT machines share the same fundamental features. Especially, both designs use (cf. Table 1):

- a superconducting linac solution to produce the required high-power proton beam, the main reasons for this choice being the power-upgrading capability of this solution and the perspectives of improvement of the beam reliability;

- a liquid metal core coolant and spallation target, the metal being pure lead for the EFIT design and lead-bismuth eutectic (LBE) for the XT-ADS, allowing for lower working temperatures.

Table 1. Baseline characteristics of the XT-ADS and EFIT machines

	XT-ADS (ADS prototype)	**EFIT (industrial transmuter)**
Goals	Demonstrate the concept Demonstrate the transmutation Provide an irradiation facility	Maximise the transmutation efficiency Ease of operation & maintenance High level of availability
Main features	50-100 MWth power	Several 100 MWth power
	K_{eff} around 0.95	K_{eff} around 0.97
	600 MeV, 2.5 mA proton beam (back-up: 350 MeV, 5 mA)	800 MeV, 20 mA proton beam
	Conventional MOX fuel	Minor actinide fuel
	Lead-bismuth eutectic coolant & target	Lead coolant & target (back-up: gas)

The reliability-oriented reference accelerator for ADS

The European Transmutation Demonstration requires a high-power proton accelerator operating in CW mode, ranging from 1.5 MW (XT-ADS operation) up to 16 MW (EFIT). The main beam specifications are shown in Table 2. At first glance, the extremely high reliability requirement (beam trip number) can immediately be identified as the main technological challenge to achieve.

The reference design for the accelerator was developed during the PDS-XADS programme [2,3] and is based on the use of a superconducting linac (see Figure 1). Such a choice allows to obtain a very modular and upgradeable machine (same concept for prototype and industrial scale), an excellent potential for reliability (see hereafter), and a high RF-to-beam efficiency thanks to superconductivity (optimised operation cost). For the injector, an ECR source with a normal conducting RFQ is used up to 3 or 5 MeV, followed by an energy booster section which uses normal conducting H-type DTL or/and superconducting CH-DTL structures up to a transition energy still under optimisation, around 20 MeV. This first part of the linac is duplicated in order to provide good reliability perspectives. Then a fully modular superconducting linac, based on different RF structures (spoke, elliptical), accelerates the beam up to the final energy (350, 600, 800 or 1 000 MeV). Finally a doubly-achromatic beam line with a redundant beam scanning system transports the beam up to the spallation target.

Table 2. XT-ADS and EFIT proton beam general specifications

	XT-ADS	EFIT
Max. beam intensity	2.5-4 mA	20 mA
Proton energy	600 MeV	800 MeV
Beam entry	Vertical from above	
Beam trip (>1 sec) number	< 5 per 3-month operation cycle	< 3 per year
Beam stability	Energy: ±1%, intensity: ±2%, size: ±10%	
Beam time structure	CW with 200 µs low frequency 0-current holes	

Figure 1. The reference accelerator scheme for the XT-ADS (and EFIT)

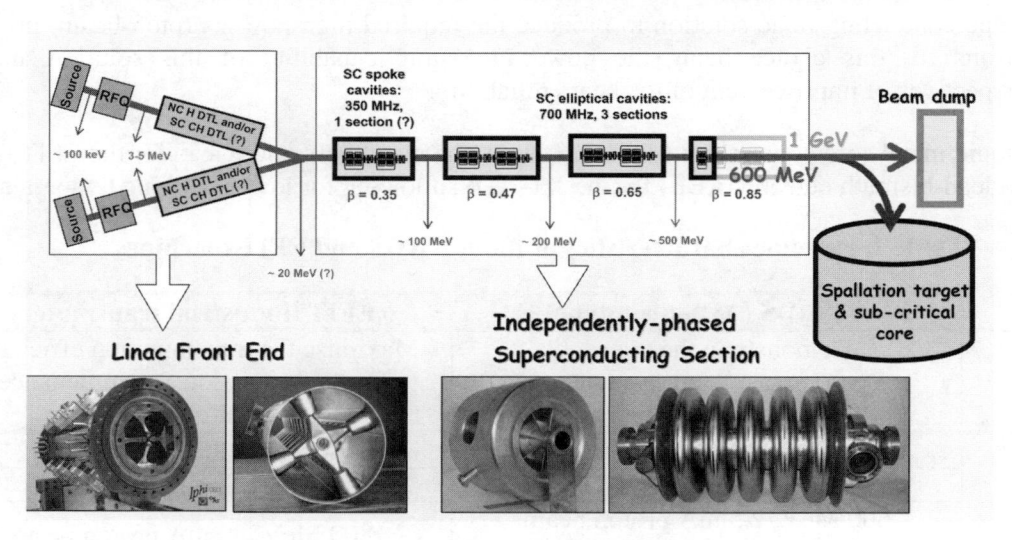

The ADS accelerator is expected – especially in the long-term EFIT scenario – to have a very limited number of unexpected beam interruptions per year, which would cause the absence of the beam on the spallation target for times longer than one second. This requirement is motivated by the fact that frequently repeated beam interruptions induce thermal stresses and fatigue on the reactor structures, the target or the fuel elements, with possible significant damages, especially on the fuel claddings; moreover these beam interruptions decrease the plant availability, implying plant shutdowns in most of the cases. Therefore, it has been estimated that beam trips in excess of one second duration should not occur more frequently than five times per three-month operation period for the XT-ADS, and three times per year for the EFIT.

To reach such an ambitious goal, which is lower than the reliability experience of typical accelerator-based user facilities by as many as 2 or 3 orders of magnitude, it is clear that design practices oriented toward reliability need to be followed from the early stage of component design. In particular:

- "Strong design" practices are needed: every linac main component has to be de-rated with respect to its technological limitation (over-design).

- A rather high degree of redundancy needs to be planned in critical areas; this is especially true for the identified "poor-reliability" components: linac injector and RF power systems, where solid-state amplifiers should be used as much as possible.

- Fault tolerance capabilities have to be introduced to the maximum extent: such a capability is expected in the highly modular superconducting RF linac, from at least 20 MeV [9].

A preliminary bottom-up reliability analysis (Failure Mode and Effects Analysis – FMEA) has been performed in order to identify the critical areas in the design in terms of impact on the overall reliability [11]. This activity confirms the choice to provide a second, redundant, proton injector stage (composed of the source, RFQ and low-energy booster), with fast switching capabilities. After the injector stage, the superconducting linac has a high degree of modularity, since the whole beam line is an array of nearly identical "periods". All components are operating well below any technological limitation in terms of potential performances, and therefore a high degree of fault tolerance with respect to cavity and magnets can be expected in the superconducting linac, where neighbouring components have the potential to provide the functions of a failing component without affecting the accelerator availability. Clearly this approach implies a reliable and sophisticated machine control system, and in particular a digital RF control system to handle the RF set points to perform fast beam recovery in the case of cavity failures.

R&D activities for ADS accelerator development

The EUROTRANS accelerator work package (WP 1.3) is split into several tasks, focused to design the main subcomponents of the ADS linac in the different energy areas, and to identify their reliability characteristics. Activities are dedicated to:

- experimental evaluation of the proton injector reliability, performed at the IPHI injector;

- assessment of the reliability performances of the intermediate-energy accelerator components, with particular attention to the comparison of different accelerating structures;

- design and experimental qualification of the reliability performances of a high-energy cryomodule tested at full power and nominal temperature;

- design and test of a prototypical RF control system intended to provide fault tolerance operation of the linear accelerator;

- update of the accelerator design, including beam dynamics issues and investigation of control strategies for fault tolerance, development of further reliability analyses and cost estimations for the XT-ADS and EFIT.

Proton injector reliability tests

The IPHI injector (see Figure 2), developed in France at Saclay by CEA and CNRS, fulfils the specifications of the ADS proton injector, with wide margins in terms of beam current capabilities. It is composed by the 95 keV SILHI ECR ion source, an already tuned low-energy beam transport line, a 3 MeV copper RFQ under final completion, and the associated diagnostic beam line [4].

In recent years, the SILHI source has been successfully used for several week-long reliability tests at currents of 30 mA, showing no beam stops and occasional sparks in the extraction region, causing no beam interruptions. In the EUROTRANS context, these tests will be extended, to include acceleration in the RFQ and propagation in the beam lines: the IPHI injector, once completed in 2008, will be used for a long-run test (two months), to demonstrate and assess, on a real scale, the reliability characteristics of this accelerator subcomponent.

Figure 2. Scheme of the IPHI injector and picture of the diagnostic beam line installation in Saclay

Moreover, the possibility to achieve the sharp 200 µs "beam holes" at low repetition rate (10^{-3} to 1 Hz), that is required for the subcriticality monitoring of the ADS core, has been successfully tested directly pulsing the SILHI source. Encouraging fall and rise time durations of less than 30 µs have been obtained. This test will also be extended to the full beam of the IPHI injector, at 3 MeV.

RF structures for the intermediate-energy section

For the intermediate-energy region after the RFQ and up to approximately 100 MeV, several cavity types are considered as valid "candidates". Studies and tests of prototypes are being developed in order to evaluate their feasibility and assess their potential reliability performances. In particular, activities concentrate on: 1) high shunt impedance copper DTL structures of the IH and CH type, using "Konus" focusing scheme (focusing elements outside the drift tubes) that provides high real estate gradients; 2) superconducting multi-gap CH structures, with "Konus" focusing scheme, ensuring both high real estate gradients and optimised RF-to-beam efficiency thanks to superconductivity; 3) superconducting spoke cavities, which are very modular, providing some fault tolerance capability, and can operate efficiently from very low energies (around 5 MeV) up to 100 MeV and more. The two first structures are meant to be part of the redundant injector front end, while the spoke structures will be part of the fault tolerant independently-phased linac. At the present phase of the project, the transition energy is around 17 MeV, still to be optimised given the results of the ongoing R&D.

The first activity is being carried out by IBA and IAP Frankfurt and is still in the design phase. Concerning the second activity, a 19-gap superconducting 350 MHz CH prototype has been successfully tested at IAP Frankfurt in vertical test (Figure 3) and is ready to be installed in an existing horizontal cryostat, with an external driven mechanical tuner for further testing [5]. Concerning the spoke activities, CNRS/IPN Orsay is equipping an existing cryomodule (Figure 3) with a 350-MHz $\beta = 0.15$ spoke cavity, fully dressed with its stainless steel helium reservoir, its cold frequency tuning mechanism and its high power (20 kW) RF coupler [6]. The cavity, which has already reached the design goals in vertical tests, once completed with its ancillary RF components, will be operated in 2008 at nominal operating conditions in long tests to determine the reliability characteristics of the components.

The high-energy cryomodule

The high-energy part of the linac uses low-beta elliptical cavities from an energy of approximately 100 MeV, using three different beta-sections (0.5, 0.65, 0.85). Such a technology is already successfully used world wide (e.g. at the SNS), but the full demonstration of the very low-beta section is not yet

Figure 3. Pictures of the SC CH-cavity during vertical test at Frankfurt (left) and of the spoke horizontal test cryomodule during its installation at Orsay (right)

fully accomplished. Excellent test results have been achieved with the $\beta = 0.47$ TRASCO cavities at INFN Milano [7], but besides the development of the bare superconducting cavity, it is important to prototype and test each auxiliary system needed for the cavity operation in a real environment (power coupler, RF source, power supply, RF control system, cryogenic system, cryostat...).

The goal of this EUROTRANS third task is thus to design, build and test before 2009 an operational prototypical cryomodule of the high-energy first section of the proton linac [8]. This cryomodule will host one of the existing $\beta = 0.47$ TRASCO cavities, equipped with a cold frequency tuner, developed by INFN, and a 150 kW CW RF coupler, developed by CNRS. This cryostat, under development at INFN and CNRS, will be assembled and tested at IPN Orsay under nominal 2 K operating conditions, and extensive tests (without beam but at full 80 kW RF power level) will be performed to qualify its reliability characteristics. Figure 4 shows a conceptual layout of the module and of the RF power coupler.

Figure 4. Conceptual layout of the single cavity module (left) and associated power coupler (right) for the testing of high-energy components

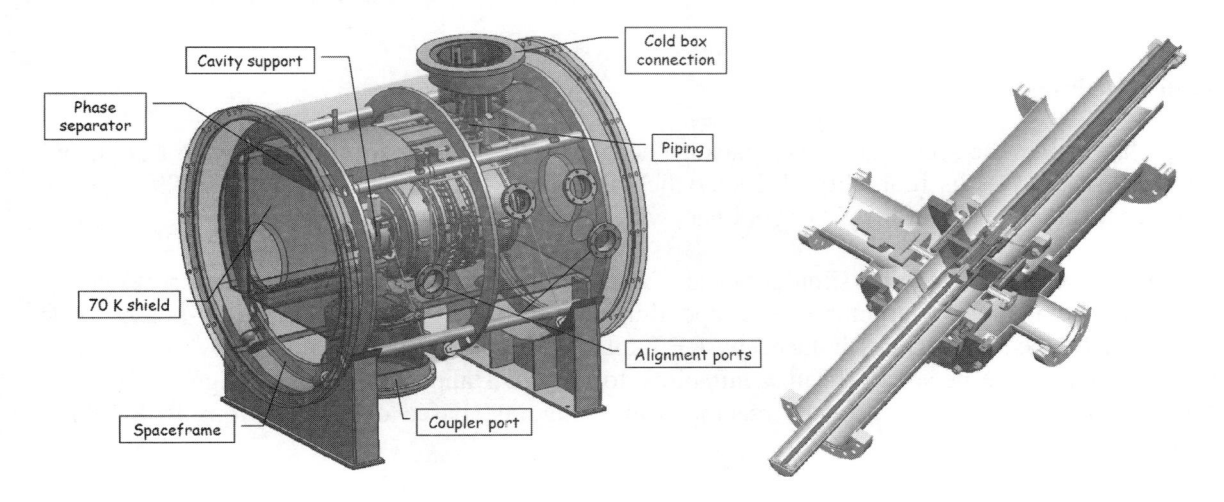

Low-level RF control system and fault-tolerance

The scope of this activity is to develop a digital low-level RF control system intended to provide the necessary field and phase stability (±0.5% in amplitude, ±0.5° in phase) and to identify and handle fast recovery scenarios for cavity failures in the superconducting linac, by means of local compensation at neighbouring cavities.

This local compensation method (cf. Figure 5), in which neighbouring components have the potential to provide the functions of a failing component, has been demonstrated on the beam dynamics point of view from 10 MeV [9], given that modular independently-phased accelerating cavities are used and that some RF power margin (up to 30%) is available. Such a fault-recovery system has even been recently demonstrated on real operation at the SNS for high-energy beams (>200 MeV) [10], using a global compensation method where all the linac components are retuned, but with the advantage of requiring very few margins in terms of RF power. The remaining step is now to identify and develop fast failure recovery scenarios to ensure that such a retuning is performed in less than 1 second. This would probably require: fast fault detection and beam shut-down; fast communication between neighbouring LLRF systems; fast update and tracking of the new field and phase set-points; adequate management of the tuner of the failed cavity; final beam recovery.

Digital techniques become necessary to meet the speed and software configuration required by such a retuning procedure. A FPGA-based digital LLRF control system, in which a number of key functionalities are implemented on a single chip, offers indeed a high-grade reliability and flexibility. Developments are going on in CEA and CNRS at 704 MHz and 352 MHz respectively, with very encouraging preliminary results [6]. Figure 5 shows a picture of a first prototype developed by CNRS (LPNHE/IPNO).

Figure 5. Principle of the local compensation method (left) and picture of a prototypical FPGA-based digital LLRF control card developed at CNRS (right)

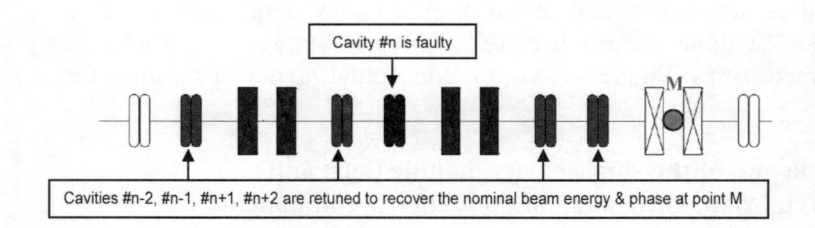

Accelerator design

The last task in the accelerator work package deals with the progress of the accelerator design and the characterisation of its reliability characteristics, the final goal being to obtain in 2009 a frozen detailed conceptual design of the XT-ADS linac, with assessed reliability and costing.

Start-to-end beam dynamics simulations are being performed, especially for the different options for the intermediate-energy structures. The modelling also includes in the independently-phased linac the effect of the beam transients induced by RF faults, and explores the possible recovery scenarios needed to provide the necessary beam availability to the ADS application. The design of the final beam line connecting the linac to the reactor is also ongoing, based on the previous PDS-XADS design guidelines.

Furthermore, an integrated reliability analysis is being performed to predict the reliability characteristics of the proposed accelerator, following the standard practices used in the nuclear energy applications (mainly failure mode and effects analysis and reliability block diagrams), and integrating the experimental results and demonstrated reliability numbers. This activity is the continuation of the PDS-XADS work, that already showed that a reliability-oriented design can, without necessarily improving the mean time between failure (MTBF) values of each sub-system, greatly improve the MTBF of the whole system by one or two orders of magnitude, especially with the implementation of redundancy, fault-tolerance and corrective maintenance strategies in critical areas [11,12].

The deduced and optimised design will then provide the basis for the cost estimation for XT-ADS and EFIT and a possible schedule for its realisation. This costing activity has already been started with parametric studies to test the influence of the final energy on the linac price, especially showing that a 600 MeV solution appears to be only 20% more expensive that a 350 MeV solution.

Conclusion

This paper describes the ADS accelerator reference solution, based on a reliability-oriented linear superconducting accelerator, and summarises the R&D presently ongoing on some prototypical accelerator components. Compared to warm temperature linacs or circular machines, a superconducting linac, much more "forgiving", has a great potential for high reliability that is the main requirement for the ADS application.

The "less than a few beam trips per year" goal is still two or three orders of magnitude below present accelerator performance, but seems to be reachable. As a matter of fact, the concept of fault-tolerance in a superconducting linac with independently-phased cavities is now proven, at least at high energies, by the SNS experience, where a beam recovery system is successfully running [10]. An integrated reliability analysis of the whole system shows that this SNS proof of principle of the fault-tolerance concept potentially improves the reliability figure by at least one order of magnitude (given that we can implement it in less than the 1-second limit) [12]. The remaining order of magnitude will be gained by redundancy, corrective maintenance plans, careful design of auxiliary systems and by improving the MTBF of individual components using the results from the ongoing R&D performed within EUROTRANS.

Acknowledgements

The principal institutions in the accelerator work package WP1.3 of EUROTRANS include CNRS (F), CEA (F), IBA (B), IAP-Frankfurt University (D) and INFN (I). Additional contributions, especially in issues dealing with the accelerator-reactor interface, are provided from ANSALDO (I), AREVA NP (D), ITN (P) and UPM (S). The programme has the financial support of the European Commission through the contract FI6W-CT-2004-516520.

REFERENCES

[1] European Technical Working Group, *The European Roadmap for Developing ADS for Nuclear Waste Incineration*, ISBN 88-8286-008-6, ENEA (2001).

[2] Mueller, A.C., "The PDS-XADS Reference Accelerator", *International Workshop on P&T and ADS Development*, SCK•CEN, Belgium, October 2003.

[3] Biarrotte, J-L., *et al.*, "A Reference Accelerator Scheme for ADS Applications", *Nuclear Instruments and Methods*, A 562, 565-661 (2006).

[4] Beauvais, P.Y., "Recent Evolutions in the Design of the French High Intensity Proton Injector (IPHI)", *Proceedings of the EPAC 2004 Conf.*, Luzern, Switzerland, 5-9 July 2004.

[5] Podlech, H., "Development of Room Temperature and Superconducting CH-structures for High Power Applications", these proceedings.

[6] Ponton, A., "Development of Superconducting Spoke Cavities for an ADS Linac", these proceedings.

[7] Panzeri, N., "Status of the Preparation of the Elliptical Cavity System for the EUROTRANS Cryomodule", these proceedings.

[8] Barbanotti, S., "Design of a Test Cryomodule for the High-energy Section of the EUROTRANS Linac", these proceedings.

[9] Biarrotte, J-L., *et al.*, "Beam Dynamics Studies for the Fault Tolerance Assessment of the DPS-XADS Linac Design", *4th OECD NEA International Workshop on Utilisation and Reliability of HPPA*, Daejon, South Korea, May 2004.

[10] Galambos, J., "Operational Experience of a Superconducting Cavity Fault Recovery System at the Spallation Neutron Source", these proceedings.

[11] Pierini, P., "ADS Reliability Activities in Europe", *4th OECD NEA International Workshop on Utilization and Reliability of HPPA*, Daejon, South Korea, May 2004.

[12] Pierini, P., "Reliability Studies for a Superconducting Driver for an ADS Linac", these proceedings.

DESIGN OF A TEST CRYOMODULE FOR THE HIGH-ENERGY SECTION OF THE EUROTRANS LINAC

Serena Barbanotti, Nicola Panzeri, Paolo Pierini
Istituto Nazionale Fisica Nucleare, Milano

Jean-Luc Biarrotte, Sébastien Bousson, Emmanuel Rampnoux, Hervé Saugnac
Institut de Physique Nucléaire, Orsay

Abstract

One task of the Accelerator Work Package of the EUROTRANS programme is dedicated to the design and realisation of a prototype cryomodule of the high-energy section of the linac, equipped with elliptical superconducting niobium cavities. We present the status of the design and the planned programme that foresees the experimental characterisation of the fully-equipped cavity and RF system under its nominal operating conditions.

Introduction

The EUROTRANS programme and its prototypical activities intended to provide an experimental evaluation of the reliability figures for the main modular components of the accelerator configuration are described in a separate contribution of these proceedings [1]. One of the tasks in the accelerator working package of the EUROTRANS project is dedicated to building and testing a full prototypical cryomodule of the high-energy section of the superconducting proton linac.

A cryomodule is the basic building block of the superconducting accelerator section and has the main role to provide to the cavities both a mechanical support and their cryogenic environment for operation. The XADS/XADT linac, the main parameters of which were briefly reviewed in Ref. [1], on the basis of the high availability/reliability requirements and maintenance considerations, needs a cryomodule designed for easy disconnection both from the beam line and the cryogenic liquids distribution plant [2]. The aim of the work briefly presented here is to deliver an operational prototype cryomodule, which can be extensively tested (without beam, but at high RF power levels as in its operating condition), to assess its main reliability characteristics.

The test cryomodule will be a prototypical module of the beta 0.5 section containing one single elliptical multi-cell superconducting niobium cavity with all its auxiliary equipments. In the high-energy section of a proton linac, consisting of a few families of elliptical cavities with gradually increasing lengths, the technological complexities decrease gradually with the increase of the cavity length (that is, with the increase of the velocity, and energy, for which the proton cavity is designed). It is felt by the accelerator designers that the lowest limit for the elliptical cavity technology is a cavity design for proton velocities at half the speed of light (beta = 0.5), which has been indeed chosen for the first family of the high-energy section of the XADS linac.

Figure 1. XT-ADS reference accelerator scheme: a doubled linac front end is followed by a fully modular spoke and elliptical cavity SC section, upgradeable from 350 MeV up to 1 GeV for the EFIT needs. Typical cavity prototypes are shown in the lower part: from left to right RFQ, CH structure, spoke, elliptical 5-cell.

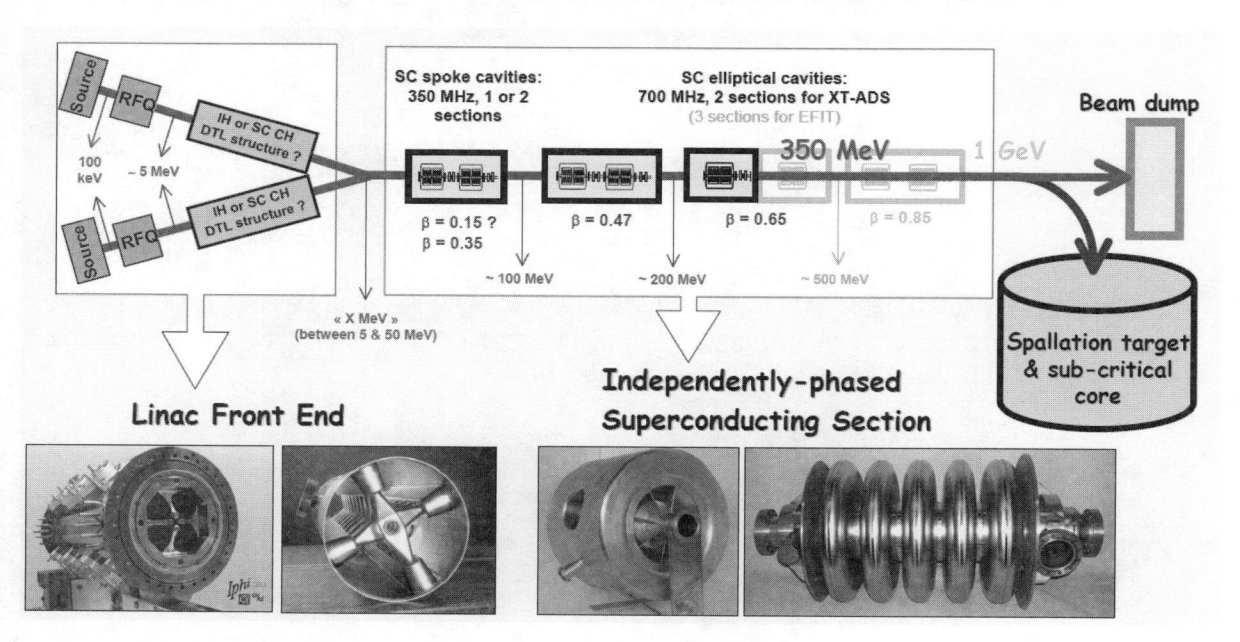

Besides the development of the bare superconducting cavities, it is important to prototype each auxiliary system needed for the cavity operation in a real environment (frequency tuner, power coupler, RF source, power supply, RF control system, cryogenic system, cryostat…), and relative procedures of assembly and alignment. The construction of a full-scale module with a $\beta = 0.5$ cavity (100-200 MeV energy range) can be considered as a rather general proof-of-principle of the technology and a test stand for determining its reliability characteristics.

The development of the cryomodule relies heavily on prior R&D results, existing infrastructures and investments of INFN, CNRS and CEA. INFN will make available its two beta 0.5 TRASCO [3] cavities equipped with the cold tuning system [4], which have outperformed the EUROTRANS specifications during vertical tests [5]. CNRS will contribute with clean room for cryomodule assembly, cryogenic infrastructure for test at 2 K, and manpower for integration, assembly and tests. CEA will contribute with cavity preparation (chemical treatment) and RF digital controls.

The final installation and testing of the cryomodule will be done at the Supratech infrastructure under preparation at IPN/Orsay, where the following hardware and facilities will be available:

- High-power RF sources:
  - a 350 MHz source (for EURISOL and EUROTRANS spoke cavities): 10 kW unit;
  - 700 MHz sources: 80 kW unit (IOT).
- A clean room for cryomodule assembly:
  - 85 m^2 clean room, with 45 m^2 of class 10/100 for cavity assembling and handling;
  - ultra-pure water production system and HPR facility.
- A helium liquefier.
- A cavity chemistry facility.

Layout of the module

The actual layout of the module is shown in Figure 2. The vessel of the module for a single cavity has a diameter of about 1.2 meters (namely 48″, as the SNS vessel) and a length of about 1.4 meters.

The design has been performed on the following reliability-based considerations:

- As a general request, easy and reliable connection interfaces to the cryogenic and RF system and fast and reliable cold mass alignment strategies are needed to guarantee a short mean time to recover in the case of a module exchange in the linac.

- As in the SNS experience, in order not to produce mechanical stresses on the warm ceramic RF window the fundamental power coupler is positioned vertically. Furthermore, in order to avoid the possibility of contamination of the inner cavity surface during the coupler assembly with dust particles, the warm RF window is positioned below the cavity.

- The actual layout simplifies the handling of the subassembly coming out of the clean room. As discussed further on, the suspension of the dressed cavity to a room temperature "space frame" with tension tie rods does not require any vertical movement of the cavity during assembly. In more details the cavity, with its ancillary components, will be supported by eight tension rods in a symmetric X pattern to a room temperature "space frame" support cage similar to the SNS concept. The space frame also acts as the support for the thermal shield that protects the cold mass at 2 K from the room temperature surfaces of the vacuum vessel, intercepting the thermal radiation at higher temperatures.

- This design allows the connection to the cold box providing the liquid cryogenic circuits at different temperature levels already developed by IPN Orsay and similar to the CM0 cold box used for the spoke cryomodule. The cold box will be located above the module, connected through a big flange on the top of the vacuum vessel.

Figure 2. Cryomodule layout

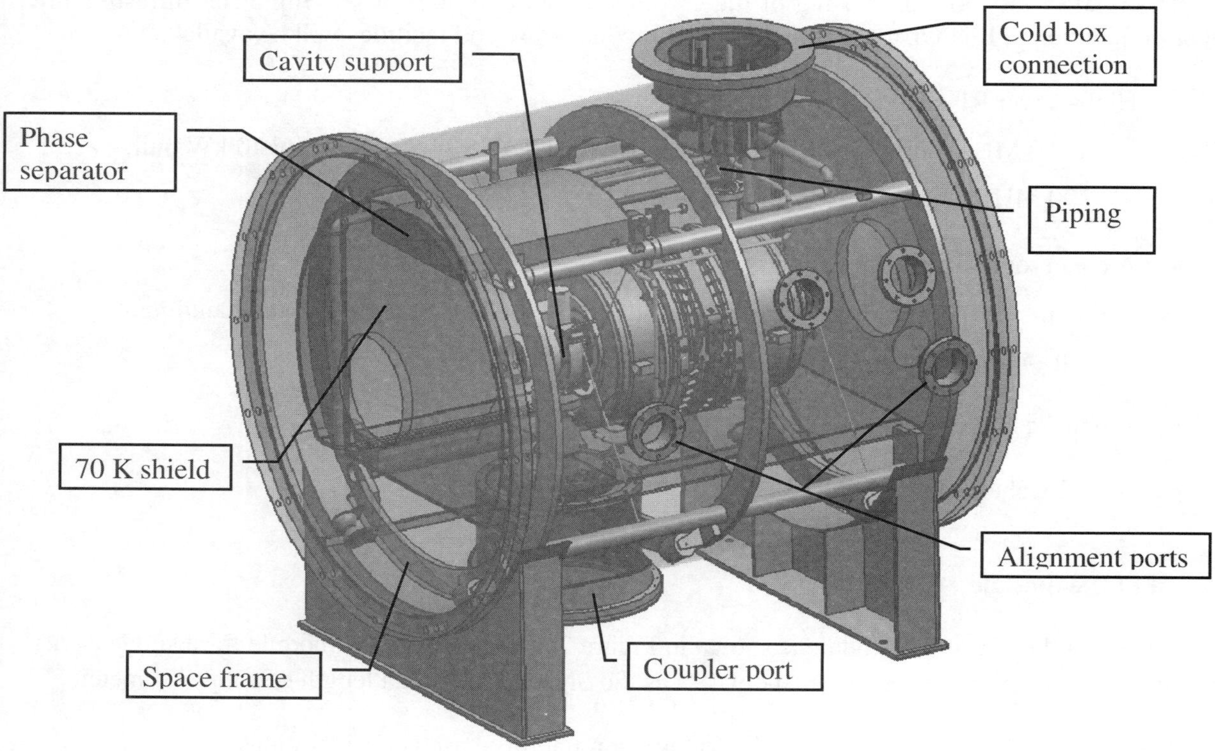

As a last consideration, it seems important to borrow, as extensively as possible, proven technologies from existing state-of-art cryomodule designs, again with the perspective of reaching the ADS reliability goals. Most of the technical solutions, and the underlying superconducting RF technology, outlined for the module layout have been derived from the huge experience accumulated by the TESLA Test Facility (TTF) and the Spallation Neutron Source (SNS) experience [6]. These technologies represent the state-of-art in the design of a large superconducting RF infrastructure with perspectives of reaching an unprecedented high operational availability, necessary for an acceptable economical cost of the underlying applied and fundamental physics experimental programmes for which both accelerators have been designed.

The "dressed" cavity

The module will be equipped with one of the two TRASCO cavities, fully "dressed" with the components required for their operation in a linac cryomodule. In particular, the cavity will be equipped with a Ti-He reservoir (which provides the low pressure bath liquid He for the operation at 2 K) and with a jacket of a high-permittivity (at low temperatures) material (Cryoperm) that provides the required shielding from the earth magnetic field in order to guarantee low values of surface resistance for RF loss minimisation. An alternative and simpler solution for using a magnetic shield at the inside of the helium reservoir is currently under investigation.

Design verification according to pressure vessel standards

An ASME calculation package has been set up to analyse the main vessel package according to pressure vessel standards. The calculation package allows to rapidly check the conformity of the parametric design as it evolves. The formal ASME procedure has been followed to verify the design data, in terms of vessel dimensioning (thickness), nozzle verification, vessel head dimensioning and stresses under dead load. Finally, all the QA/QC procedures used in the TTF module fabrication (mainly welding procedures and testing) will be used for the module fabrication.

Thermal design

One of the main issues driving the technical design of a cryomodule is the heat load budget at low temperatures. This can be either coming from a static contribution (thermal radiation, heat conduction or convective heat transfer from the room temperature environment) or from a dynamic contribution driven by the presence of the RF fields sustained by the accelerating cavities.

Static heat loads

Static losses can be handled with several technological expedients in a module, minimising the heat flows towards the 2 K He bath and intercepting it at higher temperatures, thus gaining in thermodynamic efficiency. The thermal radiation flowing from surfaces at room temperature is intercepted with a thermal shield fixed at intermediate temperature, and minimised by using multi-layer insulating (MLI) blankets (layers of doubly aluminised Mylar sheets separated by a low thermal-conducting spacer material). Ideally, a single thermal shield level in the temperature range between 35 and 50 K would be the best solution to minimise effectively these sources of static in-leak. This solution, however, requires a cryogenic system with the ability to mix warm helium gas at the module entrance, in order to produce the desired temperature coolant for the shields. The actual cold box system that will be used during testing lacks this feature and thus an independent circuit using liquid nitrogen has been assumed in the design. Direct heat conduction paths from the room temperature environment to the 2 K circuit need to be carefully intercepted at intermediate temperatures. Finally, convective effects are prevented by providing a good insulating vacuum between the cold mass and the external room temperature containing vessel.

For the estimation of the static heat load budget, and its proper handling, a thermal analysis of each direct path from the room temperature vessel to the cold mass environment has been performed, by taking into account the proper temperature-dependent material properties. Estimation based on simplified geometries and tabulated material data has been verified with finite element calculations on the model geometry.

Dynamic loads: RF losses

A superconducting cavity deposits the power P_{cav} on its walls according to the relation:

$$P_{cav} = \frac{(E_{acc} L_{active})^2}{R/Q \; Q_0}$$

where E_{acc} is the accelerating field level, L_{active} the active cavity length, Q_0 the quality factor of the resonator and R/Q is a geometrical parameter, determined only by the shape of the resonator.

For the TRASCO cavity operation at its design parameters (E_{acc} = 8.5 MV/m, $Q_0 = 10^{10}$, with L_{active} = 0.5 m and R/Q = 160 Ω), this estimation gives 11.3 W of dissipated power at each cavity. For comparison, in the SNS case, the two cryomodule types (medium beta and high beta) have a nominal RF dynamic load of 10 and 13 W, respectively. Even if the vertical tests performed on the existing structures were able to exceed these limits, for the heat load computations, a conservative value of $Q_0 = 5 \; 10^9$ has been assumed.

Heat load budget

Table 1 reviews the overall heat load budget on the module. Only the 2 K main circuit and the shield circuit have been taken into account. The overall heat load estimates, even with the uncertainties arising from a few missing contributions from the table, seems to be well within the capacity of the CM0 cold box (~ 50 W).

Table 1. Heat load budget on the module

	2 K circuit		70 K circuit	
	Static	Dynamic	Static	Dynamic
RF load @ 8.5 MV/m and 5 10^9		22.58	8.14	
Tie rod connections from shield	0.12		52.10	
Phase separator support frame	0.44		18.33	
Thermal radiation[1]	0.40		15.08	
Cabling	0.10			
Coupler[2]	1.00	6.77	1.00	
Total [W]	**2.07**	**29.35**	**94.65**	**0.00**

[1] Based on 0.1 W/m² from 77 K surfaces and 2 W/m² from 300 K.
[2] Static from SNS estimates (30% conduction, 70% radiation). Dynamic based on SNS estimate, 30% of RF load.

On the basis of the heat load assessment reported above have been calculated the required mass flow for the shield and cavity circuits. For these calculations, it has been assumed that the shield circuit is cooled by nitrogen evaporation. The calculation does not take into account the additional heat losses in the cold box. Even using a conservative estimate for the dynamic heat loads and a 30% overcapacity budget for the coolant mass flow, a total requirement of 1.7 g/s at 2 K is needed for the LHe circuit, consistently with the 2.2 g/s capabilities of the CM0 cold box.

For the moment, the estimation of the cryogenic needs does not include the cooling circuit for the coupler, since the details of this component are still under finalisation. Nonetheless, for comparison, the SNS coupler requires a flow of 5 K, 3 atm He of 0.0375 g/s, of which 2/3 is used to handle the static losses and the rest handles the dynamic losses at 6% RF duty cycle. In terms of cold box capabilities, the extra flow needed for the coupler cooling should be compatible with the existing CM0 cold box flow limitations.

The cryogenic cold box

The cryogenic cold valves box is derived from the IPN Orsay CM0 design, which has a nominal cooling capacity of 50 W at 2 K. The cold box supplies 3.3 g/s of liquid helium for the cavity cooling, 2.5 g/s of liquid helium at 4.3 K and a Joule Thomson valve for the 2 K operation (it can produce a maximum flow rate of 2.2 g/s at 1.9 K, ~50 W).

A ~20-litre helium buffer is placed inside the valve box to prevent pressure perturbations on the cavity when feeding liquid helium. A thermal shield, able to evacuate around 60 W at around 60 K, made of copper and covered with multi-layer insulation, surrounds the cold internal parts of the valve box. The valve box has no specific cryogenic loop for the power coupler cooling.

The coupler

The 704.4 MHz power coupler is under development at IPN Orsay and is based on the SNS design, adapted and optimised for the needs of this design (different frequency, more CW power, cooling included).

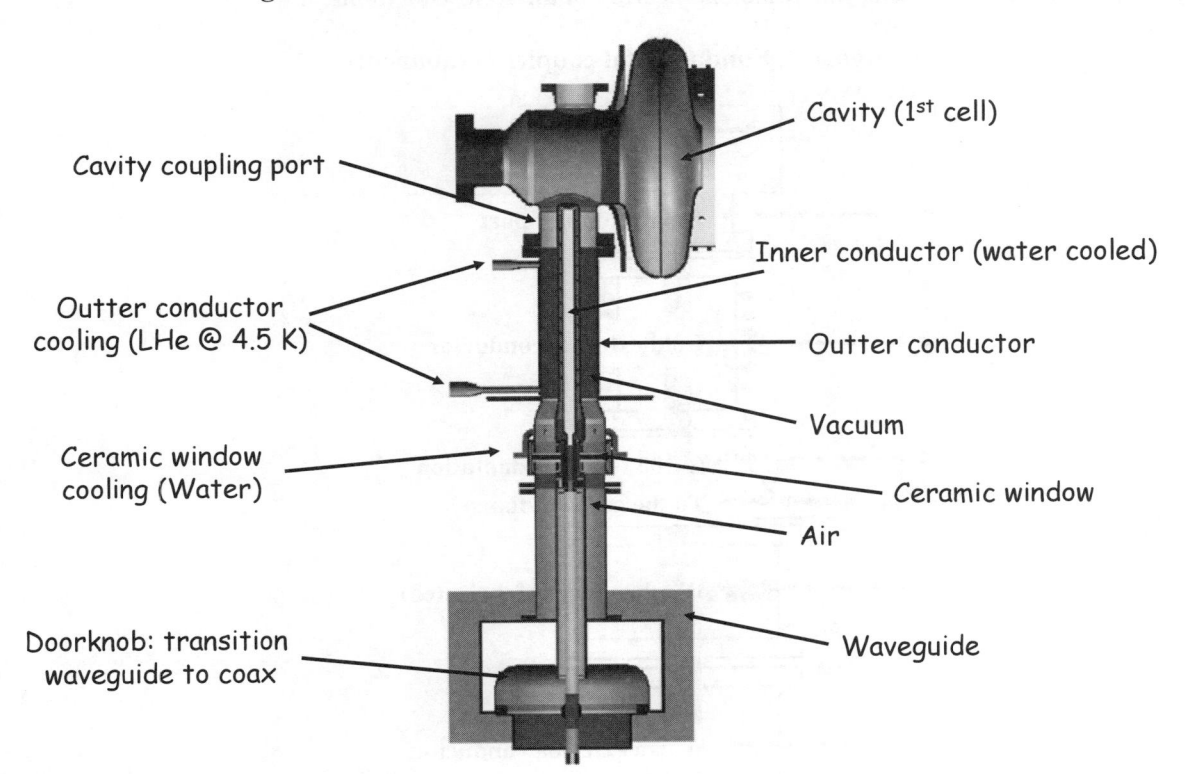

Figure 3. Scheme of the SNS fundamental power coupler

The main characteristics of the coupler are the following:

- capacitive coupling to the cavity;
- coaxial geometry of the antenna;
- ceramic window geometry: disk with chokes;
- doorknob to perform the transition from the coaxial to the waveguide geometry;
- capable to transmit 150 kW of RF power to the cavity;
- inner cooling of the antenna with room temperature water;
- window cooling by water @ 300 K;
- cooling of the outer conductor between the ceramic and the cavity by supercritical liquid helium @ 4.5 K.

Considerations about the global dimensions of the coupler

The only part of the coupler which has already been fabricated is the window. Every other coupler part has been designed, but not fabricated. A further iteration of the coupler design will be performed during the finalisation of the cryomodule engineering in the following months.

Figure 4. Fundamental coupler components

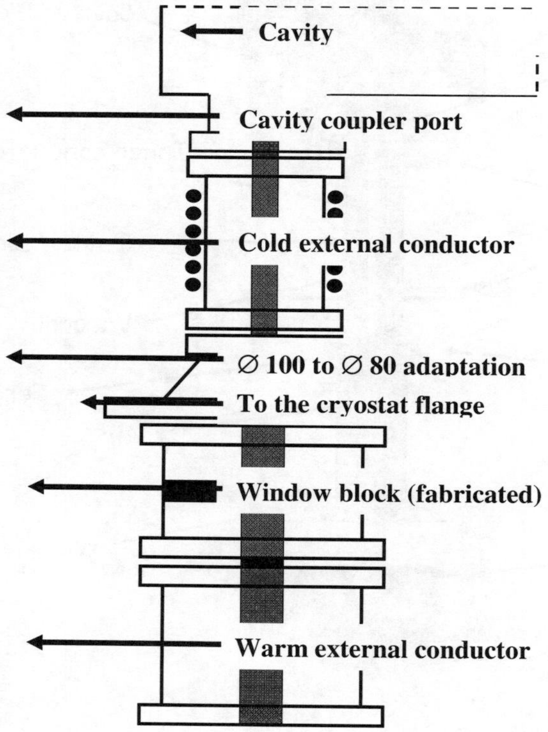

The cold external conductor and the diameter adaptation could be only one mechanical piece, as the one which has been used for the module development.

At the present stage of the study, the only constraints on the dimensions are given by:

- The already-fabricated window block (fixed dimensions).

- The height of the adaptation from $\varnothing$ 100 to $\varnothing$ 80 which has not yet been calculated (the height should be between 5 and 20 cm).

- The height of the cold external conductor has a minimum value because this part is also the thermal transition between 4 K and 300 K. A ~25-cm height should be a minimum value for this dimension, but the thermal calculation has to be performed (this answer is also linked to the cooling fluid properties). An interception point at 80 K will also be defined and placed on this coupler part.

The clean room cavity-coupler assembly

The cavity, the helium tank and the coupler window have to be mounted inside a clean room in order to seal the cavity volume and to avoid any contamination. The assembly procedure inside the cryomodule has to allow the positioning of the window outside the cryostat volume.

The following procedure is envisaged for the assembly of the cavity in the module:

- STEP 1 – Pre-assembly of the cold mass on the space frame. The cavity and cold coupler part (up to the room temperature window) are connected in the clean room and the entire assembly, translated on a temporary support, is attached to the space frame via the tension tie rods.

- STEP 2 – Rolling into the vessel: The space frame is rolled into the vessel.

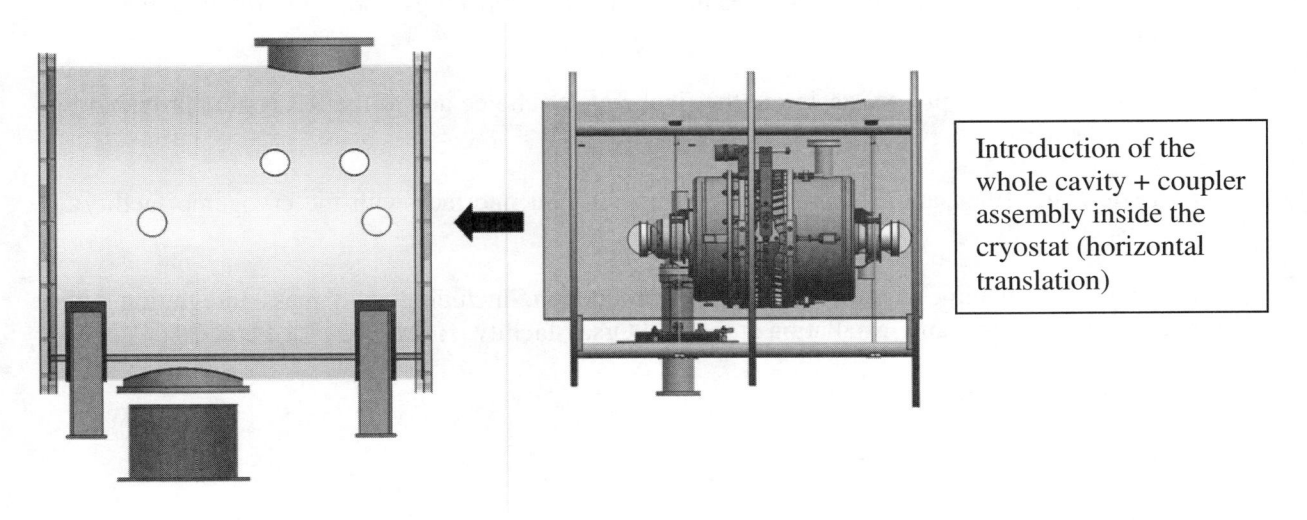

Introduction of the whole cavity + coupler assembly inside the cryostat (horizontal translation)

- STEP 3 – Vacuum enclosure: The vacuum closure which separates the warm coupler part from the cryostat vacuum is slid from below to separate the two environments.

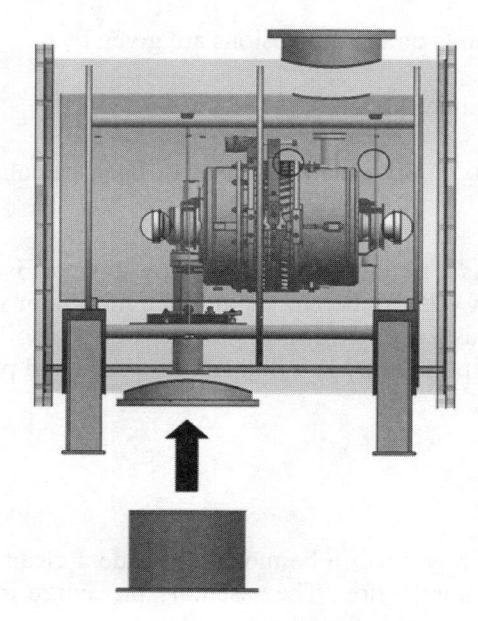

Insertion of the vacuum closure separating the warm coupler window and extension.

Future work

The next step in the design activities will be the finalisation of the following aspects in the cryomodule layout:

- final choice of beam vacuum details: metal gate valves, cavity pumping system …;

- definition of all service ports: insulation vacuum openings, pressure relief valves and alignment references;

- adaptation of the present design to the final coupler choice and to the RF system provided by IPN Orsay;

- review and finalisation of the cryogenic piping and interface with the cryogenic facility at Orsay.

The expected timeline for the cryomodule production, including cold mass integration after tuning, cleaning and tests, and installation at the IPN Orsay facility is:

- 12/2007 final engineering and start procurement;

- Fall 2008, assembly in Orsay.

Acknowledgements

This work is supported by the EURATOM 6th Framework Programme of the EC under contract FI6KW-CT-2004-516520.

REFERENCES

[1] Biarrotte, J-L., "Status of the EUROTRANS R&D Activities for ADS Accelerator Development", these proceedings.

[2] Pierini, P., C. Pagani, *Task 1.3.3. Preliminary Report: Specifications of the Prototype Cryomodule*, EUROTRANS report, Deliverable 1.15 of EU contract FI6W-516520, April 2006.

[3] Napolitano, M., *et al.*, "Status of the High Current Proton Accelerator for the TRASCO Program", *Proceedings of EPAC 2002*, Paris, France, p. 251 (2002).

[4] Pierini, P., *et al.*, *Report on Tuner Design*, CARE-Note-2006-003-HIPPI.

[5] Bosotti, A., *et al.*, "RF Tests of the Beta = 0.5 Five Cell TRASCO Cavities", *Proceedings of EPAC 2004*, Lucerne, Switzerland, p. 1024 (2004).

[6] Campisi, I., "Testing of the SNS Superconducting Cavities and Cryomodules", *Proceedings of the 2005 Particle Accelerator Conference*, Knoxville, TN, USA (2005).

PROTON ENGINEERING FRONTIER PROJECT

Yong-Sub Cho, Jun-Yeon Kim and Byung-Ho Choi
Korea Atomic Energy Research Institute, Korea

Abstract

The Proton Engineering Frontier Project (PEFP), promoted by the Korean government since 2002, has the goal to develop a 100-MeV high-current proton linear accelerator, its beam utilisation and industrial applications. The 100-MeV proton accelerator will be operated with maximum 8% duty and maximum 160 kW beam power. High-power proton beam handling technology and accelerator reliability are the key issues in the R&D for PEFP. The front-end 20-MeV accelerator has been constructed and tested at KAERI test stand. The user beam lines, which can utilise high-power proton beam for many applications, have been designed. In addition, site preparation and construction works are being carried out in co-operation with the city of Gyeongju. In the workshop, the status and the future plan of the project including the test results of the front-end accelerator, the site preparation and construction works will be presented in detail.

Introduction

The PEFP was launched by the Korean government in 2002 to realise potential applications of the high-power proton beams. Its primary goal is to develop a high-power proton linear accelerator to supply 100-MeV, 20-mA proton beams [1] and to construct user beam line facilities, with which the users can utilise proton beams of wide ranges of energies and currents for their research and development programmes. In addition, the 100-MeV accelerator can be used as a proton injector for the next-stage high-power accelerators with higher energy, such as a higher-energy linac or rapid cycling synchrotron for GeV proton beam application like a spallation neutron source.

The main concept of the PEFP proton beam facility is that a high-power proton accelerator supplies proton beam to many users simultaneously. This concept can be compared with a facility with many low-power proton accelerators for many users. Based on the user demand survey for proton beam applications, we chose a facility with a high-power accelerator. There are many types of proton accelerator for proton beam applications, such as cyclotron, synchrotron, and linear accelerator. Because the capability of high beam power and future extension is the most important feature, we had decided to choose a linear accelerator as the main accelerator of this facility. Figure 1 shows the schematic diagram of the PEFP user beam line.

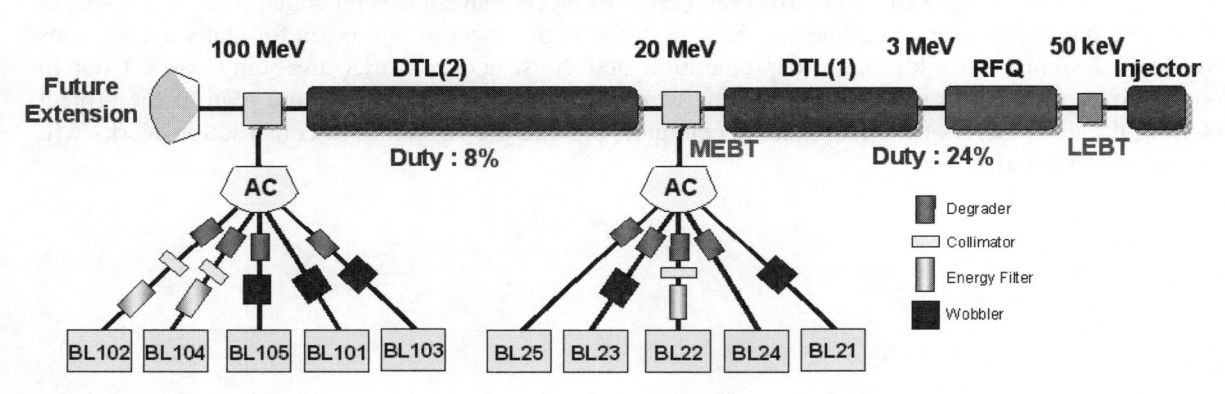

Figure 1. Schematic diagram PEFP accelerator facility

Proton beams of 100 MeV and 20 MeV will be extracted and distributed to a maximum of five users simultaneously by AC magnets with a programmable current power supply. The total power of 100 MeV beam is 160 kW and that of 20 MeV is 96 kW. We will control the beam energy step-wise with RF ON/OFF for each DTL tank. To control the beam energy continuously, we will put energy degraders and energy filters in the beam lines for special applications.

Proton beam requirements

The surveys for proton beam demands from many application fields, such as nanotechnology (NT), biotechnology (BT), space technology (ST) and radioisotope production, were undertaken through the homepage (http://www.komac.re.kr) and the user programme from 2003 to 2006. From these activities, we selected the common requirements for many applications and summarised the beam line requirements for 10 beam lines of 100 MeV and 20 MeV (see Tables 1 and 2). In the selection process, we put more weight on high beam power applications, which will be the main applications in this facility.

Table 1. 100 MeV beam line requirements

Beam line no.	Energy (MeV)	Average current	Irradiation condition	Max. target size
BL100	103	~1.8 mA	Horizontal/vacuum	Beam dump
BL101	33,45,57,69,80,92,103	30~300 μA	Horizontal/vacuum	100 mm
BL102	20~103	~10 μA (10 nA)	Vertical/external	300 mm
BL103	20~103	30~300 μA	Horizontal/external	300 mm
BL104	20~103	10 nA~10 μA	Horizontal/external	300 mm
BL105	103	30~300 μA	Horizontal/vacuum	100 mm

Table 2. 20 MeV beam line requirements

Beam line no.	Energy (MeV)	Average current	Irradiation condition	Max. target size
BL20	20 MeV	~4.8 mA	Horizontal/vacuum	Beam dump
BL21	20 MeV	120 μA~1.2 mA	Horizontal/vacuum	100 mm
BL22	3~20 MeV	10 nA~60 μA	Vertical/external	300 mm
BL23	3~20 MeV	60 μA~1.2 mA	Horizontal/external	300 mm
BL24	20 MeV	120 μA~1.2 mA	Horizontal/vacuum	100 mm
BL25	20 MeV	120 μA~1.2 mA	Horizontal/vacuum	300 mm

20-MeV accelerator at KAERI test stand

The PEFP 20-MeV proton accelerator has been developed and tested at KAERI test stand as shown in Figure 2. The main accelerator facilities at KAERI test stand are the 20-MeV accelerator which consists of a 50-keV proton injector, 3-MeV RFQ and 20-MeV DTL, two sets of 1-MW, 350-MHz RF system, two sets of –100-kV, 20-A DC high voltage power supply for the klystron, two sets of 2-MW cooling system for the cavity and RF system. The design duty of the 20-MeV accelerator is 24% and two sets of 1-MW, 350-MHz klystron are used to drive a 20-MeV accelerator, one is for RFQ and the other is for DTL. All the other ancillary facilities such as klystron power supply and cooling system were designed for 100% duty operation. During the low duty operation at KAERI test stand, the RF system is operating such that the electron beam of the klystron is CW whereas only the input RF signal is modulated for the low duty pulse operation. The purposes of the initial test of the

Figure 2. 20-MeV linac at KAERI test stand

20-MeV accelerator are to check and evaluate the overall system performance and to study the machine operation parameters for proper beam acceleration. Especially during the initial test, we should limit the average beam current because the radiation shield is not enough for full beam power at KAERI test stand. Therefore, the operation parameters for initial test are 50 μs beam pulse width with 0.1 Hz repetition rate [2].

The proton injector includes a duoplasmatron proton source and a low-energy beam transport (LEBT). The beam current extracted from the source reached up to 50 mA at a voltage of 50 keV using a 120-V, 10-A arc power. The normalised emittance of the beam from the proton source is 0.2π mm · mrad of a 90% beam current, and the proton fraction is larger than 80%. To operate the ion source in pulse mode, a high voltage semiconductor switch is installed in the high voltage power supply of which the rising and falling time are less than 50 ns. With the semiconductor switch, the pulse length and the repetition rate can be easily changed [2]. The LEBT consists of two solenoid magnets for the beam focusing, two steering magnets for the beam position and angle correction, ACCT for beam current measurement and electron trap at the entrance of the RFQ to remove the electrons. Figure 3 shows the beam signal, which is measured with the Faraday cup at the exit of the ion source.

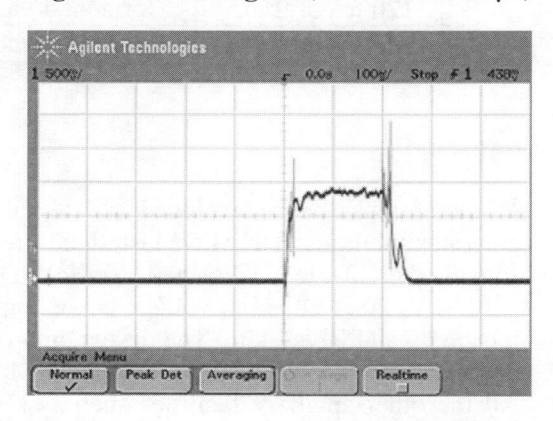

Figure 3. Beam signal (1 mA/div. 200 μs)

The output beam current from the RFQ was measured with respect to the LEBT magnet parameters and RFQ vane voltage. A current transformer developed by the Bergoz was installed at the exit of the RFQ to measure the beam current. The current transformer is tuned to the fundamental beam frequency and can measure only the bunched beam signal component of 350 MHz. During the test, the current transformer tuned to 350 MHz operated like a band-pass filter, therefore the current transformer picked up the accelerated beam signal. From the test results, the operating point of the RF amplitude could be determined, which was about 10% higher than the design value [3].

During the initial test, the beam signal was not stabilised because of the improper LLRF control. Therefore, the digital LLRF system was newly developed. The stability requirements of the RF field are ±1% in amplitude and ±1° in phase. The digital feedback control system is based on the commercial FPGA PMC board hosted in VME board as shown in Figure 4. An ICS-572B commercial board was used, which is a PMC module with a two-channel 105-MHz ADC, a two-channel 200-MHz DAC, and a four million gate onboard Xilinx FPGA. For the host system, a Motorola VME processor module, MV5100 was used. The main features of the system are a sampling rate of 40 MHz, which is four times higher than the down converted cavity signal frequency, a digital in-phase and quadrature detection, and a simple proportional-integral (PI) feedback and feed-forward algorithm. The control logic based on the feedback and feed-forward control is implemented in the FPGA by using very high speed integrated circuit hardware description language (VHDL).

Figure 4. FPGA PMC board hosted in VME board

The closed loop test results using dummy cavity showed that the stabilities in amplitude and phase were enhanced by about 10 times better than those of the open loop test. To check the pulse-to-pulse stability under high-power conditions, the high-power RF system was operated in pulse mode without beam loading. During the test, the pulse width was 200 μs and repetition rate was 0.1 Hz. The RF phase of the DTL cavity is shown in Figure 5. The measured RF amplitude and phase were kept within ±0.10% and ±0.15° respectively, with the standard deviations of 0.02% in amplitude and 0.03° in phase, which meet the requirements of the RF control system [4].

Figure 5. Pulse-to-pulse RF phase variation

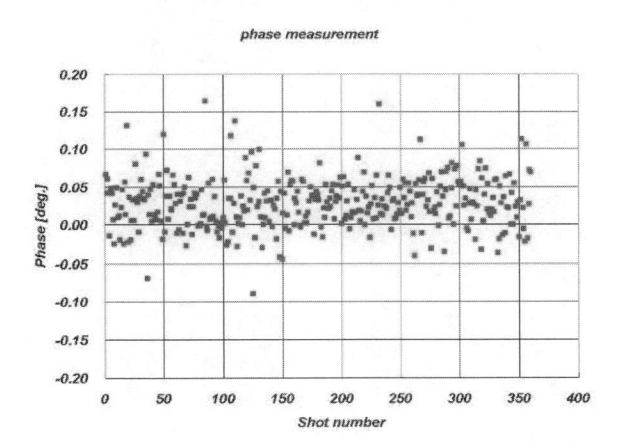

The unique characteristic of PEFP 20-MeV DTL is that one klystron drives four DTL tanks simultaneously. Also, the cavity cooling circuits for four tanks are connected in parallel from one cooling system. For this multi-cavity driving concept, the tank wall temperature control mechanism with heater was adopted, and mechanical phase shifters were also installed in each waveguide leg to the tank. By doing this, the four independent tanks can be considered as one tank. During operation, the overall frequency deviation of four tanks was compensated by controlling the coolant temperature and the resonant frequencies of each tank were finely adjusted by controlling the wall temperature.

The current transformers were installed at the entrance of tank 1 and at the exit of tanks 2 and 3. Also a Faraday cup was installed at the exit of tank 4. The beam transmission through the DTL tanks was measured using these beam diagnostic devices for tuning the machine parameters. The typical beam signal is shown in Figure 6, when the peak beam current was 2 mA. The signals of channels 1, 2 and 3 correspond to the current transformer signals of the entrance of tank 1 and at the exit of tanks 2

Figure 6. 20-MeV DTL beam signal

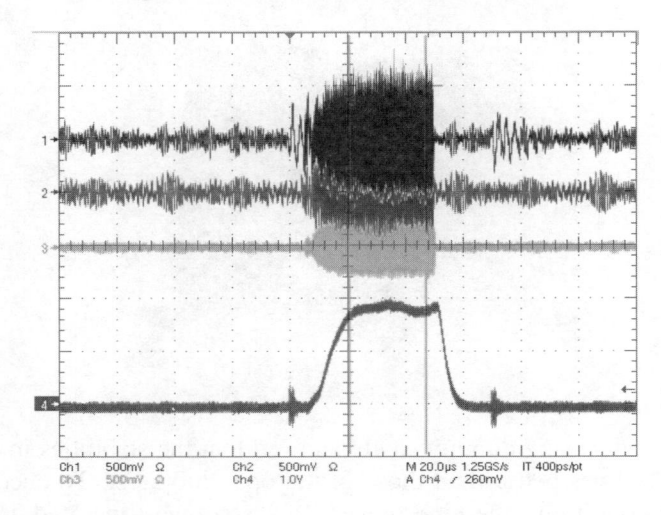

and 3, whereas the Faraday cup signal from the exit of tank 4 is shown in channel 4 of Figure 8. Recently, the 20-MeV proton beams were extracted into the air through the beam window for future beam user. The beam profile was measured using MD-55 Gafchromic film as shown in Figure 7.

Figure 7. Beam profile on the film

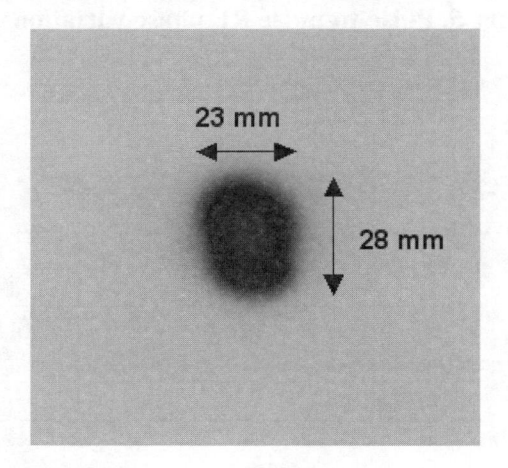

DTL2 and MEBT [5]

The low-energy part (3 MeV ~ 20 MeV), called DTL1, of DTL structures was designed for 24% beam duty. But after 20 MeV the beam duty is 8%, and we designed the higher-energy part of the linac for more efficient acceleration under the 8% beam duty. First, we determined the dimensions of the DTL2 tanks and drift tubes (DTs) by studying how the effective shunt impedance per unit length (ZTT) depends on the geometry. The sensitive geometrical parameters are the tank diameter, face angle, DT diameter and bore radius. We additionally considered the installation of the quadrupole magnet into the drift tubes when determining the DT diameter and face angles. Figure 8 shows the resulting ZTT as a function of energy. It also includes the shunt impedance of the DTL1 for completeness. The geometric parameters of DTL2 tanks are summarised in Table 3.

Figure 8. Effective shunt impedance per unit length depending on energy

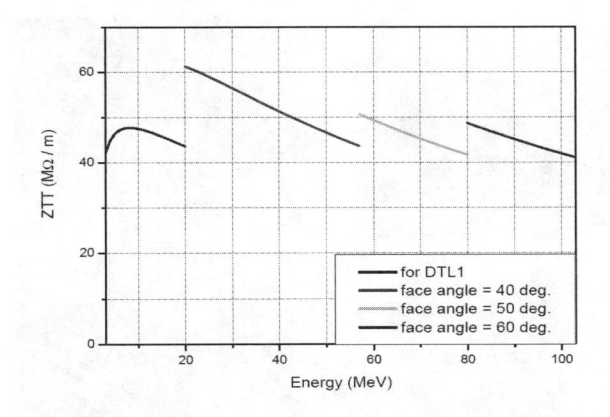

Table 3. Summary of PEFP DTL2 parameters

Parameters	Values
Tank diameter	546.657 mm*
DT diameter	135 mm
Bore radius	10 mm
Face angle	40°, 50°, 55°**
Stem diameter	40 mm
Post-coupler diameter	26 mm
Lattice	FFDD
Integrated field	1.75 T

* The value determined by including the effects of slug tuners, stems and post-couplers.

** 40° for the initial three tanks, 50° for the following two tanks, and 55° for the remaining two tanks.

We used the PARMILA code in order to determine the tank length which is less than 7 m. Each tank is divided into three sections and driven by one 1.3-MW klystron. From the engineering design point of view, the thermal stability is the most important issue. We verified the thermal stability of the tank using a thermal analysis code. RF duty was assumed as 9%. The maximum temperature increase was 12° at the bottom of the tank. The increase of the tank diameter due to thermal deformation was 20 μm. The frequency change due to the thermal deformation was expected to be about 10 kHz, which is in the controllable range. The thermal stability of the other parts such as vacuum grill and post coupler was also analysed. The maximum temperature increase was just 4° at the vacuum grill, so it was concluded that there would be no problem during normal operation.

After finishing the design of the PEFP DTL2, the fabrication of DTL tanks, drift tubes, quadrupole magnets, slug tuners and post coupler is under progress as shown in Figure 9.

One of the main characteristics of the PEFP linac is the supply of 20-MeV proton beams for low-energy beam utilisation. A 90° bending magnet which is located after the 20-MeV accelerator for the beam extraction makes a serious potential problem in beam matching at the entrance of the higher energy part (20 MeV ~ 100 MeV) called DTL2. In order to solve the matching problem, we will install a MEBT system which consists of eight quadrupole magnets and two buncher cavities. The initial four magnets are controlling the beam size in the drift space where a 90° bending magnet is located for the

Figure 9. Fabrication DTL tanks and some components

beam extraction. The following quadrupoles are matching 20-MeV proton beams into the next accelerating structure. The buncher cavities are for longitudinal matching. The system can be realised as two small DTL tanks with three cells. We used the TRACE-3D code for the beam matching between DTL1 and DTL2 through the MEBT. We used the PARMILA code for beam simulation in the linac in MEBT and DTL2. The simulated output beam of the DTL1 is used for the MEBT input beam which is obtained by the PARMILA code. The 20-MeV beam properties depend on the matching scheme between RFQ and DTL1. In this design, the beam matching between the RFQ and DTL1 was achieved by adding a quadrupole magnet between two accelerators. Figure 10 shows the 100-MeV output beam calculated with the PARMILA code.

Figure 10. DTL2 output beam: 100 MeV

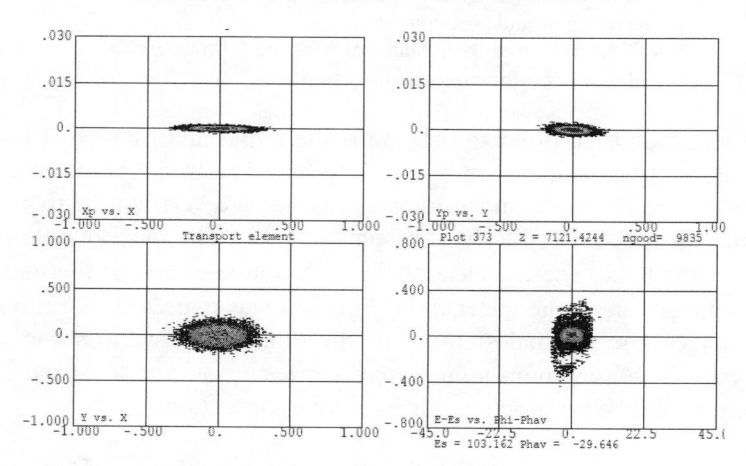

Beam line design

A 20-MeV proton beam from the drift tube linac (DTL) is transported with bending magnets and quadrupole magnets from linear accelerator tunnel to experimental hall. The first bending magnet for the user beam line is located between two buncher cavities of medium energy beam transport (MEBT) at the end of the 20-MeV linear accelerator. The basic lattice for beam transport line is FODO from the linac to the targets. Dipole magnets exited with shaped AC currents will distribute the beam from the linac to five targets simultaneously. Figure 11 shows the layout of the 20-MeV beam lines.

Figure 11. Schematics of 20-MeV beam lines

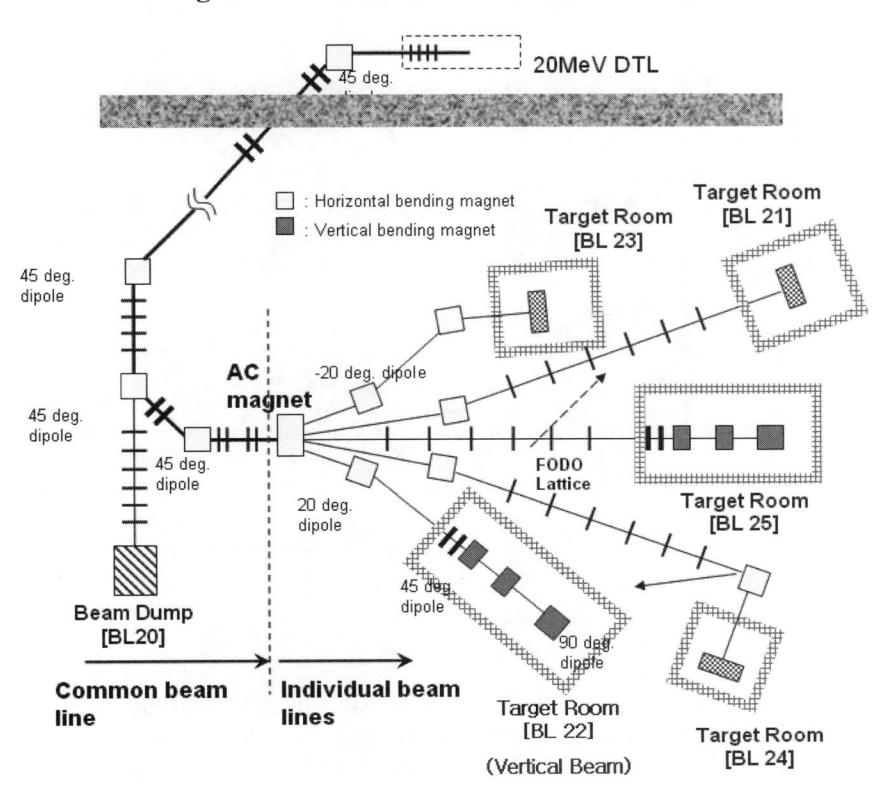

To provide flexibilities of irradiation conditions for users from many application fields, we design beam lines to the targets with wide or focused, external or in-vacuum, and horizontal or vertical beams. Figure 12 is an example of the beam optics calculation from the 20-MeV linear accelerator to target room #25 with TRACE-3D. For 100-MeV beam lines, the schematic layout is almost the same as that for the 20-MeV beam lines. A 100-MeV beam is transported through long transport line with doublet lattice from the 100-MeV DTL. Figure 13 shows the layout of the experimental hall, which is developed with the beam requirements and the other constraints.

Figure 12. Beam optics of BL25

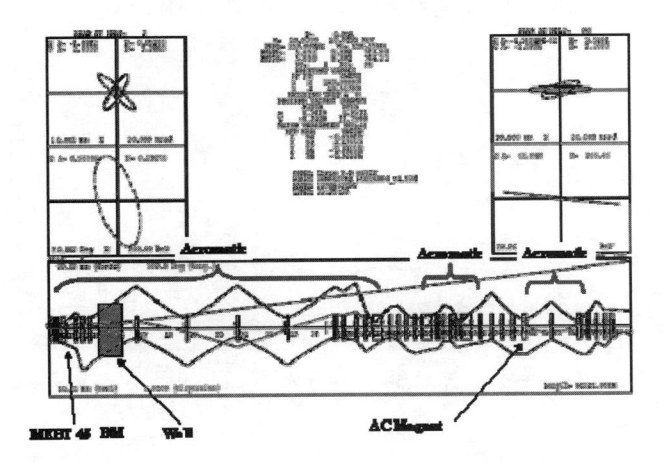

Figure 13. Layout of experimental hall

R&D for future extension

We are studying two technologies for the future extension of the PEFP accelerator programme. One is the superconducting radio frequency (SRF) cavity, and the other is rapid cycling synchrotron (RCS). A 1-GeV superconducting linac or 2-GeV RCS with 200-MeV superconducting linac are being considered, after the existing 100-MeV linac.

The SRF linac is designed to accelerate proton beams upwards of 80 MeV at a frequency of 700 MHz. It consists of three sections. We have completed a preliminary design of the cryomodule, which includes the design of superconducting RF cavity, a HOM coupler, a fundamental power coupler, a tuner, a cooling system, a magnetic shielding, a vacuum system, a cryomodule stand and a PEFP superconducting RF laboratory. Five-cell superconducting RF cavities with beta value of 0.42 and frequency of 700 MHz are shown in Figure 14. As concerns the cavity design, we studied cavity shape optimisation, HOM analysis, multipacting simulation and mechanical analysis.

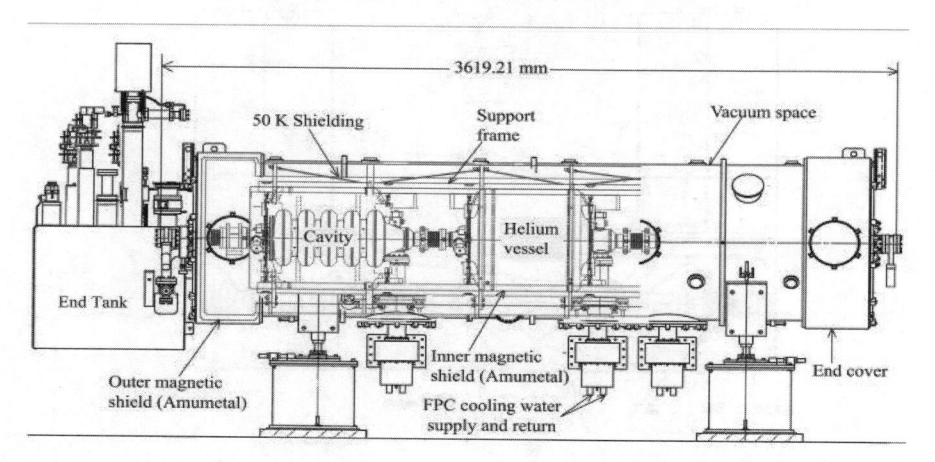

Figure 14. Superconducting linac design

A RCS is designed for nuclear physics experiments, spallation neutron source, radioisotope production, medical research, etc. The spallation neutron source requires high power and high current proton beam, but the radioisotope production and medical research facilities need relatively low energy and low current proton beam. The RCS with the extraction energy 1-2 GeV and injection energy 100-200 MeV is able to optimally support both the spallation neutron source and radioisotope production and medical research facilities. The fast extraction system of the RCS machine can be a spallation neutron source facility and the slow extraction system can be a radioisotope production and medical research facility simultaneously. The RCS is composed a of magnet lattice with main ring, injection system and extraction system. The schematic layout of the RCS machine is shown in Figure 15. The RCS is initially at 15 Hz with injection energy of 100 MeV where the target beam power is about 58 kW at 1 GeV at the first stage. In addition, the RCS is designed to have upgrade option of the beam power from 58 kW to 900 kW step by step.

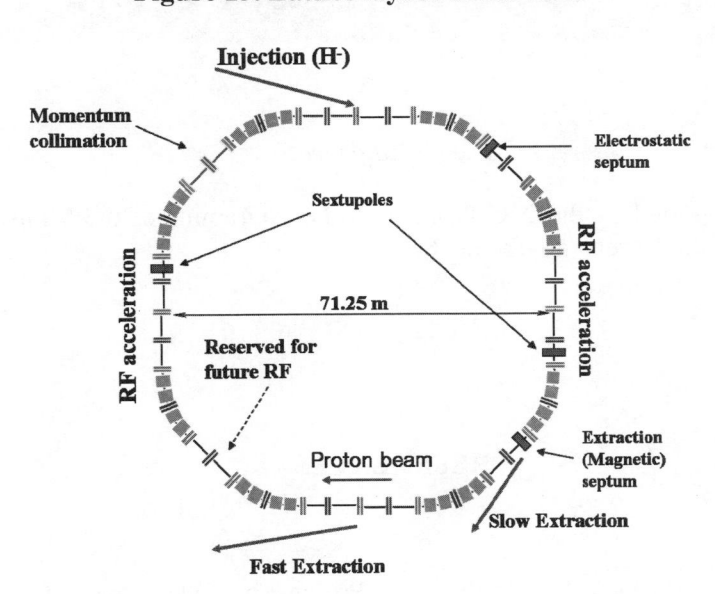

Figure 15. Lattice layout of the RCS

Construction works

The project host site was selected, in January 2006, to be Gyeongju city. After geological surveys of the site and site plan have been done, the preliminary facility layout has been undertaken, as shown in Figure 16. The architectural works of conventional facilities are under way. This work is to build up facilities to supply the research centre with electric power, water and other kinds of utilities from outward sources.

Conclusion

We have developed and tested a 20-MeV proton linac, and are fabricating a 100-MeV proton linac. User beam lines have been designed with the beam requirements from many users. The site for utilisation of high-power proton beam will be constructed in 2009 and the beam will be supplied to users from 2010. For the extension of the 100-MeV linac, we are studying SCL and RCS.

Figure 16. Bird's eye view of the PEFP research centre

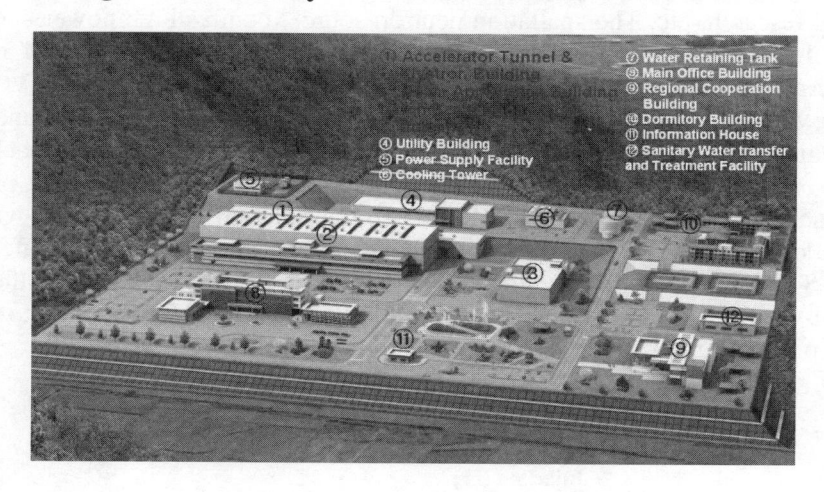

Acknowledgements

This work was supported by the 21C Frontier R&D programme of the Ministry of Science and Technology (MOST) of the Korean government.

REFERENCES

[1]   Cho, Y.S., *et al.*, "100-MeV High-duty-factor Proton Linac Development at KAERI", *Proc. of LINAC2006*, Knoxville, May 2005.

[2]   Cho, Y.S., Ky Kim, "Transient Characteristics of a Pulsed Proton Injector for the PEFP", *J. Korean Phys. Soc.*, 48. 721 (2006).

[3]   Kwon, H.J., *et al.*, "RF Set Point Determination of the PEFP 3MeV RFQ", *J. Korean Phys. Soc.*, forthcoming.

[4]   Kim, H.S., *et al.*, "Low Level RF Control System Development for the PEFP Proton Accelerator", *J. Korean Phys. Soc.*, forthcoming.

[5]   Jang, J.H., *et al.*, "Design of the PEFP MEBT", *Proc. of PAC 2005*, Knoxville, May 2005.

DEVELOPMENT OF SUPERCONDUCTING SPOKE CAVITIES FOR AN ADS LINAC

A. Ponton, J-L. Biarrotte, S. Bousson, C. Joly, N. Gandolfo,
J. Lesrel, L. Lukovac, F. Lutton, A.C. Mueller, G. Olry, E. Rampnoux
Institut de Physique Nucléaire d'Orsay, CNRS/IN2P3 Université Paris-Sud, France
E-mail: pontoon@ipno.in2p3.fr

Abstract

A high degree of reliability is required for the XADS linac in order to demonstrate the feasibility of a nuclear waste transmutation facility including a high-power proton linac, a spallation neutron source and a subcritical core reactor. IPN Orsay is strongly involved in the XADS project in the frame of which a new horizontal small-scale cryostat called CM0 has been acquired and modified in order to perform cold tests of a fully equipped spoke cavity that will compose one part of the superconducting linac. With a working operation of CM0 close to an accelerator configuration, a good reliability is foreseen for all the spoke subsystems. In this work, fully linked with reliability engineering, we report the experimental results of the spoke subsystems and describe the future work that will be completed with CM0.

Introduction

Reliability engineering is already one of the main concerns for many high-level technical branches such as aviation, automobile or medicine. An accelerator-driven system (ADS) for nuclear waste transmutation purposes must also meet a high degree of reliability that will be discussed in this paper. We will underline the consequences of beam trips and therefore the tight requirements that will be compared with data from the LANSCE accelerator. Three main guidelines can be followed to reach the wanted degree of reliability: overdesign, redundancy and the concept of fault tolerance. They will be defined and their effects on the PDS-XADS [1] design will be shown. We will present IPN Orsay activities related to spoke cavities and the link of this work with the XADS proton linac reliability. We will show current experimental spoke subsystem results and give outlooks with CM0: a new horizontal small-scale cryostat dedicated to fully equipped spoke cavities. ADS reliability has been strongly highlighted throughout this work.

Reliability for a waste transmutation ADS

Needs for reliability

In the most general case, reliability can be defined as follows: facing each component failure of a complex system, such as a driver accelerator, i.e. anticipating and monitoring the probability of these failures to achieve the nominal and required behaviour of the system.

Concerning an ADS linac, it is obviously mandatory to reach a high level of reliability. Indeed, frequent beam interruptions, in addition to reducing plant availability, can seriously damage the target, the reactor structure and the fuel elements.

Proton beam specifications

Taking into account the consequences of beam trips, the proton beam is submitted not only to classic requirements but also to acute considerations about its stability and the acceptable number of interruptions. The proton beam specifications for the PDS-XADS are summarised in Table 1 [2].

Table 1. Proton beam specifications for the PDS-XADS (data extracted from [2])

Maximum beam intensity	4 mA CW on target
Proton energy	600 MeV
Beam entry	Vertically
Beam trip number	< 5 per three-month operating cycle (exceeding 1 second)
Beam stability	Energy: ±1%, intensity: ±2%, size: ±10%

State of the art of reliability: Example of the LANSCE accelerator

A large amount of data dealing with accelerator reliability has been accumulated at LANSCE in operation for many years [3]. The facility delivers 800-MeV protons at 1 mA of beam current. What captures attention in Figure 1 is the number of trips which are quite beyond the PDS-XADS requirements: 3 515 and 1 167 beam interruptions per operating year lasting less than one minute for the H$^-$ and H+ beam respectively. Furthermore, according to Figure 2, it can be seen that the injector

Figure 1. Number of trips vs. duration of beam interruption at LANSCE

SUMMARY OF BEAM TRIPS AT LANSCE
(Normalized number of trips per operating year)

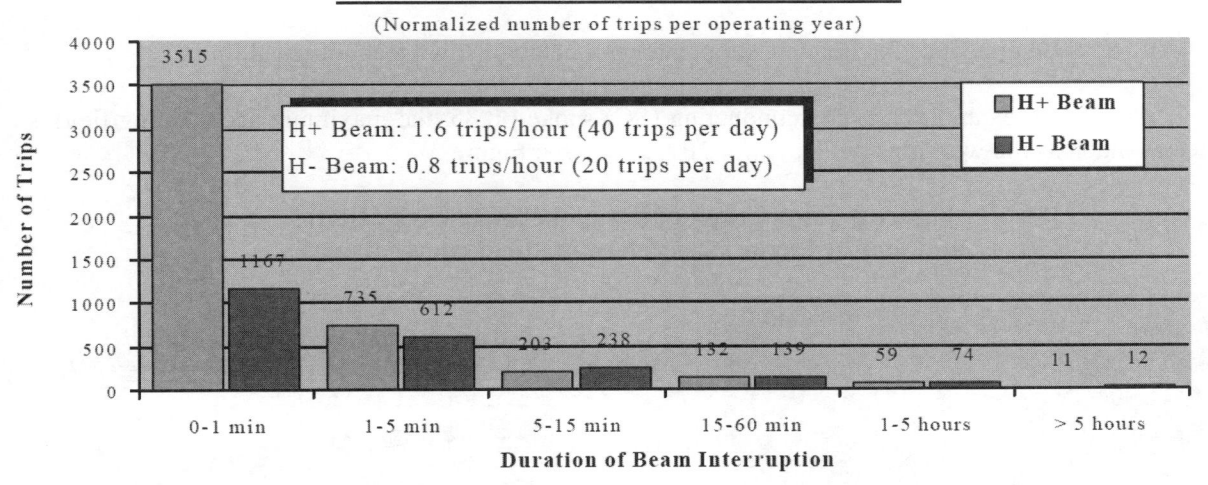

H+ Beam: 1.6 trips/hour (40 trips per day)
H- Beam: 0.8 trips/hour (20 trips per day)

Duration	H+ Beam	H- Beam
0-1 min	3515	1167
1-5 min	735	612
5-15 min	203	238
15-60 min	132	139
1-5 hours	59	74
> 5 hours	11	12

Figure 2. Histogram showing the LANSCE components accountable for most beam trips and downtime

Causes for beam trip and down time in the H+ beam
(normalized number of trips per operating year)

Summary of trips and downtime in the H+ beam:

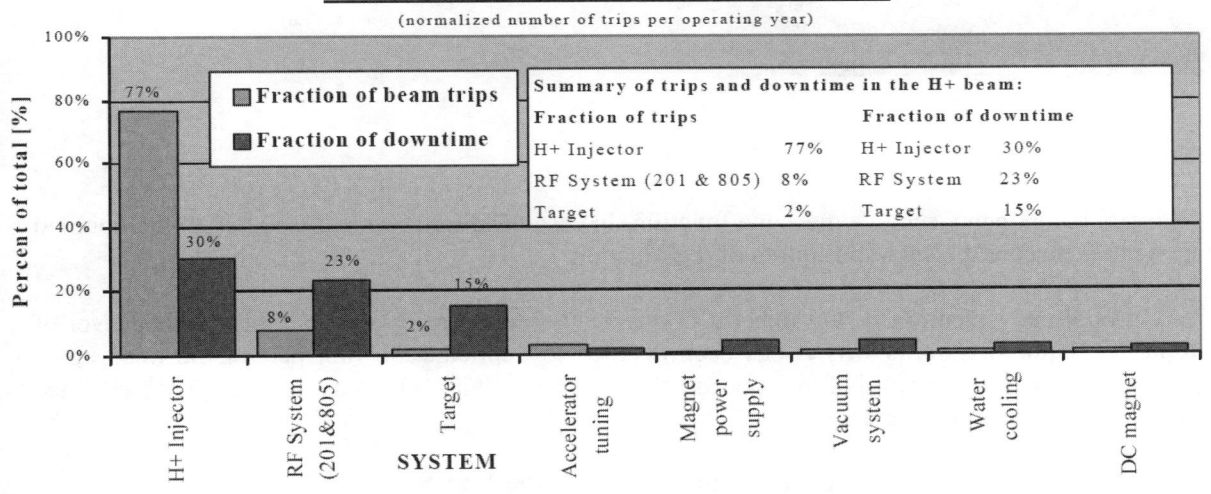

System	Fraction of trips	Fraction of downtime
H+ Injector	77%	30%
RF System (201 & 805)	8%	23%
Target	2%	15%

Systems: H+ Injector, RF System (201&805), Target, Accelerator tuning, Magnet power supply, Vacuum system, Water cooling, DC magnet

and the RF system fail very frequently and cause the most downtime. Failures to the injector constitute 77% of the trips and 30% of the downtime, while 8% of the trips and 23% of the downtime come from the RF system.

Designing the XADS linac is consequently a very challenging work and many efforts must be focused on injectors and RF systems in order to increase their reliability. In the next section the main guidelines to reach the proton beam requirements will be described and the experimental results from IPN Orsay will be shown.

Reliability guidelines for an ADS linac

Derating/overdesign

Any one component of the linac must operate well below its own technological limit.

Taking as an example the superconducting (SC) spoke β0.35, the maximum accelerating field is twice as much as the working one: $E_{acc,max} = 2E_{acc,nominal}$ (see Figure 3).

Figure 3. Photograph of the spoke β0.35 cavity (left) and quality factor vs. accelerating field curves (right)

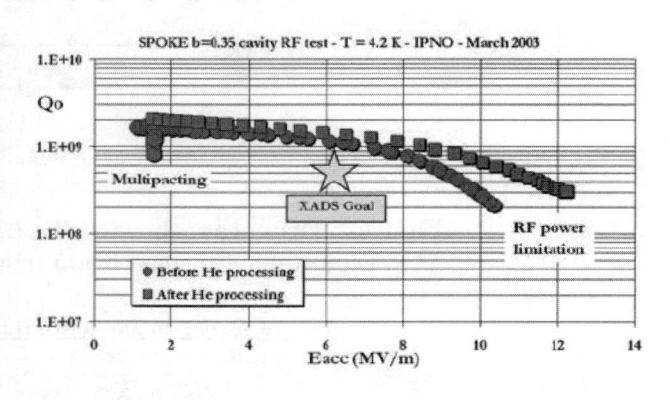

Redundancy

Several components achieve the same function. In case of failure, we could switch to the doubled components to ensure the smooth running of the function.

As was already mentioned, the injector is one of the most critical points when a high level of reliability is required. This is why it has been decided to duplicate the first part of the linac up to 17 MeV: the transition energy before entering the spoke structure. One can see the proposed linac layout for the XADS in Figure 4..

Figure 4. Linac layout of the XADS

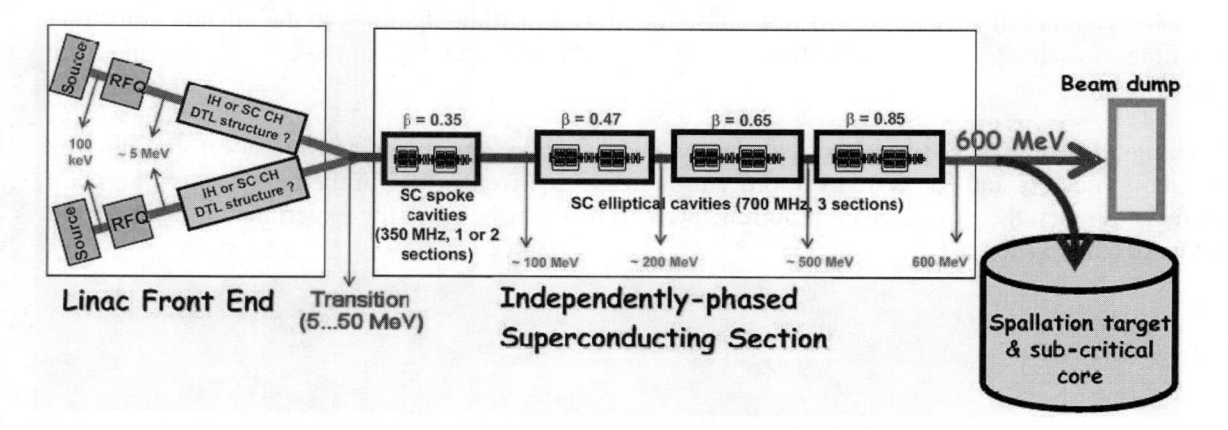

Fault tolerance

In order to achieve fault tolerance, it is necessary to monitor the system in such a way that component failures do not lead to a system failure. In this context, a beam dynamics study has been performed by J-L. Biarrotte [4] concerning the local compensation method, which is presented in Figure 5. In case of a cavity or a quadrupole failure, it has been shown that with an adequate retuning of the surrounding cavities or quadrupole doublets, the proton beam can be transported up to the high energy. There is not any beam loss and very similar envelopes, in comparison to the nominal case, can be obtained. Figure 6 shows, as an example, a multi-particle simulation when a cavity is faulty and the local compensation method is used.

Figure 5. The local compensation method for a faulty cavity

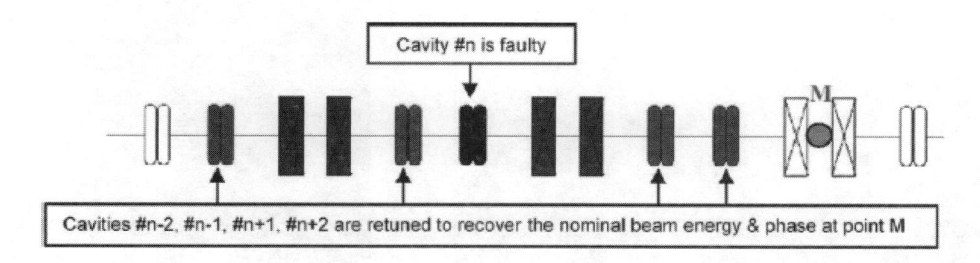

Figure 6. Multi-particle beam envelopes (x-transverse and phase) if spoke cavity #4 (5.5 MeV, 3 m) is faulty and local compensation method is performed

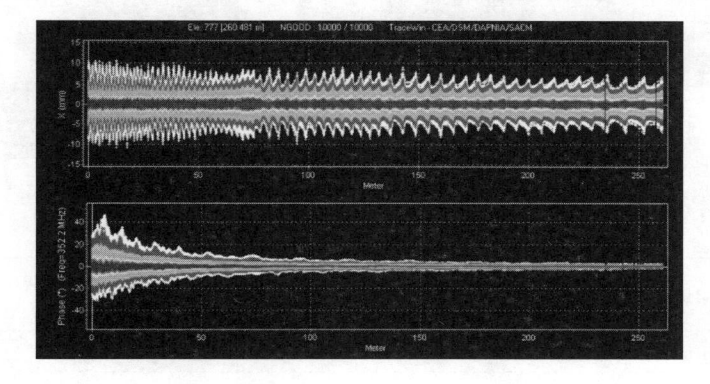

Reliability and spoke activities at IPN Orsay

Strong requirements and guidelines to reach a very high reliability level for the proton XADS linac have been presented. IPN Orsay is currently developing a horizontal small-scale cryostat, CM0, to perform cold tests of a fully equipped spoke cavity. Main focus on reliability engineering has been engaged for all the spoke subsystems such as the tuning system, the digital LLRF system, the power source and the high power coupler. In this section, we will present recent results concerning spoke activities and the outlook with CM0.

SC spoke cavities

Two prototypes of spoke cavities [5] working at 352.2 MHz with two different geometrical Betas, 0.35 and 0.15, have been designed and tested in a vertical cryostat at IPN Orsay. The β0.15 is shown in Figure 7 during test preparations. These cavities seem, under beam dynamics considerations, to be very suitable candidates for an acceleration of high intensity proton beams from 20 to 150 MeV. Since they are two-gap structures, the proton energy loss through one faulty cavity could easily be recovered by using the local compensation method already mentioned earlier. Radiation safety is also very good thanks to the large beam aperture: 50 to 60 mm. Another point deals with the performances foreseen with such a cavity. Table 2 shows that, over 20 years of spoke cavity development, in three different laboratories, mean values of 6 MV/m, 38 MV/m and 100 mT have been reached for the accelerating gradient, the peak electric field and the peak magnetic field respectively.

Figure 7. The β0.15 spoke cavity on its supporting frame without (left) and with (right) its super-insulation coat

Table 2. Summary of spoke cavity development

Labs	Spoke-type	Geometrical /Optimal betas	Eacc max* [MV/m]	Epk [MV/m]	Bpk [mT]	Voltage gain [MV]	Limitation
IPN Orsay	Single	0.15/0.20	4.77	32	69	0.81	Quench
	Single	0.35/0.36	8.15	38	104	2.49	Power
ANL	Single	0.29/0.29	8.46	40	106	2.21	Quench
	Single	0.40/0.40	7.57	46	123	2.63	Quench
	Double	0.40/0.40	8.60	40	79	4.40	Quench
	Triple	0.50/0.50	7.65	28	88	6.65	Quench
	Triple	0.63/0.63	8.61	34	104	9.40	Quench
LANL	Single	0.175/0.21	7.50	38	99	1.34	Quench

Accelerating gradient given with the βλ convention.

RF couplers and amplifiers

A 10-kW solid state amplifier (see Figure 8) has been built in strong collaboration with INFN Legnaro and will be delivered within the next two months to IPN Orsay. It is composed of several

Figure 8. The solid state amplifier at Legnaro

modules of 315 W each for which a test bench shown in Figure 9 has been realised at IPN Orsay in order to characterise their parameters. The following points, about the solid state amplifier technology, are of crucial importance from reliability point of view:

- one module failure shall not significantly affect the amplifier;

- circulators, implemented in each module, shall support the total reflected power.

Figure 9. RF module test bench principle (left) and photo (right)

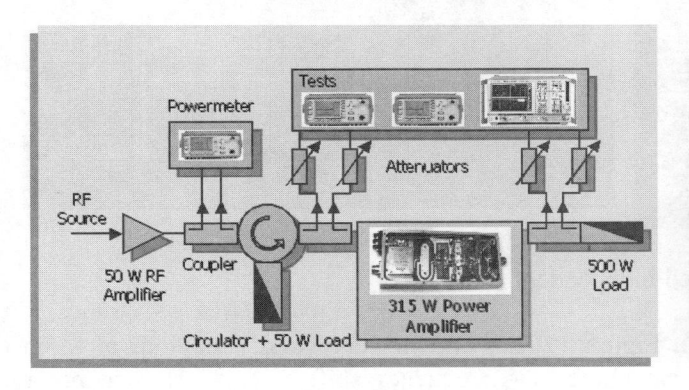

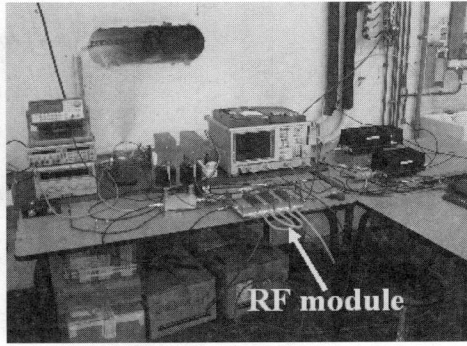

As a first step, this 10-KW solid state amplifier will be used for the high-power conditioning of RF power couplers (see Figure 10) dedicated to spoke cavities. They will be tested in transmission through a stainless steel coupling cavity at room temperature in July 2007. This power source will then be used to feed a fully equipped spoke cavity in CM0.

Figure 10. High-power conditioning test bench for spoke power couplers

Next steps with CM0

CM0 is a horizontal cryostat dedicated to fully equipped spoke cavities with a tuning system and a power coupler. It has a useful space of 690 mm in length and 490 mm in diameter while its dimensions are 1.5 m in length and 1 m in diameter as shown in Figure 11. The working operation of CM0 is close to an accelerator configuration.

Figure 11. Photograph of the cryostat CM0 closed (left) and opened with a view of the SC spoke β0.15 cavity in its supporting frame

Some components have been already tested independently from CM0:

- The cold tuning system, presented in Figure 12, has shown very good behaviour at room temperature in good agreement with calculations. A frequency shift of 1.887 Hz per motor step (2 Hz for calculations) has been found.

- The digital LLRF system has been already successfully tested in a vertical cryostat with the β0.15 spoke cavity at 4.2 K. While the XADS linac requires less than 0.1% in amplitude and less than 0.5° in phase for the regulation of the field inside the cavity, first analysis showed encouraging performances of 2% and 1.6° for the regulation of the field amplitude and phase respectively (see Figure 13).

Figure 12. Photograph of the cold tuning system on the β0.15 spoke cavity (left) and room temperature results giving frequency shift vs. motor step

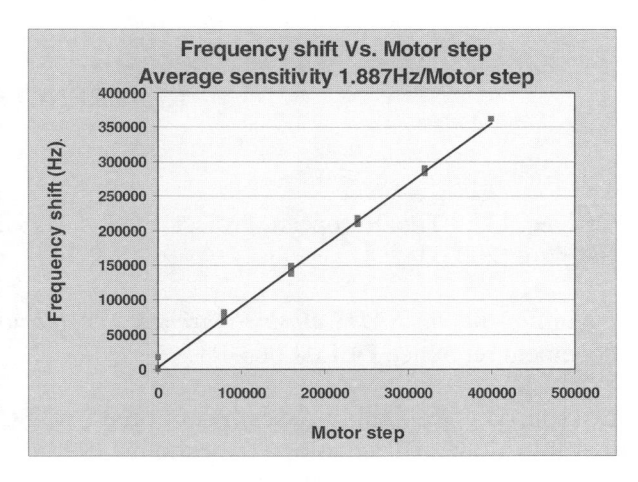

Figure 13. Phase and amplitude without (left) and with the digital LLRF system regulation (right)

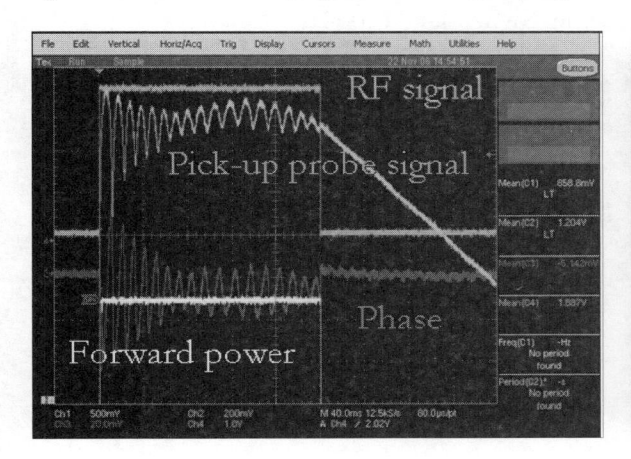

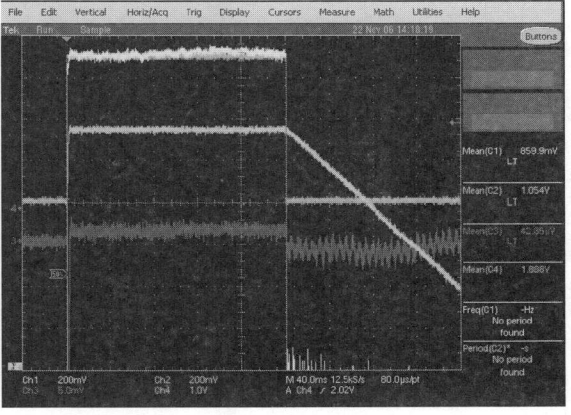

CM0 is now almost ready for the first cryogenic test in July 2007 including an upgraded digital LLRF system and the cold tuning system. Next year, long duration tests are planned with the 352 MHz RF power coupler supplied with the 10-kW solid state amplifier. An upgraded digital LLRF is currently under development and piezo actuators will be soon implemented in the present cold tuning system to face fast frequency variations.

Conclusion

A high degree of reliability is required for the XADS linac in order to enhance the availability of the nuclear plant and thereby demonstrate the concept of nuclear waste transmutation with a nuclear reactor coupled with a high-power proton linac. To reach the needed degree of reliability, derating, redundancy and the concept of fault tolerance are the main guidelines. At IPN Orsay, good results have already been obtained in the vertical cryostat for the digital LLRF and at room temperature for the cold tuning system. A new horizontal small-scale cryostat CM0 has recently been acquired and is now almost ready for the first cryogenic test of a fully equipped spoke cavity. Thanks to its working operation close to an accelerator one, every spoke component is foreseen to reach a good reliability in order to fulfil the XADS linac requirements.

REFERENCES

[1] Carluec, B., "The European Project PDS-XADS", *Int. Workshop for P&T and ACS*, Mol, Belgium (2003).

[2] *Definition of the XADS-class Reference Accelerator Concept and Needed R&D*, PDS-XADS document reference DEL04-063-R1.

[3] Erikson, M., "Reliability Assessment of the LANSCE Accelerator System", M.Sc. Thesis at the Royal Institute of Technology, Stockholm (1998).

[4] Biarrotte, J-L., "Beam Dynamics Study for the Fault Tolerance Assessment of the PDS-XADS Linac Design", *Proceedings of EPAC 2004*, Lucerne, Switzerland (2004).

[5] Olry, G., *et al.*, "Development of Spoke Cavities for the EURISOL and EUROTRANS Projects", *Physica C, 12th Int. Workshop on RF Superconductivity*, Ithaca (2006), pp. 201-204.

STATUS OF THE PREPARATION OF THE ELLIPTICAL CAVITY SYSTEM FOR THE EUROTRANS CRYOMODULE

Nicola Panzeri, Serena Barbanotti, Angelo Bosotti, Paolo Pierini
Istituto Nazionale di Fisica Nucleare, Italy

Abstract

One of the TRASCO low-beta elliptical cavities is being equipped with a helium tank and cold tuning system in order to be used in the cryomodule prototype for the high-energy section of the EUROTRANS linac. The cold tuning system has been fabricated and its performances evaluated by means of numerical simulations. An experimental test will be done in short time.

We report on the status of the cavity preparation procedures necessary for the integration in the cryomodule.

Introduction

The activities of the EUROTRANS project [1] and the activities in the task dedicated to the cryomodule design [2] are presented in dedicated contributions to these proceedings. One TRASCO low-beta elliptical cavity will be installed in the EUROTRANS test cryomodule described in Ref. [2]. This design activity can be considered as a proof-of-concept prototype for the construction of the full-scale module in the 100-200 MeV energy range (the first section of the high-energy superconducting linac). For the installation in the cryomodule the cavity has to be "dressed" with its helium vessel and tuning system. These components were properly designed at the LASA department of INFN Milano in order to obtain the required performances with a reasonable manufacturing cost and good reliability perspectives.

The manufacturing phase has just been completed and the experimental tests are foreseen in the next months to verify the correct behaviour of all the mechanical parts and motorised actuators. A complete test at low and high power is foreseen in the horizontal test cryostat CryHoLab (Saclay) at the beginning of the next year before its use in the EUROTRANS module.

In this paper the design process of the tuning system will be reviewed, pointing out the main aspects and the requirements driven by the cavity RF behaviour.

The TRASCO cavities

Two five-cell SC cavities (Z501 and Z502) were developed at INFN Milano for the TRASCO/ADS national programme and built at Zanon in 2002. Their main mechanical and RF parameters are reported in [3]. The cavities were tested at JLAB (see Figure 1) and Saclay and they fulfilled the requirements of the project quite well in terms of accelerating field [4,5], with ample margins. During these tests the Lorentz force detuning (LFD) coefficient was also derived; the values obtained were larger than expected and with a relevant spread (Figure 2), but this is mainly due to the uncertainty of the external constraints applied to the cavities during the tests, as explained in [6].

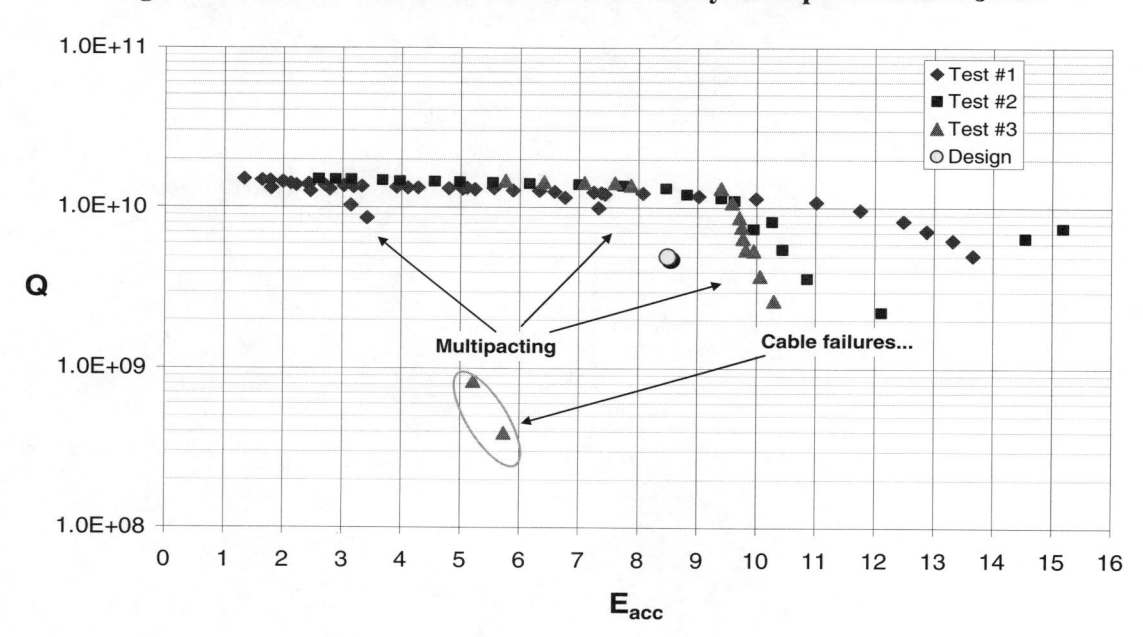

Figure 1. Results of several vertical test on cavity Z501 performed at JLAB

Figure 2. LFD coefficient from the test performed at JLAB (Z501) and Saclay (Z502)

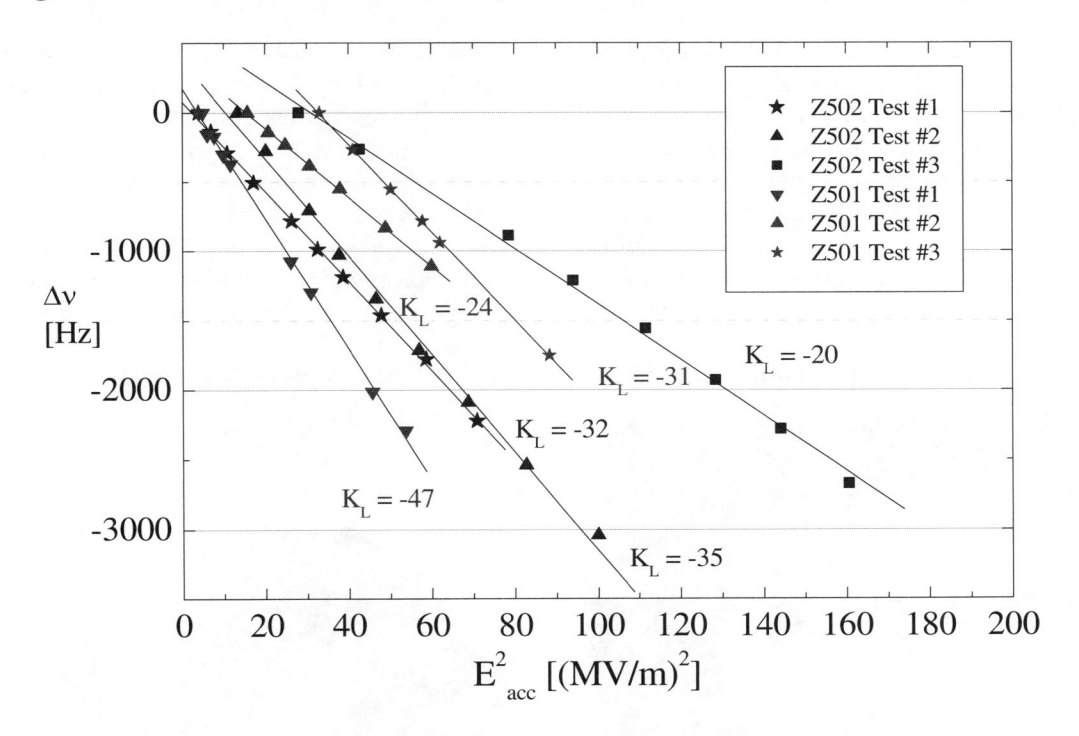

Because the LFD strongly depends on the external stiffness seen by the cavity (see Figure 3), one of the main requirements that the helium tank and tuning system must satisfy is to have a sufficiently high axial stiffness. With a design accelerating field of 8.5 MV/m, the overall K_L in the operating condition should be limited below -10 Hz/(MV/m)^2 to allow for LFD compensation by the LLRF system. In order to achieve this condition the combined stiffness of the He tank and tuning system needs to be greater than 9 kN/mm.

Figure 3. Lorentz force detuning coefficient as a function of external stiffness

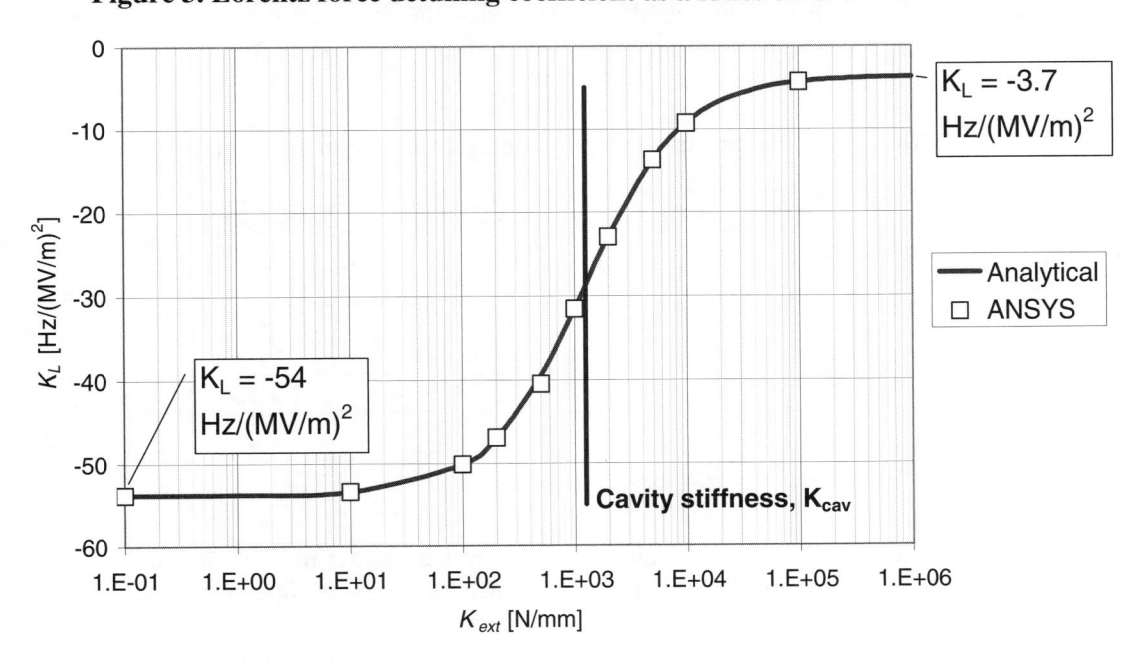

The helium tank

The helium tank is a titanium reservoir connected to the cavity end dishes (see Figure 4). In order to allow the longitudinal tuner action on the cavity, it is made of two cylindrical parts connected by a bellow in the central region. The tank is slid on the cavity at the smaller end dish side, where an adaptation ring allows adjusting the final length for the final welding to the cavity. This solution has been studied with these goals in mind: reduction of cost, easy cavity inclusion, simplest as possible execution of welds and optimisation of overall stiffness. Moreover the cylindrical parts are also used to support the tuning system: at one side a ring welded to the helium tank is used to fix the blade tuner while the piezo elements required for LFD feedforward and microphonic oscillation compensation are positioned against the pad support at the other side in such a way to not waste longitudinal space.

Figure 4. The helium tank: manufactured part and complete 3-D model

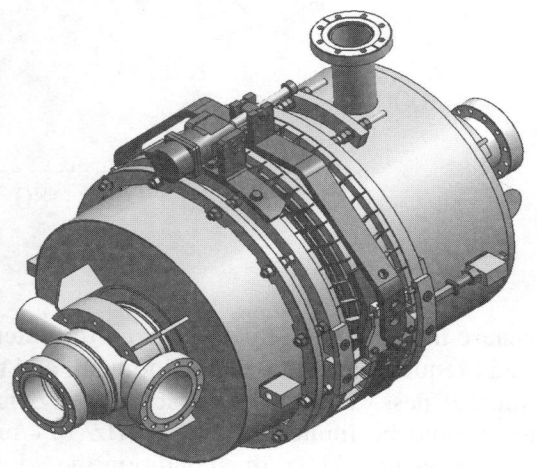

The stiffness of the helium tank, bellow and end dishes has been evaluated by means of the finite element method. The obtained values are reported in Table 1. Because these elements can be considered as spring in series, the overall compliance is equal to the sum of the three compliances. It is important to point out that, the bellow being in parallel with the tuning system, the overall stiffness of the reservoir is governed by the stiffness of the end dishes.

Table 1. Stiffnesses of the helium tank, end dishes, tuner and piezo elements

Part	Axial stiffness	c (μm/kN)	k (kN/mm)
Helium tank	K_H	1.17	856
End dishes	K_W	63.4	15.7
Bellow	K_B	3 205	0.312
Blade tuner	K_T	40.0	25.0
Piezo PIC255	K_P	4.176	2×10^5

The tuning system

Under working conditions at cold temperatures the cavity frequency must be maintained as near as possible to the nominal value of 704.4 MHz, to guarantee proper matching to the RF amplifier, avoiding power reflection at the coupler due to the extremely narrow bandwidth of the resonator.

Two different tuning actions are required:

- a slow tuning action to compensate the quasi-static frequency variations that can occur during the operation of the cavity;

- a fast tuning action to compensate the Lorentz force detuning (LFD) due to pulsed RF or compensation of microphonic oscillations.

These two actions will be provided by a coaxial piezo blade tuner, the design of which has been inspired by the concept proposed for the TESLA and ILC projects [7,8]. It should be noted that a blade tuner has already successfully been tested for the slow tuning of the TTF superstructures [9].

The tuner consists of three main components (see Figure 5, left):

- the rings-blades assembly, made of titanium, that provide the cavity slow tuning action;

- the leverage mechanism, in stainless steel 316L and brass MS58, that, equipped with a stepping motor, drives the rings-blade assembly movements;

- the piezo actuator part, that, during operation, provides the fast tuning action necessary for Lorentz Force Detuning compensation under pulsed operation.

The tuner stiffness and its elongation range are presented in the following subsections.

Figure 5. Drawing of the piezo blade tuner and the FE model used for the mechanical analyses

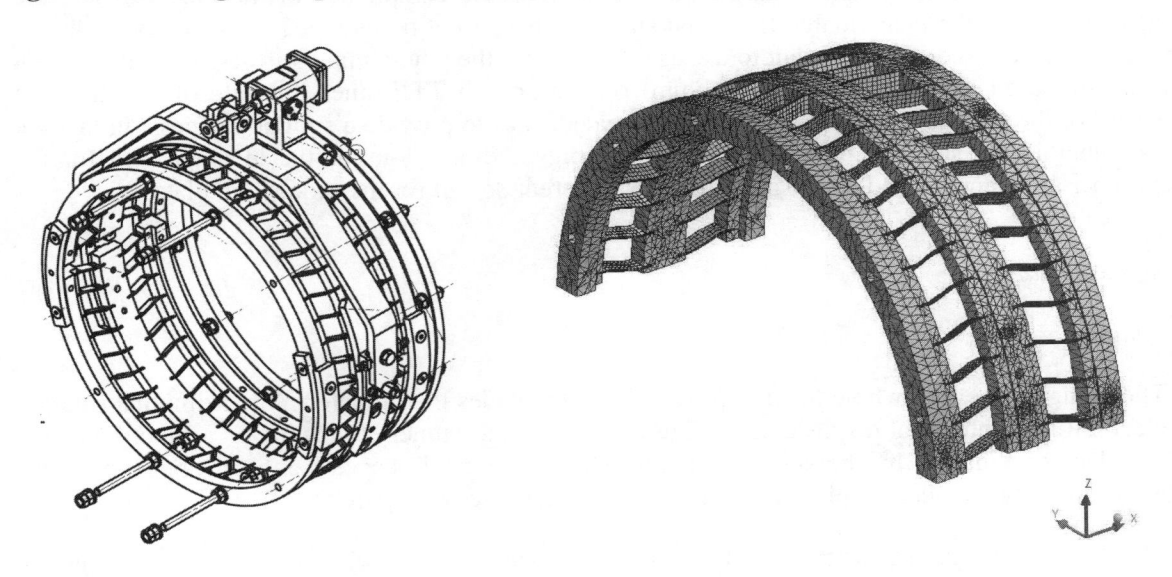

Tuning stiffness

The tuner device, although simple in its kinematics (Figure 6), is composed of many parts and the global behaviour has been obtained studying separately the leverage mechanism and the rings-blades assembly. The adopted finite element models are reported in Figures 5 and 6 (right-hand images).

Figure 6. Slow tuning working scheme and driving mechanism FE model

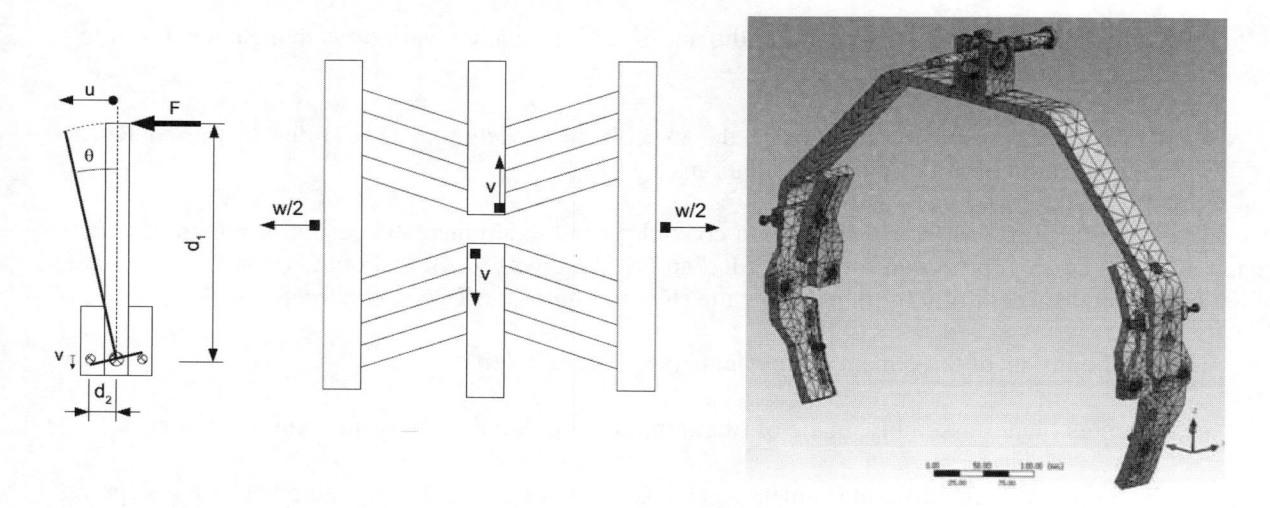

The detailed computations are described in [10]. Here we report two reference values of the axial stiffness obtained applying different boundary conditions at the central ring:

- central ring free (without leverage): K_T = 2.4 kN/mm;

- central ring with leverage (working condition): K_T = 52.3 kN/mm.

The values obtained by these analyses do not include the compliance of the motor and of the bearings. Moreover, the unavoidable lacks and slack joints can not be included consistently in the FE analysis, therefore it seems convenient to use as reference for the whole tuner stiffness the value which has been measured experimentally (25 kN/mm) on the proven TTF tuner. In spite of the different geometries of the two tuners, the design has been scaled so as to provide a similar stiffness. In fact the blade number is the same in both designs and the bigger dimensions of the leverage arm for the present tuner are compensated by the adoption of a different design for the saddle and plates.

Tuning axial range

Slow tuning

The elongation of the whole tuner is provided by the blades that transform the azimuthal rotation of the central ring into axial displacement. Figure 7 displays the tuner elongation as a function of the shaft displacement driven by the stepping motor. The "theoretical" curve has been obtained by FE analysis of the single blade, coupled with the leverage kinematics of Figure 6.

The experience gained with the TTF tuner shows that real conditions reduce the range by approximately 20% with respect to the FE estimation. This effect has been taken into account in the "expected data" curve. Thus, with the actual leverage configuration, the tuner is capable of a maximum elongation of approximately 1.3 mm.

Fast tuning

The fast tuning is provided by two piezo elements acting between the blade tuner and the pad fixed at the other side of the helium tank. Piezo of different length, up to 72 mm, can be used. One of

Figure 7. Tuner elongation as a function of shaft displacement

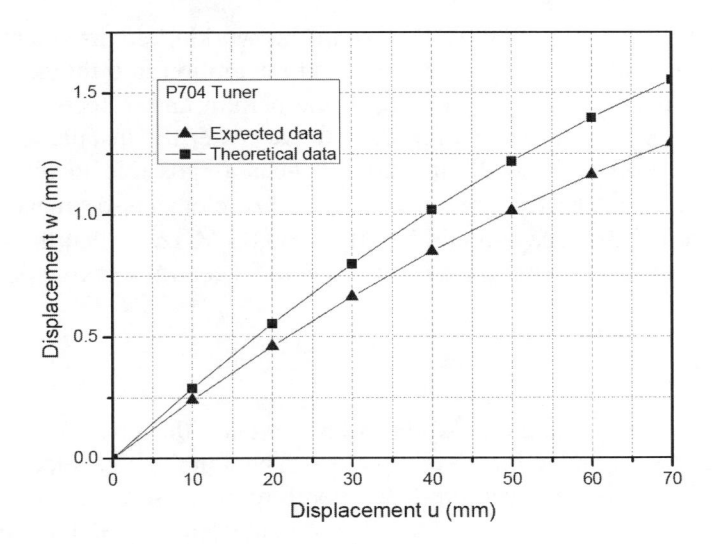

the remaining uncertainties concerning the piezo materials is their stroke capabilities at the low operating temperatures [11]. By assuming a 2 μm stroke capability for the 40-mm long piezo, a very conservative value (if compared with the data recently obtained at DESY), a 700-Hz frequency offset can be compensated during the fast tuning action (meeting the $K_L > -10$ Hz/(MV/m)^2 requirement.

Behaviour of dressed cavity

From the mechanical point of view, the dressed cavity can be described as a spring system. Two different configurations may be considered: the first (Figure 8) for the slow tuning and the second one for the fast tuning. It easy to demonstrate that not all the tuning action is transferred to the cavity, but a non-negligible part is lost in the deformations of the end dishes. In particular only 92% of the slow tuning action is transferred to the cavity, while for the fast tuning the efficiency is 85%.

Figure 8. spring model for the slow (left) and fast (right) tuning

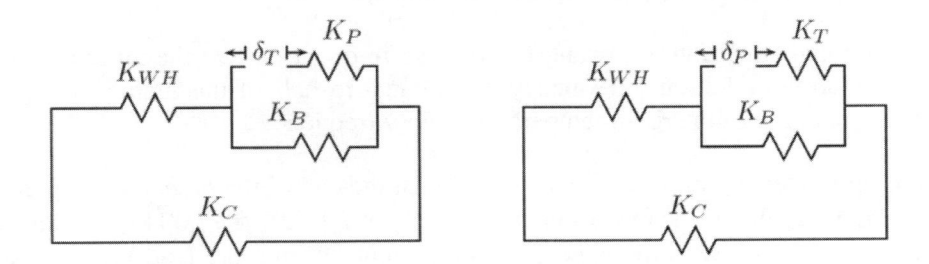

The external stiffness seen by the cavity can be easily computed by considering a series of sprigs: end dishes, helium tank, blade tuner and piezo elements. The bellow can be neglected. Assuming the use of PIC 255 piezo elements, 40 mm length with a stiffness of 105 kN/mm each, the external stiffness is:

$$1/K_{ext} = 1/K_H + 1/K_T + 1/K_W + 1/(2K_P) \rightarrow K_{ext} = 9.2 \text{ kN/mm}$$

This value is near, but acceptable, to the limit value of 9 kN/mm required for LFD and microphonic compensation capabilities.

Preparation of the cavity for the welding procedure

To obtain the nominal frequency of 704.4 MHz in the working environment at 2 K, the room temperature cavity frequency, after surface treatments and integration into the helium tank, has to be 702.8 ± 0.1 MHz. Chemical treatment and welding to the helium tank effects have to be separately considered in order to get the right frequency reported above. After the installation of the tuner (with the tuner blade in rest position and the driving arm as near as possible to the motor), the cavity frequency is tuned to 703.2 ± 0.01 MHz, preloading the piezo elements by means of their threaded supports. After cool down the frequency should be 704.2 ± 0.05 MHz, so that the working frequency of 704.4 MHz at 2 K is obtained by moving the mechanism to its central working position.

Cavity tuning philosophy

With the introduction of the dynamic capabilities on the tuner, the cavity has to be tuned at room temperature in such a way to guarantee the correct preload range on the piezoelectric elements. In fact the cavity elasticity is used to give the piezo pre-load, therefore the elastic deformation of the cavity at cold has to be determined to simultaneously obtain the nominal frequency and a good preload on each piezo. Moreover the wider tuning range has to be assured to avoid piezoelectric element breakage or their complete unload, which would result in an unusable device.

Therefore the correct cavity pre-tuning has to take in account for:

- warm/cold frequency shift;

- piezoelectric element pre-load;

- tuner range capabilities.

The application of a pre-load force on the piezo, mandatory for reliability reasons, can be exerted compressing it while stretching the cavity of the proper amount. If P [N] is the total pre-load force to apply to the piezos, and k_{cav} = 1 248 is the cavity stiffness, the cavity must be elongated by $\Delta l = P/k_{cav}$ to apply the chosen pre-load force. This operation will result in a positive frequency shift $\Delta f_P = \Delta l \, S$, where S = 353 kHz/mm is the cavity axial frequency sensitivity.

Moreover, in nominal conditions we want the working frequency when the tuner is in its middle range position. Therefore an additional frequency shift, equal to half of the tuning range, has to be considered in the computation of the room temperature cavity frequency.

Thus the cavity room temperature resonant frequency at the end of the preparation operations will be: $f_{RT} = f_0 - \Delta f_C - \Delta f_P - \Delta f_t/2$ where f_0 is the nominal frequency at 2 K, $\Delta f_c \cong$ 1 000 kHz is the warm/cold frequency shift and Δf_t = 420 kHz is the expected tuning range. With a pre-load P = 750 N for each piezo element, the goal value of 702.8 kHz at room temperature is obtained.

Cavity tuning procedure

The two cavities were originally built without planning to test them in a cryomodule environment, hence during fabrication their frequency was not adjusted, and they have now a frequency of about 699.9 MHz at room temperature, below the required value of 702.8 MHz needed to meet the nominal frequency of 704.4 MHz at cryogenic temperature with the correct tuner setting. This would be unacceptable for the horizontal test and the subsequent installation in the EUROTRANS test module. Therefore a tuning procedure, consisting of the following steps, has been set-up:

- a first tuning from 699.9 MHz to about 702 MHz followed by a leak test;
- a heat treatment, mandatory for the hydrogen degassing of the cavity;
- tuning to the final frequency of 702.8 MHz at room temperature, followed by a leak test.

The first tuning phase has successfully been carried out in our laboratory, and now the cavity is ready for the heat treatment in a vacuum furnace that will be done at CERN. This treatment is required to desorb from the niobium the hydrogen introduced by the deep etching performed after fabrication to remove the surface damage layer. When the cavity is cooled down slowly (as is the case for a horizontal cryostat or a cryomodule) a high hydrogen content would lead to the formation of lossy hydrides within the RF penetration length, and therefore to a substantial degradation of the cavity performances (so-called "Q-disease").

Figure 9. The Z502 cavity constrained in our tuning device before the leak test

Conclusions

The helium tank and tuning system have been designed to fulfil the requirements driven by the Lorentz force detuning and the total tuning range. The effect of the end dishes deformability is not negligible, indeed they are less stiff than the tuning system. Nevertheless an external stiffness higher than the minimum required has been obtained and the total tuning range is of the order of 1.2 mm, which corresponds to about 400 kHz. The construction of the helium tank and of the piezo blade tuner has just finished and we are preparing for mechanical testing and for the horizontal RF tests in CryHoLab. This last activity requires the tuning of the cavity from the actual frequency of 699.9 MHz to 702.8 MHz. The first phase of the tuning has been done, showing the possibility to reach the nominal resonance frequency without problems.

Figure 10. Assembly of the manufactured helium tank and tuning system

Acknowledgements

This work is supported by the EURATOM 6th Framework Programme of the EC under contract FI6KW-CT-2004-516520.

REFERENCES

[1] Biarrotte, J.-L., "Status of the EUROTRANS R&D Activities for ADS Accelerator Development", these proceedings.

[2] Barbanotti, S., N. Panzeri, P. Pierini, J-L. Biarrotte, S. Bousson, E. Rampnoux, H. Saugnac, "Design of a Test Cryomodule for the High-energy Section of the EUROTRANS Linac", these proceedings.

[3] Barni, D., A. Bosotti, G. Ciovati, C. Pagani, P. Pierini, "SC Cavity Design for the 700 MHZ TRASCO Linac", *Proceedings of EPAC 2000*, Vienna, Austria, p. 2019.

[4] Bosotti, A., C. Pagani, P. Pierini, P. Michelato, R. Paulon, G. Corniani, J-P. Charrier, B. Visentin, Y. Gasser, J-P. Poupeau, B. Caodou, *Report on Cavity A (Z502) Fabrication and Tests*, CARE-NOTE-2005-001-HIPPI.

[5] Bosotti, A., C. Pagani, P. Pierini, J-P. Charrier, B. Visentin, G. Ciovati, P. Kneisel, "RF Tests of the Beta = 0.5 Five Cell TRASCO Cavities", *Proceedings of EPAC04*, Lucerne, Switzerland, p. 1024.

[6] Bosotti, A., C. Pagani, N. Panzeri, P. Pierini, G. Ciovati, P. Kneisel, "Characterization of an Elliptical Low Beta Multicell Structure for Pulsed Operation", *Proceedings of 12th International Workshop on RF Superconductivity (SRF 2005)*, Ithaca, New York, USA, July 2005.

[7] Barni, D., A. Bosotti, C. Pagani, "A New Tuner for TESLA", *Proceedings of EPAC 2002*, Paris, France, p. 2205.

[8] Pagani, C., N. Panzeri, "Comparative Analysis of Blade Tuner Optimization Options for the ILC", *APAC 07*, WEPMA123, Indore, India, 29 January-2 February 2007.

[9] Sekutowicz, J., *et al.*, "Test of Two Nb Superstructure Prototypes", *Phys. RST-AB*, 7, 012002 (2004).

[10] Pierini, P., N. Panzeri, A. Bosotti, P. Michelato, C. Pagani, R. Paparella, *Report on Tuner Design*, CARE-Note-2006-003-HIPPI.

[11] Sekalski, P., *et al.*, "Static Absolute Force Measurement for Preloaded Piezo Elements Used for Active Lorentz Force Detuning System", *Proceedings of Linac 2004*, Lubeck, Germany, p. 48.

DEVELOPMENT OF ROOM TEMPERATURE AND SUPERCONDUCTING CH STRUCTURES FOR HIGH-POWER APPLICATIONS

Holger Podlech, Alexander Bechtold, Marco Busch, Gianluigi Clemente,
Horst Klein, Holger Liebermann, Rudolf Tiede, Ulrich Ratzinger, Chuan Zhang
Institut für Angewandte Physik (IAP), University of Frankfurt, Germany

Abstract

The Crossbar-H mode (CH) structure developed at the IAP in Frankfurt is a multi-cell drift tube cavity for the efficient acceleration of low- and medium-energy ions and protons. A room temperature as well as a superconducting prototype cavity have been developed and tested. These cavities could be used for high-intensity linacs like accelerator-driven systems (ADS, EUROTRANS) or the international fusion material irradiation facility (IFMIF). Additionally, the new proton injector for FAIR (324 MHz, 70 mA, 70 MeV) will use room temperature CH cavities. The development of this new type of drift tube cavity and the tests of the prototypes are presented.

Introduction

The CH structure is a multi-cell cavity and belongs to the family of H-mode cavities which are operated in an H_{n1}-mode. This new structure has been named CH structure because of its cross-bar geometry and the H_{21}-mode [1,2]. H-mode cavities have been developed and operated successfully over the past 30 years [3-7]. Considering the two lowest modes (H_{11} and H_{21}) there are four types of accelerating cavities, two RFQ structures and two drift tube cavities (IH cavity and CH cavity). Figure 1 shows the family of H-mode cavities used. The IH structure is used at lower operation frequencies between 30 MHz and 250 MHz whereas the CH structure can be used between 175 MHz and 700 MHz. The typical energy range of the CH structure is between 3 and 100 AMeV.

Figure 1. H-mode drift tube cavities: r.t. IH-DTL (217 MHz, left), r.t. CH-DTL (350 MHz, centre) and s.c. CH-DTL (350 MHz, right)

One main feature of H-mode drift tube cavities which boosts the efficiency and makes long lens free multi-cell cavities possible is the KONUS (Kombinierte Null-Grad-Struktur) beam dynamics [8]. The KONUS concept reduces the transverse rf defocusing by using zero-degree main accelerating sections. A KONUS period consists of three sections with separate functions. The first section consists of a few gaps with a negative synchronous phase of typically -35° and acts as a rebuncher. The beam is then injected into the 0° section with a surplus in energy compared with a "synchronous" particle. This multi-gap main accelerating section is then followed by transverse focusing elements which are magnetic quadrupole triplets. Superconducting CH cavities will house only the rebuncher and the main acceleration section. The transverse focusing elements will be located in the inter-tank sections. The number of cells in each section depends mainly on the rf frequency, on the gradient and on the beam current.

In general H-mode structures have high acceleration efficiencies in the low- and medium-energy range especially when using slim drift tubes without internal lenses. The IH and the CH DTL have no competitor with respect to efficiency and shunt impedance in the energy range from $\beta = 0.01$-0.3. For β-values up to 0.1 the plug power needed to operate an IH linac is competitive with superconducting solutions at identical accelerator length. Figure 2 shows the effective shunt impedance Z as a function of the particle velocity $\beta = v/c$ for different rf structures.

In many cases the rf linac efficiency can be increased significantly by the use of multi-cell cavities. For instance, in case of actual projects involving proton and light ion driver linacs with rf frequencies between 175 and 350 MHz there is an obvious lack of efficient superconducting low β cavities. In these cases efficient means a high-energy gain per cavity which leads to a low total number of individual cavities and rf systems. Due to the rf frequency and to the RFQ voltage gain which is typically between 1 MV and 5 MV the cell length $\beta\lambda$ is around 40 mm at the superconducting DTL front end. Using conventional superconducting two-gap cavities this significantly reduces the filling factor as cavities with a small number of cells imply a lot of drift spaces and increase the mechanical complexity of the linac greatly.

Figure 2. Effective shunt impedance for different rf structures including the transit time factor T and the synchronous phase ϕ as function of the particle velocity β = v/c

The black horizontal bars represent some existing IH DTLs and the red bars represent the expected shunt impedance of the rt CH cavities for the GSI-FAIR 70 MeV, 70 mA proton injector

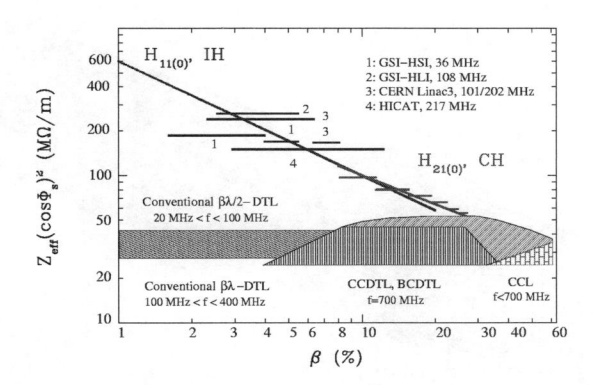

Room temperature CH cavity development for the p-linac of the FAIR facility

The Facility for Antiproton and Ion Research (FAIR), presently under design, will provide new features for experiments in the field of atomic, nuclear and particle physics. One main feature of the new facility will be the availability of intense cooled antiproton beams [9]. The antiprotons are produced by a primary beam of protons with an energy of up to 30 GeV. But the existing UNILAC as injector is unable to deliver the required beam current to fulfil the physics programme. The intensity has to be increased by more than two orders of magnitude. Therefore a dedicated proton linac is necessary. The new p-linac must provide a 70 MeV, 70 mA proton beam [10]. The linac will consist of a 3 MeV RFQ followed by r.t. six coupled CH cavity (CCH) doublets (Figure 3). One doublet consists of two CH cavities operated in the H_{21}-mode coupled by a cylindrical coupling cell operated in the E_{010}-mode. The coupling cell is housing a magnetic quadrupole lens for transverse focusing. The frequency of the linac is 325.2 MHz for which commercial 2.5 MW klystron are available. The rf power will be coupled magnetically into the coupling cell. Figure 4 shows the cavity doublet 2 of the p linac. The effective shunt impedance is between 100 MΩ/m and 45 MΩ/m. The required rf power including beam loading is around 2 MW per doublet. The rf pulse length is 70 μs, the beam pulse length is 36 μs. Together with a repetition rate of 4 Hz the rf duty cycle is very low (2.8E-4) although the pulse beam power is significant (4.9 MW). The high shunt impedance and the low duty cycle allow very high accelerating gradients between 2.5 and 6.7 MV/m. A r.t. CH prototype has been built and tested successfully. Figure 5 shows the test set-up. The measured Q-factor was 13 000, which is about 95% of the value predicted by MicrowaveStudio [11]. The available rf power of 2 kW cw could be coupled to the cavity within 10 minutes. The cavity has not shown any multi-pacting problems. Table 1 summarises the parameters of the p-linac.

Figure 3. Schematic layout of the proton-injector for the new FAIR facility. The linac will consist of an RFQ followed by 12 coupled CH-structures. The operation frequency is 325.2 MHz; the linac length is about 25 m.

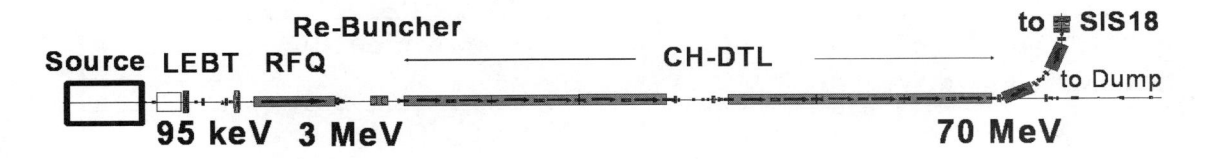

Figure 4. Schematic layout of a coupled CH structure (CCH)

Two cavities operated in the H_{21}-mode will be coupled with a coupling cell housing a magnetic triplet lens. The coupling cell is operated in the E_{010}-mode.

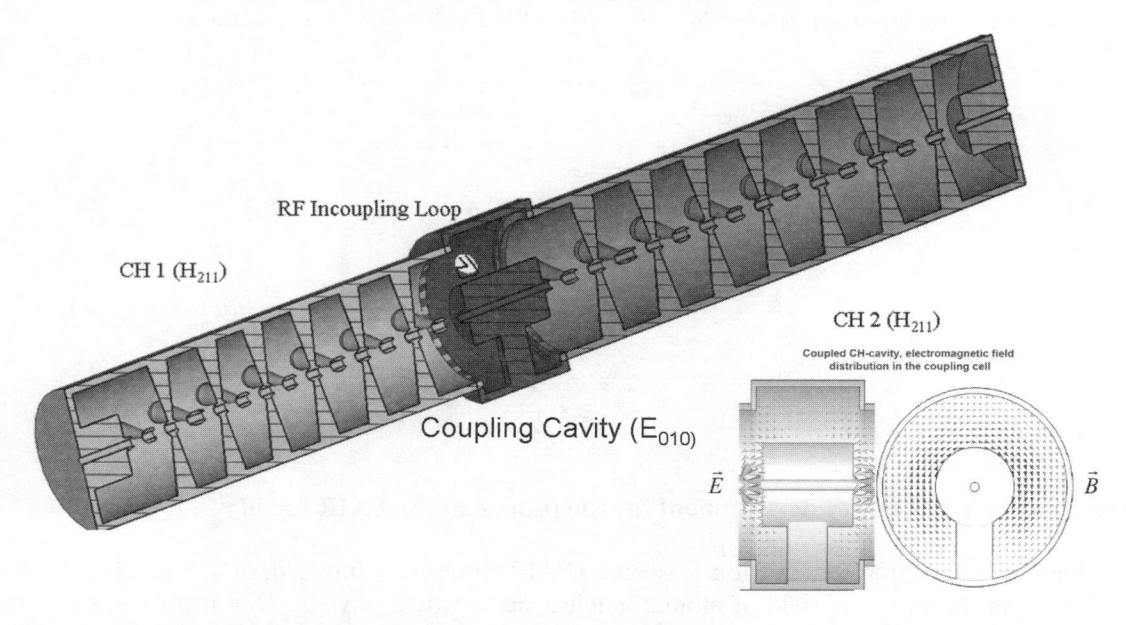

Figure 5. Test of a r.t. prototype for the proton-linac. The cavity has been tested with 2 kW cw.

Table 1. Parameters of the p-linac for the FAIR facility

Particles	Protons	RF pulse length (µs)	70
Beam current (mA)	70	Beam pulse length (µs)	36
Energy (MeV)	70	$\Delta p/p$ (%)	0.1
RF structures	CCH r.t. (6×)	ε_x (95%, norm., mm*mrad)	3.3
Frequency (MHz)	325.2	ε_y (95%, norm., mm*mrad)	3.4
Klystron power (MW)	2.5	E_a (MV/m)	6.7-2.5
Z_{eff} (MΩ/m)	100-45	Repetition rate (Hz)	4

Superconducting CH-cavity development

The CH-structure with its crossed-stem geometry provides a high mechanical stability which is required for superconducting operation. A superconducting CH prototype cavity with 19 accelerating cells (f = 350 MHz, β = v/c = 0.1) has been designed and built [12]. Figure 6 shows the prototype before the final welding of the end walls. It is the first superconducting multi-cell cavity for the low- and medium-energy range. Table 2 summarises the main parameters of the prototype cavity.

**Figure 6. View into the superconducting CH cavity,
b = 0.1, f = 350 MHz, before welding of the end walls**

Courtesy ACCEL instruments GmbH, Bergisch Gladbach, Germany

The superconducting CH-structure has been tested in the cryogenic rf laboratory at the IAP in Frankfurt. Effective gradients of up to 4.7 MV/m have been achieved which is significantly more than required [13]. This corresponds to electric peak fields of 25 MV/m and of magnetic peak fields of 26 mT. Presently the limitation is field emission due to a single field emitter. As expected the CH-structure has shown only little multi-pacting which could be conditioned within one hour. The Q-value at low field level was 5.7E8. Table 2 summarises the parameters of the prototype cavity.

Table 2. Main parameters of the superconducting CH prototype cavity

Cavity type	CH	R_a/Q_0 (Ω)	3 180
Material	Bulk niobium	G (Ω)	56
Accelerating cells	19	$R_a R_s$ (Ω^2)	180 000
β	0.1	E_a (MV/m)	4.7
Frequency (MHz)	350	U_a (MV)	3.6
Length (mm)	1 048	Q_0 (low fields)	5.7E8

Figure 7. Measured unloaded quality factor Q_0 as function of the acceleration gradient E_a. The maximum achieved gradient of 4.7 MV/m corresponds to an effective voltage of 3.7 MV.

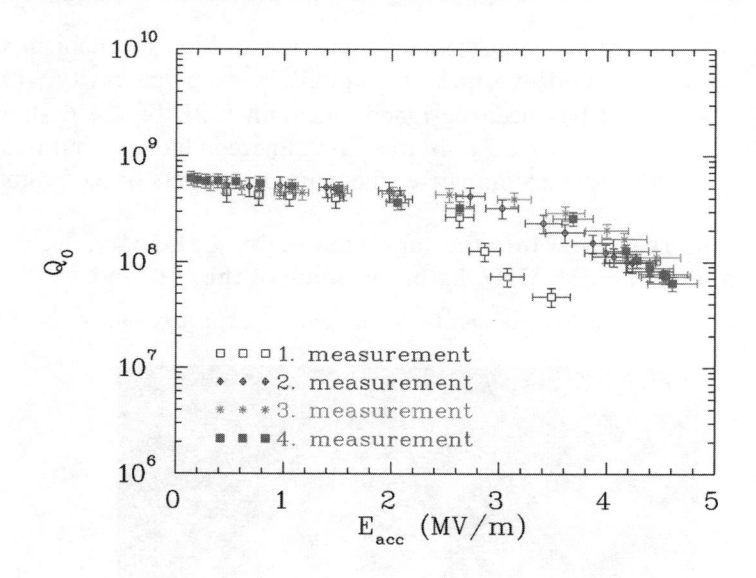

EUROTRANS

For future accelerator-driven systems (ADS) efficient and reliable high-power proton drivers have to be developed. Within the XADS project (eXperimental ADS) the feasibility of an ADS for nuclear waste transmutation has been studied [15]. EUROTRANS is the continuation of XADS and it has been launched by the EU in 2006. One goal of EUROTRANS is to build, to test and to assess different accelerator structures for a driver linac of a European Transmutation Demonstrator (ETD). The linac parameters are mainly inspired from the MYRRHA project [16]. The beam current is 3.5 mA with a final proton energy of 600 MeV. The beam with a power of 2 MW will hit a spallation target within a subcritical core.

The most challenging feature of the transmuter driver linac is the extremely high reliability requirement. Beam trips longer than 1 s will cause severe stress in the material of the target and of the core due to thermal shocks. Therefore the number of trips with t > 1 s has to be limited to about 10 per year for an ETD. In case of a future industrial transmuter with a beam power of 25 MW (25 mA, 1 GeV) the number of beam trips should not exceed three per year. Beam trips will most likely be caused in the low- and intermediate-energy part of the linac. At higher energies (i.e. several 10 MeV) the failure of a cavity or magnet will not cause a beam trip: It is planned to raise the field and to readjust the phases in the cavities neighbouring a failing cavity. But even using redundancy with respect to power supplies, control systems and amplifiers it seems to be questionable to fulfil the reliability requirements for the low-energy part. Therefore a possibility is to use two identical injectors up to a certain energy, which may be 17 MeV or higher. Injector 1 delivers the beam while Injector 2 is on-line and ready to deliver the beam in case of a failure in Injector 1. Due to the required cw operation the main part of the linac will be superconducting. In the design which has been proposed by IAP the beam will be delivered by an ECR source followed by a 352 MHz four-vane RFQ with a final energy of 3 MeV. A r.t. CH structure accelerated the beam to 5 MeV. The energy range between 5 and 17 MeV could be covered by four superconducting 352 MHz CH cavities. The energy gain in each cavity is typically 3 MeV. Three quadrupole triplets will be placed between the tanks for transverse focusing. The total length of the superconducting CH linac is about 6 m. The average beam load in case of the ETD is 15 kW per cavity.

**Figure 8. Schematic layout of the EUROTRANS
driver linac for a European transmutation demonstrator**

*The IAP proposal for EUROTRANS consists of an RFQ with an energy of 3 MeV followed by a
room temperature CH structure which delivers 2 MV effective voltage. The energy range from
5 to 17 MeV could be covered by four superconducting CH structures. Between 17 and 100 MeV
single-spoke cavities are foreseen. The high-energy section will consist of elliptical cavities.*

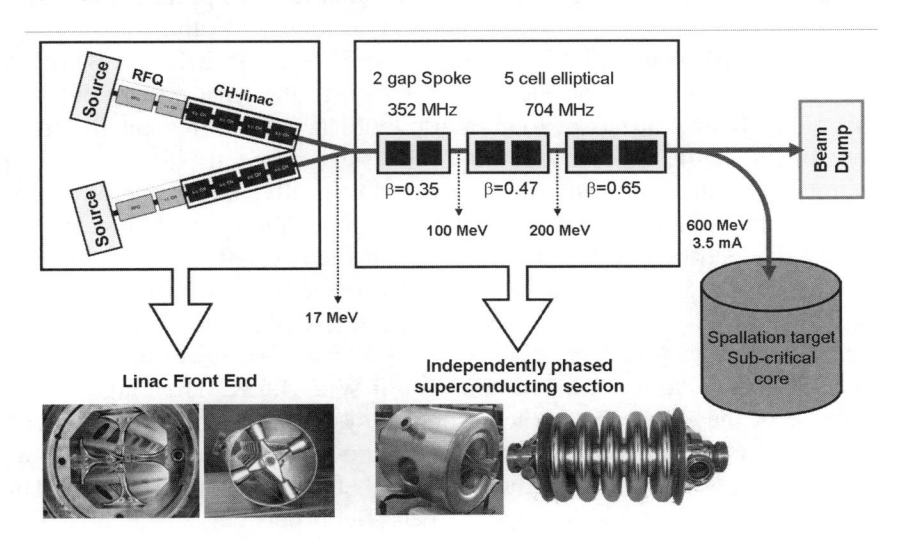

Table 3. Parameters of the CH injector linac for EUROTRANS

RFQ energy (MeV)	3	Nr. of s.c. CH-cavities	4
r.t. CH energy (MeV)	3-5	Length CH-linac 3-17 MeV(m)	9
s.c. CH (MeV)	5-17	Number of cryomodules	1

IFMIF

Future fusion reactors using the D-T-reaction need inner walls which can stand very high fluxes of fast neutrons of up to 14 MeV. This is especially true for the first wall of the vacuum chamber. A certain fraction of this flux will be absorbed in the material causing displacements of lattice atoms. This will result in a fatigue of the used material. To extend the lifetime and to limit the activation damages of the first wall it is necessary to find new alloys with sufficient robustness against neutron radiation. Presently there is not any high flux source of fast neutrons which can probe the conditions in future reactors. The International Fusion Material Irradiation Facility (IFMIF) which is under design will provide the neutron flux needed to develop new reactor materials [16].

**Figure 9. Schematic layout of the 175 MHz IFMIF linac proposed
by the IAP Frankfurt. The DTL consists of a room temperature
IH or CH structure and eight superconducting CH structures.**

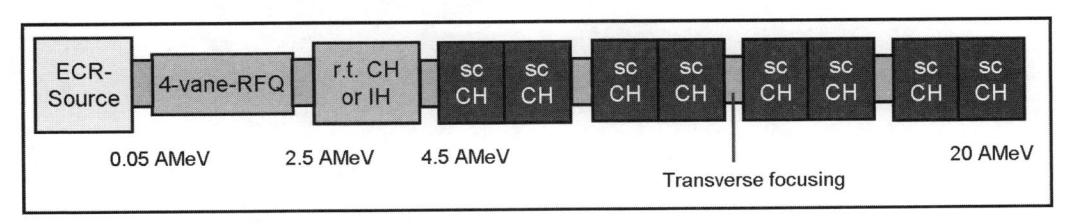

The neutrons will be produced from a 40 MeV deuteron beam hitting a liquid lithium target. The total required beam current is 250 mA with cw operation delivered by two linacs in parallel operation. The beam power is 10 MW. The loss rate along the accelerator must be very low, about the order of 1 W/m to avoid activation and to guarantee hands-on maintenance. The front end of each IFMIF accelerator will consist of an ECR source injecting into a 175 MHz four-vane RFQ. The reference design of the DTL is an Alvarez-type accelerator. The IAP Frankfurt has proposed an alternative DTL concept using room temperature and superconducting CH structures. Within this concept the beam will be accelerated after the RFQ from 2.5 AMeV to 4.7 AMeV by a 175 MHz r.t. IH or CH structure. A following chain of s.c. CH structures accelerates the beam to the final energy of 20 AMeV. The superconducting part of the linac consists of four cavity doublets each corresponding to one KONUS period. One main advantage of the superconducting option is significantly higher accelerating fields, as the limiting cooling power aspect is not valid here. Additionally, larger apertures up to 8 cm in diameter become possible, providing an additional safety margin against particle losses. A third surplus is the significant saving in electricity costs. The rf power for the two Alvarez-type linacs is estimated to 4.1 MW without beam loading. Assuming an amplifier efficiency of 60% the required plug power due to Ohmic losses in the DTL is 6.2 MW. The two r.t. CH cavities need 240 kW rf power and 400 kW plug power, respectively. A conservative estimation for the rf power in one superconducting cavity is 30 W. Both s.c. linacs require a maximum of 480 W at 4.5 K. The plug power is typically 300 times higher because of the efficiency of the cryogenic system. The total plug power for two CH linacs is less than 0.6 MW. Assuming 8 000 hours of operation time per year a s.c. solution could save 45 million kWh per year. The accelerating gradient will be between 3.5 and 4 MV/m. This results in electric peak fields of about 20 MV/m and in magnetic peak fields between 25 and 30 mT.

Table 4. Main parameters of the proposed CH-linac for IFMIF

Cavity type	CH	Particles	Deuterons
Energy range (AMeV)	2.5-20	Beam current (mA)	2 × 125
r.t. CH (AMeV)	2.5-4.5	RF power per cavity (kW)	400-500
Number of s.c. CH-cavities	8	E_a r.t. CH (MV/m)	2.0
Frequency (MHz)	175	E_a s.c. CH (MV/m)	4.0

Summary and outlook

The CH structure operated in the H_{21}-mode is a novel multi-cell drift tube cavity which can be realised for room temperature as well as for superconducting operation. The r.t. CH structure will be used for the new 70 MeV proton injector for the FAIR facility. A prototype has been built and tested successfully. The superconducting CH structure which could be used for cw operated linacs like EUROTRANS and IFMIF has been built and tested at cryogenic temperatures. Acceleration gradients of 4.7 MV/m have been achieved. Presently the CH structure is being optimised with respect to high rf power up to 500 kW per cavity.

Acknowledgements

This work has been supported by Gesellschaft für Schwerionenforschung (GSI), BMBF contract No. 06F134I. and EU contract No. 516520-FI6W. We also acknowledge the support of the European Community-Research Infrastructure Activity under the FP6 "Structuring the European Research Area" program (CARE, contract No. RII3-CT-2003-506395) and EU contract No. EFDA/99-507ERB5005 CT990061 between EURATOM/FZ Karlsruhe IAP-FU. The work was carried out within the framework of the European Fusion Development Agreement. The views and opinions expressed herein do not necessarily reflect those of the European Commission. The authors would like to thank the company ACCEL for the excellent work on technical drawings as well as in the fabrication of the superconducting prototype cavity. In addition, the authors would like to thank the technical staff of the IAP in Frankfurt, especially D. Bänsch, I. Müller, G. Hausen and S. Reploeg.

REFERENCES

[1] Ratzinger, U., R. Tiede, *Nucl. Instr. and Methods in Phys. Research A*, 415, 229-235 (28-1998)

[2] Podlech, H., *Proc. 2004 Linear Accelerator Conference*, pp. 28-32 (2004).

[3] Amendola, G., *et al.*, *Proc. 1992 European Particle Accelerator Conference*, pp. 536-538 (1992).

[4] Ratzinger, U., *et al.*, *Proc. 1987 Particle Accelerator Conference*, pp. 367-368 (1987).

[5] Ratzinger, U., *et al.*, *Proc. 2000 European Particle Accelerator Conference*, pp. 98-102 (2000).

[6] Schneider, J.D., *Proc. 2000 European Particle Accelerator Conference*, pp. 118-122 (2000).

[7] Angert, N., *et al.*, *Proc. 1992 European Particle Accelerator Conference*, pp. 167-169 (1992).

[8] Ratzinger, U., *Proc. 1991 Particle Accelerator Conference*, pp. 567-571 (1991).

[9] Spiller, P., G. Franchetti, *Nucl. Instr. and Methods A*, 561, 305-309 (2006).

[10] Clemente, G., *et al.*, *Proc. 2006 European Particle Accelerator Conference*, pp. 1283-1285 (2006).

[11] http://www.cst.de

[12] Podlech, H., *et al.*, *AIP Conference Proc.*, 773 ICFA-HB, pp. 107-109 (2004).

[13] Podlech, H., *et al.*, *Proc. 2006 European Particle Accelerator Conference*, pp. 1588-1590 (2006).

[14] Junquera, T., *Proc. of the International Workshop on P&T and ADS Development*, Mol, Belgium (2003).

[15] Abderrahim, H.A., *et al.*, *Nucl. Instr. and Methods A*, 463, 487-494 (2001).

[16] IFMIF International Team, *IFMIF Comprehensive Design Report*, International Energy Agency (2004).

UPGRADE OF THE PSI PROTON ACCELERATOR FACILITY TO 1.8 MW

Pierre A. Schmelzbach
Paul Scherrer Institute, Switzerland

Abstract

The PSI proton accelerator currently delivers a 590-MeV beam with an intensity close to 2 mA. The upgrade programme aiming at boosting the beam power from 1.2 to 1.8 MW includes the commissioning of new bunchers in the transfer lines to the injector cyclotron and between injector and ring cyclotron, the ongoing replacement of the four Al cavities of the ring cyclotron by Cu cavities operated at 1 MV, and the future installation of two additional accelerating cavities in the injector cyclotron. In parallel to the implementation of this new equipment simulation studies are under way to improve our understanding of the space charge effects at the different stages of acceleration. The present status of the project will be presented.

Introduction

A high-power proton beam of 590 MeV is produced at PSI using two consecutive cyclotrons, and delivered to two meson production targets and a neutron spallation source. The production beam current has increased over the years to 1.9-2.0 mA. After the passage of the graphite targets M and E, ~1.3 mA of beam is transported to the spallation neutron source SINQ, which uses as target solid lead rods with a steel or Zircaloy cladding. A liquid metal target (lead-bismuth eutectic) has recently been tested in the frame of the MEGAPIE project [1]. In addition to the main use, up to 100 µA of the beam from the injector cyclotron are split and directed to an isotope production facility. An ultra-cold neutron facility which will use the full high-energy beam during about 8 seconds every 15 minutes is in construction. The medical applications previously using a few µA split off the high-energy beam have recently been transferred to a dedicated facility equipped with a compact, superconducting 250-MeV cyclotron.

Beam is delivered for ~5 000 hours per year. The evolution of the performance of the accelerator complex is illustrated in Figure 1. High-power operation became possible after the commissioning of a high-intensity injector cyclotron in 1985 and the rebuilding of the Target E station in 1990. Since the facility is in permanent development and some systems are operated at the limit of the technology (HF) or are inherently critical in the particular environment (electrostatic elements), occasional failures are not unexpected. Other problems (e.g. site power, cooling) are not specific to the high-power operation. Altogether, an availability of 85-90% is achieved.

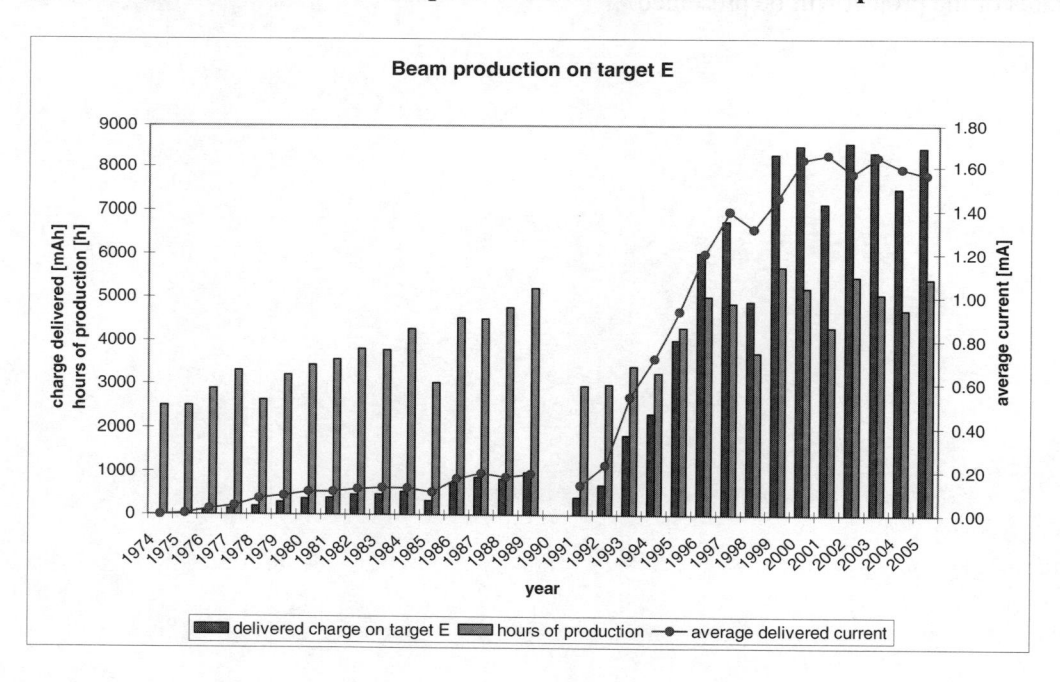

Figure 1. Performance of the PSI proton accelerator since the first operation in 1974

Since the SINQ target is designed for a load of 2 mA the main proton beam should be increased to 2.7 to 2.8 mA to fully exploit the potential of the facility. For higher currents an improved design of the target is requested. The present project aims at an increase of the main beam intensity to 3 mA.

The current limit of a cyclotron is essentially determined by the losses at the injection and extraction septa. A large turn separation, i.e. a high-energy gain per turn in comparison to the energy spread induced by the space charge effects, is requested to keep the losses at an acceptable level. For routine operation a loss of 0.5 µA for each element (extraction from the injector cyclotron, injection into and extraction from the ring cyclotron) is a reasonable value with regard to the maintainability of the accelerators. With present technologies powers in the range of 10-15 MW could be achieved with a cyclotron, e.g. a 10 mA proton beam could be accelerated to 1-1.5 GeV, where the relativistic mass increase sets a limit on the achievable turn separation. The 590-MeV PSI ring cyclotron is equipped with four accelerating cavities. Their upgrade to 1 MV peak voltage will boost the current limit to 5-6 mA. The maximum beam current at our facility will no longer be given by the performance of the accelerators, but by conditions encountered at other components, e.g. the maximum temperature of the meson production Target E.

The ring cyclotron was originally equipped with Al cavities. The peak voltage has been increased over the years up to 750 kV. A new Cu cavity designed for operation at 1 MV was tested up to 1.4 MeV on the test bench. Two such cavities are now installed in the accelerator. They have been reliably operated above 900 kV in 2006. The replacement of all four new cavities will be completed in 2008.

The injector cyclotron is now able to deliver beam currents up to 2.2 mA and the "round beam" operation mode (i.e. acceleration of a beam prepared such that a stable bunch structure of very small phase width is conserved) is well established. The upgrade of this machine can be realised in several steps. While some improvements are achievable in a short range at relatively low costs, the final step will request substantial investments for two additional accelerating cavities in order to increase the turn separation at the extraction.

In the concept of our facility, the meson production Target E plays a crucial role for the operation of the spallation neutron source. It scatters the proton beam in such a way that a suitable beam density distribution can be achieved after transportation to the spallation target. While in the past failures of Target E were the cause of two to three unscheduled maintenances per year, the present performance is excellent. However, the particular design of this device limits the proton beam current to 3 mA. A reconstruction of the target station would be very costly and would necessitate a one-year shutdown of the facility. This is not foreseen in this project.

Several components enabling beam tests and production above the presently allowed 2 mA are already installed. We expect that the supervising authorities will accord permission for a current increase in the next few months.

Beam simulations

The prediction of the performance of a high-power accelerator is a difficult task since the relevant factors are not accessible by usual beam dynamical calculations. The current limit is given by the losses due to tails and halos several orders of magnitude smaller than the beam itself. In routine operation the injection and extraction losses are in the range of 0.02% of the beam intensity. A reliable beam simulation requests tracking of millions of particles, a good knowledge of the initial conditions, the consideration of higher order effects, and detailed beam diagnostics for comparison and validation of the calculations.

The development of the computational tools needed for such simulations (MAD9P, efficient parallel programming) is in progress [2], and the results obtained so far are very promising. Good results have been obtained in the simulation of the beam transport from the source to the injector cyclotron. Simulations of the bunching of the space charge dominated beams have been performed as well. Figure 2 shows a simulation of the cleaning of the 870-kV beam entering the injector cyclotron. The calculated and measured loads on the collimators used for this purpose agree within 10%.

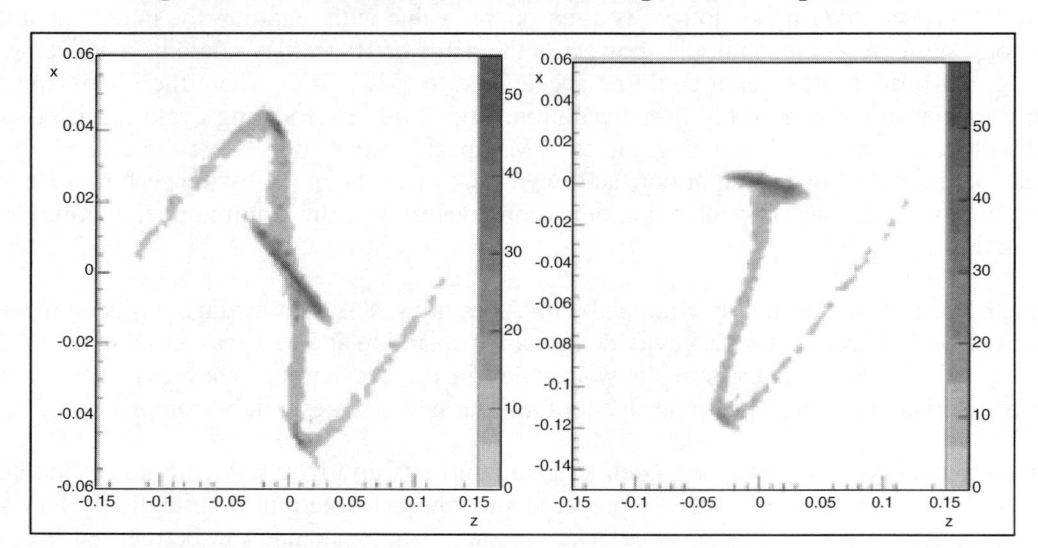

Figure 2. Simulation of the beam cleaning at the first phase slit

The calculated rms beam parameters in the injector cyclotron agree well with the measured ones and the formation of the short, stable bunches which are the key to the high current operation are well described. It is however a long way until start-to-end simulations will be able to predict losses at the 10^{-3} level. For the ring cyclotron the work is less advanced. At present, projections must still rely on extrapolation and scaling based on the performances observed at different steps of the development of the facility.

The investigation of the influence of beam-induced high order modes in the ring cyclotron shows that such effects remain small at the beam current considered in this upgrade.

Upgrade of the injector cyclotron

The quality of the beam extracted from the injector cyclotron depends crucially on the initial conditions at injection. In the past few years, significant progress has been made (in fact an increase of the beam current from 1.5 to 2 mA) mainly by better handling the space charge effects on the bunching of the 870 kV beam from the Cockroft-Walton pre-accelerator. Pursuing along this line a second buncher operated at the third harmonic (150 MHz) has been installed during the 2006 shutdown. The expected increase of the beam intensity in the phase space defined by the collimators at the injection and the generation of the conditions required for acceleration in the "round beam" mode have been achieved. It is possible to inject 3.5 mA with a DC beam of 10 mA, compared with the previous 2.2 mA at 12.5 mA DC. Beam intensities of up to 2.7 mA (limit of the available beam dump) have been extracted. While in the present stage of the development the extraction losses at this current are too high for routine operation, the observed beam quality suggests that up to 2.4 mA could already be used for injection into the ring cyclotron.

In order to improve the stability and reduce the frequency of maintenance work we plan to replace the multi-cusp source by a device with a longer lifetime and greater proton efficiency. A compact, permanent magnet, microwave source has been constructed. The ongoing tests are very promising. The new source will be installed during the 2008 shutdown. The reduced load and a new stabilising system for the HV generator of the Cockroft-Walton pre-accelerator will result in a reduction of the beam jitter, thus helping to lower the losses in the whole facility.

The ultimate step to achieve maximum performance with the injector cyclotron will be the installation of two new resonators in order to replace the two flat-top resonators (which are obsolete in the "round beam" acceleration mode where the phase width is as low as 2° HF) by 50 MHz accelerating systems. Figure 3 shows measured and estimated beam widths at the extraction of the injector cyclotron as a function of the beam current, and the limits given by the present and future acceleration system. To keep the losses at the present level, the turn separation at the extraction should be at least 7 σ (beam).

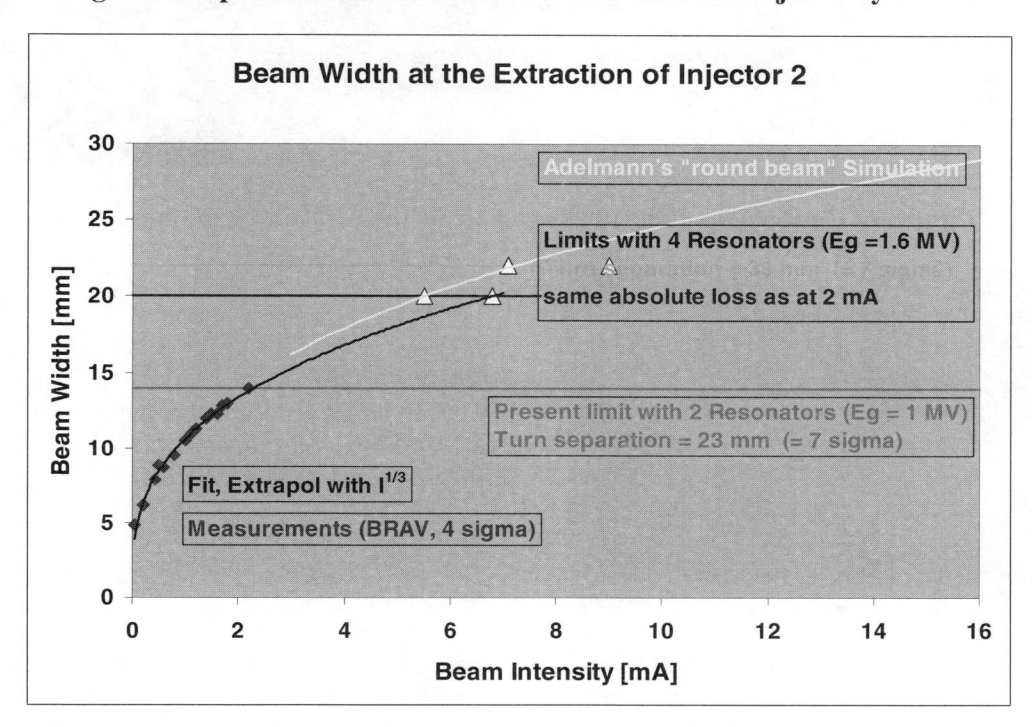

Figure 3. Expected beam width at the extraction of the injector cyclotron

The concept study led to the choice of single-gap cavities, as discussed in Ref. [3]. The engineering phase has just been completed and the installation is planned in 2009/2010. A sketch of the new resonator is shown in Figure 4 together with the sector containing the flat-top cavity to be replaced. The technical requirements set by this upgrade have already been considered in the ongoing renewal of the vacuum system of the injector cyclotron. For economical reason (optimisation of the cooling power by operating the old resonator at a lower level) the energy gain will be somewhat lower than quoted in the figure. Nevertheless the production of a 3- to 3.5-mA beam will definitely be possible, even below the present loss limit.

Improved injection into the ring cyclotron

Due to the space charge the phase width of the bunches increases significantly during the transport to the ring cyclotron. Simulations shows that a "superbuncher" working at the 10th harmonic (500 MHz) and located in the middle of the 50-m long transfer line can be used for bunching or phase rotating the 72-MeV beam prior to injection into the ring cyclotron. The design chosen after a comparative study is a double-gap drift tube cavity [4]. The dissipated power of less than 30 kW is moderate for the infrastructure in the beam line vault. The buncher is ready for the power tests and the necessary infrastructure in the vault has been installed. Some of its components are shown in Figure 5.

Figure 4. Accelerating (future) and flat-top (present) cavities in the injector cyclotron

Figure 5. Double-gap drift tube buncher for the 72-MeV beam line

Technical data:
506-MHz two-gap drift tube cavity
218-kVpp RF-voltage per gap
30-kW power (op. 10 kW)

Preliminary calculations show that space charge compensated bunching is applicable to generate a variety of starting conditions at the ring cyclotron. A full simulation of the beam injection is in preparation and should demonstrate if the "round beam" acceleration mode is practicable in this accelerator. If this technique does not work, the "superbuncher" can alternatively be used as phase rotator to improve the matching to the phase acceptance of the ring cyclotron.

Upgrade of the ring cyclotron

The current limit observed during the previous stepwise increase of the accelerating voltage shows a third power dependence on the number of turns, which confirms the rule proposed a long time ago by W. Joho [5] for the strength of the longitudinal space charge effects in a cyclotron. The emittance term entering the calculation of the beam width in the ring contributes the same way to the current limit. Since the emittance of the injected beam is known, both parts can be disentangled, and a calibration is obtained for the space charge contribution itself. With this knowledge, the beam width for any combination of emittance, space charge distribution and accelerating voltage shape can be estimated.

For identical conditions for the compensation of the space charge effect by means of a tilted flat-top, the maximum current achievable (with the present beam quality delivered to the ring) will be about 5 mA with four accelerating cavities at 1 MV. This feature is illustrated in Figure 6. However, the potential of improvement of the present flat-top system is limited and it is questionable if it would be able to provide the optimal correction when the new cavities are operated at the rated voltage.

Figure 6. Observed and extrapolated current limit for the ring cyclotron

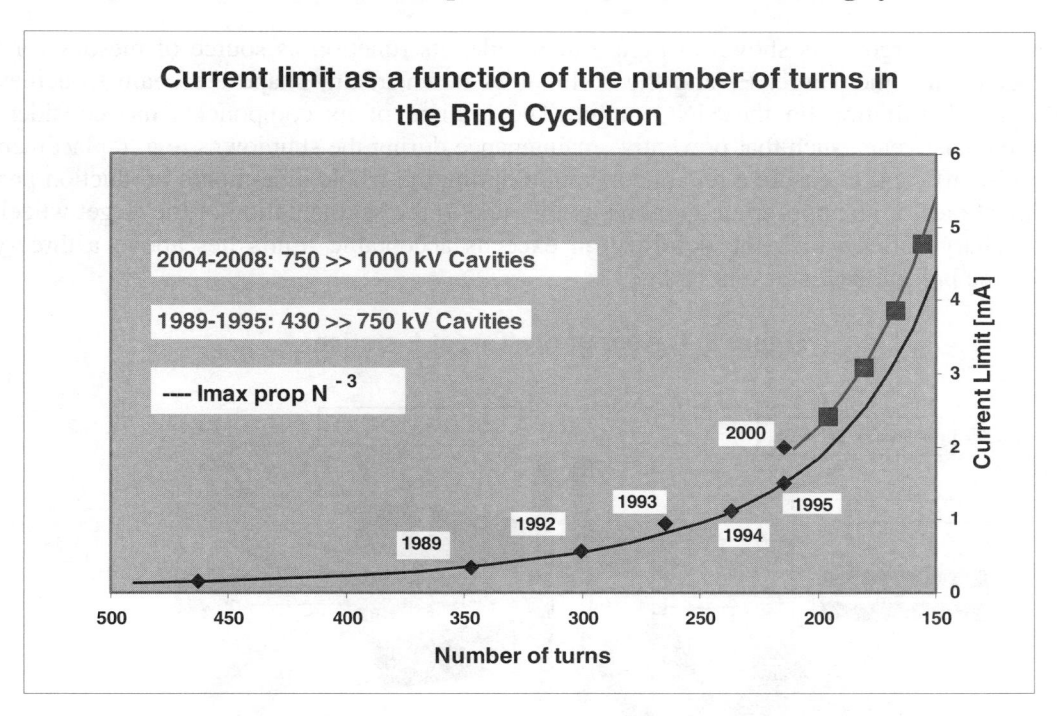

Current limit as a function of the number of turns in the Ring Cyclotron

- 2004-2008: 750 >> 1000 kV Cavities
- 1989-1995: 430 >> 750 kV Cavities
- ---- Imax prop N^{-3}

Data points labeled: 1989, 1992, 1993, 1994, 1995, 2000.
X-axis: Number of turns (500 to 150)
Y-axis: Current Limit [mA] (0 to 6)

In the "round beam" acceleration mode a flat-topping cavity would no longer be required in the ring cyclotron. However, since results from a reliable simulation are not at hand at the moment, the further use of the flat-top system should also be considered. Using a very simplified model to evaluate the shape of the longitudinal phase space one can estimate that beam currents up to 2.7 mA are achievable with the present flat-top performances in combination with the "superbuncher". Therefore, the decision whether the development of a new flat-top system should be undertaken can be postponed until results of the simulation and/or first experimental tests confirm or deny the applicability of the "round beam" technique in the ring cyclotron.

Figure 7. Installation of the new cavity in the ring cyclotron

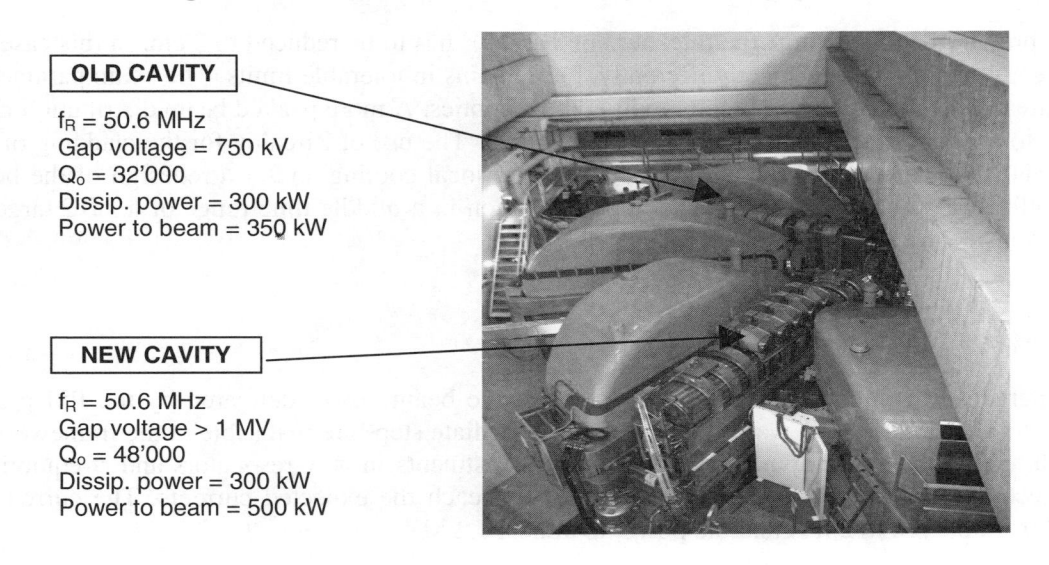

OLD CAVITY
f_R = 50.6 MHz
Gap voltage = 750 kV
Q_o = 32'000
Dissip. power = 300 kW
Power to beam = 350 kW

NEW CAVITY
f_R = 50.6 MHz
Gap voltage > 1 MV
Q_o = 48'000
Dissip. power = 300 kW
Power to beam = 500 kW

Target E and beam transfer to the SINQ target

The layout of Target E is shown in Figure 8. Besides its function as source of mesons for five secondary beam lines the Target E (and the collimators downstream) shape the beam to achieve a suitable density distribution on the SINQ target. The lifetime of its components has considerably improved in the past years, such that preventive maintenance during the shutdowns (e.g. replacement of the bearings) is sufficient to ensure a reliable operation during the whole nine-month production period. Thanks to the choice of an appropriate graphite quality and to the segmentation of the target wheel the accumulated charge after which the deformation exceeds acceptable limits lies above a three-year production at the present beam current level.

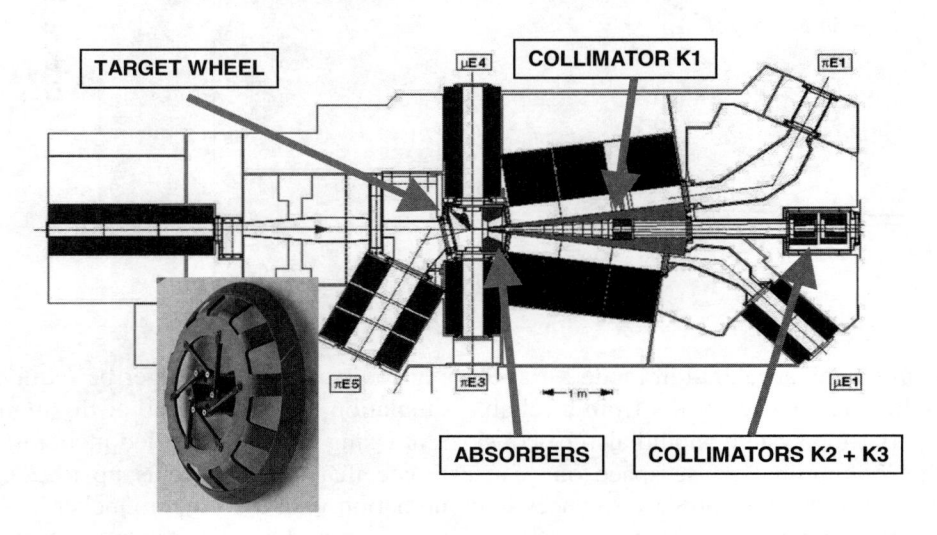

Figure 8. Layout of the Target E station

The 3-mA load limit on Target E is set by the temperature at which it starts to evaporate. No problems are expected with the collimators up to a beam current of 2.6 mA with the 4-cm thick target. The second collimator defining the beam to SINQ will reach its limit. The beam optics remain the same. The current on SINQ is about 1.8 mA and the maximum density is 55 $\mu A/cm^2$.

For a primary beam of 3 mA the thickness of Target E has to be reduced to 3 cm. In this case the load on the collimators is distributed differently, but remains in tolerable limits. The beam parameters on SINQ are slightly changed due to the modified beam optics. A more peaked beam distribution must be taken into account in a new design of the SINQ target. The use of Zircaloy for the cladding of the lead rods and a higher LD_2O flow allowing for a better local cooling and/or a redesign of the beam shaping collimators are in discussion. The implementation in a middle time range of a LBE target is also being considered.

Conclusion

An upgrade programme aiming at an increase of the beam power delivered by the PSI proton accelerator to 1.8 MW is under way. While some intermediate steps are realisable in the framework of the smooth, continuous development of the facility, investments in new resonators and rf-equipment for the injector cyclotron will be needed to eventually reach the expected currents. The current on SINQ will be improved to 2 mA and the power to almost 1.2 MW.

REFERENCES

[1] Groeschel, F., "MEGAPIE – Irradiation Experience of the First Megawatt Liquid Metal Spallation Target", these proceedings.

[2] Adelmann, A., *et al.*, "Beam Dynamic Activities at PSI's High Intensity Cyclotrons", ICFA BD-Newsletter No. 37, August 2005.

[3] Bopp, M., *et al.*, "Upgrade Concepts of the PSI Accelerator RF Systems for a Projected 3 mA Operation", *Proc. 16th Conf. on Cyclotrons and their Applications*, East Lansing, p. 300 (2001).

[4] Raguin, J.Y., *et al.*, "Comparative Design Studies of a Super Buncher for the 72 MeV Injection Line of the PSI Main Cyclotron", *Proc. 9th European Particle Accelerator Conference (EPAC'04)*, Lucerne, p. 1162 (2004).

[5] Joho, W., "High Intensity Problems in Cyclotrons", *Proc. 5th Int. Conf. on Cyclotrons and Their Applications*, Caen, p. 337 (1981).

ERAWAST – A NEW PRODUCTION ROUTE FOR EXOTIC LONG-LIVED RADIONUCLIDES

Dorothea Schumann, Jörg Neuhausen
Paul Scherrer InstituteVilligen, CH-5232 Villigen PSI, Switzerland

Abstract

Exploitation of highly-activated accelerator waste as a source for long-lived exotic radionuclides for application in several scientific and technological fields is presented as a possible new production route. In the framework of an international, ESF-funded exploratory workshop, held at PSI in November 2006, the demands concerning such isotopes for research in nuclear structure, nuclear astrophysics, basic nuclear physics, radiopharmacy, accelerator mass spectrometry and related research were defined. Thirty (30) participants from 12 countries decided to create an international network collaborating on this subject and launched a proposal for a Research Networking Programme (RNP) at ESF (European Science Foundation). Due to the availability of considerable amounts of proton-activated material (mainly from the surrounding of the 590-MeV ring cyclotron), PSI acts as unique production site for the radionuclides of interest for this collaboration. It has already been demonstrated that: a) radionuclides like ^{60}Fe, ^{26}Al, ^{44}Ti and ^{10}Be are available in the required amounts; b) in principle, separation of these isotopes is possible. Several samples in laboratory-scale quantity have already been provided for some selected experiments.

Introduction

Long-lived exotic radionuclides such as ^{60}Fe, ^{26}Al, ^{44}Ti or ^{10}Be are of great interest in several fields of research like radiopharmacy, basic nuclear physics, astrophysics, accelerator mass spectrometry and/or radioactive ion beam techniques. Some examples for the use of such rare isotopes are the potential medical application of ^{44}Ti/^{44}Sc as a generator system for Positron Emission Tomography (PET) as well as investigations of core collapse supernovae or studies of several neutron capture reactions on radioactive isotopes like the ^{60}Fe(n,γ)^{61}Fe reaction at stellar energies.

The production of all these nuclides in sufficient amounts is very time consuming and extremely expensive. Conventional techniques in commercial radioisotope production – restricted mainly to reactor-based or accelerator-driven production routes – are approaching their limitations. Consequently, alternative production possibilities and means of co-operation in large basic-physics facilities are discussed. One of these possibilities is the exploitation of accelerator waste.

Until very recently, no concerted effort had been made to reclaim such radionuclides for future use from activated components of particle accelerators. This was due, in large part, to the small amounts of suitably irradiated material. Nowadays, this situation has changed drastically with the construction of new, large accelerator facilities, spallation neutron sources and radioactive beam facilities.

The Paul Scherrer Institute operates the most powerful spallation neutron source (SINQ) in Europe, driven by a high-power proton accelerator, the 590-MeV ring cyclotron with a proton beam current up to 1.8 mA. Previous radiochemical analyses showed that beam dumps, shielding and target materials from these facilities contain long-lived radionuclides in such high amounts that chemical separation for several applications seems to be attractive. It was found that the isotopes ^{60}Fe, ^{26}Al, ^{10}Be, ^{44}Ti and probably many others can now be separated in amounts of 10^{16}-10^{18} atoms [1].

Due to this potential, we decided to establish an international collaboration of users and producers of such long-lived exotic radionuclides called ERAWAST – Exotic Radionuclides from Accelerator Waste for Science and Technology. On behalf of this, an ESF-funded workshop was held at PSI in November 2006 with 30 participants from 12 countries [2]. The participants agreed to the launch of a Research Network Programme at ESF to support their activities. Short-term as well as long-lasting working fields were defined:

1. *Existing accelerator waste material.* From a copper beam dump, which was irradiated at the 590-MeV proton beam station at PSI and was dismounted about 15 years ago, nuclides like ^{26}Al, ^{59}Ni, ^{53}Mn, ^{60}Fe, or ^{44}Ti can be separated. Other irradiated materials, i.e. carbon (source for ^{10}Be), stainless steel, or concrete are also available.

2. *Target material from the SINQ facility.* Two irradiated lead targets from the spallation source are available as well, which can be used to extract heavier isotopes, such as ^{182}Hf or several rare earth nuclides (^{146}Sm and several Dy isotopes). In principle, lead targets from the SINQ will be available every second year.

3. *Nuclide extraction from mercury.* Similar to lead, mercury can also serve as a spallation target for neutron production. In the framework of the EURISOL programme, studies for isotope separation from such targets are foreseen. Two already-irradiated rods filled with Hg are available for preparing small samples of interesting isotopes. An extension of this possibility is expected with a working EURISOL facility.

4. *Special irradiations.* The SINQ facility offers the possibility to irradiate materials with 590 MeV protons at special positions, allowing the production of specific isotopes.

In the following, an overview is given on short-term (2-4 years) actions concerning chemical separations as well as some examples for potential application. Due to the partially high γ dose rate of the starting material, the amount allowed to be handled in a laboratory without special radiation safety equipment is limited. Therefore, only relatively small amounts of the interesting isotopes can be separated at the moment. An adaptation of the developed separation techniques to large-scale quantities using hot cells and remote technique is planned.

Available material

At the moment, PSI can provide proton-irradiated copper stemming from a former beam dump, as well as irradiated graphite targets with a high content of ^{7}Be and ^{10}Be. Additionally, the first lead targets irradiated at the spallation neutron source, SINQ, will be ready for use in the near future.

After extended chemical analyses, the radial and length distribution of radionuclides within the copper beam dump could be determined [3]. From this data, the available amount of the isotopes of interest could be roughly estimated and 500 g of material were drilled out containing around 80% of the total activity. Additionally, several graphite targets were analysed concerning their content of ^{7}Be and ^{10}Be [4]. Table 1 shows data for expected amounts of selected isotopes.

Table 1. Estimation of the available amount of selected isotopes (without chemical separation)

	^{10}Be	^{7}Be	^{60}Fe	^{53}Mn	^{26}Al	^{44}Ti
Activity [kBq]	2	10^8	50	500	7	10^5
Atoms	10^{18}-10^{20} per target	10^{17}-10^{18} per target	10^{18}*	10^{19}	10^{17}	10^{17}

* Preliminary estimation.

Results for chemical separation and potential application

^{60}Fe

For all separations emanating from the copper beam dump, the copper samples were dissolved in half-concentrated nitric acid. Iron can then easily be separated by extraction from HCl-containing solution. Due to the high contamination with ^{60}Co within the copper beam dump – also a residue nucleus produced via the spallation reaction – extremely high separation factors of around 10^7 both from copper and cobalt, are necessary. Details of the chemical procedures for iron are described elsewhere [5,6]. Depending on the specific requirements of the envisaged experiments, the separation technique can differ slightly. Samples were prepared for the following applications:

- 10^{15} atoms (5 mg carrier added) for the determination of the half-life of ^{60}Fe in collaboration with Technical University Munich, Germany.

- 10^{16} atoms, carrier-free for studies of the neutron capture reaction ^{60}Fe(n,γ)^{61}Fe in collaboration with Forschungszentrum Karlsruhe, Germany.

^{44}Ti

Titanium forms – similar to its heavier homologues Zr and Hf – soluble anionic fluoride compounds. This property can be used for its isolation from the copper beam dump using ion exchange methods. For a detailed description see Ref. [6]. Two carrier-free samples were prepared:

- 1 MBq aimed to develop a $^{44}Ti/^{44}Sc$ generator for radiopharmaceutical use in collaboration with the University of Mainz, Germany.

- 3.5 MBq for studies of core collapse supernovae reactions, carrier-free in collaboration with the University of Edinburgh, UK.

^{26}Al

Aluminium can be precipitated as a hydroxide using ammonia solution. For carrier-free separation a non-isotopic carrier should be added, such as iron or lanthanum. A final purification is achieved by ion exchange on strongly acidic cation exchangers and following elution with diluted HCl solution [6]. Two samples could be obtained:

- 10^{14} atoms with carrier as standard material for AMS in collaboration with ETH Zürich, Switzerland, and University of Vienna, Austria.

- 10^{13} atoms in carrier-free form for laser spectrometry (RIMS) experiments in collaboration with the University of Mainz, Germany.

^{10}Be

Whereas the radionuclides described above were separated from the copper beam dump, beryllium isotopes were obtained from irradiated carbon targets, which requires a completely different dissolution technique. A mixture of $HNO_3/HClO_4/H_2SO_4$ had to be used for complete dissolution. Carbon dioxide was caught with a $Ba(OH)_2$ trap due to the contamination with ^{14}C. In the following, the separation and purification could be carried out similar to that of aluminium. A detailed description of the separation scheme can be found in [4]. At the moment, two applications for ^{10}Be are foreseen:

- 10^{17} atoms in carrier-free form for the development of a radioactive ion beam at the Université Catholique de Louvain-la-Neuve, Belgium.

- 10^{13} atoms in carrier-free from for laser spectrometry in collaboration with GSI Darmstadt, Germany.

Experiments using the samples described above are underway or planned for the near future. An extension of the application fields is expected with the availability of hot cell technique.

Outlook

The development of the hot cell/remote technique is expected to be complete within the next two to three years. This will enable us to provide the requested radionuclides in amounts one magnitude larger than as at present. Moreover, detailed analytical studies of the lead targets from the SINQ facility are ongoing. This material will enlarge the spectrum of radionuclides, e.g. ^{182}Hf or selected lanthanides, which are of interest in several fields of application, too. Separation techniques can then also be adapted for remote-controlled hot cells. Since lead targets from the SINQ facility will be available in principle every second year, a routine production of exotic long-lived radionuclides seems to be prospective.

REFERENCES

[1] Atchison, F., D. Schumann, R. Weinreich, TM-85-04-16, 6 December 2004.

[2] http://lch.web.psi.ch/radwaste/workshop/index.html

[3] Schumann, D., J. Neuhausen, S. Horn, P. Kubik, H-A. Synal, S. Köchli, G. Korschinek, *LCH-PSI Annual Report 2006*, 38.

[4] Schumann, D., J. Neuhausen, S. Horn, P.W. Kubik, I. Günther-Leopold, *Radiochim. Acta*, submitted.

[5] Schumann, D., R. Michel, G. Korschinek, K. Knie, J-Ch. David, *NIM A*, 562 1057 (2006).

[6] Schumann, D., R. Weinreich, *LCH-PSI Annual Report 2005*, 40.

150-MeV FFAG ACCELERATOR COMPLEX AS A NEUTRON PRODUCTION DRIVER FOR ADS STUDY

Minoru Tanigaki, Yoshiharu Mori, Tomohiro Uesugi, Kota Okabe*, Masamitsu Aiba, Yoshihiro Ishi**, Kaichiro Mishima, Seiji Shiroya, Makoto Inoue

Research Reactor Institute, Kyoto University, Japan
*Faculty of Engineering, Fukui University, Japan
**Mitsubishi Electric Corporation, Japan

Abstract

The Kumatori Accelerator-driven Reactor Test (KART) project, aiming to demonstrate the basic feasibility of ADS, has started at Kyoto University Research Reactor Institute (KURRI) from the fiscal year of 2002. The proton FFAG accelerator complex as a neutron production driver for this project is now in the final stage of the test operation. The developments and the current status of this FFAG accelerator complex, including the current status of this project, will be presented.

Introduction

As a substitute for the 5-MW reactor at Kyoto University (KUR), a neutron source based on the accelerator-driven subcritical reactor system (ADSR) concept was proposed in 1996 [1]. The conceptual design study on ADSR using the MCNPX code clarified the lack of reliable effective multiplication factor k_{eff} in the proton energy region between 20 MeV and 150 MeV. Since our current experimental studies are limited to those with a 300-keV Cockcroft-Walton accelerator [2,3], a proton beam source which covers between 20 MeV and 150 MeV is required to extend our study on ADSR system.

A fixed-field alternating gradient (FFAG) accelerator originally proposed by Ohkawa 40 years ago [4] has a lot of advantages compared to synchrotrons such as a large acceptance, a possible fast repetition rate because of no active feedback in the acceleration. In ADS system, the stability of beam acceleration is directly connected to the stability of reactor itself. In such a context, FFAG accelerators can be a good candidate for the neutron source driver in ADS system.

While an FFAG has a potential to a neutron source driver, there are still technical difficulties in FFAG accelerators, such as the lack of wide band high voltage RF cavity or the short straight section insufficient for beam injection and extraction. Mori has demonstrated that these difficulties can be overcome by recent developments [5,6] with the successes of a 500-keV PoP FFAG and an 150-MeV FFAG with RF.

On the basis of our study and the technical developments of FFAG, the KART project was approved and started from the fiscal year of 2002. In this project, a practical proton FFAG accelerator complex of E_p = 20~150 MeV as a proton driver for ADSR is constructed in KURRI and the basic feasibility of ADSR system and the multiplication factor k_{eff} in the energy region of E_p = 20 ~150 MeV will be studied.

FFAG accelerator complex

The FFAG accelerator complex for the KART project consists of the FFAG injector with an induction acceleration and two FFAGs with RF as the booster and main accelerators, respectively. The basic specifications for the FFAG complex are summarised in Table 1. The layout of these FFAG accelerators in the accelerator room is shown in Figure 1. All of these accelerators will be in pulse operation with a repetition rate of up to 120 Hz. The beam energy of the current FFAG complex can be varied between 20~150 MeV by the change of beam energy from the injector.

Table 1. Specification of FFAG complex

	Injector	Booster	Main
Focusing	Spiral	Radial	Radial
Acceleration	Induction	RF	RF
k	2.5	2.45	7.6
E_{inj}	100 keV	2.5 MeV	20 MeV
E_{ext}	2.5 MeV	20 MeV	150 MeV
p_{inj}/p_{ext}	5.00	2.84	2.83
r_{inj}	0.60 m	1.42 m	4.54 m
r_{ext}	0.99 m	1.71 m	5.12 m

Figure 1. FFAG complex at KURRI

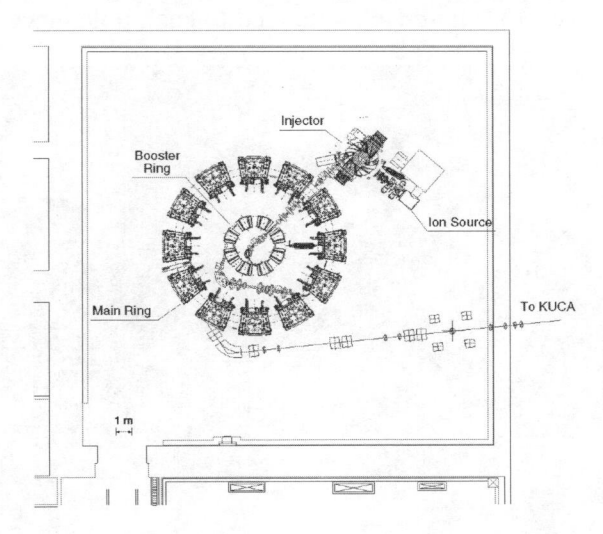

Ion source and injector FFAG with induction acceleration

H$^+$ ions are extracted from the typical multi-cusp-type ion source and accelerated to 100 keV, then transported to the injector. Since all of the FFAG complex is operated in the pulse mode, the ion source itself is also operated in the pulse mode for less power consumption. The arc voltage is pulsed at the duty of ~10%, then the pulsed beam is shaped to ~50 μs at the beam chopper placed in the transport line between the ion source and the following injector.

The injector FFAG (Figure 2) is a 2.5-MeV FFAG with induction acceleration scheme. Eight (8) spiral sector magnets with the spiral angle of 42° produces the FFAG magnetic field. The spiral sector type is chosen to this injector because of its rather long straight section and its higher packing factor. The index k for the FFAG field is defined by 32 trim coils placed on the pole face along radius direction (Figures 3, 4). The beam energy of the current FFAG complex can be varied through the change of this k by supplying the proper current set for trim coils. Induction acceleration scheme is adopted in this injector because the d.c. acceleration is more tolerant towards higher beam current. Typical induction voltage is 2~3 kV for ~300 μs.

Figure 2. FFAG injector

**Figure 3. Spiral sector magnets assembled in the factory.
Thirty-two (32) trim coils attached to each pole piece.**

**Figure 4. A typical FFAG magnetic field of the injector
ring magnet with trim coils obtained by the field measurement**

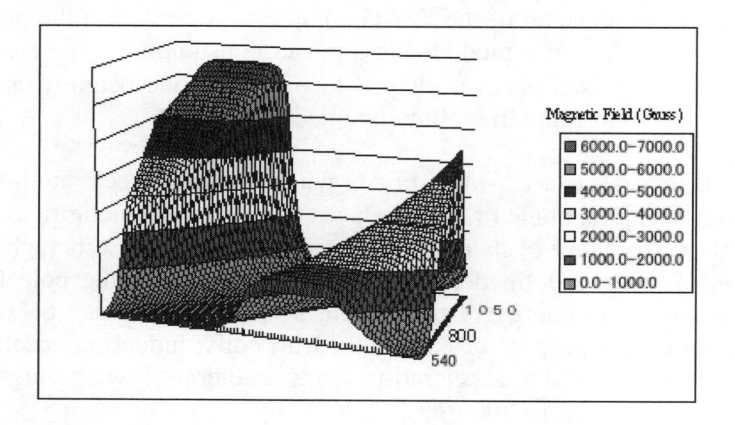

Booster FFAG with RF

The beam from the injector is then accelerated up to 20 MeV in this booster ring. This FFAG with RF is the radial sector type, consisting of 8 cells of DFD magnets. The FFAG magnetic field is produced by its pole shape with the half gap proportional to $(r/r_0)^k$. In this booster ring, $k = 2.45$ is chosen to minimise the beam excursion and the resonance variation. The window-frame type magnetic shield is attached to each side of the magnet to reduce the fringing field at the straight sections. The fringing field at the centre of the straight section is confirmed to be less than 100 Gauss from the magnetic field measurement. The beam injection is performed by a septum magnet, an inflector electrode and a pair of bump magnets. Acceleration voltage is supplied by an RF cavity with very low Q (~1) for the flat output voltage over the acceleration frequency range. The beam extraction is performed by a kicker and septum magnets. All the components for beam injection/extraction except the inflector electrode for beam injection are in pulse operation. The booster ring under construction is shown in Figure 5.

Figure 5. The booster ring under construction before the bump magnet for beam injection and the magnetic septum for beam extraction are installed

They will be installed in the straight section just before the extraction channel. The beam from the injector is injected to the ring through the magnetic septum, electrostatic inflector and a π-bump magnets. Beam extraction is performed by a kicker magnet and a magnetic septum.

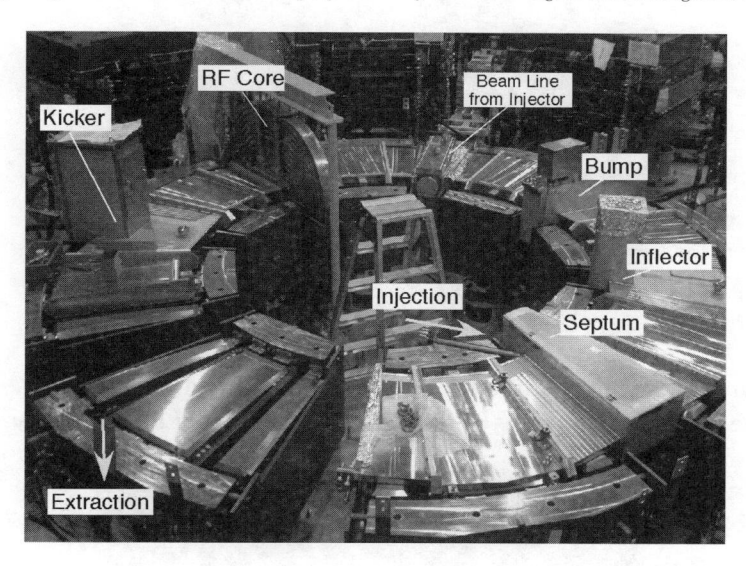

Main FFAG with RF

The main accelerator is basically identical to the 150-MeV FFAG in KEK. Detailed discussions are available in Ref. [6]. This time, the purity of iron in the magnets are increased to accept a high magnetic flux required for acceleration at 200 MeV, aiming the beam energy upgrade by the reinforcement of power supplies in near future. In our 150-MeV FFAG, a conventional amorphous core, not the finemet core, is used in the RF cavity. The beam injection process is performed by a septum magnet, a static-electric inflector and a kicker on the closed orbit. The saturation of RF core by rather large fringing field in the straight sections is expected in the case of "yoke-free" magnets, the RF cavity is installed in a magnetic shield. This magnetic shield becomes the additional COD source, so a pair of dipole magnet is installed in both sides of this RF cavity to minimise the disturbance.

Current status and future prospects

The construction of "Innovation Research Laboratory" was completed in March 2004. This building is not only for FFAG accelerator complex, but also for the multi-purpose usage of the beam from the FFAG complex in future, such as nuclear physics, chemistry, material science and cancer therapy. Currently, the construction of our accelerator complex itself is almost completed (Figure 6).

We successfully completed the first acceleration and beam extraction at the injector without trim coils in June 2005 (Figure 7). The energy of the extracted beam was 250 keV. It is quite interesting to find that the beam came over the integer resonance in this case. We plan to start investigating on this acceleration. The first acceleration and extraction with trim coils was successfully completed in January 2006 (Figure 8).

Figure 6. Current status of FFAG complex at KURRI

**Figure 7. First extraction of the beam from the injector FFAG.
Half of the injected beam was obtained as the extracted beam over $\nu_\chi = 1$.**

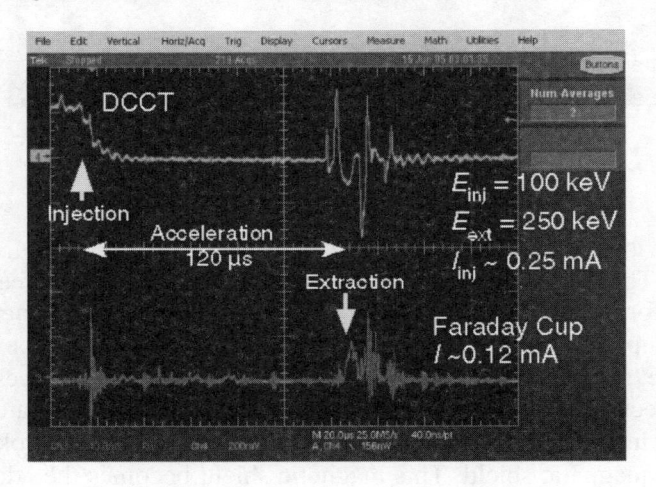

Figure 8. First extracted beam from the injector FFAG with trim coils

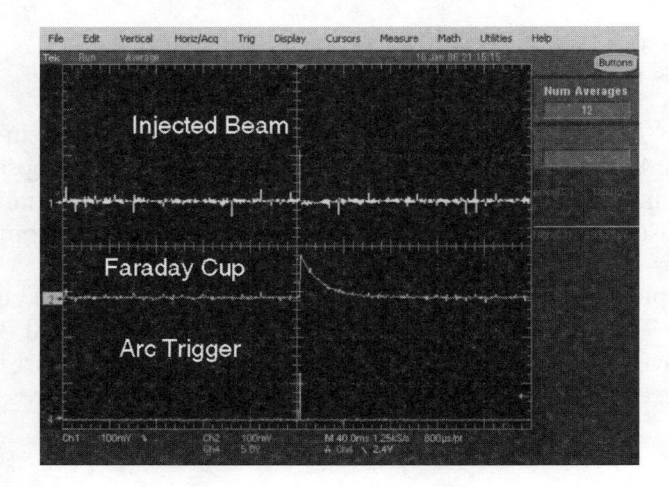

The first acceleration at the booster ring was employed in June 2006 (Figure 9). As the injection energy was limited ~1.1 MeV for stable operation of the injector, the RF frequency for the injection period was set to 1.6 MHz, then swept up to 3.6 MHz corresponding to the extraction energy of ~9 MeV. The beam inside the ring was monitored by a bunch monitor and confirmed to reach the extraction radius by the radial probe. The first beam extraction (Figure 10) was delayed to September 2006, mainly due to the termination of beam commissioning during the construction of main ring.

Figure 9. First beam acceleration at booster ring. Yellow line and magenta line correspond to observed beam and applied RF, respectively. Beam lost point in the picture corresponds to the position of radial probe in the ring.

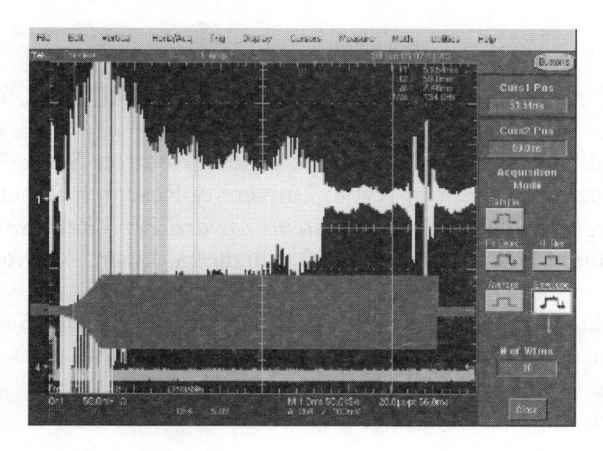

Figure 10. First extraction of beam from booster ring observed by the bunch monitor and profile monitor just after the exit of booster ring

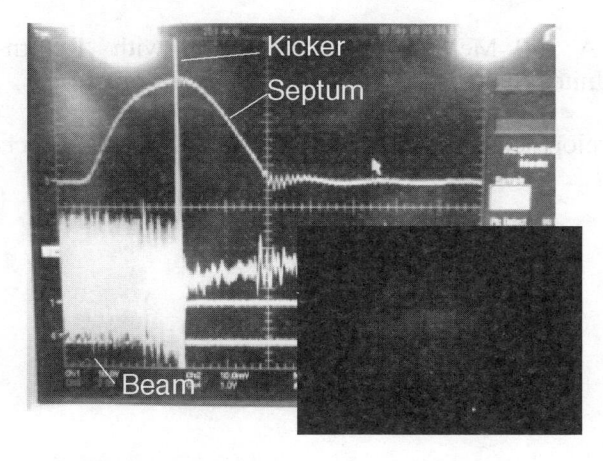

As of April 2007, the injector and booster are in stable operation and studies for beam injection and acceleration in the main ring are now in progress. The average beam current from the injector is ~20 nA, and the average beam current inside the booster ring reaches over 0.5 nA. These beam intensities are limited by the regulations of radiation protection applied to the current restricted area, not by the technical problem. The first beam from this FFAG complex is expected by the summer of 2007. The preparation of the subcritical core and the target for the neutron production is almost completed and the inspection for the actual operation will be held as soon as the accelerator complex is approved by the government.

This paper contains some results obtained within the task "Feasibility Study on ADSR Using Fixed-field Alternating Gradient (FFAG) Synchrotron as an Energy Amplifier" entrusted from the Ministry of Education, Culture, Sports, Science and Technology of Japan.

REFERENCES

[1] Kawase, K., M. Inoue, "Neutron Factory Project at KURRI", *APAC 1998*, Tsukuba, Japan, p. 104.

[2] Shiroya, S., H. Unesaki, *et al.*, "Neutronics of Future Neutron Source Based on Accelerator Driven Subcritical Reactor Concept in Kyoto University Research Reactor Institute (KURRI)", *Int. Seminar on Advanced Nuclear Energy Systems toward Zero Release of Radioactive Wastes, 2nd Fujiwara Int. Seminar*, 6-9 November 2000, Shizuoka, Japan, Abstracts p. 58.

[3] Shiroya, S., H. Unesaki, *et al.*, *Trans. Am. Nucl. Soc. 2001 Annual Meeting*, 17-21 June 2001, Milwaukee, Wisconsin, p. 78.

[4] Ohkawa, T., *Proc. of Annual Meeting of JPS* (1953).

[5] Mori, Y., *et al.*, "A New Type of RF Cavity for High Intensity Proton Synchrotron Using High Permeability Magnetic Alloy", *EPAC 1998*, p. 299.

[6] Adachi, T., *et al.*, "A 150 MeV FFAG Synchrotron with 'Return-yoke Free' Magnet", *PAC 2001*, Chicago, United States, p. 3254.

[7] Aiba, M., *et al.*, "Development of a FFAG Proton Synchrotron", *Proceedings of EPAC 2000*, Vienna, Austria, p. 581.

HYBRID PBG STRUCTURE: A CHALLENGING CAVITY SYSTEM FOR HIGH-POWER, HIGH-INTENSITY ACCELERATORS

V.G. Vaccaro[1], S. Albanese[1], A. Andreone[2], E. Di Gennaro[2], M.R. Masullo[3], M. Panniello[1]

[1]Dipartimento di Scienze Fisiche and I.N.F.N. Sezione di Napoli, Università *Federico II*, Napoli, Italy
[2]CNISM and Dipartimento di Scienze Fisiche, Università *Federico II*, Napoli, Italy
[3]I.N.F.N. Sezione di Napoli, Italy

Abstract

The presence of higher order modes (HOM) is one of the main limitations in the performance of high-intensity r.f. accelerators, since HOM degrade the beam quality. Accelerating cavities require HOM suppression while maintaining a high quality factor (Q) for the operational mode. Furthermore, at high power the r.f. breakdown can strongly influence the acceleration process and cause irreversible damages. This behaviour can be surmounted, raising the operation frequency f, since the Kilpatrick law states that the r.f. breakdown threshold increases with f. All these requirements, however, can be hardly be met in closed metallic cavities, especially at high frequencies where the configurations become cumbersome and technically unviable. The study presented here concerns the possible use of hybrid (metallo-dielectric) "open resonators" based on the photonic band gap (PBG) concept for operation in future high-intensity, high-power accelerators and working in the low GHz region. A number of prototypes have been designed, fabricated and tested at room temperature. All hybrid (copper or niobium + sapphire cylinders) structures under test show a good monomodal behaviour, in perfect agreement with simulations. In order to limit radiation problems, typical of hybrid resonators, we realised and measured a new TM_{02} mode cavity, at about 13 GHz, based on the simulation results. The geometry can easily be scaled to work at different frequencies. Measurements at low temperatures (1.5 K) are foreseen.

Introduction

The design of high-intensity and high-energy particle accelerators always faces a number of different problems. This paper will deal with two of them, namely the presence of higher order modes (HOM) interacting and degrading the quality of high-intensity beams, and r.f. breakdown phenomena, that easily show up in structures where large accelerating field gradients have to be achieved.

The beam loading of an accelerating cavity increases linearly with beam intensity. This is due to the energy transfer from the beam into the cavity where the e.m. field will reach a level different from the one without beam. The harmonic content in the bunch is the driving source of this transfer. Moreover, a high acceleration is generally connected with short bunches: this implies an increase of the higher order harmonics. The presence of these harmonics has a bad influence on the beam dynamics. This phenomenon is depicted in Figure 1, where the discontinuity introduced by the cavity in the vacuum chamber captures a part of the e.m. field which stays into the cavity until it is damped. This field consists of a spectrum of modes which includes the fundamental one and the higher order modes (HOM). The fundamental one can be compensated by varying the amplitude and the phase of the feeder voltage. The higher modes are more or less damped inside the cavity according to the Q of the relevant frequency. However, in general this Q is not sufficiently low to damp them in a fast way. It is therefore necessary to resort to *ad hoc* conceived devices to remove them from the cavity. Many efforts have been spent to reach this goal at no expense in terms of cavity performances. At relatively low operation frequency f, the removal of HOM is obtained connecting waveguides of various cut-off frequencies to the cavity. As an example, in Figure 2 the picture of the accelerating cavity of the main ring in DAFNE, running at 368 MHz, is displayed [1]. In order to reduce the interaction of the beam with the HOM of the cavity, the latter is equipped with long tapered beam tubes and three waveguides to couple out the parasitic modes. To suppress HOM other configurations can also be used [2], however all of them become cumbersome at high f.

Figure 1. Electromagnetic field due to the interaction between the beam and the accelerating cavity

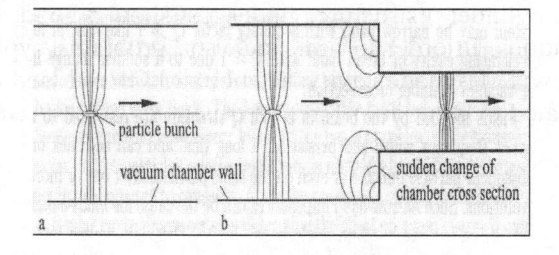

Figure 2. A DAFNE cavity

As mentioned above, in the design of high-energy particle accelerators a further priority requirement is to achieve large energy gradient in order to increase the final particle energy. In linear colliders, and more in general in linear accelerators (linacs) this translates in gradient values up to hundreds of MV/m [3]. This goal is limited by two phenomena: electron field emission and r.f. breakdown.

The first phenomenon can itself be a precursor of r.f. breakdown, and causes "dark currents". Further, it may lead to Q-value degradation for superconducting cavities.

The r.f. breakdown influences the acceleration process and can cause irreversible damage to the structure [4]. This phenomenon as well as its frequency dependence is not yet entirely understood, and all criteria adopted up to now for the design of cavities are empirically based. Experiments have demonstrated that the key parameter in the evaluation of the gradient limit value is the maximum achievable surface electric field. For normal-conducting accelerating structures, following the Kilpatrick criterion [5], the choice of high operation frequency enables higher surface electric field, and so higher gradients, without breakdown effects. The adopted Kilpatrick scaling law gives yield values which are quite conservative [3], but still reasonable for the design of the accelerator. Furthermore, in the case of long linacs, the choice of high r.f. working frequency limits the problem of cumulative currents due to secondary electrons. In fact, the on-axis capture of these currents, which is detrimental for the beam stability, depends on the accelerating field amplitude whose yield value rises with the frequency of operation. The choice of the cavity geometry and of the material can also improve the peak surface field limit, as experiments have demonstrated [3].

The emission of secondary electrons can also be reduced resorting to a cavity coating without free conducting electrons such as dielectric overlayers on metallic walls or bulk dielectric materials. It has recently been shown, in fact, that breakdown and charging can be limited in dielectric accelerating structures [6]. The shunt impedance per unit length, r_s, relates the accelerating gradient and the power required for establishing it through the field attenuation parameter, which is determined by the loss tangent of the dielectric, the resistive loss of the conductor, and the group velocity. As reported in [6], high shunt impedance values can be achieved in low-loss dielectric materials.

Dielectric structures based on photonic band gap (PBG) concepts [7,8] can be designed to be naturally mode selective, providing an operation mode with a high Q-factor and an effective damping of all unwanted higher order modes. This means that for a selected operating mode the losses can be lowered, enhancing the shunt impedance. In addition, they act as low electron emitters. For all these reasons hybrid PBG monomodal cavities can be an appealing choice as basic elements of a new generation of high-power, high-intensity accelerators.

A PBG structure is based on a periodic alignment of macroscopic objects, like metallic or dielectric cylinders, sandwiched between metallic plates [8,9]. Compared to a traditional cavity, the structure exhibits frequency band gaps (known as stop bands) that prevent the propagation of e.m. radiation along the periodicity directions in the structure itself. In such a system a cavity can be realised by removing one or more cylinders. In fact, when the lattice contains such "defects", new modes can exist in the frequency stop bands, being localised in the "cavity" and decaying exponentially in all directions away from the defect site. In this way, one can obtain a resonant structure with a very high quality factor, Q, at the fundamental mode. By choosing the appropriate lattice geometry HOM will fall into the frequency pass bands, exhibiting very low Q's. Therefore, they will propagate along the structure and can be easily damped using an absorber.

By means of proper electromagnetic codes, the design of such structures can be optimised according to the frequency requirements. The working mode (localised in the stop band) can be chosen to be a quasi-TM_{0n} mode, so that the electric field is parallel to the beam direction in the centre of the cavity (parallel to the cylinders axis). By adequately shaping the geometry around the defect, this mode can be optimised as the operating accelerating mode. By employing superconducting materials, in principle Q for this mode can be extraordinarily enhanced [9]. A "hybrid" configuration consisting of dielectric cylinders and metallic or superconducting plates can therefore be used as accelerating cavity.

Our work on PBG cavities started with all metallic structures [10] from the normal-conducting to the superconducting ones. Previous experiments, devoted to the r.f. characterisation of the cavities, have been performed on prototypes that have been numerically studied and designed by means of simulation e.m. codes. In the following we will present numerical simulations and experimental measurements carried out on different hybrid configuration structures.

Cavity design

A PBG structure working in the GHz region, or simply a microwave photonic crystal, represents a lattice of macroscopic pieces, usually with the same size and geometry (for example, rods), of metal or dielectric. Scattering of the e.m. waves inside such a structure can be described in the same way as electrons in the atomic potential of a real crystal. According to the Bloch theorem, the e.m. field in the lattice can be decomposed into a set of plane waves with the wave vectors multiple of the reciprocal lattice vectors. In a 2-D PBG structure, the wave field can be divided into two independent sets of modes: the transverse electric (TE) modes and the transverse magnetic (TM) modes. Using this approach, one can calculate the dispersion characteristics of the microwave photonic crystal under test, in order to evaluate the full band gap chart for both TE and TM modes. The presence of a global photonic band gap allows constructing a PBG resonator, simply removing one or more "atoms" (rods) in the periodic lattice. In fact, a "defect" mode having a frequency inside a full band gap will not be able to propagate through the bulk of the structure and will be spatially localised in the defect region.

Thereafter, the electromagnetic behaviour of a number of PBG based cavities can be studied by means of the MWS 3-D e.m. code (CST™) using both the eigen-mode and the transient-mode solvers in order to find the confined mode and to evaluate the related losses. In all simulation analyses the measurement antennas have been included, in order to predict the actual microwave response of the structure under study. All cavities work in the GHz range and exhibit a transverse periodicity based on a triangular lattice. This geometry provides a much better azimuthal symmetry compared to a square lattice. The resonant accelerating cavity is defined by the perimeter of the defects (number and size of the missing rods) realised in the centre of the structure. In this position a hole is created on both conducting plates allowing for the beam clearance. The walls of the cavity are formed by the surrounding periodic lattice.

In these conditions, the 3-D code allows to evaluate the mode field patterns in the whole PBG volume and outside the resonator itself. To properly design the cavity, the properties and the amplitude distribution of both the accelerating mode and HOM have been calculated along particular curves inside the volume. Because the real PBG prototype cannot be infinite, the field dependence on the number of rods around the defect have been analysed for the different configurations. The final cavity design also has to take into account the size restrictions dictated by the cryogenic experimental set-up.

In order to estimate the sensitivity of the operation frequency on the geometrical parameters, tolerance studies have been performed on a number of them. The lattice parameter b has a strong influence, as well as the radius of the central hole and the position of the first surrounding rows of cylinders. This last dependence is quite distinctive of PBG structures, underlining that maximum attention must be paid in the design and positioning of the rods closer to the defect (1st and 2nd rows, mainly), since they set the boundary conditions of the localised e.m. field [12]. Moreover, looking at the field pattern, one can exploit the different distribution of the accelerating mode and of higher order modes to determine the position of *ad hoc* absorbers and of the coupling system to the power feeder.

Experimental characterisation

The experimental r.f. characterisation has been carried out using a vectorial network analyser through the measurement of the scattering matrix at room and cryogenic temperature. The pick-ups for the experimental characterisation have been coupled to the cavity through the central holes and/or laterally from one open side through the columns. Because of the small dimensions of the input and output antennas, the coupling with the cavity has was carefully chosen, paying particular attention to avoid cross-talk and overcoupling.

Experiments at cryogenic temperatures are realised inserting the PBG structure under test in a liquid helium vessel and connecting it to the room temperature r.f. apparatus via semi-rigid coaxial cables. Measurements are carried out as a function of temperature as well, varying the position of the PBG cavity inside the cryostat (from 300 K down to 4 K) or reducing the liquid helium pressure (from 4 K down to 1.5 K) by means of a mechanical pump.

E.m. simulations have been performed modelling the structure in the same range of frequencies (10-20 GHz) investigated through measurements. There is always a very good frequency agreement (usually within 1%) between the operation frequency evaluated using the e.m. code and the experimental value. An example is reported in Figure 3, where the S_{21} scattering parameter measured at 300 and 77 K for an all metallic structure (36 metallic rods of height 4.6 mm, radius 1.5 mm, lattice constant 8.6 mm, for an overall diameter of 60 mm) excited in the fundamental mode (TM_{01}-like mode) at 14.478 GHz is compared with numerical results. The measured shunt impedance for this prototype is 70 MΩ/m, underlining the applicability of such resonant structures as accelerating cavities.

Figure 3. Metallic PBG cavity: comparison between experimental (cryogenic and room temperature) and simulated S_{21} scattering parameter

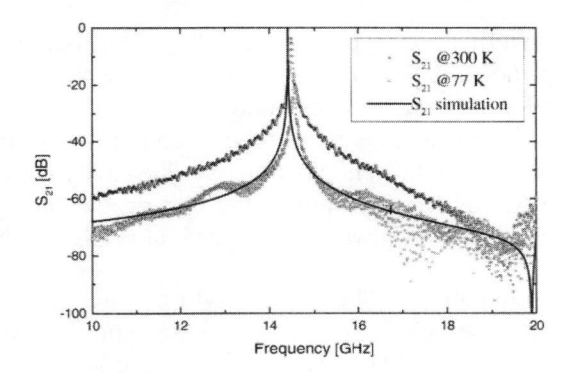

PBG hybrid resonant structures

PBG all metallic or superconducting cavities offer the dual advantage of intrinsic monomodal characteristics together with potentially very high quality factors. Previous measurements [13] on a compact (54 mm external diameter) Nb structure working at 16 GHz in the TM_{01} mode were very promising, showing in the superconducting state at the lowest temperature (1.5 K) a quality factor $Q = 1.2 \times 10^5$. Compared to a similar copper prototype, Q is more than one order of magnitude higher.

However, this value is still much lower than that expected taking into account the extremely low surface losses of superconducting Nb. The reason lies in the "open" character of a finite-size PBG resonator, where e.m. fields, even when well localised around the defective region, unavoidably tend to propagate outside the structure.

The radiation limited performance of all metallic PBG cavities points to a different approach, where metallic rods can be replaced by low-loss dielectric cylinders sandwiched between two conducting or superconducting plates. There are two main reasons behind this choice: first, a reduced complexity in the mechanical construction, that can be an important parameter especially when multi-cell cavities are considered; second, an increased flexibility in the selection of the superconducting materials and the operation temperature, provided that conductive losses remain lower than copper. In this context, high-temperature superconductors (with critical temperature T_C values ranging between 90 and 135 K) or the newly discovered magnesium diboride (T_C = 40 K) can be a valid alternative to niobium. Besides that, the planar geometry of the cavity walls can be exploited for easily replacing the bulk superconducting walls with copper walls coated with a superconducting (thin or thick) film.

Another advantage in the choice of a dielectric PBG structure is the absence of the cut-off in the photonic band structure. This means that adequately shaping the defect region we can localise, in the band gap, modes having an order higher than the TM_{01}, which on the contrary would be always confined in metallic PBG cavities [14]. As a consequence, it is possible to design an accelerating dielectric cavity working in an overmoded configuration.

Starting from the work on the metallic PBG cavities, hybrid structures, consisting of single-crystal c-axis oriented sapphire (dielectric constant ε_c = 9.7) cylinders sandwiched between conducting plates, have been designed and constructed to operate in a range of frequencies suitable for our microwave measurements. The choice of sapphire is motivated by the extremely low losses shown by this dielectric, which decrease further decreasing the operation temperature (from 10^{-4} at room temperature to 10^{-6} or even below at 4 K).

The TM_{01} dielectric structure

Tests at room temperature on a first prototype excited in the TM_{01}-like fundamental mode operating at 15.5 GHz confirmed the mono-modal behaviour. The PBG structure has an overall diameter of 55 mm and consists of three rows of dielectric cylinders surrounding a defective area of only one missing rod (region diameter ~13 mm). The mode was transversely localised but the confinement properties of the resonator were very poor, with an e.m. field extending well over the cavity dimensions (the ratio between the maximum value of the electric field E_{max} in the centre and on the third rod circle is roughly 2). A very low Q of roughly 10^3 was measured, lower than expected because of the high radiation losses.

Simulations indicate that the radiation pattern of this hybrid structure can be improved by adding rods. Using this new configuration, the ratio between E_{max} in the centre and on the third rod circle can increase to 50. Metallic rods could be added as well in order to confine the e.m. field more efficiently, simultaneously maintaining the high field gradient of the dielectric cavity. Of course, this can be done only at the expenses of the compactness and mechanical simplicity of the PBG real structure.

The TM_{02} dielectric structure

Following the work of Shapiro, *et al*. [15], we considered a new cavity design where a TM_{02}-like mode is excited. This mode is expected to be more confined in the centre in comparison to the TM_{01}-like mode. The new cavity (overall diameter 120 mm) is obtained removing from the previous structure two circles of cylinders in the centre region. In this way the mode falling in the first stop-band is the TM_{02}. In fact the wavelength λ of the confined mode is roughly equal to the half of the defect region diameter ($6b-2a$, with a cylinder radius). A TM_{01} mode with a wavelength equal to the defect diameter cannot be localised since the corresponding frequency does not fall in the dielectric stop band, as mentioned before. For the realisation of our prototype, we studied configurations with cylinders from the third to the sixth or seventh row, taking into account the overall dimension of the structure. The cavity presents a central hole with a diameter of 4.1 mm. The constructive parameters are reported in Table 1. In Figure 4, a picture of this structure with copper plates is shown.

Table 1. Geometrical parameters of a PBG cavity with a TM_{02}-like mode operating at 13 GHz

Cylinder radius (mm)	Cylinder height (mm)	Lattice constant (mm)	"Defect" region diameter (mm)	ε
1.5	6	8	45	9.7

Figure 4. The hybrid metallic-dielectric PBG cavity, based on single-crystal sapphire cylinders between two copper plates. The external clamping system is also shown.

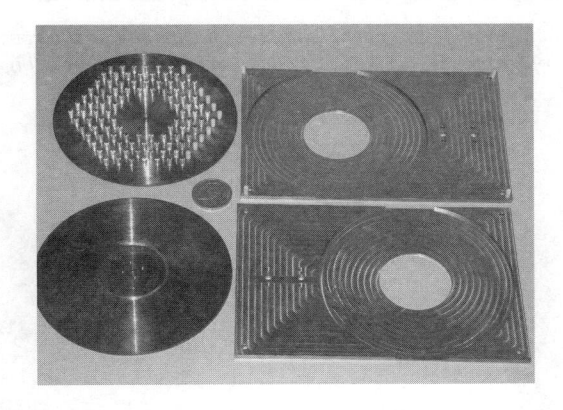

In Figure 5 the dispersion diagram of a perfect and infinite triangular lattice having the same geometrical parameters and calculated with the MBP freeware code [16] is plotted along the boundary of the first Brillouin zone (see the inset). Fully-vectorial eigenmodes of Maxwell's equations with periodic boundary conditions were computed by reconditioned conjugate-gradient minimisation of the block Rayleigh quotient in a planewave basis. In the same figure, the reflection coefficient S_{11} vs. the frequency experimentally determined, once the first two cylinder rows are removed, is also shown (black line). As expected, the TM_{02}-like mode at 12.98 GHz falls in the band gap for TM polarisation; unfortunately a second undesired resonance is also present at 17.8 GHz.

Figure 6 shows the amplitude distribution inside the cavity of the longitudinal electric field for the first two modes, excited at 12.98 GHz and at 17.81 GHz respectively. The latter mode presents a field less confined, with a maximum value in the cavity centre 30% less than in the former mode. Moreover, its frequency is not a multiple of the operational mode. Nevertheless, the presence of such resonance can be detrimental for the beam dynamics.

A way to damp it without affecting the mode at 13 GHz can be the removal of some cylinders from the fourth circle on. The creation of aisles, acting as internal waveguides, allows for propagation of the field through the cavity as shown in Figure 7. The mode is clearly propagating outside the cavity and can be easily damped using absorbers. Damping this mode also improves the effectiveness of the 13 GHz mode, whose maximum field increases to about a factor two in the cavity centre.

Figure 5. Band structure of a triangular photonic crystal having the lattice parameters reported in Table 1

In the inset the directions around the contour in the reciprocal lattice are indicated. The experimental S_{11} (black curve) related to a hybrid PBG cavity with the same structure and a central defective region is also shown.

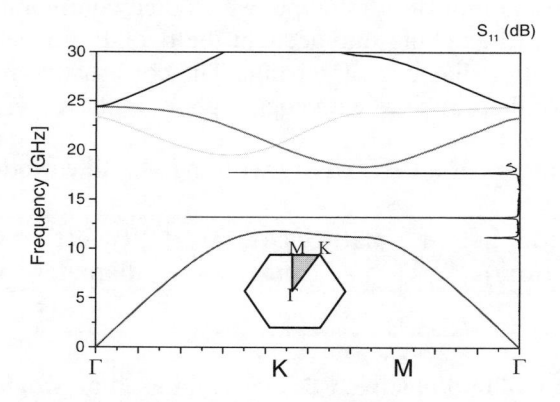

Figure 6. TM_{02}-like hybrid cavity simulation

The grey nuances code indicates the different magnitude of the longitudinal electric field excited for the fundamental mode (operating at ~12.98 GHz) and a second mode (operating at ~17.81 GHz)

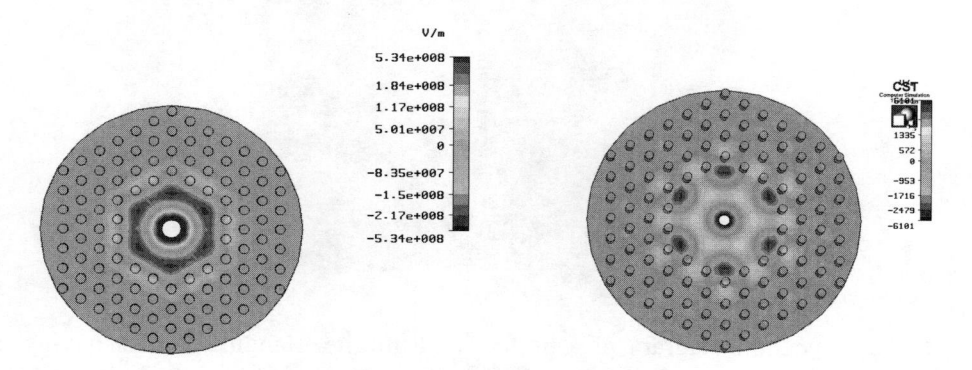

Figure 7. Electrical longitudinal field on a transverse cavity plane for the mode at ~17.15 GHz in the TM_{02}-like hybrid cavity, after the removal of some external cylinders

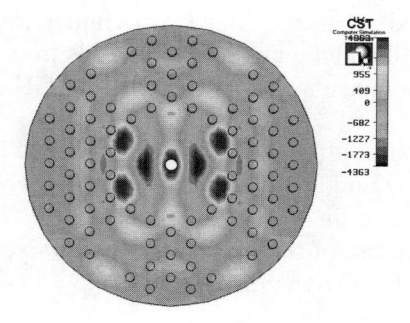

Pick-ups positioned out of the cavity allow the measurement of the field radiating outside the structure. The comparison between the S_{21} parameter evaluated inside and outside the modified PBG cavity is displayed in Figure 8. The difference between the two curves clearly shows that the fundamental mode stays much more confined inside the cavity.

Figure 8. Transmission scattering parameters inside (black curve) and outside (blue curve) the cavity for the hybrid PBG structure shown in Figure 7

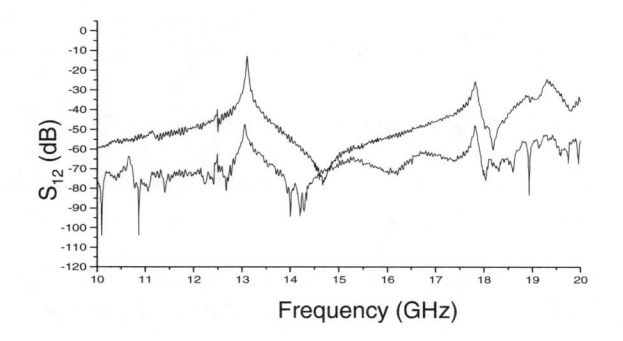

Conclusions

In this work, simulations and measurements on a number of metallo-dielectric PBG cavities based on a triangular lattice have been presented, in order to study their r.f. behaviour as accelerating cavities. The hybrid structures, consisting of single-crystal c-axis-oriented sapphire cylinders sandwiched between conducting plates, show the expected mono-modal behaviour and a good frequency agreement between the simulated and measured scattering parameters. Radiation seems to be a major problem for the hybrid resonator excited in the TM_{01}–like mode, therefore a TM_{02} hybrid resonator has also been preliminarily studied. This mode is expected to be more confined in the central region, where the hole for the beam clearance has to be realised. Simulations performed using both a freely available software package and a commercial 3-D e.m. code support this idea.

REFERENCES

[1] Bartalucci, S., Alii, "DAFNE Accelerating Cavity: R&D", *3rd European Particle Accelerator Conference*, Berlin, March 1992, p. 1263; LNF-92/033 (P), 24/04/1992.

[2] Shintake, T., "The Choke Mode Cavity", *Jpn. J. Appl. Phys.*, 31, p.1567-1570 (1992).

[3] Braun, H., *et al.*, "Frequency and Temperature Dependence of Electrical Breakdown at 21, 30, and 39 GHz", *Phys. Rev. Lett.*, 90, 22 (2003).

[4] Wang, J.W., G.A. Loew, "Field-emission and RF Breakdown in High-gradient Room-temperature Linac Structures", *Proceedings of the Joint US-CERN-Japan School, RF Engineering for Accelerators*, IOP, SLAC-PUB-7684 (1997).

[5] Kilpatrick, W.D., "Criterion for Vacuum Sparking Designed to Include Both rf and dc", *Rev. Sci. Instrum.*, 28, Issue 10, p.824 (1957).

[6] Hill, M., "High-gradient Millimetre Wave Accelerator on a Planar Dielectric Substrate", *Phys. Rev. Lett.*, 87, 94801 (2001).

[7] Shapiro, *et al.*, "Theoretical Analysis of Overmoded Dielectric Photonic Band Gap Structures for Accelerator Applications, *Proceedings of the Particle Accelerator Conference*, p. 1255 (2003).

[8] Yablonovitch, J., "Photonic Band-gap Structures", *J. Opt. Soc. Am. B*, 10, 283 (1993).

[9] Joannoupolous, J.D., R.D. Meade, J.N. Winn, "Photonic Crystals: Molding the Flow of Light", Princeton University Press, Princeton, NJ (1995).

[10] Andreone, A., *et al.*, "A Study on a Mono-modal Accelerating Cavity Based on Photonic Band-gap Concepts", *Workshop on Physics with a Multi-MW Proton Source*, CERN, Geneva (2004).

[11] Andreone, A., *et al.*, "Approaching to a Mono-modal Accelerating Cavity Based on Photonic Band-gap Concepts", *Proceedings of the 2004 European Particle Accelerator Conference*, p. 1309. © 2006 Wiley Periodicals, Inc.

[12] Akahane, Y., T. Asano, B.S. Song, S. Noda, "High-Q Photonic Nanocavity in a Two-dimensional Photonic Crystal, *Nature*, 425, 944 (2003).

[13] Masullo, M.R., *et al.*, "PBG Superconducting Resonant Structures", *Proceedings of EPAC 2006*, p.454.

[14] Smith, D.R., S. Schultz, *et al.*, "Experimental and Theoretical Results for a Two-dimensional Metal Photonic Band Gap Cavity", *Appl. Phys. Lett.*, 65 (5), p. 645 (1994).

[15] Shapiro, M.A., E.I. Smirnova, C. Chen, R.J. Temkin, "Theoretical Analysis of Overmoded Dielectric Photonic Band Gap Structures for Accelerator Applications", *PAC 2003*, p. 1255.

[16] Johnson, S.G., J.D. Joannopoulos, "Block-iterative Frequency-domain Methods for Maxwell's Equations in a Planewave Basis", *Optics Express*, 8, 173 (2001).

SESSION II

Accelerator Reliability

Chairs: B.H. Choi, J-L. Biarrotte

OPERATIONAL EXPERIENCE OF A SUPERCONDUCTING CAVITY FAULT RECOVERY SYSTEM AT THE SPALLATION NEUTRON SOURCE*

J. Galambos, S. Henderson, A. Shishlo, Y. Zhang
Spallation Neutron Source, Oak Ridge National Laboratory, USA

Abstract

The Spallation Neutron Source accelerator complex includes a superconducting RF linac that accelerates a proton beam from 186 MeV to 1 000 MeV, at a design power level of 1.5 MW. It includes 81 independently powered cavities. One of the advantages of the many independently powered cavities is rapid fault recovery in the event of a problem with a single cavity or any of its subsystems. We have developed a system to automatically calculate the new downstream cavity phase set-points, in the event of an upstream cavity failure (or any change in its setting). The system requires an initial setting of each cavity using a beam measurement, and uses a model to predict changes in cavity arrival times due to cavity failures. It has been successfully tested with up to 20 simultaneous changes in cavity amplitude and phase set-points, and is used regularly. The principles behind the scheme and results will be presented.

---

* ORNL/SNS is managed by UT-Battelle, LLC, for the US Department of Energy under contract number DE-AC05-00OR22725.

Introduction

The Spallation Neutron Source (SNS) is an accelerator-based source of high-intensity, short pulse neutrons, produced at 60 Hz [1]. SNS will be the world's highest-power pulsed, neutron scattering user facility. It is designed to operate at an average accelerator beam power of 1.4 MW, making it the highest-power pulsed accelerator. Beam commissioning was completed one year ago, and presently SNS is transitioning to operations with a scheduled three-year power ramp-up. To date, beam powers have been run up to 60 kW for sustained run periods of days, and to 90 kW for a sustained period of hours.

Beam acceleration is accomplished in a linac composed of warm and superconducting structures. The warm structure is composed of drift tube linac (DTL) and coupled cavity linac (CCL) structures and provides acceleration up to 186 MeV. The superconducting cavity linac (SCL) is designed to provide acceleration from 186 MeV to 1 000 MeV. The SCL linac is composed of two families of cavities: 1) medium beta cavities design for acceleration from 186 to roughly 400 MeV; 2) high beta cavities designed to accelerate the beam from 400 MeV to 1 000 MeV. There are 33 medium beta cavities and 48 high beta cavities, each separately powered by an individual 550 kW klystron, and each cavity has 6 cells.

In addition to operating at high power, operation at high reliability is crucial. Many users will be scheduled for beam time for only a few days, and loss of beam for any fraction of a day would have a significant impact. A key concern for reliability is the large number of klystrons and RF control equipment in the SCL. If a single cavity fails, the downstream cavities in the SCL cavities must be retuned and if the final energy changes, the transport lines and storage ring must be retuned – a time-consuming affair. From the initial proposal to use many superconducting cavities for SNS acceleration, it was noted that some sort of fast cavity retuning would be needed to realise high availability. Although SNS does not yet have outside users, the need to rapidly retune the cavities was imperative even during the beam commissioning period. SCL cavity operational limits have changed as they become better understood, and the need to rapidly reset cavities was needed to optimally utilise beam commissioning time. We have developed a model-based technique for SCL cavity fault recovery, and implemented it as a control room application. The method is used regularly in the control room to recover from changes in the SCL operating parameters. This technique and example applications of it are discussed in this paper.

Longitudinal beam modelling

A key ingredient in setting the cavity phases and scaling them in the event of a cavity failure is use of a model of the longitudinal dynamics. This is a simple tracking model of the synchronous particle. The longitudinal dynamics of the beam through a multi-gap cavity are modelled using standard first order equations [1,2] This model uses a drift-kick-drift approach, where the beam drifts through the first half of the gap, and then the effects of the RF are applied as a kick, and the beam drifts though the latter half of the gap (see Figure 1). This approach is applied successively through each gap. The beam phase and energy are tracked using:

$$\Delta E_n = qE_0 L_n T_n \cos(\phi_n)$$

$$\phi_n = 2 \cdot \pi \cdot f \cdot (\tau_n - \tau_0) + \varphi_0$$

$$\tau_n = \tau_{n-1} + \frac{L_{n-1}/2 - \delta g_{n-1}}{\beta_{n-1}^{(f)} \cdot c} + \frac{L_n/2 + \delta g_n}{\beta_n^{(i)} \cdot c}$$

$$\beta_{n-1}^{(f)} = \beta_n^{(i)}$$

$$\delta\tau_n = \frac{-E_0 L_n \beta_{n-1} T_n' \sin(\phi_n)}{\beta_{n-1}^2 \gamma_{n-1}^3 E_r 2\pi f} + \delta g_n \left(\frac{1}{\beta_{n-1}c} - \frac{1}{\beta_n c}\right)$$

Here ΔE_n is the energy gain at gap n, and ϕ_n is the synchronous phase of the beam relative to the RF at gap n. E_0 is the RF field on axis, L_n is the cell length of gap n, and T_n is the transit time factor of gap n. The synchronous phase is specified at the first gap centre (ϕ_0), where the tracking starts. τ_n represents the time the beam spends travelling from the first gap centre to the centre of gap n. δg_n is the offset distance of the electrical centre of the gap (where the kick is applied) relative to the geometrical centre. The term $\delta\tau_n$ is the phase correction term applied at the point of the kick. In this term, the T' term is $\partial T/\partial\beta$. Other terms used are: f is the cavity frequency, β and γ are the usual relativistic factors and E_r is the beam rest mass. For the SNS SCL cavities, the end cells for each cavity type have different cell lengths, electric field levels and transit time factors than the inner cells. Also, the end cells have non-zero gap offsets whereas the inner cells do not. This information comes from Superfish [3] modelling of the cavities and is tabulated separately for each cell.

Figure 1. Schematic of the longitudinal acceleration model

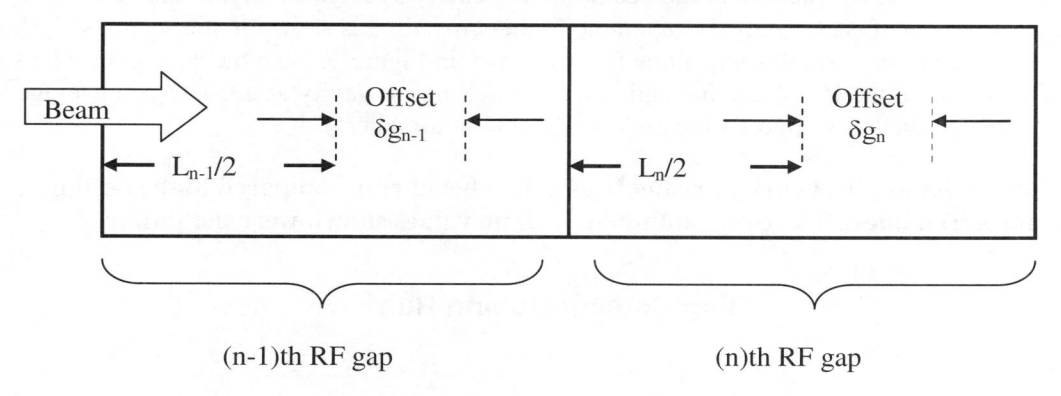

The synchronous phase of the SNS at design conditions is shown in Figure 2 at each gap of the 33 medium SCL beta cavities. Note that there are large excursions from the design phase for cavities where the beam energy deviates from the cavity geometrical beta (the medium beta cavities geometrical beta is 0.61). This "phase slip" has ramifications in the phase scaling technique described later.

Figure 2. Beam synchronous phase throughout each cell of the 33 medium beta SNS superconducting cavities for nominal conditions. Average phase is about -20° for each cavity.

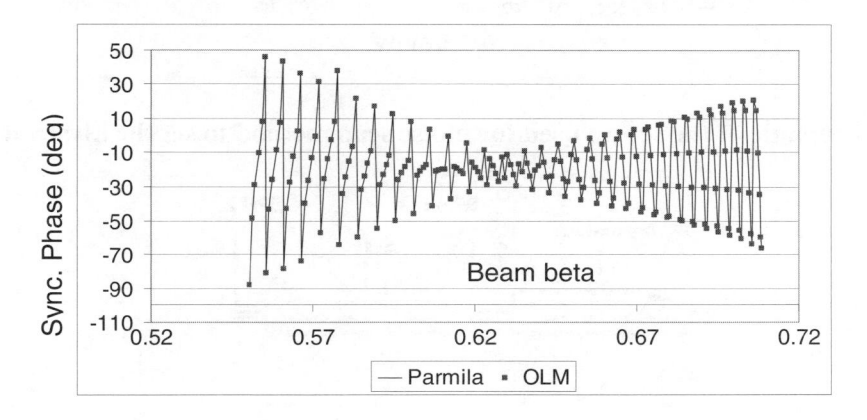

Superconducting linac cavity phase setting

The SCL cavities are operated in a mode where each cavity is run at as high a gradient as deemed safe. Cavity operational gradients are limited by different phenomena (e.g. field emission, quench protection, RF power limitation, etc.), and we typically run cavities with a wide range of gradients (including some which are not even operational), as shown in Figure 3. The average cavity RF phase relative to the beam is usually picked to roughly provide constant focusing ($E_0 \sin <\phi_0>$ is constant, where $<\phi_0>$ is the average synchronous phase of a cavity). Setting a klystron phase to get a desired $<\phi_0>$ requires some sort of measurement with beam. We use an RF phase scan technique [4] which is depicted schematically in Figure 4. In this technique, the phase of the cavity is varied 360°, and the beam arrival time is measured at two downstream points. All downstream cavities are un-powered during the measurement so as to not affect the beam. In addition the procedure is done with a short pulse low intensity beam so as to not excite any intervening cavities between the one being tuned and the devices that measure arrival time [5]. The longitudinal tracking model is used to simulate beam acceleration through the cavity and predict the time of flight (TOF) between the two beam phase detectors (the relative positions of the BPMs is known to within ~1 mm). Input to the model for: 1) initial beam energy, 2) cavity field, 3) offset between the klystron RF phase and ϕ_0 are varied to best match the measured TOF values over the 360° RF phase sweep. This matching provides the relationship between the RF phase and the beam, the RF cavity average field, and the beam energy at the entrance (and exit) of each cavity. An example of this procedure is shown in the screen snapshot of the application used to perform this technique for one cavity in Figure 5. Note the near sinusoidal shape of the TOF response to the RF phase, indicating that modelling the cavity as a single-cell average kick would likely be sufficient (as opposed the multi-cell tracking used here).

Figure 3. Example of cavity gradients used in a beam run, compared to the design values (red lines). The cavity numbers with no values shown were not powered.

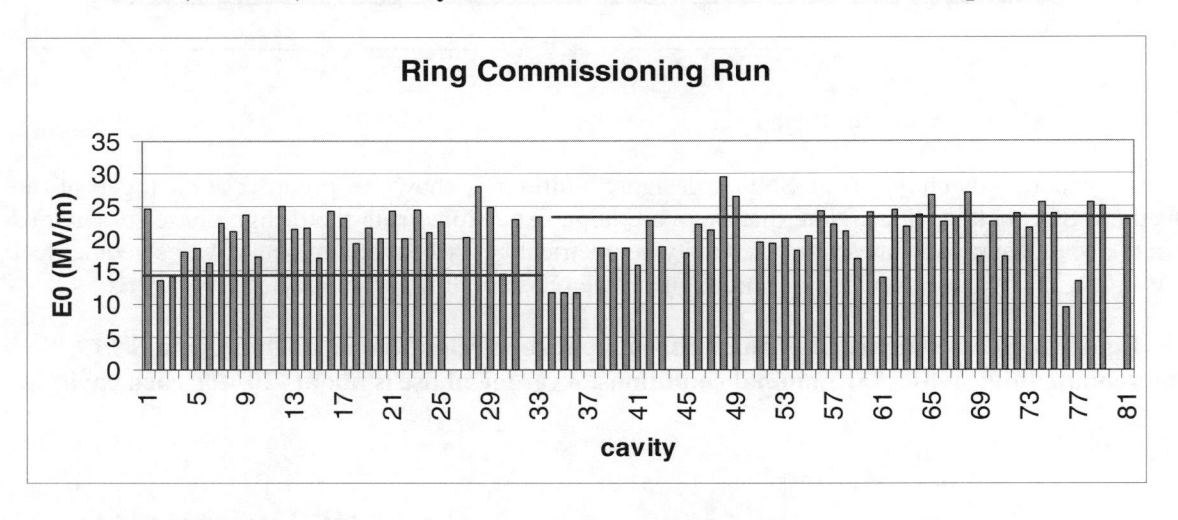

Figure 4. Schematic of the set-up used for phase scan method to set the klystron phase

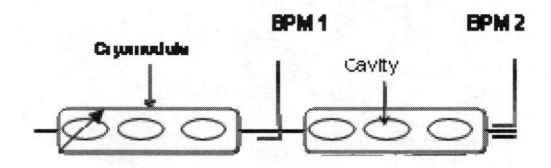

Figure 5. A phase scan matching example for an SCL cavity. Dots are model predictions, the black line is a fit of the BPM measurement and the red line is a cosine fit of the measurement.

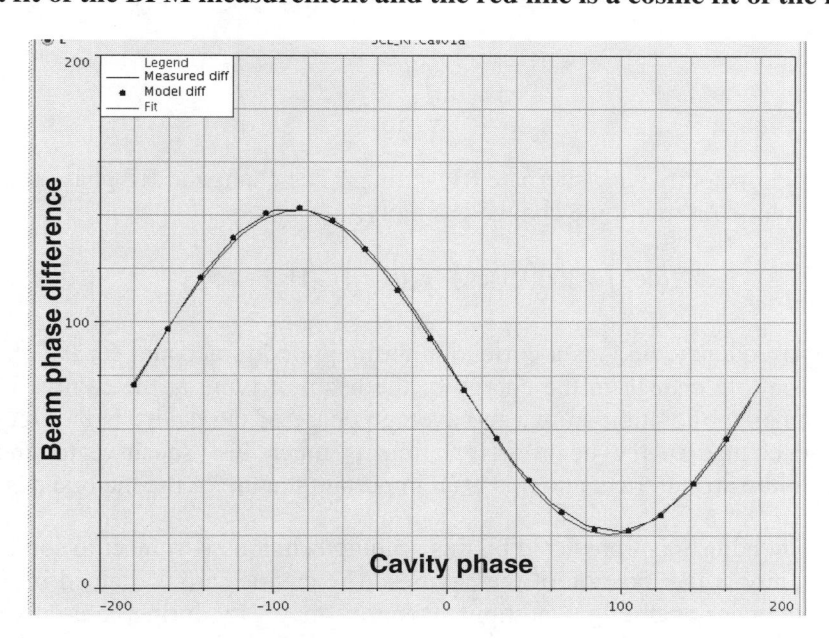

The scanning technique described above takes only 1-2 minutes per cavity with all equipment set up properly for the cavity phase scan. However, other activities such as turning on the cavity being scanned, adjusting quadrupole focusing magnets as the energy is increased, etc., can take considerable time. Initially the procedure took two to four days to set up the entire SCL. A big improvement has recently been realised by turning on all the SCL cavities simultaneously at the operational repetition rate (say 15 Hz), and blanking the RF pulses in all cavities downstream of the one being scanned so as to not affect the beam. Since beam is only run at 1 Hz for the scanning procedure, the cavities are effectively on the entire time, eliminating the time consuming procedure of turning them on one at a time as each cavities is set. This improvement allows tuning the entire SCL linac in six hours. Further optimisation may reduce this to three to four hours, but it is difficult to imagine any beam-based technique happening any faster. It was recognised early in the beam commissioning that a need to rapidly reset cavity phase set-points was required when, for example, SCL cavity amplitudes have to be reduced for more reliable operation or cavities must be turned off for various reasons.

Cavity phase set-point scaling method

A method has been developed to scale the cavity phase set-points arrived at with the above technique, in the event that one or more cavity amplitudes or phases must be modified. The first step in predicting new cavity phase set-points is to use the model to calculate beam arrival times at the first gap of each cavity for a configuration in which a complete set of beam-based phase set-points have been determined, i.e. the arrival time for each cavity "n" at position $L =$

$$t_n = \int_0^L \frac{dl}{v}$$

where $v(l)$ is the particle velocity with nominal RF settings. The position integration starts at the first cavity gap. The time spent traversing a cavity is calculated using the equations in the section above entitled *Longitudinal beam modelling*, and the time spent between cavities is a simple drift at the exit speed from the upstream cavity. Then, new beam arrival times are calculated including a change in one or more upstream cavity's amplitude or phase, i.e.:

$$t_{n_n}^* = \int_0^L \frac{dl}{v_*}$$

where $v_*(l)$ is the particle velocity with alternate RF settings. The change in RF phase at cavity "n" due to an upstream cavity phase or amplitude change is simply:

$$d\phi_n(\deg) = 360(t_n^*(\sec) - t_n(\sec))f_n(\text{Hz}) + \delta\phi_{slip}$$

where f_n is the cavity frequency. $\delta\phi_{slip}$ is a correction term applied to account for the change in phase slip across a cavity due to a change in the energy of the beam arriving at the cavity. This correction factor is found by numerical iteration so that the average phase of the cavity is that requested by the user. Calculating the change in the beam arrival time is much less sensitive to uncertainty than calculation of the absolute arrival time, and this is an important feature of the method described here.

The phase set-point adjustment predicted by this scaling technique is subject to some uncertainties, however. One uncertainty is the location of the cavities. The cavities are contained in a cryostat, and not directly observable under operating conditions. Positions may vary during cool-down to cryogenic temperature, but we believe the relative positions between cavities is known to within a few mm. Figure 6 shows the phase set-point error resulting from uncertainties in the cavity relative positions. Curves are shown for cavity separation uncertainties of one and two cm. Also curves are shown for energy gains in a cavity of 10 and 20 MeV – typical of the SNS cavities. This error estimate is for changing a single cavity – not a cumulative error from multiple cavity set-point changes. We expect less than 1° error in the phase set-point of any particular downstream cavity from the cavity position uncertainty. Another source of error is the uncertainty in the energy gain of a particular cavity, as shown in Figure 7. Here phase set-point errors are indicated for uncertainties in the energy gain of 1 and 2 MeV. Curves are also shown for different cavity separations. Based on experience using the beam-based phase scan technique described above, we believe that we can predict the energy gain in a cavity to within a few hundred keV (less than the energy gain uncertainty show in Figure 7). The distance between cavities is less than 1 m for sequential cavities in a cryostat, but may be up to 6-8 m if an entire cryostat is removed. For these parameters, the error in phase set-point from the energy gain uncertainty is also < 2-3° (for the error involved in changing a single cavity – not a cumulative error). The phase scaling errors are much smaller at higher energies, where the beam beta changes less.

This phase scaling technique is incorporated in a control room software application at SNS [6]. In this application the phase set-points, cavity voltages and energy gains derived from beam-based phase scan measurements are stored for all the cavities during a beam set-up typically done after any extended shutdown. The longitudinal acceleration model is run using the measured cavity voltages and synchronous phase set-points for all the cavities and the beam arrival times tabulated at the start of each cavity. The user can input changes to one or more cavity amplitudes and/or synchronous phase set-points. The beam model is rerun and new RF phase set-points are calculated based on the change in arrival time as described above, and can be sent to the machine. The model and application are part of the XAL [7] programming infrastructure. This infrastructure provides convenient methods for finding cavity locations, a longitudinal transport model, arrival time at cavities, and a GUI framework to build the application, among many other things.

Figure 6. Scaled cavity set-point error of a downstream cavity that arises from uncertainty in the cavity position. Phase set-point errors for cavity position uncertainties of 1 and 2 cm, and energy gains in the upstream cavity of 10 MeV and 20 MeV are shown.

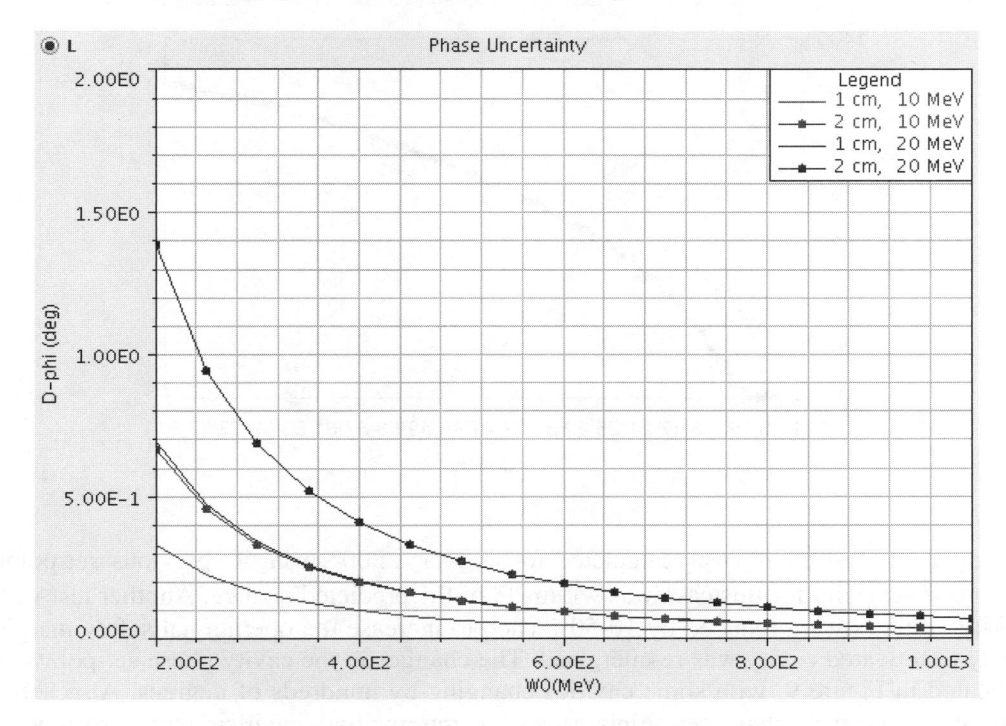

Figure 7. Scaled cavity set-point error of a downstream cavity that arises from uncertainty in the energy gain in an upstream cavity. Phase set-point errors are shown for energy gain uncertainties of 0.5 and 1 MeV, and drifts between cavities of 1.5 and 4.5 m.

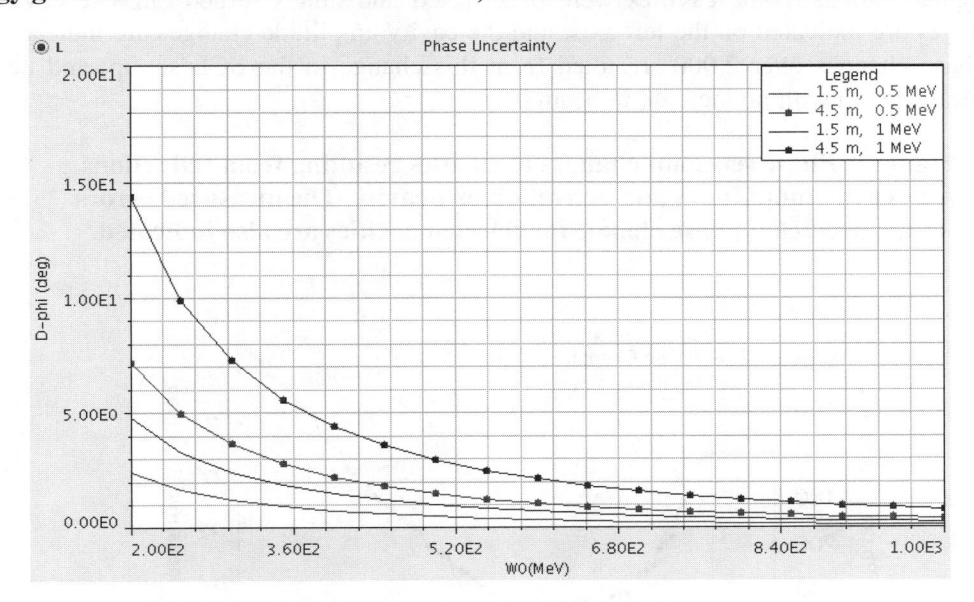

The application was originally tested by turning off single cavities, calculating and propagating the resulting change in downstream cavity phase set-points, and then checking the phase set-point of the last cavity (which is affected the most) with a beam-based phase scan technique. An example check is indicated in Figure 8. Here, the seventh (out of 81) cavity was turned off, and the change in

Figure 8. The phase set-point change of cavities in the SNS linac resulting from the deactivation of cavity number 7

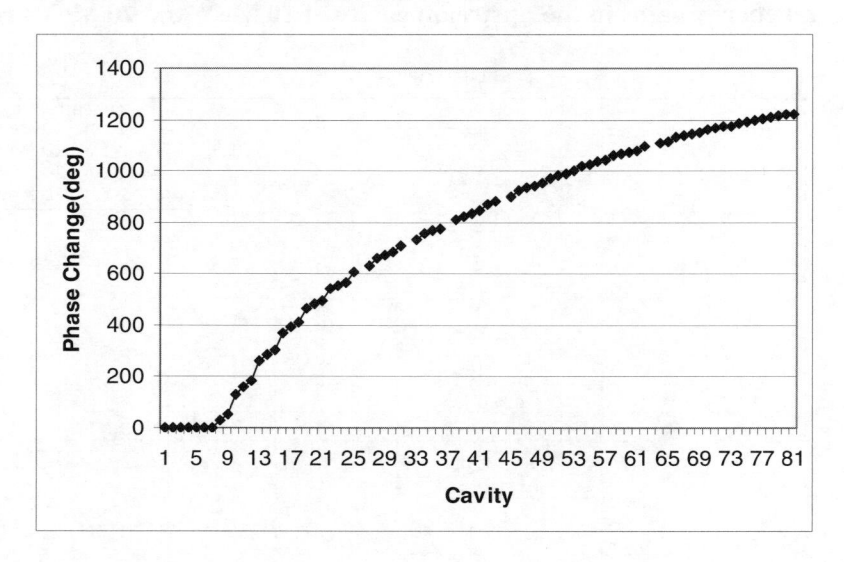

phase set-point of the last cavity was calculated to be over 1 200° from its previous set-point. The check of the phase set-point for this case was within 1° of the predicted change. Another test was done when 11 cavity voltages were reduced by 10-50% each to increase the operational safety margin, and one previously deactivated cavity was re-energised. The changes in the cavity phase set-points for this case are indicated in Figure 9, with some cavities changing by hundreds of degrees. Also shown are checks of the predicted new phase set-points at several intermediate cavities, with errors being less than ±4°. A final example of this technique is shown in Figure 10, for a case when the operating temperature of the SCL was changed from 4.5 to 2 K. Many cavities' amplitudes had to be changed from the original settings, some cavities were deactivated and others turned on. The cavity phase set-point changes are indicated on the left axis, and the cavity amplitude changes are indicated on the right axis. Phase changes over 2 000° resulted from this change in the SCL set-up, and beam was immediately transported with no increase in beam loss.

Figure 9. Phase set-point changes in cavities resulting from reductions in 11 cavity amplitudes and activation of 1 cavity. The measured errors of the predicted phase change for selected cavities are also indicated.

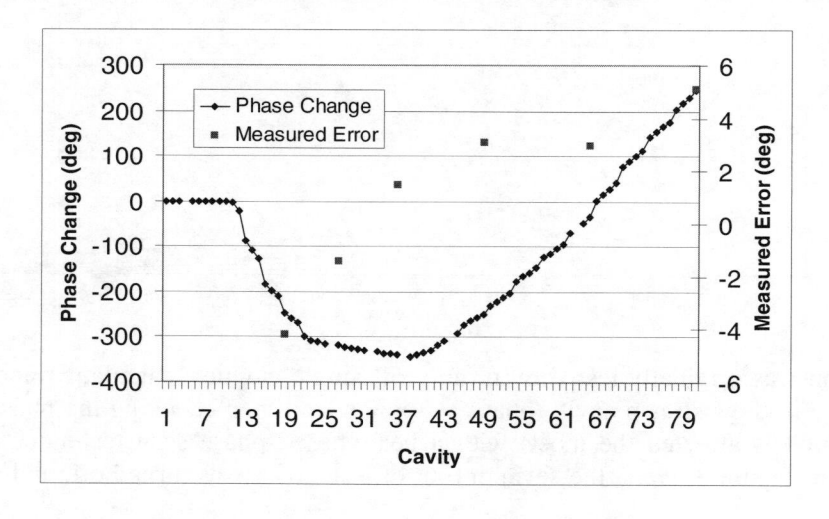

Figure 10. Cavity amplitude changes (right axis) and resultant phase set-point changes from the phase scaling method (left axis), during a recent transition from 4.2 K to 2 K SCL cryogenic temperature.

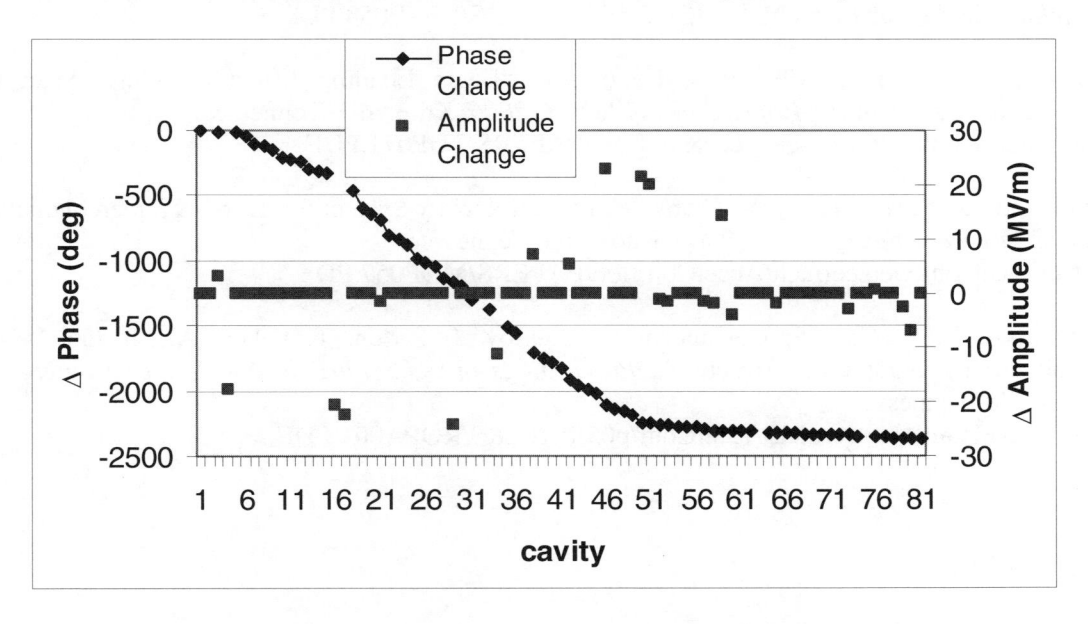

Summary

As the SCL cavity behaviour becomes better understood, safe operating levels change. Also sometimes cavities become disabled for unanticipated reasons. Whenever a cavity amplitude changes, the downstream cavity phase set-points must be reset. Using beam-based techniques to set cavities individually is a time-consuming endeavour, with 81 SCL cavities at SNS. The use of the model-based scaling technique allows quick resetting of downstream cavities without the need for any beam measurements, has become a standard practice at SNS. It has been used three times to accommodate wholesale changes in the SCL set-up. Beam is quickly re-established, in a state with comparable loss as the initial state.

REFERENCES

[1] Wangler, T., "Principles of RF Linear Accelerators", John Wiley & Sons (1998).

[2] Lapostolle, P., M. Weiss, *Formulae and Procedures Useful for Design of Linear Accelerators*, CERN-PS-2000-001(DR) (2000).

[3] Billen, J., L. Young, *POISSON SUPERFISH*, LA-UR-96-1834, LANL (2000).

[4] Henderson, S., I. Campisi, J. Galambos, D. Jeon, Y. Zhang, "Spallation Neutron Source Superconducting Linac Commissioning Algorithms", *Proceedings of 2005 Particle Accelerator Conference*, Knoxville, Tennessee.
http://accelconf.web.cern.ch/AccelConf/p05/PAPERS/FPAE058.PDF

[5] Jeon, D., S. Kim, Y. Zhang, S. Henderson, "Beam Loading Effect on Phase Scan for Superconducting Linac", *Proceedings of LINAC 2006*, Knoxville, Tennessee.
http://accelconf.web.cern.ch/AccelConf/l06/PAPERS/TUP071.PDF

[6] Galambos, J., S. Henderson, Y. Zhang, "A Fault Recovery System for the SNS Superconducting Linac", *Proceedings of LINAC 2006*, Knoxville, Tennessee.
http://accelconf.web.cern.ch/AccelConf/l06/PAPERS/MOP057.PDF

[7] Galambos, J., C. Chu, S. Cousineau, V. Danilov, J. Patton, T. Pelaia, A. Shishlo, "XAL Application Programming Structure", *Proceedings of 2005 Particle Accelerator Conference*, Knoxville, Tennessee.
http://accelconf.web.cern.ch/AccelConf/p05/PAPERS/ROPA001.PDF

RELIABILITY STUDIES FOR A SUPERCONDUCTING DRIVER FOR AN ADS LINAC

Paolo Pierini
Istituto Nazionale di Fisica Nucleare, Sezione di Milano - LASA, Italy

Luciano Burgazzi
Ente Nazionale Energie Alternative, Bologna, Italy

Abstract

The reliability studies presented at the 4th HPPA workshop in Daejon and the RBD models developed are being extended to investigate the sensitivity to the MTBF assumed for the lumped components used in the estimations. Given the absence of precise experimental data to feed into the simulations this study is intended to better quantify the uncertainty of the reliability estimations that can be obtained with these models and the perspectives of reaching a highly reliable design.

Introduction

At the HPPA4 in Daejon we summarised [1] the reliability-oriented guidelines that have been followed to select the reference accelerator design for the PDS-XADS project [2] and presented some exploratory reliability estimates for such a system. In this and other papers [3,4] we have already described the situation in the community of accelerator designers, where reliability-oriented design considerations have been introduced in recent years both for user facilities (as the synchrotron radiation sources or neutron spallation sources, where availability is a concern) and for the assessments of the next huge high-energy physics colliders (LHC and the linear collider).

The scarce practice in the application of formal reliability analysis procedures in accelerator design implies that no consistent reliability characteristics database of accelerator component exists or can be easily gathered in the community. At each major accelerator facility the detailed fault reporting needed to infer the reliability characteristics of the components in the machine is performed differently and in most cases is not sufficiently detailed to analyse failure modes and fault frequencies of the components. Moreover, most of the key components in each new accelerator are often *ad hoc* components, entered in production after a usually short prototyping phase meant to demonstrate the possibility of reaching the technologically challenging nominal performance specifications.

In recent years a few significant accelerator subsystems have been thoroughly examined, modelled and ultimately optimised by means of formal reliability methodologies. This is the case, for example, of the LHC quench protection system (aimed at protecting damage to the hardware during the magnet quenches by properly detecting and driving gracefully the energy discharge following a quench) [5] and for the LHC machine protection system (aimed at minimising the probability of not detecting beam losses while keeping a high machine availability) [6]. A great modelling and optimisation effort has been dedicated to these mission critical subcomponents of the LHC, which is going to be the largest and most significant accelerator operating in the next few decades.

The accelerator configuration for the EUROTRANS programme (described elsewhere in these proceedings [7]) follows the design that was chosen in the PDS-XADS programme very closely, adapted to the slightly modified energy and current specifications. The reliability of the PDS-XADS accelerator design has been investigated both with a qualitative FMEA activity (aimed at the failure mode identification of the system) [8] and by means of a lumped-component Reliability Block Diagram (RBD) analysis used to identify possible configurations that maximise the overall system reliability meeting the programme goals [9].

A comprehensive review of the reliability investigation activities started with PDS-XADS has been presented in [3]. Due to the aforementioned large uncertainties in the component data that can be used in accelerator reliability modelling, our studies have been concentrated on system design issues and perspectives of fault tolerance implementation in the choice of the reference solution for the accelerator layout. The degree of fault tolerance assumed in the reliability analysis is also complemented by parallel studies on the beam dynamics aspects in the accelerator [10]. Moreover, a specific task within EUROTRANS is dedicated to the analysis of a low-level RF control system strategy allowing the implementation of the fault tolerance capabilities in the linac [11].

In this contribution we will review the reliability characteristics of the EUROTRANS accelerator, extending its analysis to considerations that take into account the system sensitivity to component reliability characteristics.

Reliability goals

The XT-ADS operation foresees a cycle of three months continuous irradiation followed by a month-long shutdown for fuel pin operations. Longer irradiation times will be required in the EFIT scenario, but for the exploratory reliability studies we have set a mission time of three months (2 190 h), with a maximum of three faults per mission time. Given that we allow a minimum number of faults during the mission time, the reliability parameter R, associated with the survival probability at mission end and generally required close to unity in all mission-critical systems (satellites, airplanes, etc.), has not been constrained in the analysis.

The role of corrective maintenance rules on the subcomponents has been taken into account both in the evaluation of the different configurations and in the fault tolerance implementation. Indeed, components located in the accelerator tunnel can be repaired only during a beam shutdown and this suggests the use of either high reliability components or a good fault tolerance strategy for this class of components.

The block layout of the PDS-XADS and EUROTRANS linac is shown in Figure 1.

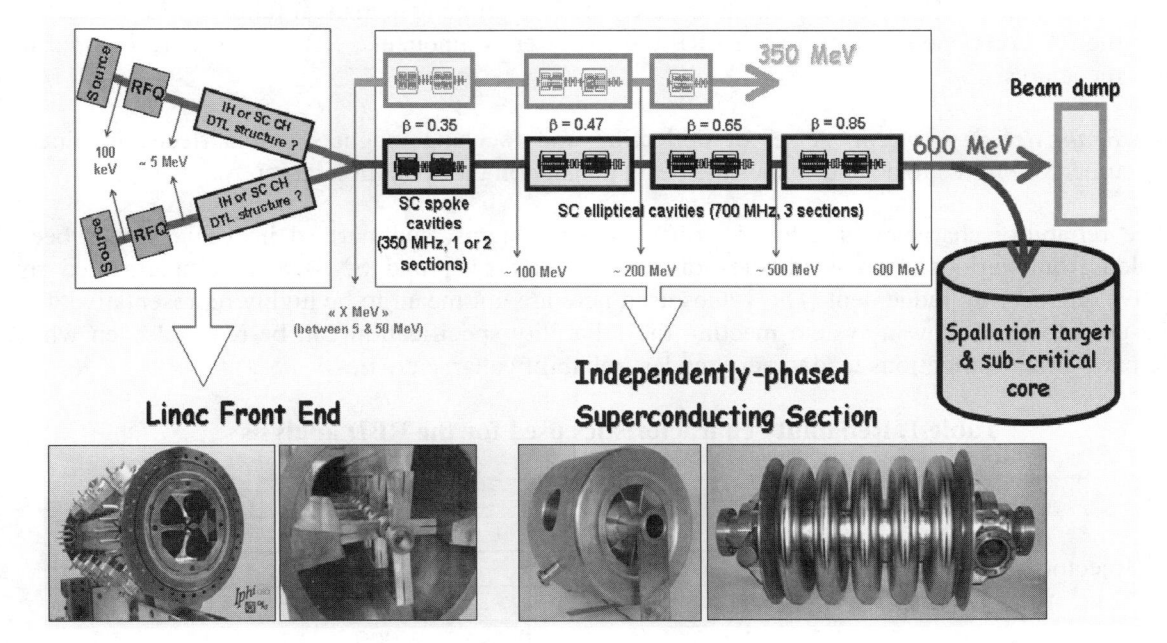

Figure 1. The layout of the ADS linac for PDS-XADS and EUROTRANS

Reliability modelling

To investigate the RBDs of different redundant configurations of the ADS accelerator we have used a commercially available reliability modelling tool, based on Monte Carlo evaluation methods. The package allows elaborate connection configurations, with all kind of possible implementation of parallelism (hot, standby, warm, "k/n"). The tools also have extensive options for maintenance schemes and preventive and corrective actions (i.e. perform repairs only at system failure or at each component failure, …) and can easily account for maintenance cost and repair and spare logistics (but we did not model these aspects).

In order to devise RBD schemes for various accelerator configurations, we had to reduce the complexity of the accelerator to a simpler system composed of "lumped" elements. In doing so we have divided the accelerator system in the following structure:

- *Injector.* A system composed of three lumped accelerator components, the proton source, the radio-frequency quadrupole (RFQ) and the (either normal-conducting or superconducting) energy booster, raising the energy to that of the superconducting spoke linac (around 20 MeV).

- *Support systems.* In addition to the site power (not considered here) the accelerator needs three large conventional support systems, the cryogenic system intended to provide the cold fluids for the cavity and magnets operation, a water-cooling system for power dissipation in the normal-conducting structures and in the cryosystem and a (hardware and software implemented) control system for the operation of the plants and of the linac itself.

- *Superconducting linac sections.* A sequence of nearly identical RF units that power the cavities providing beam acceleration. The linac is split in sections with cavities of different shape and geometry (spoke cavities or elliptical cavities, with different geometries matched for different beam energies), but from the point of view of reliability analysis, it can be seen as a series of identical RF units (96 and 92 in the spoke/elliptical sections). Also, the mechanical components (RF cavity, cryomodules, etc.) are less prone to statistical failures with respect to the RF components (klystrons, LLRF units, power components, …) and thus are ignored in the analysis.

- *Beam delivery system.* A set of deflecting and focusing magnets that deflects the linear trajectory in the linac and shapes the beam density on the spallation target.

The reliability characteristics for the different subsystems, summarised in Table 1, has been assembled from various sources (studies of IFMIF, APT [12] and SNS) and complemented by subjective engineering judgement. The values used here are not meant to be highly representative, but will be used to show how a system meeting the reliability specification can be realised even when simple parts count estimations lead to very modest reliability characteristics.

Table 1. Reliability characteristics used for the RBD analysis

System	Subsystem	MTBF (h)	MTTR (h)
Injector	Proton source	1 000	2
	RFQ	1 200	4
	NC DTL	1 000	2
Support systems	Cryoplant	3 000	10
	Cooling system	3 000	2
	Control system	3 000	2
RF unit	High voltage PS	30 000	4
	Low-level RF	100 000	4
	Transmitters	10 000	4
	Amplifier	50 000	4
	Power components	100 000	12
Beam delivery system	Magnets	1 000 000	1
	Power supplies	100 000	1

The experimental activities within the EUROTRANS programme will provide more knowledge on some of the reliability characteristics of key components listed in the above table, and will allow progresses in reliability analyses on a more firm component database.

Furthermore, the experience which is being gained with the operation of the Spallation Neutron Source at Oak Ridge in the US will be extremely relevant to further studies of ADS driver systems. The SNS accelerator, apart from its pulsed mode of operation, shares many components with the machine foreseen for an ADS driver.

It is important to mention here that, in spite of their importance, a conservative MTBF has been chosen for each support system (for example, large cryogenic systems usually have better reliability characteristics), only moderately compatible with the foreseen mission time.

Parts count and all-series connections

By far the simplest possible configuration for which reliability estimations can be derived analytically is the all-series connection of the nominal number of components required to achieve the desired beam parameters at the target. The simple reliability diagram corresponding to this situation, indicating the number of elements of each type is shown in Figure 2.

Figure 2. All series connection

Start → Assy: SOURCE, MTBF: 1000, Qty: 1, MCT: 2, hr → Assy: RFQ, MTBF: 1200, Qty: 1, MCT: 4, hr → Assy: NC DTL, MTBF: 1000, Qty: 1, MCT: 2, hr →

Assy: CRYOPLANT, MTBF: 3000, Qty: 1, MCT: 10, hr → Assy: CONTROL SYSTEM, MTBF: 3000, Qty: 1, MCT: 2, hr → Assy: COOLING, MTBF: 3000, Qty: 1, MCT: 10, hr →

Assy: SPOKE RF ELEMENT, MTBF: 5700, Qty: 96, MCT: 4, hr → Assy: ELLIPTICAL RF ELEMENT, MTBF: 5700, Qty: 92, MCT: 4, hr →

Assy: BDS MAGNETS, MTBF: 1000000, Qty: 20, MCT: 1, hr → Assy: BDS MAG PS, MTBF: 100000, Qty: 20, MCT: 1, hr → End 1::1

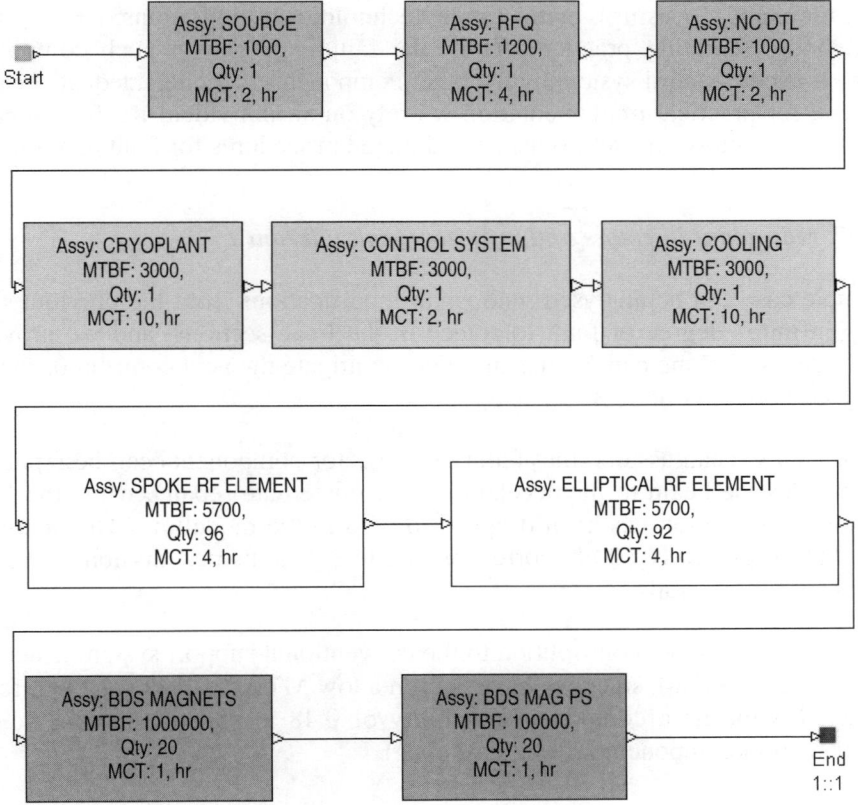

Such a configuration, given the large number of components in the system, leads to a short MTBF of approximately 28 h, and the faults are distributed in the accelerator sections according to Table 2. Of course, the fault rate is dominated mainly by the high number of RF components in the two linac sections and by the reliability characteristics of the RF unit (i.e. the RF unit MTBF of 5 700 needs to be further divided by the number of units, 188, according to the nature of the series connection, resulting in a linac section MTBF of 30.32 h).

Table 2. Fault rate contribution of the accelerator systems, for the RBD shown in Figure 2

System	Fault rate contribution
Injector	7.7%
Support systems	2.7%
Spoke linac	45.4%
Elliptical linac	43.5%
Beam delivery system	0.6%

This simple analysis already suggests the strategies to implement redundancy and fault tolerance in the linac design. Clearly, some degree of fault tolerance is desired in the superconducting linac section, where the system should be able to work in a "n out of k" parallelism, allowing as much repairs as possible to happen to failing equipment when the accelerator is functioning.

Luckily the SC linac has ideal perspectives for introducing tolerance to RF faults, since it is composed by a highly modular pattern of repeated components, all providing the same functions (i.e. beam acceleration and focusing), with ample technological limitations with respect to their maximum performances (derating practice). Thus, the failure of one of such components can be compensated with an expert control system by the other components, keeping adequate beam conditions on target. Of course, for practical implementation we rely on an individual RF feed scheme for each cavity, LLRF digital regulation with set-points and tabulated procedures for fault handling.

"Dream machine", redundant injectors and infinitely fault tolerant

A second simple case can be analysed with small modifications from the previous limiting case. If we assume an "infinite" degree of fault tolerance in the linac sections, and we provide a second injector stage with ideal switching capabilities in order to mitigate the next contributor to the fault rate, we come to the RBD shown in Figure 3.

In the injector area we can assume that failure on injector components can be repaired while the second injector provides the beam to the accelerator (i.e. immediate repair), while the failure rate of the superconducting linac section has been dropped to zero in the modelling. The analysis of such a scheme yields an MTBF of nearly 800 h, corresponding to 2.7 faults per mission. Clearly this meets the reliability goals set for our study.

It should be reiterated that the contribution to the conventional support system is not negligible at this level (2.19 faults per mission), since we have chosen a low MTBF value of 3 000 h for each of the major components (this means a component reliability of 0.48 at end of mission, i.e. 50% of the mission times each support component fails at least once).

A realistic compromise between a pure series connection and an infinitely fault tolerance assumption for the superconducting linac sections must be clearly identified. We begin by first noting that most RF components in the RF units can be placed outside the accelerator tunnel, and therefore

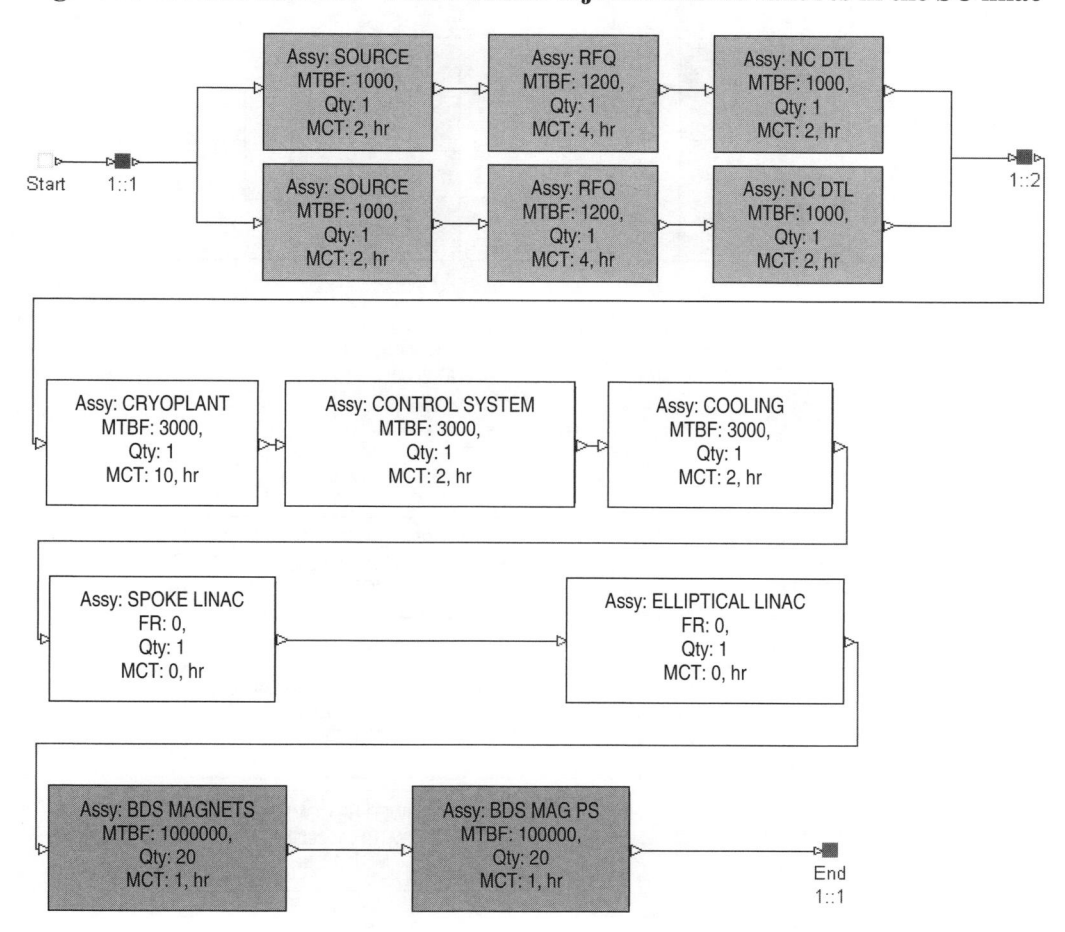

Figure 3. "Dream machine" with a double injector and no failures in the SC linac

can be immediately repaired without switching the system off, and only the power RF components (waveguides, circulators, attenuators, ...) are placed in the controlled environment of the accelerator tunnel (and therefore can be repaired only following a system shutdown). The current MTBF of the out-of-tunnel components is 6 100 h.

Final fault tolerant scheme, split RF systems

In Figure 4 we show a fault tolerance accelerator configuration based on a split repair provision for the RF unit components.

In both linac sections it has been assumed that at any time the failure of two out-of-tunnel RF components (but not more...) can be tolerated, and only one in-tunnel RF component. The components that are out-of-tunnel are immediately repaired, while the repairs on the in-tunnel component are performed only during failures that lead to the system stop.

The analysis of such a system results in a MTBF of 550 hours (corresponding to 3.8 faults per mission), which is 20% shorter than required. However, now that we have implemented fault tolerance strategies, we can use the system sensitivity to single components to find the right components to which we have to require additional engineering efforts in order to increase their MTBF and contribute to reach the system goals.

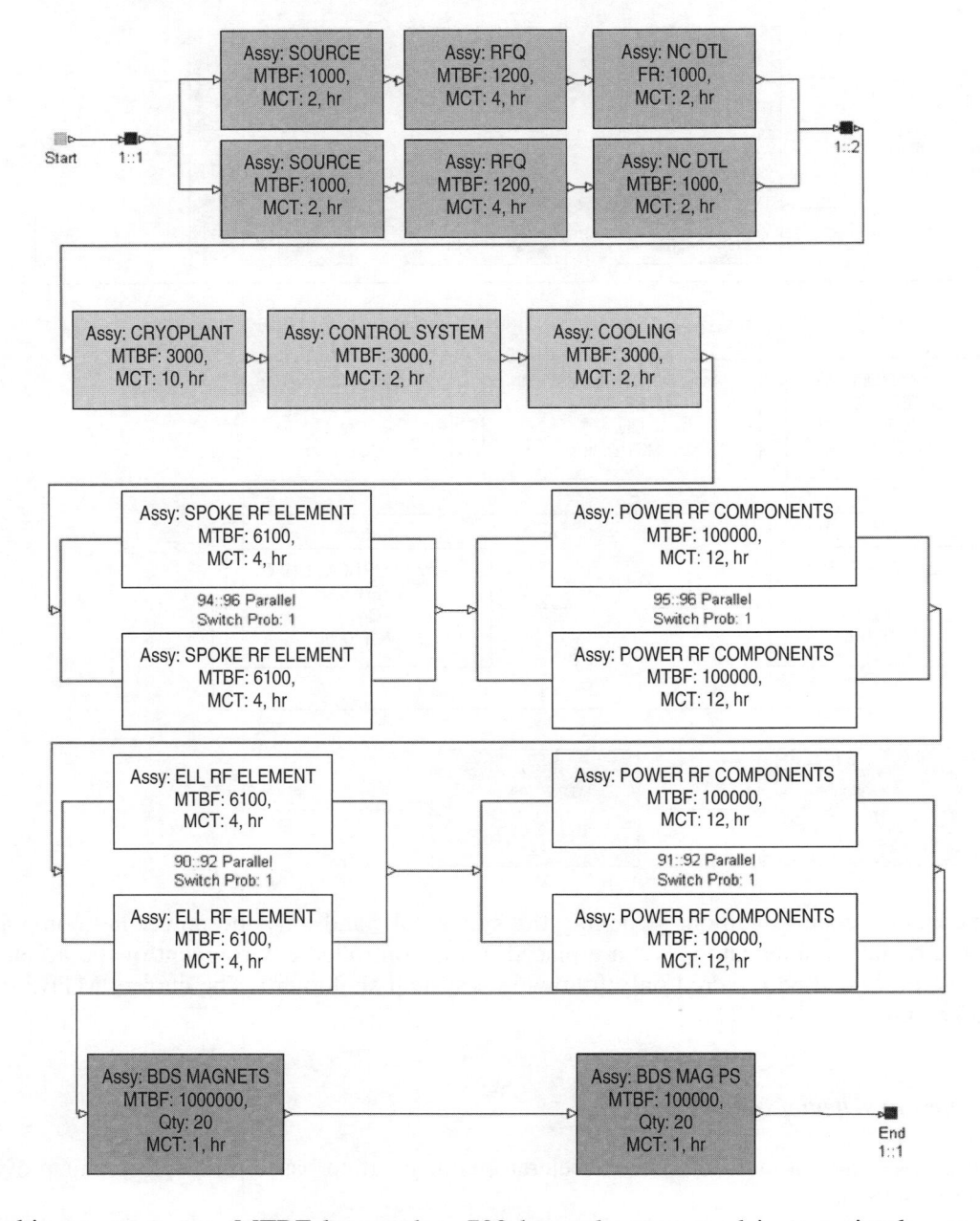

Figure 4. Final scheme for a split RF system fault tolerant linac

In this case a system MTBF longer than 700 h can be recovered in two simple ways. Either doubling the MTBF of the conventional support systems (thus asking for more reliable support systems, which indeed makes sense since they provide mission critical functions) or increasing by a factor of 10 the in-tunnel RF components (i.e. increasing their reliability due to their repair policy).

Lessons learned

The procedure followed in the previous paragraphs in order to meet the ADS reliability goals by altering the nature of the connection between moderately reliable components is not intended towards predicting the reliability characteristics of the linac proposed for EUROTRANS, but is intended rather at pointing out the fault tolerance features that need to be implemented in the accelerator design. Moreover, a similar procedure allows choosing the components for which high reliability characteristics must be guaranteed, due to their criticality or simply due to the impossibility to perform maintenance procedure.

The evolution of the design modification from the simple series connection to the inclusion of realistic fault tolerant capabilities is summarised in Table 3, which lists the number of injectors, the degree of fault tolerance assumed in the calculation, the RF units repair provisions, and the outcome of the analysis, in terms of system MTBF. The goal MTBF has been achieved mainly by changing the nature of connections, with minimal tweaking of the components reliability characteristics.

Table 3. Evolution of the system design modifications to provide realistic fault tolerance capabilities

# Inj.	Fault tolerance degree	RF unit repair	MTBF
1	None, all in series	At system stop	28
2	Infinite	Immediate	797
2	94/96 in spoke, 90/92 in all are needed	Immediate	758
2	94/96 in spoke, 90/92 in all are needed, more realistic correction provisions, by splitting the RF system	Immediate for out of tunnel at system stop for in tunnel	558
2	94/96 in spoke, 90/92 in all are needed, split RF SUPPORT SYSTEM MTBF * 2	Immediate for out of tunnel at system stop for in tunnel	720
2	94/96 in spoke, 90/92 in ell are needed, split RF IN-TUNNEL MTBF * 10	Immediate for out of tunnel at system stop for in tunnel	760

Of course, the analysis reported here is still crude. As an example, all the MTTR are inserted mainly for demonstration purposes, but require more careful analysis (some failure modes may need to require long times for vacuum venting, conditioning, cool-down/warm-up). Moreover, any decay time needed to access active components has been neglected.

The analysis assumes, still, a high degree of fault tolerance in the accelerator, where the failure of an RF unit is automatically recovered without inducing beam trips on target in time scales of the order of 1 s. This is clearly a challenging issue in LLRF and beam control systems.

Acknowledgements

This work is supported by the EURATOM 6th Framework Programme of the EC under contract FI6KW-CT-2004-516520.

REFERENCES

[1] Pierini, P., L. Burgazzi, "ADS Accelerator Reliability Activities in Europe", *4th International Workshop on the Utilisation and Reliability of High-power Proton Accelerators*, Daejeon, Republic of Korea, 16-19 May 2004, p. 137.

[2] D'Hondt, P., "European PDS-XADS", *4th International Workshop on the Utilisation and Reliability of High-power Proton Accelerators*, Daejeon, Republic of Korea, 16-19 May 2004.

[3] Burgazzi, L., P. Pierini, "Reliability Studies of a High-power Proton Accelerator for Accelerator-driven System Applications for Nuclear Waste Transmutation", *Reliab. Eng. and Sys. Safety*, 92, 449-462 (2007).

[4] Barni, D., et al., "Basis for the Reliability Analysis of the Proton Linac for an ADS Program", *Proceedings of 2003 Particle Accelerator Conference*, Portland, Oregon, p. 1506.

[5] Vergara-Fernandez, A., F. Rodríguez-Mateos, "Reliability of the Quench Protection System for the LHC Superconducting Elements", *Nuclear Instruments and Methods in Physics Research Section A: Accelerators, Spectrometers, Detectors and Associated Equipment*, Volume 525, Issue 3, 11 June 2004, Pages 439-446.

[6] Filippini, R., B. Dehning, G. Guaglio, F. Rodriguez-Mateos, R. Schmidt, B. Todd, J. Uythoven, A. Vergara-Fernandez, M. Zerlauth, "Reliability Assessment of the LHC Machine Protection System", *Proceedings of 2005 Particle Accelerator Conference*, Knoxville, Tennessee, p 1257.

[7] Biarrotte, J-L., "Status of the EUROTRANS R&D Activities for ADS Accelerator Development", these proceedings.

[8] *Potential for Reliability Improvement and Cost Optimization of Linac and Cyclotron Accelerators*, PDS-XADS Contractual Deliverable, DEL-03-057, P. Pierini (Ed.), July 2003.

[9] *Definition of the XADS-class Reference Accelerator Concept and Needed R&D*, PDS-XADS Contractual Deliverable, DEL-04-063, J-L. Biarrotte (Ed.), September 2004.

[10] Biarrotte, J-L., M. Novati, P. Pierini, H. Safa, D. Uriot, "Beam Dynamics Studies for the Fault Tolerance Assessment of the PDS-XADS Linac Design", *Proceedings of EPAC2004*, Lucerne, Switzerland, p. 1282.

[11] Biarrotte, J-L., C. Joly, M. Luong, *Preliminary RF Control Specifications*, EUROTRANS Contractual Deliverable, D1.17/D.1.3.10, November 2005.

[12] Burgazzi, L., "Accelerator Systems Reliability Issues", *Proceedings of the ESREL 2001*, Torino, Italy, E. Zio, M. Demichela, N. Piccinini (Ed.), p. 1163.

COMPARISON OF BEAM TRIP FREQUENCIES BETWEEN ESTIMATION FROM CURRENT EXPERIMENTAL DATA OF ACCELERATORS AND REQUIREMENT FROM ADS TRANSIENT ANALYSES

Hayanori Takei, Kazufumi Tsujimoto, Nobuo Ouchi, Hiroyuki Oigawa
Japan Atomic Energy Agency, Japan

Motoharu Mizumoto
Tokyo Institute of Technology, Japan

Kazuro Furukawa, Yujiro Ogawa, Yoshiharu Yano
High Energy Accelerator Research Organization, Japan

Abstract

Frequent beam trips as experienced in existing high-power proton accelerators may cause thermal fatigue problems in ADS components which may lead to degradation of their structural integrity and reduction of their lifetime. Thermal transient analyses were performed to investigate the effects of beam trips on the reactor components, with the objective of formulating ADS design considerations, and to determine the requirements for accelerator reliability. These analyses were made on the thermal responses of three parts of the reactor components; the beam window, the inner barrel and the reactor vessel. Our results indicate that the acceptable frequency of beam trips ranges from 50 to 2.5×10^4 times per year, depending on the beam trip duration. In order to measure the effect of reducing beam trips on the high-power accelerator for ADS, we compared the difference between the acceptable frequency for beam trips and the operation data of existing accelerators. The result of this comparison shows that for typical conditions the beam trip frequency for durations of 10 seconds or less is within the acceptable level, while that exceeding 10 seconds should be reduced by about 1/30 to satisfy the thermal stress conditions.

Introduction

To realise effective transmutation of minor actinides (MA) by an accelerator-driven subcritical system (ADS), a high-power spallation target should be installed at the centre of the core. In the case of JAEA's reference ADS [1], proton beam power of ~30 MW is necessary to keep the thermal power at the subcritical core at 800 MW. Frequent beam trips, as experienced in existing high-power proton accelerators, may cause thermal fatigue problems in ADS components which may lead to degradation of their structural integrity and reduction of their lifetime.

In general, beam trips are caused by two reasons: one is the failure of the accelerator components, and the other is the interruption by a Machine Protection System (MPS) to protect the accelerators against failures. For reference, the beam trip frequency caused by the "dead component" is assessed on the basis of techniques such as Failure Mode and Effect Analysis (FMEA) [2]. For these assessments, the reliability parameters of accelerator components, such as a failure rate, are usually used. However, the influence for the thermal shock damage on the ADS reactor system caused by beam trips has not been evaluated sufficiently. Conversely, it is not yet clear what times in the ADS reactor system are acceptable for the beam trips. The purpose of the present study is to ascertain the present level of accelerator technology by comparing beam trip frequencies estimated from operation data of existing accelerators and the requirement from transient analyses of the ADS reactor system.

Some definitions and useful acronyms

Let us first review some words or acronyms that will be used extensively in this paper.

- "*Frequency of the beam trip*" is the total number of beam trips divided by the total scheduled beam time.

- "*The censored event*" is defined as the termination of normal operation of the system by operators, e.g. this event corresponds to the manual termination for regular maintenance operations and the deactivating request from systems other than the operating one.

- "*Failure of the klystron*" is defined as a condition in which a radio frequency (RF) wave could not be output from the klystron. This condition corresponds not only to failure of the actual klystron and its relevant systems but also to a short interrupt caused by the MPS.

- "*Mean Time Between Failures (MTBF)*" is a basic measure of reliability for repairable items. It can be described as the number of hours of operation before a component, assembly or system fails. MTBF can be calculated as the inverse of the failure rate for constant failure rate systems.

- "*Mean Time To Repair (MTTR)*" is the sum of corrective maintenance/repair time divided by the total number of failures during a given time interval.

Conceptual design of ADS

General scheme

JAEA's reference design of ADS is a tank-type subcritical reactor, where lead-bismuth eutectic (LBE) is used as both the primary coolant and the spallation target, as shown in Figure 1. The subcritical core is composed of MA nitride fuel, with plutonium added at the initial loading of the first

Figure 1. Concept of 800 MWth, LBE-cooled, tank-type ADS

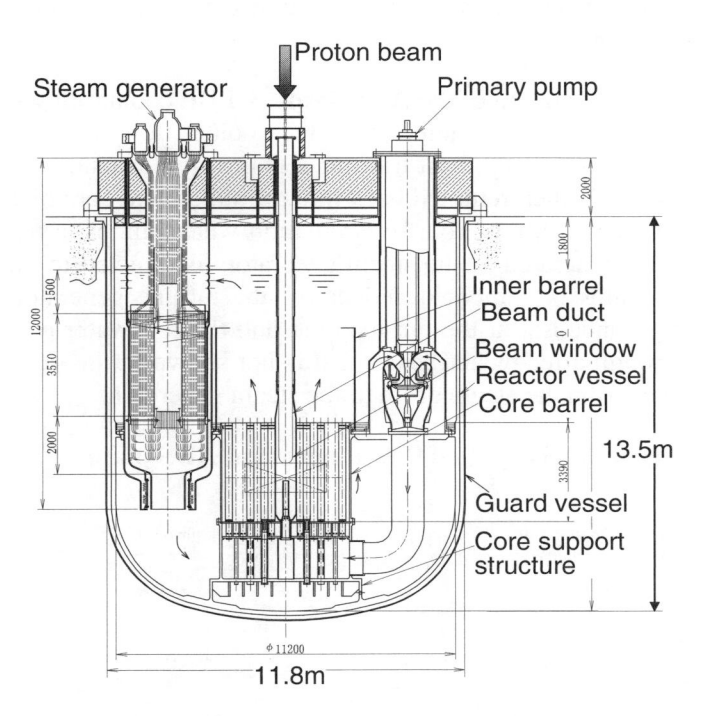

cycle to reduce the burn-up swing reactivity. The proton accelerator for the ADS must have high power intensity, more than 20 MW, with good economic efficiency and reliability. To realise such an accelerator, energy efficiency must be enhanced to assure the self-sustainability for electricity of the whole system. Taking account of these requirements, the superconducting linac (SC linac) is regarded as the most promising choice. Considering the production efficiency of the spallation neutrons in LBE, the accelerated energy of the SC linac was set at 1.5 GeV. This value will be optimised in the future taking into account the trade-off between the cost of the accelerator for higher energy and the engineering difficulty associated with higher current. In order to keep the thermal power at 800 MW, the beam current was adjusted from 8 to 18 mA (i.e. 12 to 27 MW) depending on the effective multiplication factor (k_{eff}). Taking into account these requirements, the maximum beam current of the SC linac was set at 20 mA.

Basic parameters of the SC linac, such as the number of cryomodules, the output power of the klystron and the total length of the SC linac, were optimised for accelerated energies from 100 MeV to 1.5 GeV [3]. The SC linac consists of a series of 89 cryomodules, which were designed for a 972-MHz RF wave. One klystron was provided for each cryomodule giving a total of 89 klystrons. These klystrons were classified into three categories, according to rated output power: 197, 425 and 750 kW. The total length of the SC linac was estimated as 472 m, using the effective length of equipment, such as the quadrupole magnet and the cryomodule, for the superconducting linac of the J-PARC project as reference.

Cooling system

The ADS plant was assumed to have a primary LBE cooling system and a water/steam system for power conversion through a saturated steam cycle. As shown in Figure 1, all other components of the primary system, including four steam generators, two primary pumps and auxiliary heat exchangers, are accommodated within the reactor vessel. The heat generated in the target and the subcritical core is removed by forced convection of the primary LBE, and transferred through the steam generators to water/steam for power conversion.

Figure 2 is a simplified flow diagram of the ADS. Primary LBE coolant flows upward through the subcritical core and exits into an upper plenum. Then, the coolant flow divides equally among four steam generators. The flow passes through heat transfer sections in the steam generators and enters a lower plenum. The LBE coolant is then returned back into the subcritical core by primary pumps. The water/steam system has four steam drums, four re-circulation water pumps, a turbine generator unit, and a feedwater pump. Steam generated in the steam generators is directed to the steam drums. The steam from all four steam drums is combined and drives the turbine generator unit. The steam is cooled down into water in a condenser at the turbine exit, and the feedwater pump returns the water back to the steam drums through a feedwater heater that is not shown in the figure. The recirculation water pump delivers the water from the steam drum to the steam generator.

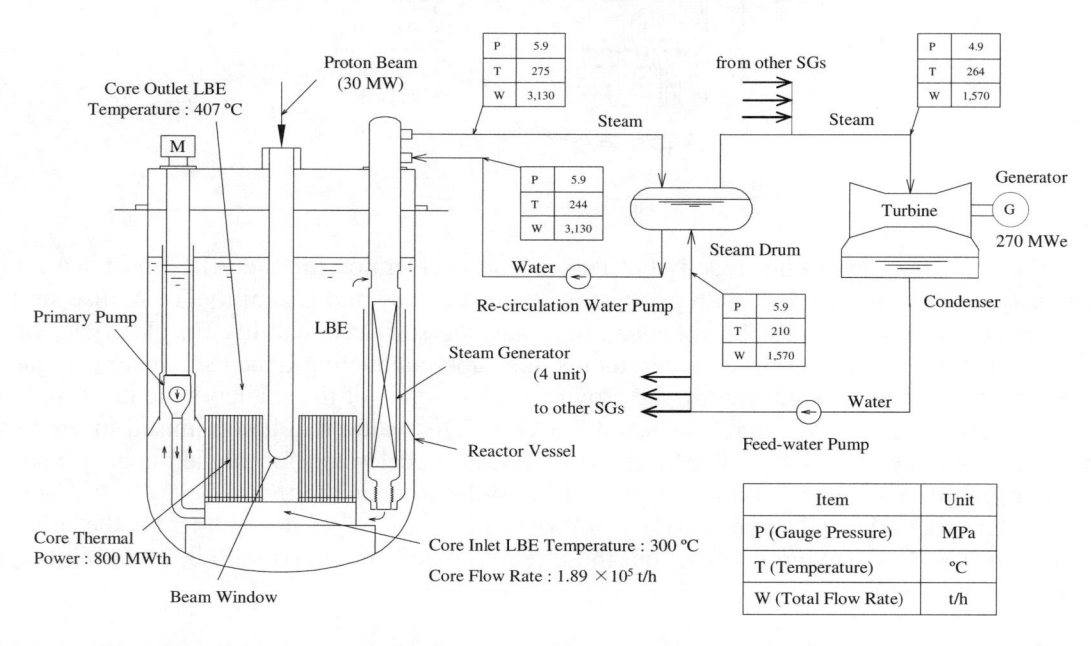

Figure 2. Simplified flow diagram of ADS plant

Item	Unit
P (Gauge Pressure)	MPa
T (Temperature)	°C
W (Total Flow Rate)	t/h

Preliminary analysis was performed for transients in the ADS plant caused by the beam trip [4]. The primary LBE cooling system and the water/steam power conversion system were modelled with a simple one-dimensional flow network. As a result, about 400 seconds of turbine operation may be possible without the beam. When the duration of the beam trip exceeds this limit, the pressure of the steam drum and the LBE temperature become too low to prevent its freezing.

Estimation of acceptable beam trip frequency

Restriction from the subcritical reactor

As shown in Figure 3, three parts of the reactor component, i.e. the beam window, the inner barrel and the reactor vessel, were picked as representative to discuss the influence of thermal shock. In the following discussion, the expected lifetime of the subcritical reactor is defined as 40 years and no replacement of the inner barrel and the reactor vessel is assumed during this period. On the other hand, the replacement of the beam window is assumed once every two years.

Figure 3. Influence of beam trip transient on reactor structure

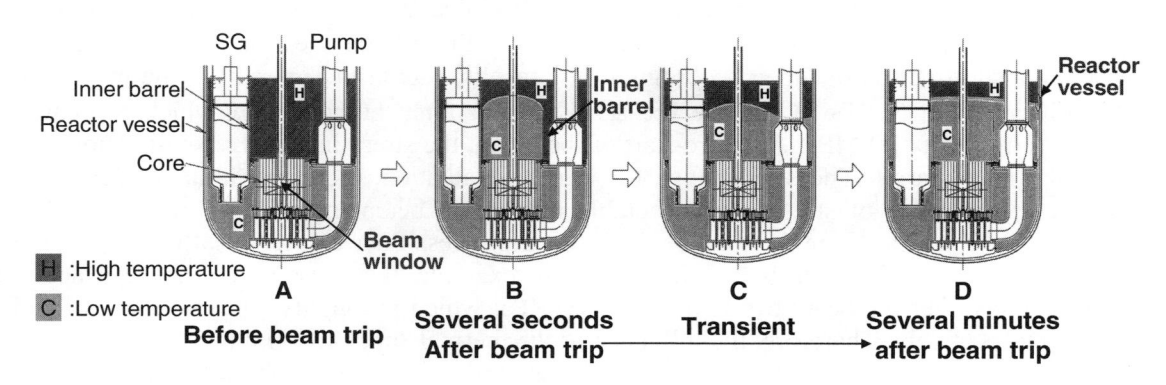

As for the beam window, a beam trip causes a rapid temperature drop, which in turn causes thermal stress because of the temperature difference between the inner and the outer surfaces of the beam window. The temperature response of the beam window (inner diameter: 450 mm, thickness: 2 mm, material: 9Cr-1Mo steel, beam power: 30 MW) was evaluated using the finite element method FINAS code [5]. In the evaluation, the beam is assumed to be restarted five seconds after the beam trip. Figure 4 shows the temperature change for the apical region of the beam window. On the beam trip at $t = 0$, the surface temperature starts to drop rapidly and asymptotically approaches the coolant temperature within about three seconds. The maximum thermal stress of 179 MPa is expected at 0.5 seconds after the beam trip. This thermal stress is much lower than what would cause buckling failure. The acceptable number of these thermal shocks is estimated at about 10^5, which means that several beam trips per hour may be acceptable for two years of the expected lifetime of the beam window (about 15 000 hours). It should be noted that this estimate is based on the material data without radiation damage, and therefore experimental verification for the effect of the proton and neutron irradiation is indispensable.

Figure 4. Change of the surface temperature for the beam window after the beam trip

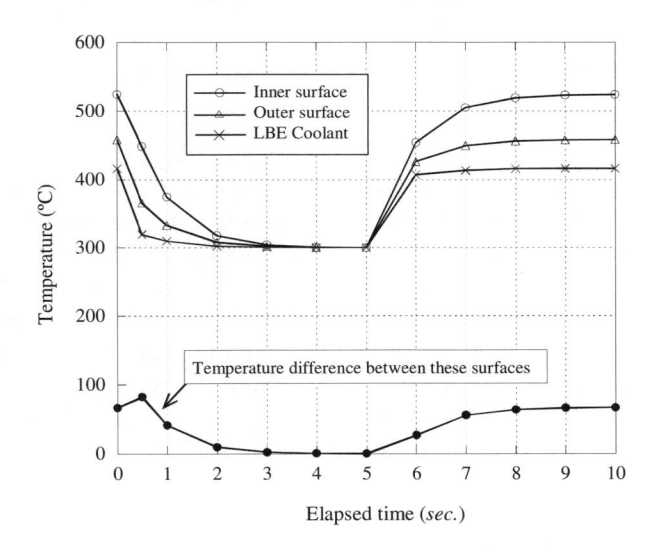

The inner barrel is a cylindrical structure, as shown in Figure 3, made of 3-cm thick 9Cr-1Mo steel and installed to straighten the LBE flow above the subcritical core. During normal operation, it is kept at the average outlet temperature 407°C. In the case of the beam trip, the inner surface of the cylinder is immediately cooled by the cold LBE (about 300°C), causing a temperature difference between the inner and outer surfaces. The temperature and stress change for 120 seconds after the beam trip was calculated by means of the FINAS code. In the calculation, the temperature for the outer surface is taken to be the steady value, 407°C, because it is difficult for the cold LBE to reach the outer surface in a short period of time. According to the results, shown in Figure 5(a), the maximum temperature difference of 60°C will be observed at 24 seconds after the beam trip, which will cause maximum stress of about 130 MPa. At the re-start of the beam, the stress in the inverse direction will be added to the inner barrel. The stress range for fatigue evaluation, therefore, is about 260 MPa. Considering this magnitude of stress, the acceptable number of beam trips for the inner barrel is estimated at about 10^4 times. If the duration of the beam trip is less than 24 seconds, the stress to the inner surface becomes smaller than the maximum. In this example, the stress at the inner surface at five and ten seconds after the beam trip is 61 and 108 MPa, which is roughly equivalent to 10^6 and 10^5 beam trips, respectively. Therefore, the influence of the thermal shock to the inner barrel could be reduced, provided that the beam was re-injected into the subcritical core within about ten seconds after the beam trip.

Figure 5. Change of physical values after the beam trip

(a) Surface temperature and stress for the inner barrel
(b) Surface temperature for the reactor vessel

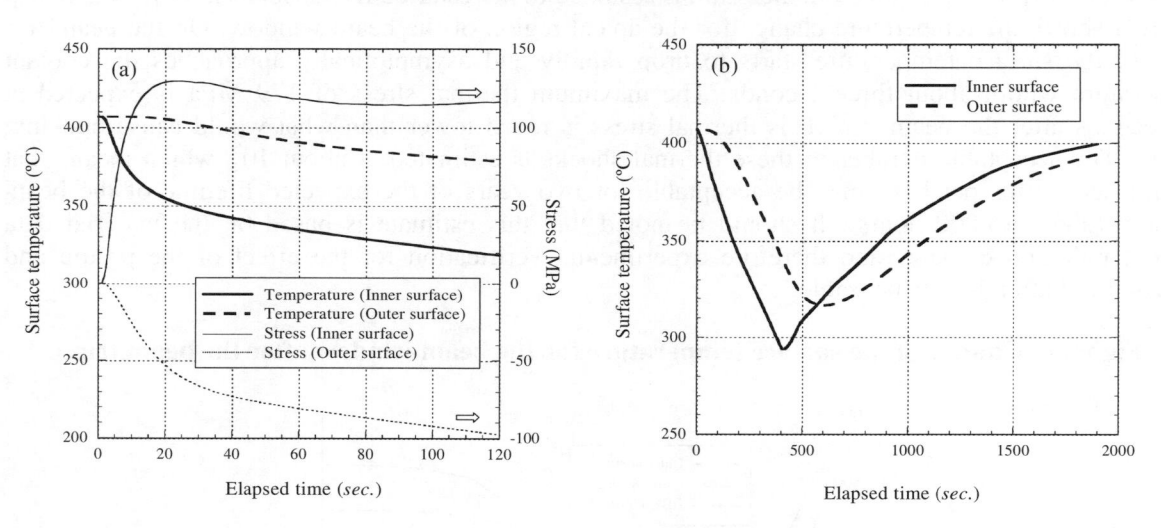

As for the reactor vessel, the beam trip causes a temperature change between the inner and outer surfaces similar to that for the inner barrel, and the formation of temperature stratification and the lowering the LBE level because of thermal shrinkage both cause thermal stress. The stress range for the fatigue evaluation was estimated for the reactor vessel made of 5-cm thick 9Cr-1Mo steel under the assumption that the beam is re-started 400 seconds after the end of the beam trip. The result is shown in Figure 5(b). We observed that the peak value of the temperature difference between the surfaces occurred just before the beam re-start. The maximum stress range calculated from the axial and circumferential stress was about 270 MPa. Considering this magnitude of stress, the acceptable number of beam trips for the reactor vessel is estimated at about 10^4 times.

Acceptable number of beam trips per year

Taking into account the above discussions, the acceptable number of beam trips for each component is summarised in Table 1. In this table, the elapsed time when the stress range is maximised after the beam trip (T_{max}) and the expected lifetime for each component are also listed. The acceptable frequency

Table 1. Acceptable number of beam trips for each component

Component	T_{max}	Acceptable number	Expected lifetime (years)
Beam window	0.5 sec.	10^5	2
Inner barrel	24 sec.	10^6 (5 sec.)* 10^5 (10 sec.)* 10^4 (24 sec.)*	40
Reactor vessel	~400 sec.	10^4	40

* The value in parentheses represents the beam trip duration.

of beam trips, obtained by dividing the acceptable number of beam trips by the expected lifetime for each component, ranges from 250 to 2.5×10^4 times per year. A requirement for ADS is that the acceptable frequency of beam trips with duration over five minutes be limited to 250 times per year. The influence of the beam trip on the power generation system is also important from the viewpoint of the availability of the system, because the re-start procedure of the system usually takes a long period of time, typically several hours, once the power generation turbine stops. As previously discussed, 400 seconds of turbine operation may be possible without the beam. Then, the acceptable frequency of beam trips is assumed to be once a week for this long beam trip.

Therefore, the acceptable frequency of beam trips was classified by four criteria, according to the beam trip duration, as shown in Table 2. The acceptable frequency of beam trips for the inner barrel was adopted for the criteria for a beam trip of five seconds or less, because the acceptable frequency of beam trips for the inner barrel was lower than that for the beam window.

This acceptable frequency of beam trips is compared with the current experimental data presented in the next section.

Table 2. Distribution of the beam trip duration

Criteria	Acceptable value (times/year)	Current operation data	
		LANSCE (N_{inj})	KEKB** (N_{rf})
0 sec. $\leq T \leq$ 5 sec.	25 000	98%*	37% (78%)
5 sec. $< T \leq$ 10 sec.	2 500		44% (6%)
10 sec. $< T \leq$ 5 min.	250		7% (5%)
$T >$ 5 min.	43	2%	12% (11%)

* The value for the beam trip duration of one minute or less is 84%.
** The value in parentheses represents the ratio for the shorten VSWR assumption.

Estimation of the beam trip frequency based on the current experimental data

In order to develop measures for reducing beam trip frequency, it is important to know the present level of accelerator technology. First, the beam trip frequency of the JAEA's SC linac is estimated from operation data of existing accelerators in this section. Pioneering experimental data obtained from the Los Alamos Neutron Science Center (LANSCE) and the High Energy Accelerator Research Organisation (KEK) was used for this estimation. Next, comparing the difference between the acceptable frequency of beam trips and the current experimental data, strategies to overcome the beam trip problem on ADS are discussed.

LANSCE accelerator facility and the beam trip frequency caused by the injector, N_{inj}

LANSCE is a spallation neutron source based on a linac [6]. The accelerator facility is one of the most powerful linear proton accelerators in the world. The proton linac delivers two proton beams at 800 MeV: the H⁺ and the H⁻ beam. The H⁺ beam can deliver 1.25 mA and the H⁻ beam can deliver 70 μA. Each injector system includes a 750-keV Cockcroft-Walton type generator. The heart of the RF system is the klystron system which provides RF power at 805 MHz to the side coupled linac. Both ions are accelerated simultaneously within the same structure. Two types of klystrons are used with both klystrons producing 1.25 MW pulses of RF power with a time duty factor of up to 12%. After acceleration, the H⁺ and H⁻ beams are separated. The H⁻ beam is injected into a proton storage ring for accumulation and delivery to the Neutron Scattering Centre or for weapons neutron research.

A thorough failure analysis has been conducted for years at LANSCE. The average value of beam trips for H⁺ was 1.62 trips/h whilst this value was 0.78 for the H⁻ beam. Detailed statistics give no room for doubt; the H⁺ injector was responsible for 86% of beam trips whose duration was less than one minute and responsible for 77% of all beam trips. Next to the injector is the RF system, responsible for 8% of all beam trips linked to the H⁺ beam. Detailed information can be found in Ref. [6].

Because 85% of all beam trips were caused by the injector and the RF system of LANSCE, the beam trip frequency of the JAEA's SC linac, N_{ads}, was roughly estimated as:

$$N_{ads} \sim N_{inj} + N_{rf} \tag{1}$$

where the N_{inj} and N_{rf} are the beam trip frequency caused by the injector and the RF system of the SC linac, respectively. And N_{inj} is estimated as:

$$N_{inj} = \frac{T_s}{\Lambda_L^{-1} + \tau_L} \tag{2}$$

where T_s is the total scheduled beam time per year for the ADS plant (7 300 h/year), Λ_L^{-1} and τ_L are the inverse of the beam trip frequency and the MTTR obtained from the data of the LANSCE H⁺ injector: 0.77 h and 3.4×10^{-2} h, respectively. For reference, Eqs. (1) and (2) were obtained assuming that the MTTR were considerably smaller than the time interval as measured by the inverse of the beam trip frequency. The distribution of beam trips durations at the LANSCE H⁺ injector is also tabulated in Table 2. Substituting these values into Eq. (2), N_{inj} was 9 100 times per year.

The beam trip frequency obtained from the LANSCE H⁺ injector was about 100 times larger than the latest experimental data. For instance, a CEA ion source for the IPHI HPPA-project (SILHI), which is the microwave ion source, produced a 75-mA proton beam for a 104-hour test with a single failure of only 2.5 minutes [7]. As the operation time of SILHI was considerably smaller than T_s, this example was not adopted for the present estimation.

Necessity of operation data of the KEKB injector linac

It is inappropriate, however, to estimate N_{rf} using the LANSCE experimental data for the following reasons:

- Although the LANSCE experimental data for beam trips of duration less than one minute could be analysed, the frequency distribution for this period was not shown.

- The extent of censored events in the LANSCE experimental data was not described in Ref. [6].

In this study, N_{rf} was estimated using the data of the RF system of the KEKB injector linac for an electron/positron beam. This is because the number of klystrons and the total scheduled beam time per year for the KEKB injector linac were closer to the specifications for the ADS linac and detailed analysis could be performed considering censoring events. It should be noted that the estimation of N_{rf} using the data for the KEKB injector linac is conservative because the accidental interrupt frequency of the KEK klystron is about 30 times greater than that of LANSCE.

KEKB injector linac and experimental data

Before estimating N_{rf}, we consider a brief review of the KEKB injector linac [8]. An electron/positron injector linac at KEK, capable of providing electrons at energies up to 8 GeV and positrons up to 3.5 GeV, is used as a multi-purpose injector not only for the KEK B Factory (KEKB), but also for the Photon Factory (PF). At present, it delivers full-energy beams of 8-GeV electrons to the KEKB High-energy Ring and 3.5-GeV positrons to the KEKB Low-energy Ring. The linac was originally constructed as a 2.5-GeV injector for the PF in 1982. However, this old linac was extensively reconstructed for the B Factory project during 1994-1998. The whole linac consists of 55 S-band (2 856 MHz) main accelerator modules. Pulse klystrons of 50 MW with a time duty factor of 0.02% were developed in conjunction with the energy upgrade of the linac for the B Factory project. These tubes were designed to be compatible with the old version of the 30-MW tubes in that the major dimension of the new tube does not need to be changed.

In order to operate the linac stably, RF monitoring systems for power level and phase were installed in all high-power klystrons and are controlled with a VMEbus eXtensions for Instrumentation (VXI)-based controller. Each klystron is controlled with a Programmable Logic Controller (PLC)-based controller, which monitors interlock status and vacuum status, and also provides a historical record of the operation data. To protect the RF window from failure, the klystron operations are terminated by the VSWR interlock which acts as a kind of MPS that trips if it is over the limiting value of 1.4. Operations are automatically restarted after five seconds. (Hereafter, this termination is referred to as "VSWR event.")

The total operation time of the KEKB injector linac has reached over 120 000 h in the FY1982 to FY2005 time period. Statistics on interruptions and the down-time distribution of KEK klystrons were obtained from the operation data in FY2005. The scheduled beam time and the down-time for the whole system was 6 815 h and 52.5 h, respectively. The total number of interruptions was 16 421, including 13 453 accidental interruptions and 2 968 censored events. Accidental interruptions were caused by failure of the klystron. Censored events were the manual terminations for regular maintenance every two weeks. Figure 6(a) shows the distribution of the operation time between the interruptions. The right arrow in Figure 6(a) shows that the last bin is the sum of all interruptions of not less than 100 hours. The solid histogram represents the operation time for accidental interruptions. On the other hand, the dotted histogram represents that for censored events. Figure 6(b) shows the number of accidental interruptions as a function of the down-time. The VSWR events accounted for 78% of all accidental interruptions. The average value of the down-time, that is, the MTTR of the klystron (τ_K) was 260 seconds. On the other hand, the one for all VSWR events was 20 seconds. The distribution of the down-time is also tabulated in Table 2.

Figure 6. Distribution of (a) the operation time for KEK klystrons, and (b) the down-time for accidental interruptions

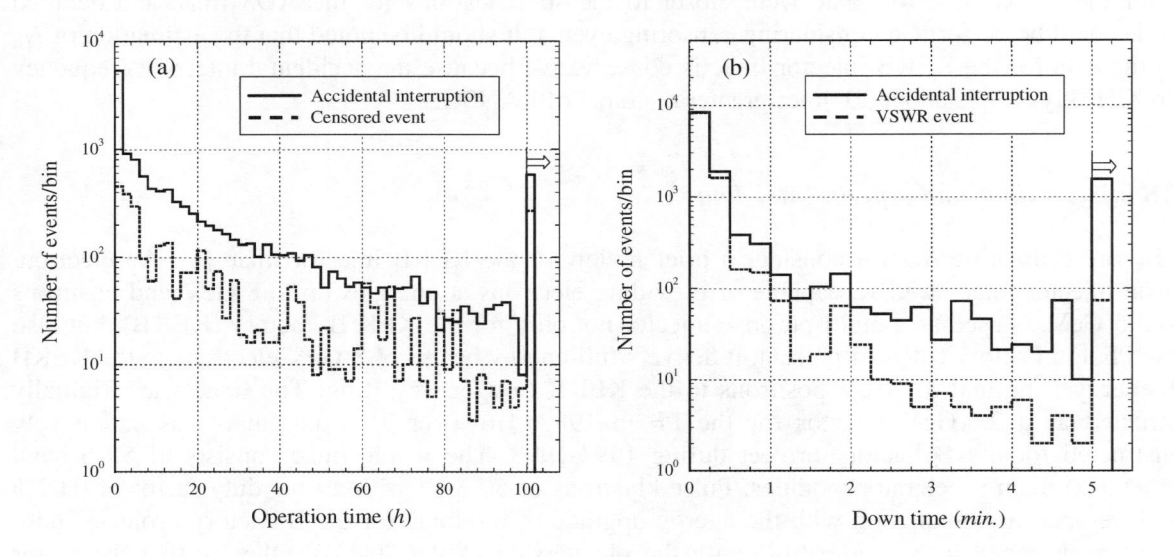

Estimation of N_{rf} based on the KEK experimental data

To estimate N_{rf}, the following assumptions were made:

- •The beam trip was merely caused by the accidental interruption of the klystron.

- •The RF system was equivalent to 60 klystrons connected in series for the KEKB injector linac and 89 klystrons for the JAEA's SC linac.

- •Each klystron was independent: the operation or failure of one klystron did not affect the operation or failure of any other klystron.

- •The accidental interrupt frequency for each klystron of the JAEA's SC linac, $\bar{\lambda}_K$, was the same. This value was obtained from the KEK experimental data and related to that for the whole RF system, Λ_K, as follows:

$$\Lambda_K = 89 \bar{\lambda}_K \qquad (3)$$

- The klystrons would be brought back to their initial state by applying the rated high voltage to the collector of the klystrons at the re-start of operation. Therefore, the klystron was deemed to be a repairable component because applying the rated high voltage to the klystron affected the repair. And the interrupt frequency could be calculated as the inverse of MTBF.

- The proportion failing as a function of operation time for each klystron, $F(t)$, followed a Weibull distribution [9], which is defined as:

$$F(t) = 1 - \exp\left[-\left(\frac{t}{\alpha}\right)^{\beta}\right] \qquad (4)$$

where $t > 0$ is operation time, $\alpha > 0$ is a scale parameter, and $\beta > 0$ is a shape parameter. Using these two parameters, the interrupt frequency, λ, is obtained as follows:

$$\frac{1}{\lambda} = \alpha \Gamma \left(1 + \frac{1}{\beta}\right) \tag{5}$$

where $\Gamma(x) = \int_0^\infty z^{x-1} \exp(-z) dz$ is the gamma function.

Similar to Eq. (2), the N_{rf} was obtained using the KEK experimental data:

$$N_{rf} = \frac{T_s}{\Lambda_K^{-1} + \tau_K} \tag{6}$$

where Λ_K^{-1} is the inverse of Λ_K. Λ_K, that is $\overline{\lambda}_K$, was estimated according to the following steps:

- *Step 1.* Obtain the cumulative number of accidental interruptions and censored events as a function of operation time for each klystron.

- *Step 2.* Estimate the proportion failing from these cumulative data using the Kaplan-Meier estimator [9].

- *Step 3.* Estimate the scale and shape parameter of the Weibull distribution, α_i and β_i, for i-th KEK klystron ($i=1, ..., 60$) by means of the statistical software package JMP [10].

- *Step 4.* Obtain $\overline{\lambda}_K$ from the accidental interrupt frequency, λ_{Ki}, for i-th KEK klystron using Eq. (5) and the following equation:

$$60 \overline{\lambda}_K = \sum_{i=1}^{60} \lambda_{Ki} \tag{7}$$

For reference, Figure 7 shows the proportion failing as a function of operation time for the typical KEK klystron. The cumulative number of accidental interruptions and the censored events is also shown in the same figure. Using the above-mentioned assumptions and the KEK experimental data, λ_{Ki} ranged from 6.1×10^{-4} to 7.9×10^{-2} times/h, and $\overline{\lambda}_K$ was estimated as 1.7×10^{-2} times/h/klystron. Finally, we got the value of N_{rf} as 9 900 times/year.

Comparison and discussion

From the above analysis, the value of N_{ads} is 19 000 times/year and using the distribution of beam trip durations, the down-time distribution of the JAEA's SC linac is a straight line as shown in Figure 8. Here, since the time distribution of events for beam trip durations of less than one minute is not clearly shown in the LANSCE experimental data, the distribution of the LANSCE data was assumed to be the same as that of the KEKB injector linac. Further, the acceptable frequency of beam trips is shown in this figure as a histogram. Comparing the two shows that even at the present technological level of accelerators, the beam trip frequency for durations of five seconds or less is within the acceptable level.

Figure 7. Proportion failing as a function of operation time for the typical KEK klystron

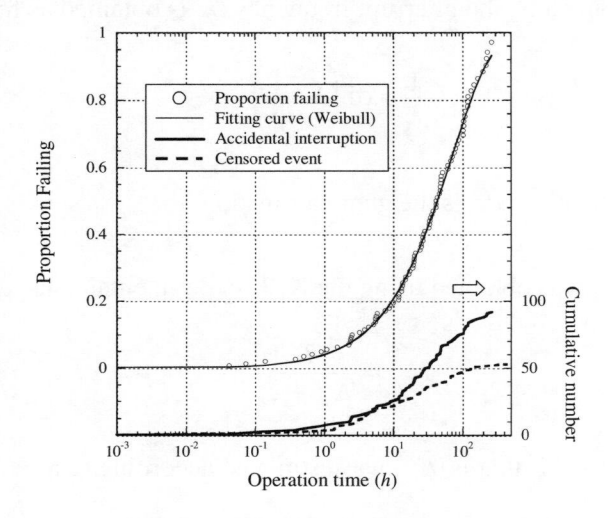

Figure 8. Comparison of the acceptable frequency of beam trips and the estimated frequency of the JAEA's SC linac

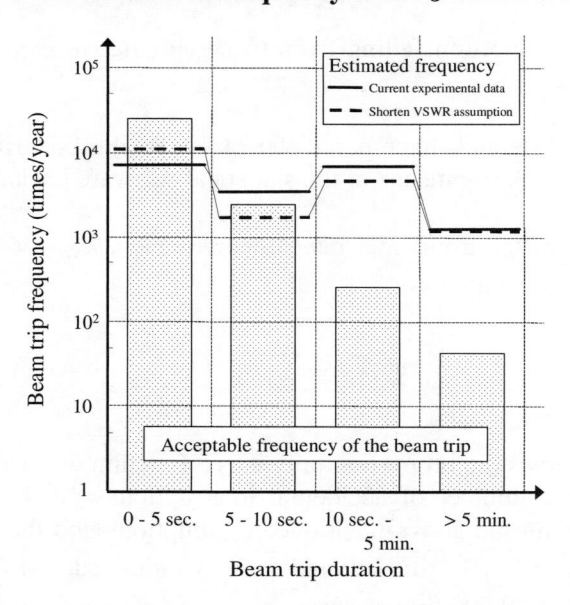

By the way, for experimental data of the KEK, we focused on the result that 78% of all accidental interruptions were VSWR events and devised measures for reducing beam trips of the high-power accelerator used for ADS. That is, since the bulk of the down-time for VSWR events is less than five seconds, we considered the down-time distribution for the case where the down-time of all VSWR events longer than five seconds can be reduced to five seconds. Under this assumption, the distribution of the N_{rf} was calculated and the results are shown in the "shorten VSWR assumption" of Table 2. The dotted line in Figure 8 shows the down-time distribution of JAEA's SC linac assuming this distribution. A comparison of the dotted line and histogram results shows that by reducing the down-time of all the VSWR events to five seconds or less, the beam trip frequency for durations of 10 seconds or less satisfies the acceptable frequency of beam trips. In this case, for the beam trip frequency with duration greater than 10 seconds, it is necessary to improve the technology level to about 30 times of present accelerator technology.

Conclusion

Frequent beam trips as experienced in existing high-power proton accelerators may cause thermal fatigue problems in ADS components which may lead to degradation of their structural integrity and reduction of their lifetime. Thermal transient analyses were performed to investigate the effects of beam trips on the reactor components. These analyses were made on the thermal responses of three parts of the reactor components: the beam window, the inner barrel and the reactor vessel. Our results indicate that the acceptable frequency of beam trips ranges from 50 to 2.5×10^4 times per year, depending on the beam trip duration.

In order to measure the effect of reducing beam trips on the high-power accelerator for ADS, it is important to know the present level of accelerator technology. We compared the difference between the acceptable frequency for beam trips and the operation data of existing accelerators. The result of this comparison shows that even at the present technological level of accelerators, the beam trip frequency for durations of five seconds or less is within the acceptable level. Under the "shorten VSWR assumption", the beam trip frequency for durations of 10 seconds or less is within the acceptable level, while that exceeding 10 seconds should be reduced by about 1/30 to satisfy the thermal stress conditions. Further, apart from accidental interruptions it is also necessary to include censored events in the analysis of the accelerator components related to the beam trip frequency.

As the acceptable frequency of beam trips is different for each ADS reactor system design, this solution for the beam trip problem is not valid for all ADS accelerators. However, we believe that this evaluation technique of the beam trip is of potential utility in ADS system planning. In the future, it will be necessary to include the following items in the study of beam trip problems.

- study the hardware methods for reducing VSWR events to five seconds or less;

- analysis of the frequency of beam trips with consideration of the frequency with which broken components are exchanged.

Acknowledgement

The authors would like to thank to Mr. K. Suzuki of Mitsubishi Electric System & Service Co., Ltd. for his help in verifying data of the KEKB injector linac.

REFERENCES

[1] Tsujimoto, K., *et al.*, "Research and Development Program on Accelerator Driven Subcritical System in JAEA", *Journal of Nuclear Science and Technology*, 44, 483 (2007).

[2] Burgazzi, L., P. Pierini, "Reliability Studies of a High-power Proton Accelerator for Accelerator-driven System Applications for Nuclear Waste Transmutation", *Reliability Engineering and System Safety*, 92, 449 (2007).

[3] Ouchi, N., *et al.*, "Development of a Superconducting Proton Linac for ADS", *Proc. of 4th Workshop on Utilisation and Reliability of High-power Proton Accelerators*, 175 (2004).

[4] Takizuka, T., *et al.*, "Responses of ADS Plant to Accelerator Beam Transients", *Proc. of 2nd Workshop on Utilisation and Reliability of High-power Proton Accelerators*, 321 (1999).

[5] *Finite Element Nonlinear Structural Analysis System (FINAS)*, ITOCHU Techno-Solutions Co., 2-5, Kasumigaseki 3-chome, Chiyoda-ku, Tokyo 100-6080, Japan.

[6] Eriksson, M., "Reliability Assessment of the LANSCE Accelerator System", M.Sc. thesis at the Royal Institute of Technology, Stockholm, Sweden (1998).

[7] Gobin, R., *et al.*, "Saclay High Intensity Light Ion Source Status", *Proc. of EPAC 2002*, 1712 (2002).

[8] Abe, I., *et al.*, "The KEKB Injector Linac", *Nuclear Instruments and Methods in Physics Research A*, 499, 167 (2003).

[9] Meeker, W.Q., L.A. Escobar, "Statistical Methods for Reliability", Wiley-Interscience Publication, New York (1998).

[10] JMP, SAS Institute Inc., Cary, NC, 27513.

CAVITY DETUNING DUE TO POWER DISSIPATION: A NEW NUMERICAL APPROACH COMBINING THE THERMO-MECHANICAL AND THE E-M CODES

V.G. Vaccaro[1], A. D'Elia[2], C. Serpico[2], M. Fraldi[3], L. Nunziante[3]
[1]Dipartimento di Scienze Fisiche and I.N.F.N. Sezione di Napoli, Università *Federico II*, Napoli, Italy
[2]CERN, Geneva, Switzerland
[3]Dipartimento di Scienze delle Costruzioni, Università *Federico II*, Napoli, Italy

Abstract

Temperature rise due to power dissipation in accelerating cavity walls may produce variations of its geometrical parameters. As a consequence the cavity resonant frequency will change. In some cases this may produce dangerous detuning which has to be known in advance with a very high accuracy. The combined use of thermo-mechanical structure analysis end e-m numerical codes needs some skill. We present a new approach to this problem: by means of specific finite element codes. Starting from an electromagnetic analysis of the structure, one may get the thermal load on the surface which is used for calculating the mechanical deformations of the structure. Instead of resorting again to the e-m code used for the non-deformed structure and to recalculate the frequency of the deformed one, much better results may be obtained from the application of the theorem which states the balance of electric and magnetic energy at resonance. The theorem is directly applied, by means of a numerical code realised *ad hoc*, to the final output of the thermo-mechanical CAD. An application of this new approach to PALME linac cavities and a comparison with other techniques will be reported.

Introduction

This work is devoted to the analysis of the resonance frequency drift in accelerating cavities. Particular attention has been devoted to side-coupled linac (SCL) cavities, even if the procedure adopted is most general. This frequency variation, induced by the thermal deformation of the metal wall of the cavity, deserves a particular attention in order to prevent the possible discord between the power feeder frequency and the cavity resonance frequency. The ultimate goal of this work is to yield a fast, light and high precision procedure for calculating this discord. A very precise evaluation of the discord has particular importance to put in operation the RF system. Indeed the design and fabrication procedure foresees that the cavity has a detuning opposite to the frequency shift that will be produced by the thermal deformation.

An SCL is formed by a bi-periodical set of resonating cavities coupled via slots. Its configuration is illustrated in Figure 1 by an example and by a diagrammatical drawing. In this accelerator two types of cavities are identified: accelerating cavities and coupling cavities, which resonate on two slightly different frequencies. However these cavities, once coupled, they lose their individuality and resonate on frequencies which are characteristic of the system (system modes).

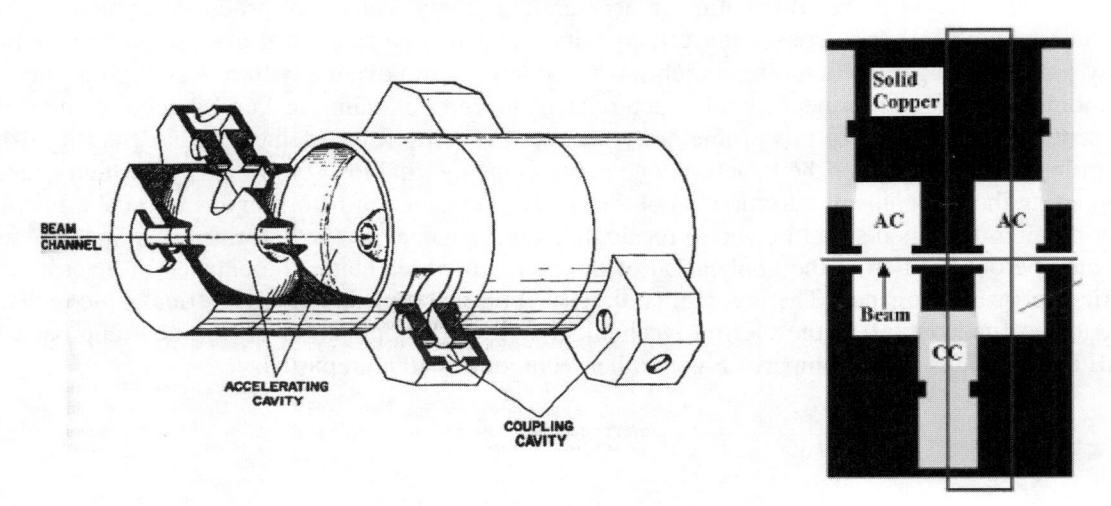

Figure 1. An SCL and its schematic representation

In Figure 2 one can find the frequency spectrum of an SCL formed by 13 cavities, 6 coupling cavities and 7 accelerating cavities. In general, the chosen working mode is the $\pi/2$ one. This choice allows to make null the electromagnetic field inside coupling cells, which are offset with respect to the beam axis and do not contribute to the acceleration. This choice has many advantages. The $\pi/2$ mode is the most distant from the adjacent ones. It is the least sensitive to the fabrication errors and, obviously, it needs the least consumption of RF power.

Furthermore because of the field behaviour, Figure 3, in the cavities it requires that the distance between the accelerating cavity gaps is such to guarantee the synchronism between the face of the electric field and the passage of the bunch.

SCL are in use as accelerators for low and high intensity beams. As an example, the PALME project and LIBO project adopted the SCL scheme for acceleration of low-intensity and low-energy proton beams for deep seated cancer radiotherapy. Likewise LINAC4, which is designed for doubling the beam "luminosity" and intensity at the exit of PS booster, foresees a section (90-160 MeV) formed by a short number of SCL modules.

Figure 2. Frequency spectrum of a chain of 13 coupled cavities

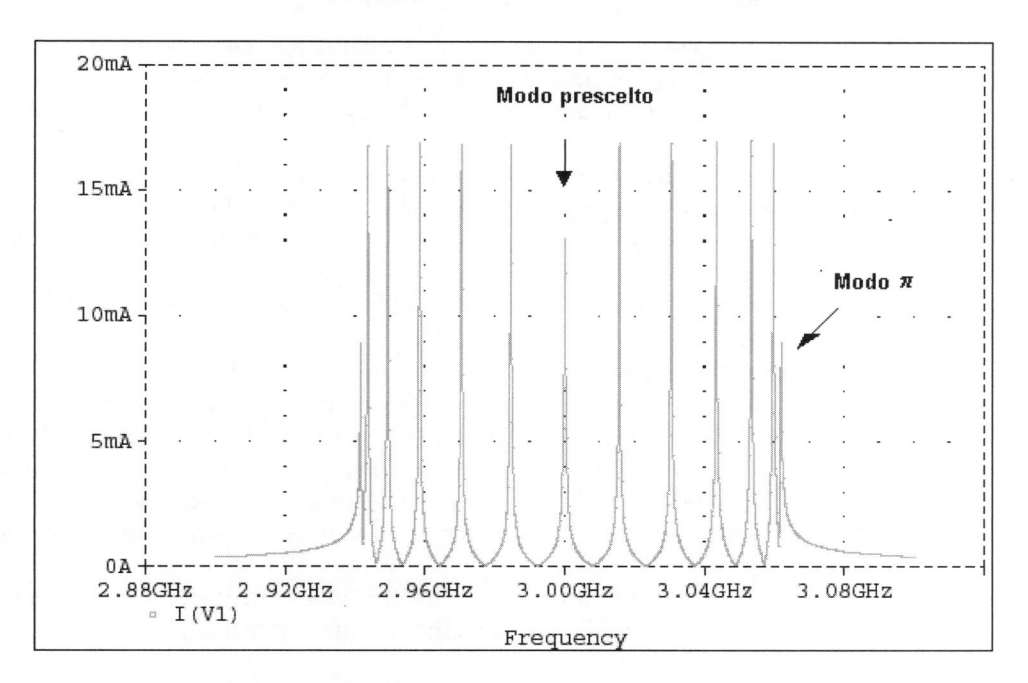

Figure 3. Field behaviour in the cavities of an SCL at π/2 mode

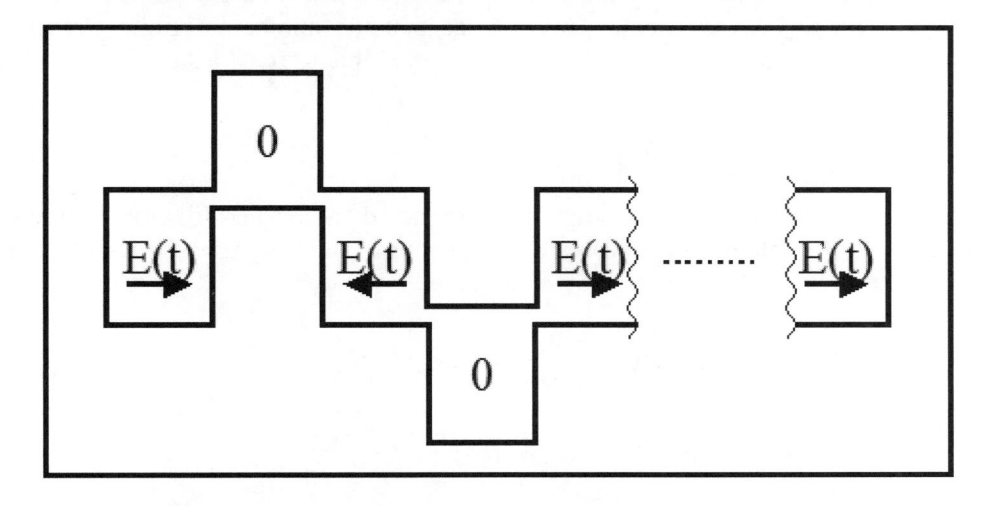

The PALME and LIBO modules are designed to be fed at 3 GHz with a duty cycle of about 0.2%. The SCL section of LINAC4 (Figure 4) is designed to be fed at 704 MHz with a duty cycle ranging between 1.5% and 10%. We will focus our attention on these two cases.

Figure 4. LINAC4 sections

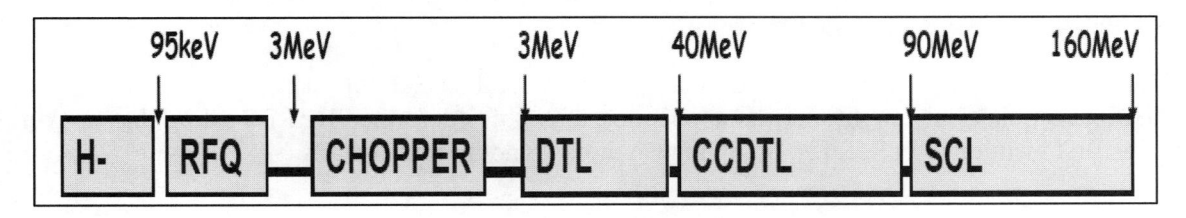

The analysis of the problem

The starting point for the evaluation of the frequency shift phenomenon is the calculation of thermal strain induced by power dissipation on the surfaces. The deformed profile of the cavity is obtained by a sequence of analyses implemented by means of the use of finite elements ANSYS CAD.

In order to minimise the complexity of the procedure, it is important to identify the smallest subset in the structure on which one could apply the sequence of analyses. This has to be done with skill: one must "cut" the subset not breaking the electromagnetic, thermal and mechanical symmetries.

A first analysis was performed in order to detect the electromagnetic parameters relevant to the non-deformed cavity: resonance frequency, form and amplitude of the EM fields, mean electric field along the axis. Furthermore, by imposing an appropriate value to wall conductivity, it is possible to calculate the power density dissipated on the metal walls of the cavity. This power is then used as the "load" for the implementation of the following thermal analysis, associated to the assigned boundary conditions. The computer code gives back as output the thermal distribution in the structure. As a final step, the previous output is the input for a thermo-mechanical analysis which, with the relevant boundary conditions (structural constrains), gives the deformation induced into the subset under test.

The "standard" procedure consists in going back to the electromagnetic CAD where the geometrical boundary conditions are given by the previous thermo-mechanical analysis.

This way to tackle the problem imposes a very high accuracy in evaluation of the frequencies of the deformed and of the non-deformed cavity. Indeed, the accuracy has to be compared to the frequency shift which may happen to be very small. This implies that the number of finite elements to be considered is exceedingly large, especially if it has not been possible to "cut" a subset of simple form.

The new method

The new method has its roots in application of the so-called "Slater Perturbation Theorem". This theorem states that it is possible to directly obtain the amount of the frequency shift in the cavity by means of following formula:

$$\frac{\Delta f_0}{f_0} = \frac{\frac{1}{4}\iiint_{\Delta V}(\mu_0 H^2 - \varepsilon_0 E^2)dV}{\frac{1}{4}\iiint_{V}(\mu_0 H^2 + \varepsilon_0 E^2)dV} \cong \frac{\frac{1}{4}\iint_{\partial V}(\mu_0 H^2 - \varepsilon_0 E^2)*(\hat{n}\cdot\bar{v})dS}{\frac{1}{4}\iiint_{V}(\mu_0 H^2 + \varepsilon_0 E^2)dV} = \frac{\Delta U_m - \Delta U_e}{U}$$

where $\hat{n}$ is the ingoing normal vector on the cavity surface and $\bar{v}$ is the displacement vector of the surface and S is surface of the cavity. The interpretation of the Slater formula is rather simple: at the resonance frequency there is a balance between the electric and magnetic energy in the cavity. If the deformation unbalances this equilibrium (see the numerator), the resonance frequency must change in order to produce a new field configuration able to restore the equilibrium. The frequency shift is the proportional to the unbalance in the two energies.

The Slater Theorem is currently used in connection with the bead pull method for measuring the field distribution along a straight line. Its application to cavities deformed by RF heating due to skin effect was first mentioned in Ref. [2], though the authors do not provide details.

The main problem for the implementation of this formula is the orientation of the normal vector $\hat{n}$ as it can be derived from the mesh data obtained from the thermo-mechanical codes. Most of these codes do not explicitly give the normal vector, which can be derived only from the co-ordinates of the mesh points. This fact produces an ambiguity on the direction of the normal, which depends on the sequence of the points. An algorithm has been found which allows making the normal vectors all ingoing (outgoing) by an appropriate change of their sign.

The integral in the above formula were implemented as finite sums in Matlab workbench by exporting the outputs of the e-m CAD and of the thermo-mechanical CAD. In order that the algorithm may give a correct result, one must check that the normal vectors be oriented in a coherent way. In general, this does not happen for most of numerical CADs. Therefore, before the execution of the algorithm, it is necessary to correct the orientation of these vectors. ANSYS itself contains a special tool by which it is possible to visualise the orientation of the normal vector to the cavity surface meshes. Once the orientations of normal vectors have been detected, the correction can be done resorting to another tool available in ANSYS.

Applications to LINAC4 and to PALME

The first application was done to the first module of LINAC4. As a first step, the subset under test was simplified. Only the accelerating cavity was taken into account. The first analysis was performed in order to detect the electromagnetic parameters relevant to the non-deformed cavity; the electromagnetic field was normalised to get the assigned mean field. The boundary conditions were fixed as follows:

- electromagnetic and thermal symmetry planes orthogonal to the axis as indicated by the red lines in Figure 1;

- thermally adiabatic walls on upper and lower walls;

- constant temperature on the lateral walls;

- the symmetry planes are assumed to be PEC;

- the symmetry planes are thermally adiabatic.

The thermal distribution, obtained according to the procedure previously mentioned is reported in Figure 5, where one may appreciate a temperature rise of 7°C in the cavity noses.

From the temperature distribution, the thermo-mechanical routine of ANSYS gave strain and stress distribution, the representation of which is in Figure 6-7.

In this treatment a particular attention has been devoted to the confrontation of the results obtained by two different methods. It has to be noted that because of the particular symmetry, the analysis could be implemented to a small fraction (1/8) of the subset. This behaviour allowed to make the implementation with a very high equivalent number of meshes. First, an analysis was done at the highest possible level of accuracy from the available hardware system. In this condition the old method and the new algorithm gave the same result, -66 kHz. The discrepancy between the two discords was less than 1%.

Figure 5. Temperature pattern in the simplified subset of LINAC4 side-coupled linac

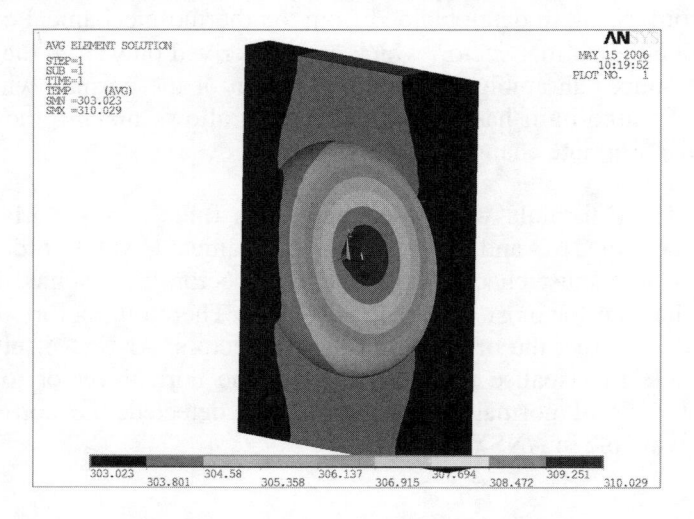

Figure 6. Strain pattern in the noses

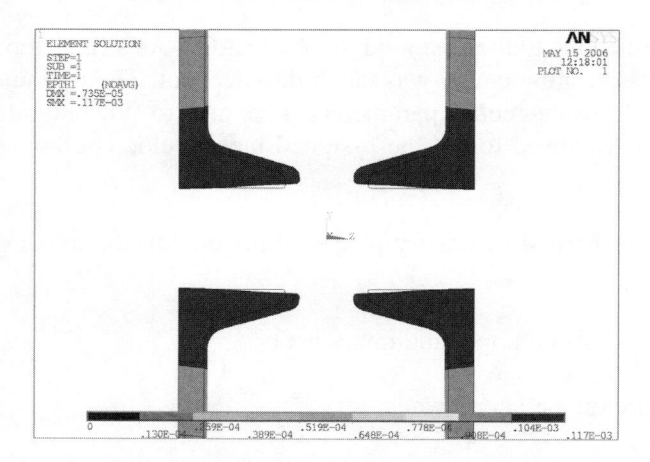

Figure 7. Stress pattern in the noses

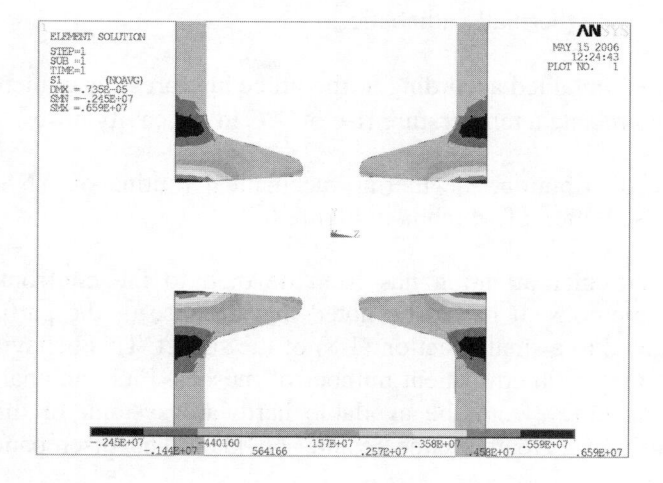

Figure 8. Comparison between the frequency shifts calculated with the new algorithm (rhombs) and the old procedure (triangles) as a function of the mesh number

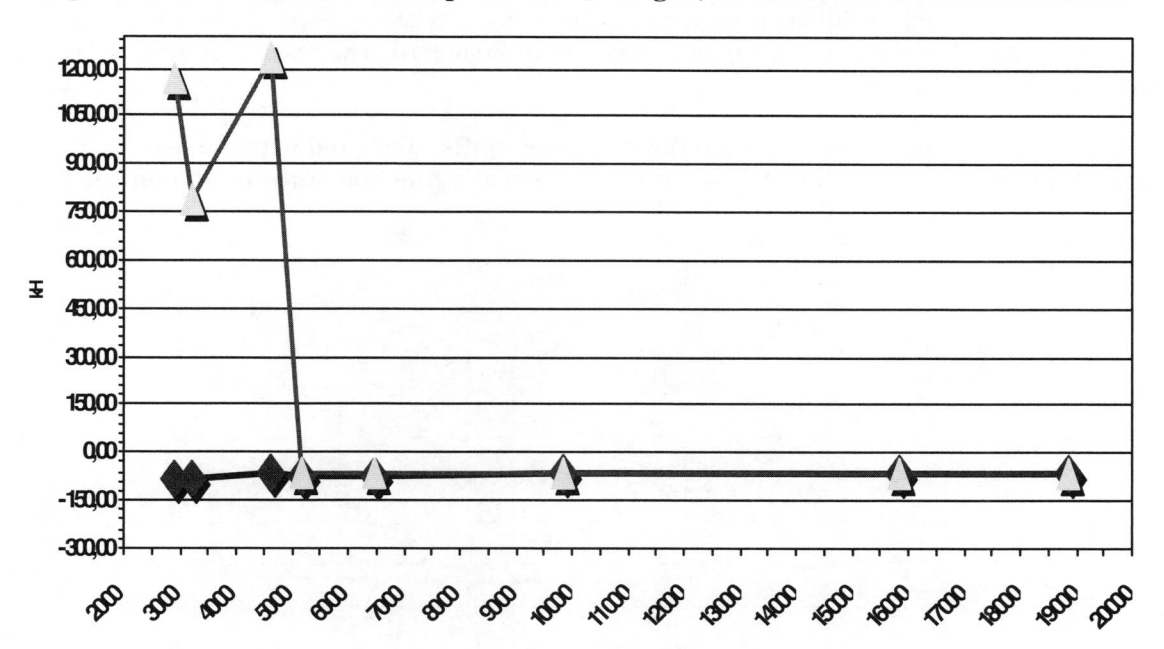

Once the accuracy was verified the next step was to implement the analysis with larger and larger meshes, i.e. decreasing the mesh number. The picture shows that the new algorithm yields results which are very stable down to small mesh numbers, where the old procedure becomes quite unstable.

When the geometry is complex, if one resorts to the old method, the need of very dense meshes may involve heavy computational load sometimes non-affordable. This limitation makes very difficult to obtain reliable results for the resonance frequency shift. Therefore the availability of an alternative tool is an essential resort for achieving an accurate result. This is well illustrated by situation of the Figure 9 where the subset which complies the real symmetries of an SCL is depicted.

Figure 9. The subset

In this case we have an increase of complexity because of the presence of two half coupling cells and of two slots connecting the adjacent cavities. In addition to this we no longer have the planes of symmetries which allowed partitioning the subset as beforehand. The comparison between the results of the old method and of the new algorithm is reported in Figure 10. The results are relevant to the 30-35 MeV module of PALME.

Figure 10. Comparison between the frequency shifts calculated with the new algorithm (triangles) and the old procedure (circles) as a function of the mesh number

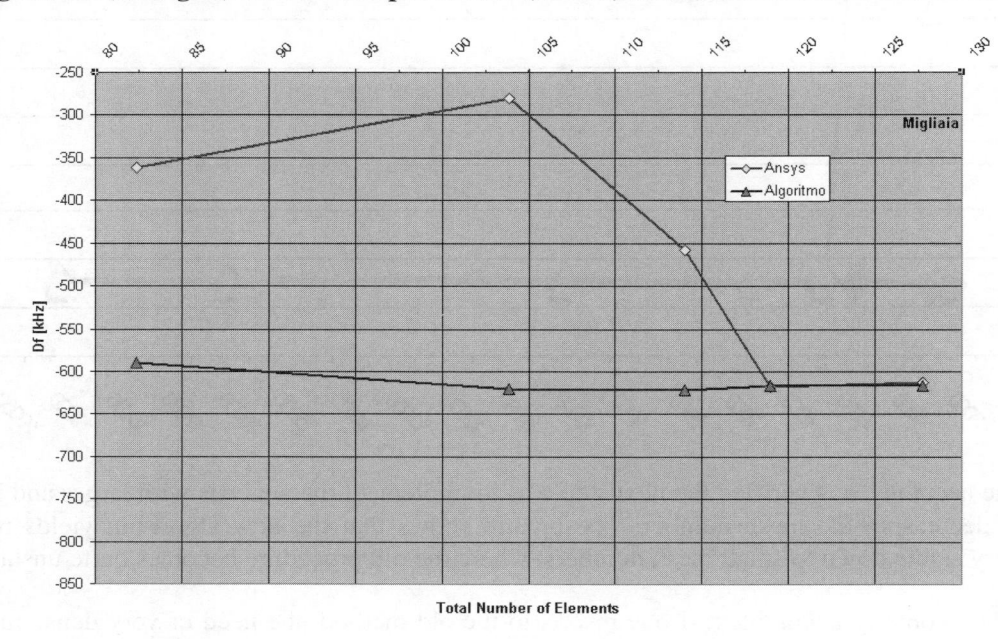

Conclusions

It is apparent from the displayed results that the new algorithm gives answers largely more stable than the old method. In addition to this, the convergence of the two approaches is shifted to larger number of meshes. This aspect confirms that an increase of the complexity of the cavity precludes resorting to the old method.

REFERENCES

[1] D'Elia, A., "System Mode Sounding (SMS) Tuning: A New Tuning Methodology for Resonant Coupling Structure", Tesi di Dottorato, Università di Napoli "Federico II", Facoltà di Ingegneria (2004).

[2] Amaldi, U., B. Szeless, M. Vretenar, E. Wilson, K. Crandall, J. Stovall, M. Weiss, "LIBO – A 3 GHz Proton Linac Booster of 200 MeV for Cancer Treatment". /AccelConf/l98/PAPERS/TU4098.

SESSION III

Spallation Target Development and Coolant Technology

Chairs: H. Oigawa, Y. Gohar

WEBEXPIR: WINDOWLESS TARGET ELECTRON BEAM EXPERIMENTAL IRRADIATION

Jan Heyse, Hamid Aït Abderrahim, Thierry Aoust, Marc Dierckx, Kris Rosseel, Paul Schuurmans, Katrien Van Tichelen
SCK•CEN, Boeretang 200, BE-2400 Mol, Belgium

Arnaud Guertin, Jean-Michel Buhour, Arnaud Cadiou
SUBATECH, *École des Mines*, 4 rue Alfred Kastler, BP 20722, F-44307 Nantes Cedex 3

Michel Abs, Benoit Nactergal, Dirk Vandeplassche
IBA, Chemin du Cyclotron 3, BE-1348 Louvain-la-Neuve, Belgium

Abstract

The WEBEXPIR (Windowless target Electron Beam EXPerimental IRradiation) programme was set up as part of the MYRRHA/XT-ADS R&D efforts on the spallation target design, in order to answer different questions concerning the interaction of a proton beam with a liquid lead-bismuth eutectic (LBE) free surface. An experiment was conceived at the IBA TT-1000 Rhodotron, a 7-MeV electron accelerator which produces beam currents of up to 100 mA. Due to the small penetration depth of the 7-MeV electron beam and the high beam currents available, the TT-1000 allows to imitate the high power deposition at the MYRRHA/XT-ADS LBE free surface. The main goals of the experiment were to assess possible free surface distortion or shockwave effects under nominal conditions and during sudden beam on/off transient situations, as well as possible enhanced evaporation. The geometry and the LBE flow characteristics in the WEBEXPIR set-up were made as representative as possible of the actual situation in the MYRRHA/XT-ADS spallation target. Irradiation experiments were carried out at beam currents of up to 10 mA, corresponding to 40 times the nominal beam current necessary to reproduce the MYRRHA/XT-ADS conditions. As a preliminary general conclusion, it can be stated that the WEBEXPIR free surface flow was not disturbed by the interaction with the electron beam and that vacuum conditions stayed well within the design specifications.

Introduction

The design of the MYRRHA/XT-ADS, the European eXperimental Accelerator-Driven System for the demonstration of Transmutation [1], includes a high-power windowless spallation target operating with liquid lead-bismuth eutectic (LBE) that will be irradiated with a 600-MeV proton beam at currents of up to 4 mA. When considering such a high power windowless target design, a number of questions need to be addressed, such as the stability of the free surface flow and its ability to remove the power deposited by the proton beam by (forced) convection, the compatibility of a large hot LBE reservoir with the beam line vacuum and the outgassing of the LBE in the spallation target circuit. These issues have been studied during previous experiments supported by numerical simulations [1].

Another crucial point in the development of the spallation target is the demonstration of safe and stable operation of the free LBE surface under the irradiation with a high-power proton beam. As a first step in this programme, the WEBEXPIR (Windowless target Electron Beam EXPerimental IRradiation) experiment was set-up. Its purpose was to investigate the influence of LBE surface heating caused by a charged particle beam in a situation representative of the MYRRHA/XT-ADS. For this purpose a compact LBE loop was constructed with flow characteristics and geometry at the irradiation point that are very similar to the MYRRHA/XT-ADS case. The loop was connected to a 7-MeV electron accelerator. During irradiation the effects of the surface power deposition on the flow stability and LBE evaporation were monitored.

Experimental set-up

Interaction chamber

The WEBEXPIR interaction chamber was constructed in such a way as to have interaction conditions that are very representative of the conditions in the actual MYRRHA/XT-ADS target nozzle (see Figure 1). In MYRRHA/XT-ADS the proton beam passes through the inner of two hollow concentric vertical cylinders with the LBE flowing in between the inner and the outer cylinder. At the bottom of this downcomer section, the outer cylinder changes into a nozzle, resulting in a confluent LBE target. In order to avoid hitting the LBE recirculation zone which is present in the centre of the nozzle, the proton beam is painted along a circular path in between the recirculation zone and the inner cylinder, forming an annulus-shaped footprint.

Figure 1. Schematic cutaway view of the MYRRHA/XT-ADS spallation target and core

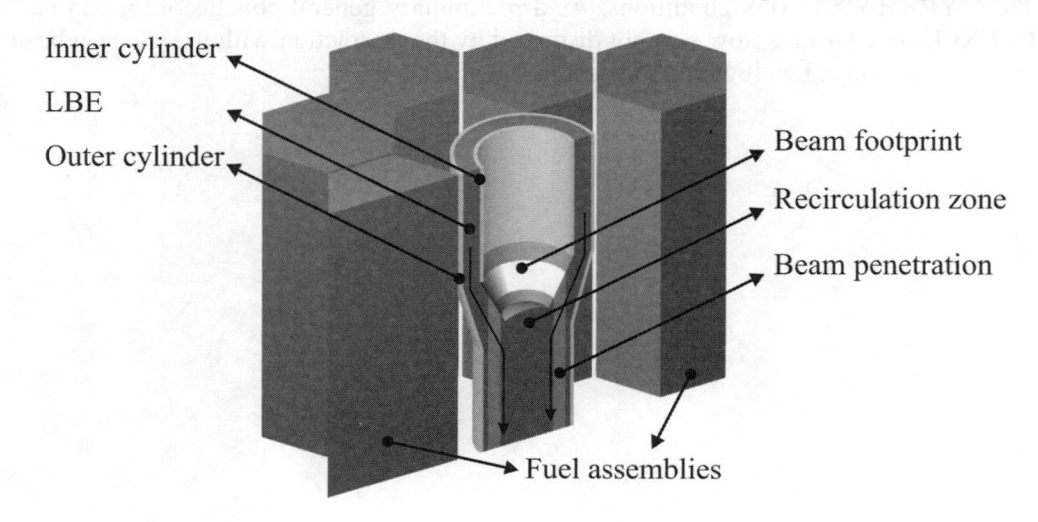

To simulate these conditions during the WEBEXPIR experiment, a geometry that can be seen as an unfolded slice of the MYRRHA/XT-ADS nozzle was chosen. It consists of a rectangular slide with its thickness corresponding to the nozzle gap thickness of 14.35 mm and with an angle of inclination of 26°, equal to the angle between the proton beam and the LBE surface. The width of the gap was taken such that the LBE flow velocity was almost constant at the beam interaction point and equal to the MYRRHA/XT-ADS nozzle velocity of 2.5 m/s.

The part of the slide upstream of the interaction point is rib-roughened to prevent fluid acceleration and tearing. At the beam interaction point, the LBE flow is confined so that the impact of the beam can only be compensated in the direction perpendicular to the free surface.

Loop layout

The interaction chamber was integrated in a compact LBE loop so as to achieve flow characteristics at the beam interaction point that were comparable to those in the MYRRHA/XT-ADS spallation target design. Apart from that, a number of spatial boundary conditions imposed by the configuration at IBA had to be taken into account.

Figure 2 shows a schematic drawing of the loop. The LBE is circulated by a 10-l/s, 5-bar magnetically coupled centrifugal pump (1). Due to space considerations, the LBE/oil heat-exchanger (2), which ensures evacuation of the heat deposited by the electron beam, has been positioned on the exit pressure side of the pump. A valve (3) has been placed behind the heat-exchanger in order to assure a stable pressure at the exit of the pump. A vortex-type flow meter (4), which requires sufficient inlet and outlet tube length to obtain the necessary accuracy, is connected to a box at the top of the interaction chamber slide (5). In the interaction chamber (6), a number of viewports have been provided to monitor the interaction of the electron beam with the LBE flow. At the drain of the interaction chamber an expansion volume (7) has been added to prevent the flooding of the interaction chamber when the pump is not running. The top of the interaction chamber is connected to the beam line (8), just at the exit of the 270° magnet that bends the electron beam coming from the accelerator into the vertical direction. The connection consists of a flexible bellows, a gate valve and a beam collimator. Expansion joints have been incorporated at several positions in the loop to relieve thermal strain and to correct for misalignment errors.

Accelerator

The irradiation was conducted at the TT-1000 Rhodotron [2], which is a very high-power electron accelerator available at Ion Beam Applications s.a. (IBA, Louvain-la-Neuve, Belgium). Table 1 shows the main accelerator characteristics. The use of a 7-MeV electron beam allows to imitate the effect of a 350- or 600-MeV proton beam hitting the LBE surface. Moreover, the use of 7-MeV electrons avoids problems with induced radioactivity, safety and waste that would occur when using protons.

Figure 3 compares the linear power deposition in LBE for a proton beam with an energy of 350 MeV and 600 MeV, respectively [3], and a 7.5-MeV electron beam, as calculated by Monte Carlo simulations. The inset of this figure clearly shows that the heating by the electron beam is concentrated near the lead-bismuth surface: about 65% of the beam power is deposited in the first 2 mm below the LBE surface. This figure also shows that the linear power density per mA at the surface is well comparable to that of a 350-MeV or 600-MeV proton beam.

Figure 2. WEBEXPIR loop layout – numbers are referred to in the text

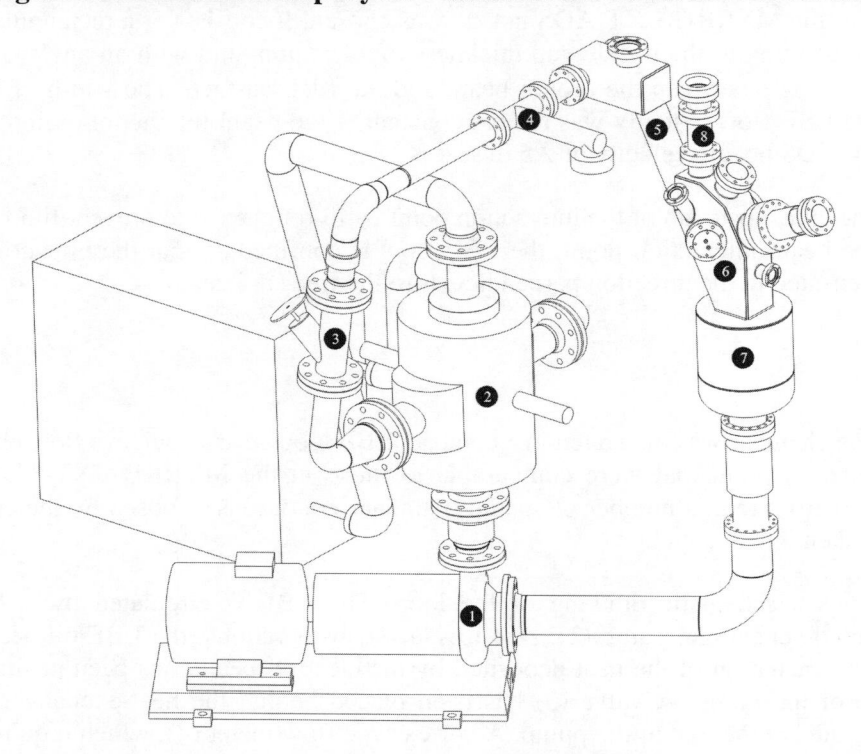

Table 1. TT-1000 Rhodotron characteristics

Energy	7 MeV e⁻
Beam power range	0.5-700 kW
Beam spot	10 mm × 3 mm FWHM
Beam entrance	Vertical from top
Diameter	3.0 m
Height	3.3 m

Figure 3. Linear power deposition

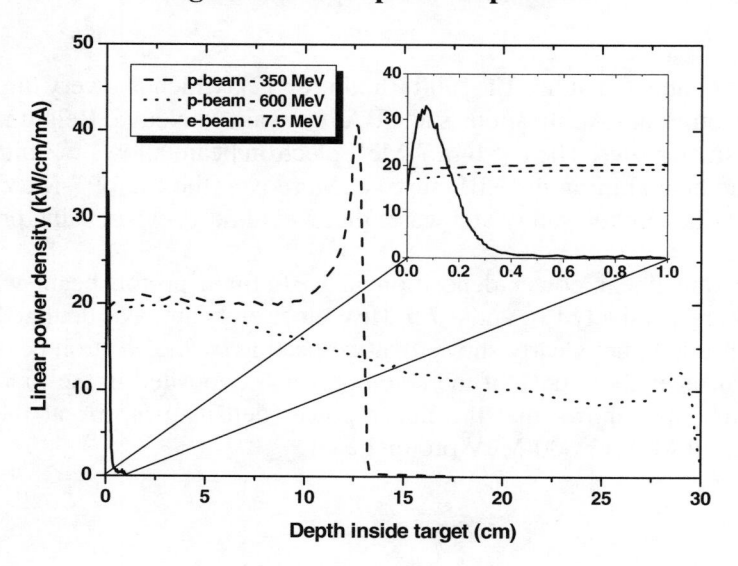

Table 2 compares the characteristics of the flow and the beam profile for the MYRRHA/XT-ADS and the WEBEXPIR case. In order to obtain at least the same surface heating conditions during the WEBEXPIR experiment, the nominal beam current was set to 0.25 mA.

Table 2. Comparison of flow and beam characteristics

	XT-ADS target	WEBEPXIR target
Target material	LBE	LBE
Beam energy	600 MeV p	7 MeV e$^-$
Average power deposition < 2 mm	1.8 kW/mm/mA	2.2 kW/mm/mA
Flow velocity	2.5 m/s	2.5 m/s
Beam shape	~ Gaussian (10 mm FWHM) painted along annulus (54 mm $\varnothing$)	~ Gaussian (3 mm FWHM $\perp$ flow, 10 mm FWHM // flow)
Beam current	4 mA	0.25 mA

Instrumentation

A number of diagnostic devices have been installed on the interaction chamber in order to study the two main effects mentioned before, i.e. free surface distortion and enhanced evaporation. Additional devices were used to monitor the LBE and beam characteristics.

- *Optical camera*: used for visual observation of the beam interaction point, both perpendicular and parallel to the LBE surface.

- *Infrared camera*: used for infrared observation of the beam interaction point and for measuring the surface heating at and downstream of the beam interaction point without interfering with the LBE flow.

- *Thermocouple array*: a set of thermocouples, inserted in the LBE flow downstream of the beam interaction point for measuring the surface heating caused by the beam.

- *Pt-100 resistive sensors*: used to accurately measure the LBE bulk temperature before and after beam interaction.

- *Thermocouples on the loop*: used to measure the temperature of the main loop components, the temperature of the top and sides of the interaction chamber and the temperature of the beam collimator.

- *Dedicated nude active ionisation pressure gauge*: used for determining the enhanced evaporation at the beam interaction point.

- *Additional Bayard-Alpert pressure gauges*: used for monitoring the vacuum conditions in the interaction chamber and the beam line.

- *Pressure transducers*: used to monitor shockwaves underneath the LBE flow and to measure the LBE pressure at the entrance and exit of the main pump.

- *Flow meter*: used to measure the LBE flow rate.

- *Beam position monitor*: Al_2O_3-coated and water-cooled copper support, used to verify the position of the electron beam in between irradiations.

- *Beam collimator current read-out*: used to monitor which fraction of the beam is stopped by the beam collimator.

Measurements and discussion

A set of measurements were performed at different electron beam currents to determine the experimental conditions and to address the beam-LBE interaction issues mentioned before. A detailed analysis of these measurements is ongoing, the results of which will be reported in a forthcoming publication. Table 3 shows the nominal LBE conditions with the electron beam turned off.

Table 3. Nominal LBE conditions with beam off

LBE flow	3.8 l/s
LBE speed at interaction point	2.5 m/s
LBE bulk temperature	200°C
Total LBE mass in loop	800 kg
Vacuum in interaction chamber	5×10^{-6} mbar

Beam profile measurements

The distribution of the current along the beam profile was determined prior to the actual irradiation experiments. A stack of electrically insulated collimators with decreasing diameters was mounted at the exit of the 270° bending magnet. The fraction of the beam current caught by each of the collimators was read out. The remaining fraction of the beam was deposited in a water-cooled beam dump. Since the profile of the beam might change as a function of the total beam current, the beam profile was measured for currents of 100 µA up to 4 mA.

The beam position and profile were also visually checked by inserting an Al_2O_3-coated and water-cooled copper support in the electron beam at the position of the beam interaction point (see Figure 4). The interaction with the 7-MeV electrons produces visual radiation which can be detected with an optical camera.

The combination of both measurements resulted in a beam profile with an average full width at half maximum (FWHM) of 10 mm in the direction parallel to the LBE flow and 3 mm perpendicular to the LBE flow.

Visual inspection

An optical camera was positioned both nearly perpendicular and parallel to the LBE surface, focussing on the beam interaction point. Visual images were recorded for beam currents of 100 µA up to 5 mA. No obvious distortion of the LBE surface was detected for any of the beam currents.

Figure 4. Beam spot on Al₂O₃

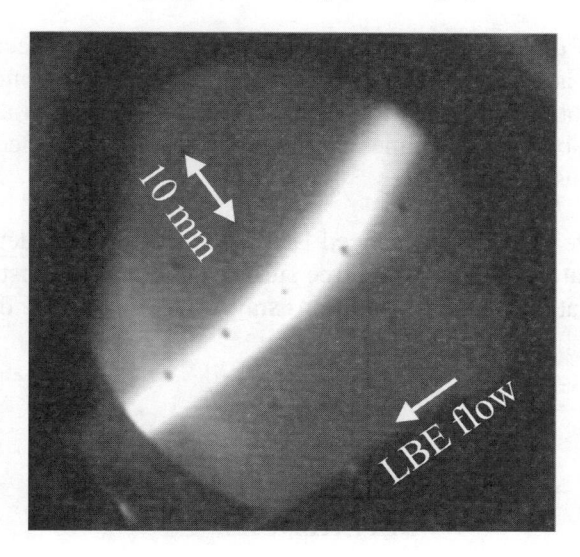

Infrared inspection

The infrared camera was positioned perpendicular to the LBE surface, focusing on the LBE surface just downstream of the beam interaction point. The camera was used to measure the surface heating of the LBE for beam currents of 100 μA up to 7.5 mA.

Enhanced evaporation

The enhanced evaporation of the LBE as a result of surface heating by the electron beam was measured with a nude Bayard-Alpert vacuum gauge positioned close to the beam interaction point. The increase in pressure was monitored for beam currents of 100 μA up to 10 mA. Even at 10 mA, the pressure in the interaction chamber stayed well below 10^{-4} mbar.

Surface samples were taken at a number of cold positions in the interaction chamber for off-line determination of the amount of LBE condensation.

Figure 5. Infrared image of beam spot on LBE at a beam current of 1.5 mA

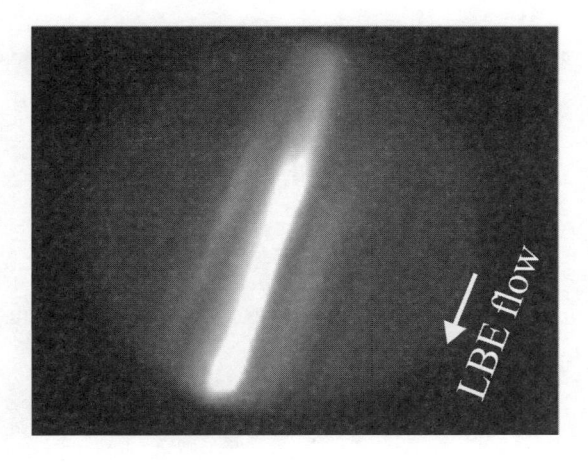

Conclusion

The aim of the WEBEXPIR experiment was to check for any LBE surface distortion or enhanced evaporation as a result of surface heating with a 7-MeV electron beam in conditions representative of the MYRRHA/XT-ADS spallation target. Various tests have been carried out at electron beam currents of up to 10 mA, which corresponds to 40 times the nominal beam current necessary to reproduce the MYRRHA/XT-ADS conditions.

No significant shockwave effects or enhanced LBE evaporation were detected. As a preliminary conclusion it can be stated that the WEBEXPIR free surface flow was not disturbed by the interaction with the electron beam and that the vacuum conditions stayed well within the design specifications.

REFERENCES

[1] Schuurmans, P., *et al.*, "Design and R&D Support of the XT-ADS Spallation Target Loop", these proceedings.

[2] Abs, M., Y. Jongen, E. Poncelet and J-L. Bol, *Radiat. Phys. Chem.*, 71, 1-2, pp. 287-290 (2004).

[3] Aït Abderrahim, H. *et al.*, *MYRRHA Draft 2*, Chapter 4, "Neutronics", RF&M/EM/em.34.B043200/85/05-26 (2005).

RELAP MODEL OF THE MYRRHA draft 2 SPALLATION LOOP

S. Heusdains
SCK•CEN

Abstract

In the aims of the XADS MYRRHA draft 2, the nominal and accidental analysis of the machine performance requires a model of the spallation loop to analyse its behaviour.

This model was made to run with the code RELAP5 mod 3.2 β, a version adapted for the use of liquid lead-bismuth eutectic by *Ansaldo Nucleare* (Genova, Italy).

In the first version, the spallation loop was modelled separately from the machine to simulate its own working comportment. Later it will be included in the total machine model. The primary circuit of the loop with the two free surfaces, the mechanical pump and the heat exchanger (HX) were modelled following design requirements. A simple secondary HX circuit reproduces its initial and final conditions.

The working simulation of the loop is a challenging undertaking with RELAP. Space limitations in the high-performance subcritical core led to proton current densities exceeding 150 µA/cm^2, requiring the use of a windowless target. Therefore, the free surface above the target is in direct connection with the accelerator vacuum. The pressure control above it is designed not to exceed 10^{-3} mbar, that is 0.1 Pa, which is not usual with RELAP. The heat flux in the target must be evacuated very quickly; if this is not done a large amount of vapour is produced and swept into the other part of the loop. The position of the target free surface is given by the balance between the nozzle in- and outflow. The nominal flow is controlled with a classic centrifugal pump velocity, easily modelled in RELAP. But the control of the free surface stability is as for controlled by a magneto-hydrodynamic (MHD) pump. It is not a classic pump; she acts on the fluid velocity to slow down or to accelerate it by Laplace forces. There is no appropriate model in RELAP to simulate its work therefore tricks were tested to respect at the best its behaviour. This paper contains the description of the design loop and the RELAP model, the requirements, how some problems were solved or not and which adaptations were made. Results in nominal conditions and a transient with loss of flow are described.

Introduction

In the scope of the MYRRHA safety analysis, a RELAP model of the spallation loop was built to simulate its behaviour under nominal and transient conditions.

Version 3.2 of RELAP5, which normally deals with water as coolant, was modified by *Ansaldo Nucleare* (Genova, Italy) for Pb-Bi applications.

In the first version, the spallation loop was modelled stand-alone to simulate its own behaviour and to test the target model. Later it will be included in the whole machine model.

There are some peculiarities in this RELAP application. Since there is no window to isolate the loop from the accelerator system, the free surface above the target is in direct connection with the accelerator vacuum. The pressure control above the free surface is designed not to exceed 10^{-3} mbar kept by an integrated vacuum system; this is not usual in the RELAP applications.

The heat flux in the target is very high; a heat rate of 1.43 MW is deposited in the LBE spallation target. The target temperature must not exceed 450°C for multiple reasons. The heat is evacuated by a heat exchanger placed in the loop downstream from the target.

The position of the target free surface is given by the balance between the nozzle in- and outflows, in relationship with the nominal flow of the loop. It is fit by means of the mechanical pump velocity around the required design value of 10 l/s. The level must be kept within a margin of a few millimetres under all circumstances. For this reason, the little level variations are reduced by the action of a magneto-hydrodynamic pump (MHD). In briefly, the MHD pump accelerates or slows down the fluid. The same effect is obtained by increasing or decreasing the pressure drop in the loop. In RELAP, the possibility to model a negative or positive DP is normally done by means of a pump. Different models exist to simulate centrifugal pumps in their various functioning modes with automatic or manual velocity control. However there is no model for MHD pumps. Several tricks were tested to approach the real working conditions.

The circuit of the spallation loop, the mechanical pump and the heat exchanger (HX) were modelled according to the design requirements. A simple circuit reproduces the initial and final conditions at the secondary side of the HX.

This paper describes the spallation loop design, its corresponding RELAP model and the requirements. It explains how some problems were solved and which adaptations were accordingly made. Unresolved problems are also mentioned. Results under nominal and transient (loss of flow) conditions are provided.

Spallation loop description

A complete description of the spallation loop is given in Ref. [1]. The different parts of the spallation unit, taken off outside the reactor, are given in Figure 1. In operation, the annular target is placed on the subcritical core axis; the LBE circulation is maintained by a mechanical pump and the heat is removed by a heat exchanger cooled by the lower plenum of the reactor vessel.

One of the main tasks of the spallation loop is to circulate the LBE to adequately remove the heat that is generated by the spallation reaction in the target. The LBE in the main reservoir of the spallation loop (free surface 1) flows almost solely by gravity into the target zone where it is heated by the proton

Figure 1. Spallation unit

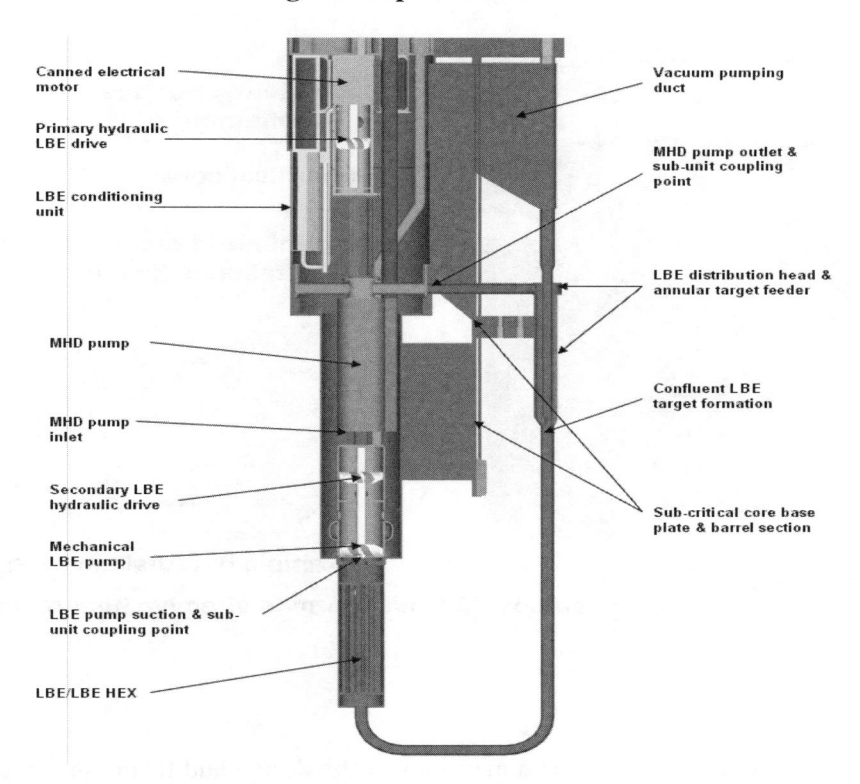

beam. The heat is removed by heat exchanger that uses the main vessel coolant LBE on its secondary side. A mechanical circulation pump pushes the LBE back into the reservoir of the spallation loop. An additional magneto-hydrodynamic pump (MHD) in the inlet branch of the target allows fine-tuning of the height of the target's free surface (free surface 2) by regulating the inlet flow rate. There are some particular points:

- The beam line does not contain any window and needs to be under vacuum; the spallation unit is connected to the central beam line via a duct and is being pumped by an integrated vacuum system to a pressure of less than 10^{-3} mbar.

- The position of the target free surface is given by the balance between the nozzle in- and outflows. It needs to be measured and kept within a margin of a few millimetres under all circumstances. The level is particularly difficult to maintain under the thermo-hydraulic dynamics of beam trips and during the start-up and shutdown procedures. For that reason a second free surface in the spallation loop main unit is maintained at about 2.3 m above the target free surface

- A mechanical pump lifts the LBE that has passed the heat exchanger to the level of the second free surface. From this second surface, the LBE flows through the go-line to the target nozzle by gravity although some small bias may be provided by an MHD pump. The three feeders are drag limited, exceeding the minimum drag of 1 bar/m necessary to compensate for the hydrostatic pressure, to prevent the spallation LBE to tear off and reach the target nozzle in free fall.

- The target level surface is sensed by a LIDAR (LIght Detection And Ranging) in line with the proton beam. Its signal is fed back to the MHD pump controller.

Figure 2. Schematics of the axi-symmetric windowless target configuration

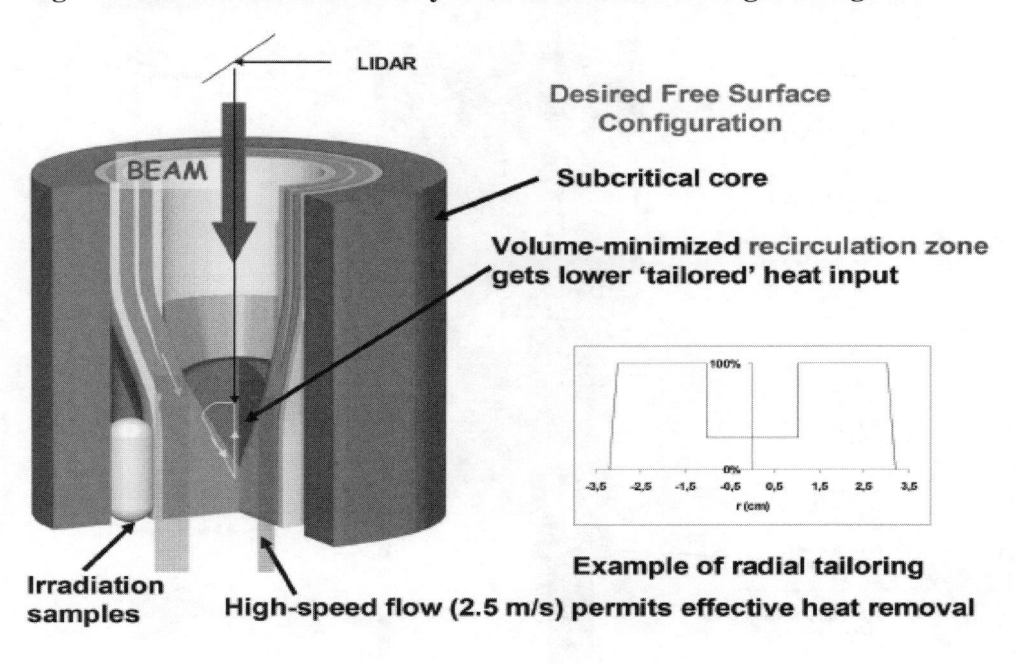

The RELAP model

The model was built according to the dimensions of the design and taking into account the level difference between the components and the reference level in the MYRRHA machine. The main vessel is represented by a vertical pipe whose large top volume contains the first LBE free surface. The LBE feeds the target by gravity via horizontal and vertical pipes. The last vertical pipe represents the annular feeder gap. The connection between the target and the loop is modelled by a junction with an adapted pressure drop representing the turbulence under the target free surface. The downstream flow is cooled in the heat exchangers and is ensured by means of the mechanical pump after the HX. The HX secondary circuit is simulated by a pipe with imposed initial conditions [Figures 3(a) & 3(b)] and a surface of heat exchange in accordance with the design.

The vacuum pumping duct is represented by a pipe with the second free surface, which can move upwards or downwards in function of the flow balance.

The conditions above the free surfaces were defined by means of a time-dependant volume (TDV). In a first approach, a pressure of 1 bar was imposed. In principle, provided that the pressure is the same above the two free surfaces, the loop behaviour should not be influenced by the absolute value of the pressure. In "shutdown" conditions, i.e. when the fluid is at rest and no power is deposited in the target, the two free surfaces are at the same level.

Flux distribution

The linear power density in the target below the free surface is axially distributed according to the curve in Figure 4. The maximum value is reached at a depth of 13 cm. The heat is uniformly deposited within the first 10 cm at a rate of 1 MW. The power delivered in the last 3 cm is 430 kW.

Figure 3. RELAP model and spallation loop sketch relation

Spallation loop

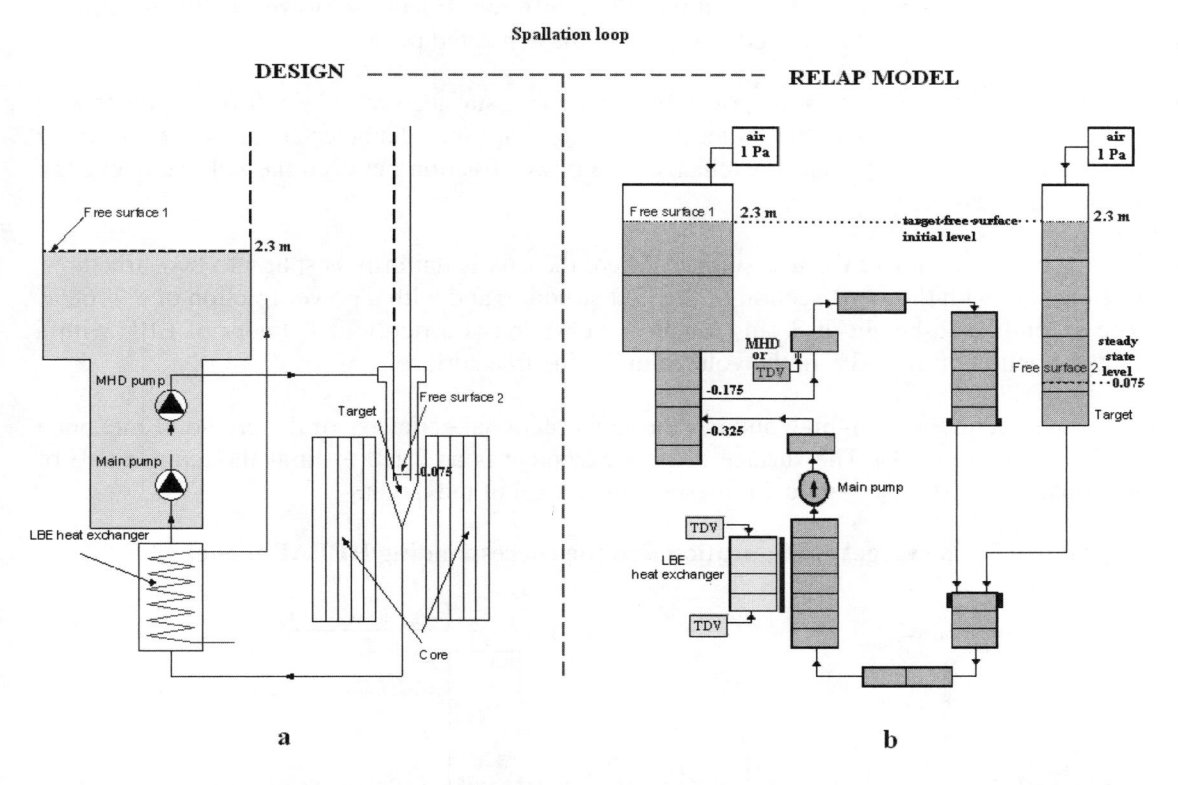

DESIGN — RELAP MODEL

a   b

Figure 4. Distribution of the linear density in function of the depth inside the target

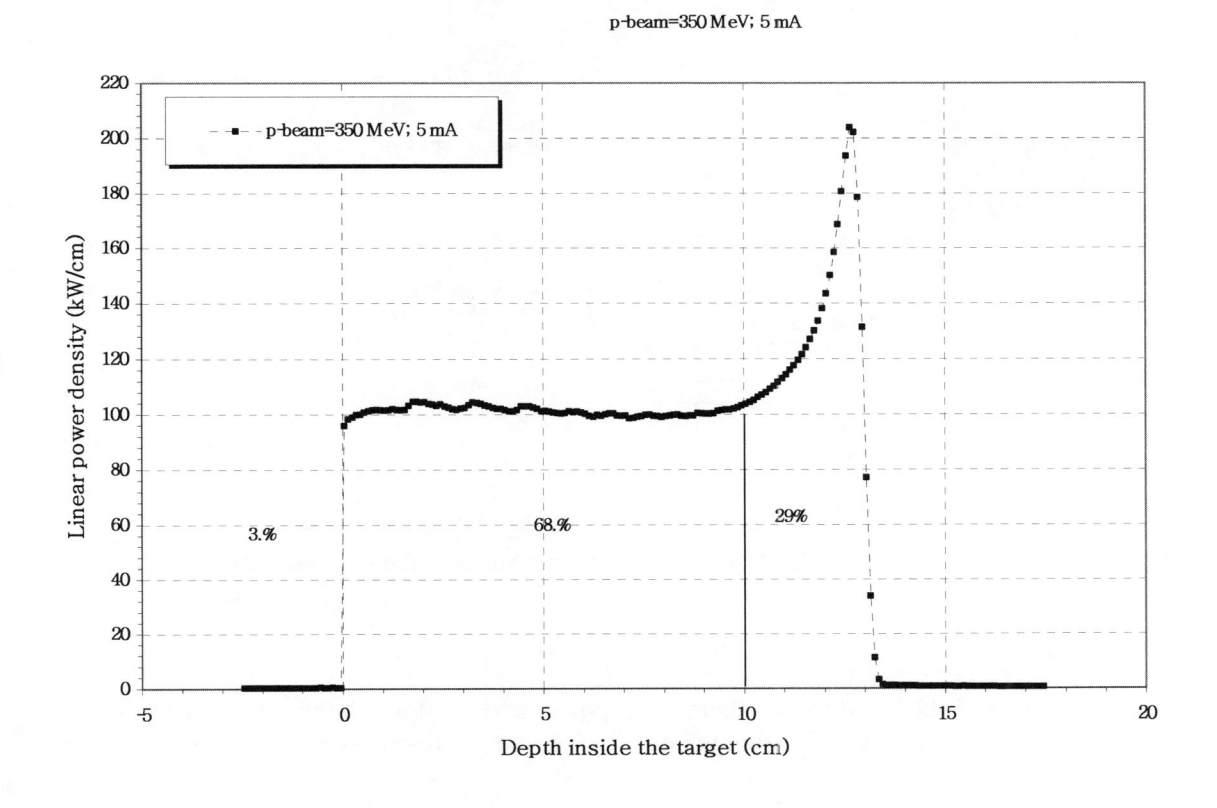

p-beam=350 MeV; 5 mA

In the RELAP model, the flux is distributed in four fictitious structures. In principle three should be sufficient because the power is subdivided into three different fractions. However, for the sake of the RELAP model, four structures are used to represent the deposited power:

- The first structure represents the power above the free surface level (3%). It is a cylinder with an inner diameter equal to the target diameter and an outer diameter equal to the gap inner diameter. It is 3 cm high and it exchanges the power fraction between the volume above the free surface and the gap.

- The 10-cm region under the free surface, where the flux is uniform, is split into two structures, one of 7 cm with the same section of the first structure and with a power fraction of 0.7*68%. The second has a height of 3 cm (10 cm – 7 cm) and is a fictitious cylinder of LBE with a power fraction of 0.3*68% in the volume under the free surface.

- The last structure is 3 cm high and has the same sectional geometry of the previous one, but a power fraction of 29%. The surface of heat exchange is adjusted to limit the temperature of this surface (0.116 m^2). Figure 5 represents the model of the target.

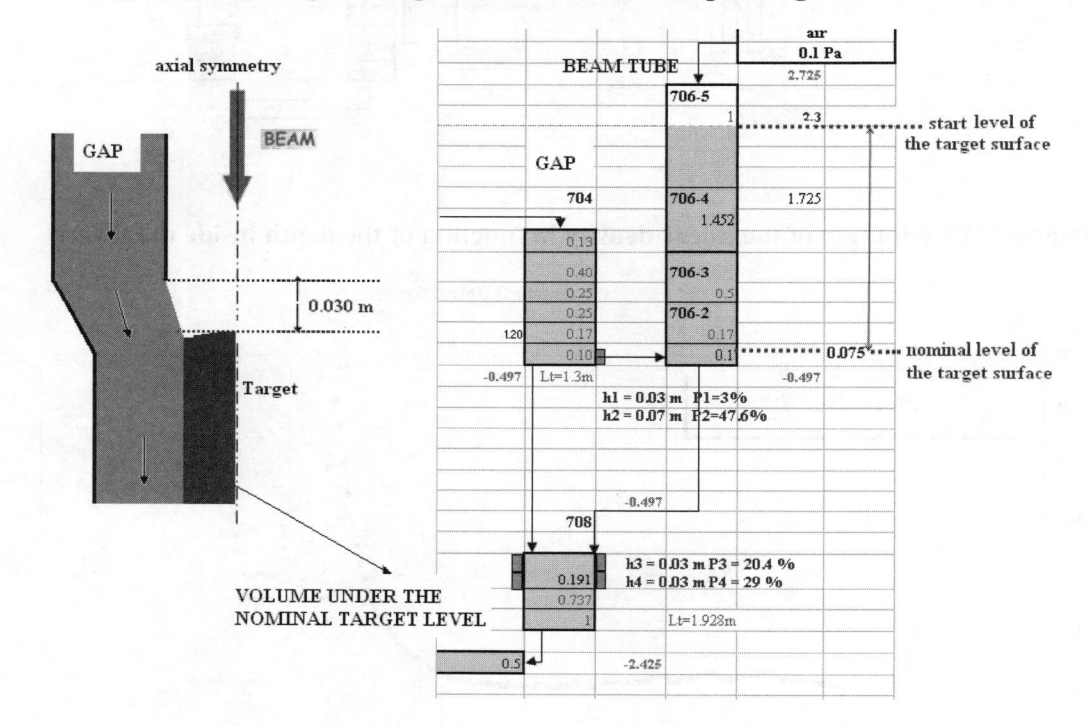

Figure 5. The target configuration and the corresponding RELAP model

Free surface level control

There are two possibilities to control the level of the free surface: the cold and warm controls.

Control without power

At the beginning of the steady state, there is no power and no flow; the two free surfaces are at the same level. This level is set by adjusting the quality (i.e. the gas fraction) in the top volumes of the feeder tank and the beam ("700-1 and 706-5). The initial pressure above the free surface is set at 1 bar (0.1 MPa) and slowly decreased by the TDV above the free surface conditions up to the nominal pressure of 0.1 Pa. When the equilibrium is reached, the pump is started. The flow in the loop increases with the pump velocity. Consequently the level of the target free surface decreases to a value close to its nominal value (0.075 m).

A time-dependent junction located at the outlet of the feeder tank adds or removes a small amount of LBE to get the exact nominal level required in the target pipe.

Control with the power

In practice the target free surface level is measured by a LIDAR (laser light) acting on a MHD pump to give a fine tuning of the gravity-fed inflow in acceleration/deceleration operation.

The modelling of the MHD pump action is not evident in RELAP. It needs to reproduce a negative or positive pressure drop in function of the target level.

Some possibilities were analysed:

- The easiest way consists of keeping the time-dependent junction localised at the loop inlet in order to inject or remove a small amount of LBE in function of the target level.

- Another possibility makes use of a servo valve normally half-open when the free surface is at its nominal position (0.075m); the valve closes or opens respectively when the free surface level is lower or higher than the nominal level. Figure 6 shows the calculations with Idel'cik data: one can have a positive DP-DP50% if the valve is close and a negative value if the valve opens, relatively to the reference opening area (50%) in nominal conditions. This option is not recommended.

Figure 6. Evolution of the pressure drop in a valve, in function of its open area

valve Idel'cik		v=2m/s	
Fh/F0	kf	DP	DP-DP50%
0	∞		
0.16	97.8	2.05E+06	1.96E+06
0.25	35	7.34E+05	6.38E+05
0.38	10	2.10E+05	1.13E+05
0.5	4.6	9.65E+04	
0.6	2.06	4.32E+04	-5.33E+04
0.7	0.98	2.06E+04	-7.59E+04
0.8	0.44	9.23E+03	-8.73E+04
0.9	0.17	3.57E+03	-9.29E+04
0.96	0.06	1.26E+03	-9.52E+04
1	0	0.00E+00	-9.65E+04

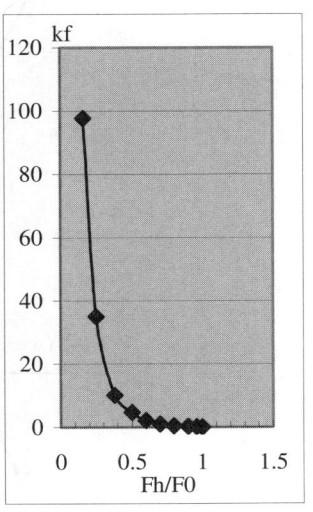

- The best way to introduce a positive or negative pressure drop is a pump. In the RELAP code a pump is modelled by the homologous curves which give the pump head in all working conditions (Figures 7 and 8). RELAP includes two centrifugal pump models: a Westinghouse pump and a Bingham pump (homologous curves). However they are not appropriate to model the MHD pump behaviour.

Nevertheless it is possible to introduce no-dimensional pump characteristics in the input deck. The fictitious homologous curves were calculated taking into account a constant velocity. The curves so obtained are unusual (Figures 9 and 10) and there is some doubt on their validity.

Figure 7. Typical centrifugal pump characteristic curve

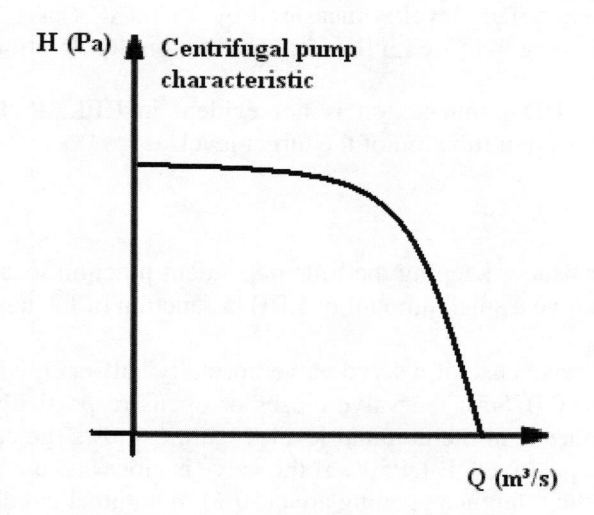

Figure 8. Typical centrifugal pump homologous curves

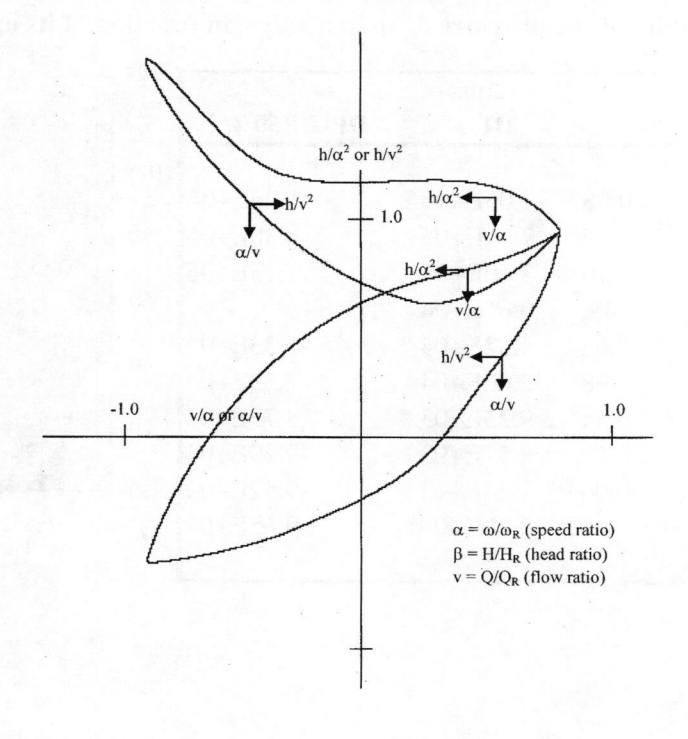

$\alpha = \omega/\omega_R$ (speed ratio)
$\beta = H/H_R$ (head ratio)
$v = Q/Q_R$ (flow ratio)

Figure 9. MHD pump characteristic curve

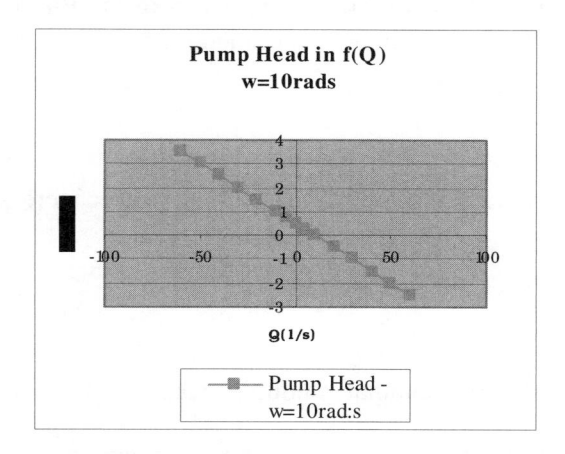

Figure 10. Calculated MHD pump homologous curves

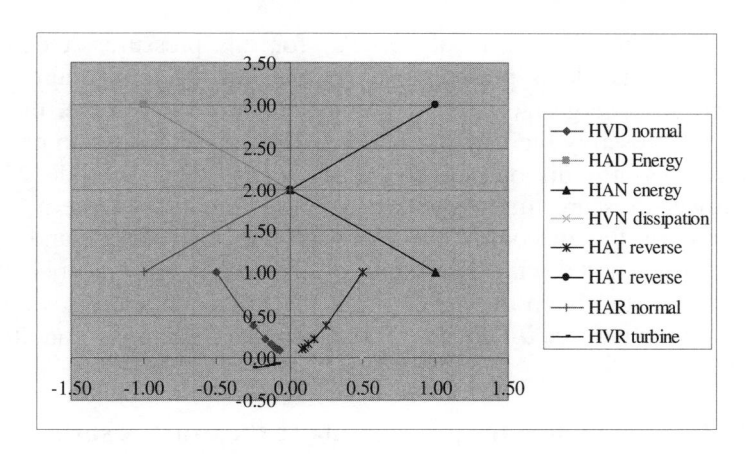

Regulation

The LBE free surface level is measured by a LIDAR, whose laser light is reflected from the target free surface. Its output controls the MHD pump via a PID regulator (Figure 11).

Figure 11. PID schema

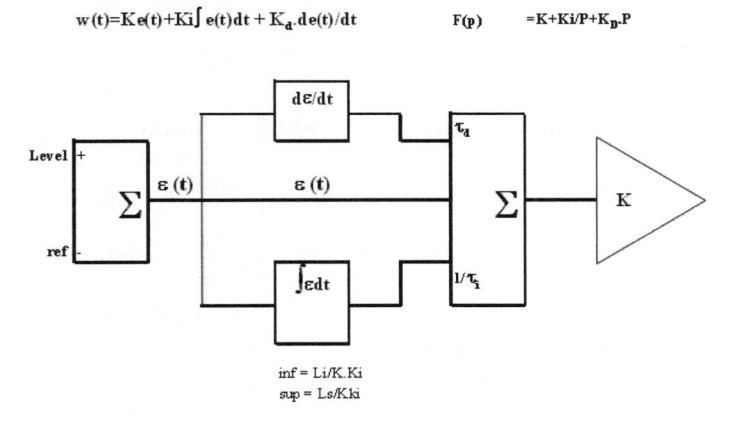

$w(t) = K e(t) + K_i \int e(t)dt + K_d \cdot de(t)/dt \qquad F(p) = K + K_i/P + K_D \cdot P$

inf = Li/K.Ki
sup = Ls/K.ki

Using this PID it is then possible to regulate the control of the target free surface level via the time-dependent junction used in steady state. Some parameters of the PID function, such as the time constants τd and τI and the gain K, have yet to be determined.

First results

In a first approach, a steady state and a transient without power were performed to analyse the physical model validity.

Steady state without power

The first step consisted to reach the nominal conditions when no power is deposited in the target.

The main challenge was to obtain the lowest possible non-condensable pressure that the code RELAP accepts, above the free surfaces.

The design imposes a value of 10^{-3} mbar (0.1 Pa) for this pressure. According to the code manual [3], when a non-condensable is present in a volume, all the temperatures are set to equal values and depend on the gas energy. As a fluid is injected into the volume, the temperature, the partial pressure of steam and possibly the vapour energy will abruptly change to new values based on the calculated conditions. Too-large discontinuities are not accepted by the code. Moreover the code manual recommends a lower pressure limit equal to 1.0 Pa. The nature of the non-condensable gas also influences the degree of the discontinuities between the gas energy and the LBE energy. Therefore the steady state was started with a pressure of air equal to an atmosphere: 105 Pa and the pressure was slowly decreased to obtain the lowest limit that the code accepts. Figure 12 shows the evolution of the pressure which reaches 0.1 Pa after 3 000 sec. An imperative condition is the use of a very small time step (0.001 s.).

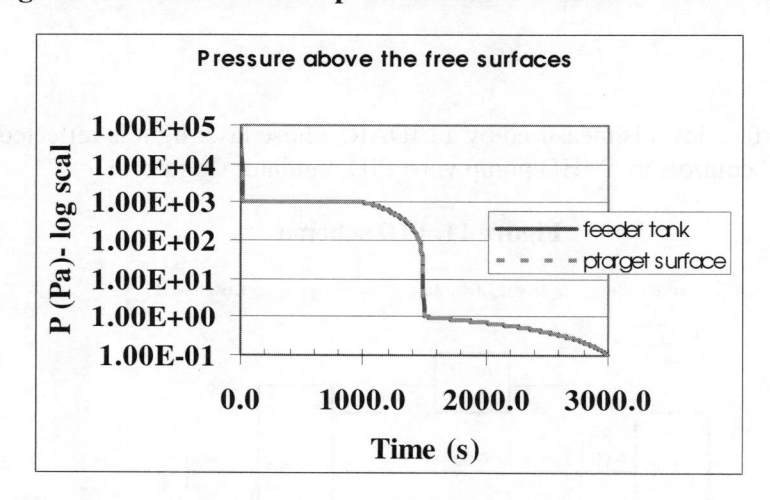

Figure 12. Evolution of the pressure above the two free surfaces

While the pressure is decreasing, the mechanical pump is started at about 400 s.

The fluid velocity increases and the LBE is seeped under the target level. Consequently the target free surface level decreases. When the nominal flow is obtained, the absolute target level must be equal to 0.075 m against the core middle plane of MYRRHA.

Figure 13 shows the evolution of the flow rate in the loop (first axis) and in the same time the lowering of the target free surface level (secondary axis). The stabilisation of the flow and the level are obtained after 1 000 s.

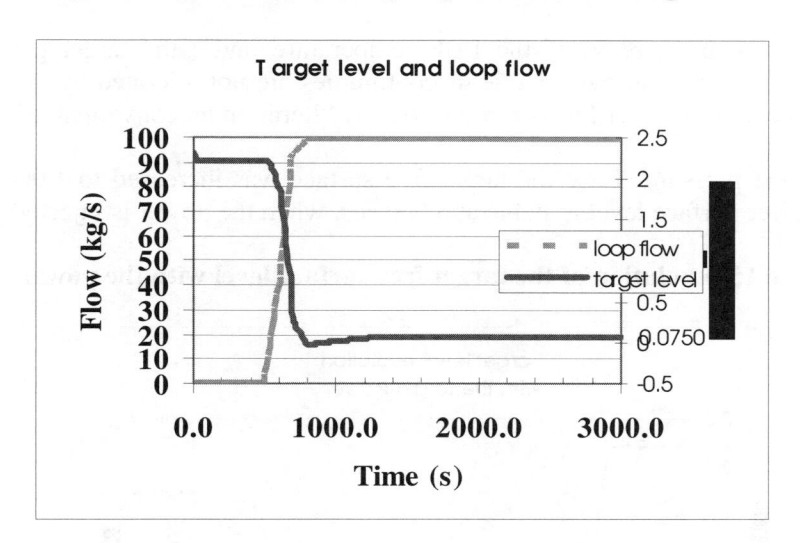

Figure 13. Evolution of the pump flow and the target level

Transient without power

This transient was simulated to study the evolution of the target free surface level when the pump is stopped. Normally the level must reach its starting position.

The pump was stopped in 10 seconds at time 2 400 s. We can see in Figure 14 that the level increases immediately after the pump stop and reaches the initial level at about 2.3 m. Nevertheless some fluctuations appear during ~ 150 s, but they are very small, i.e. some millimetres. The regulation with a PID regulator will certainly improve the stability.

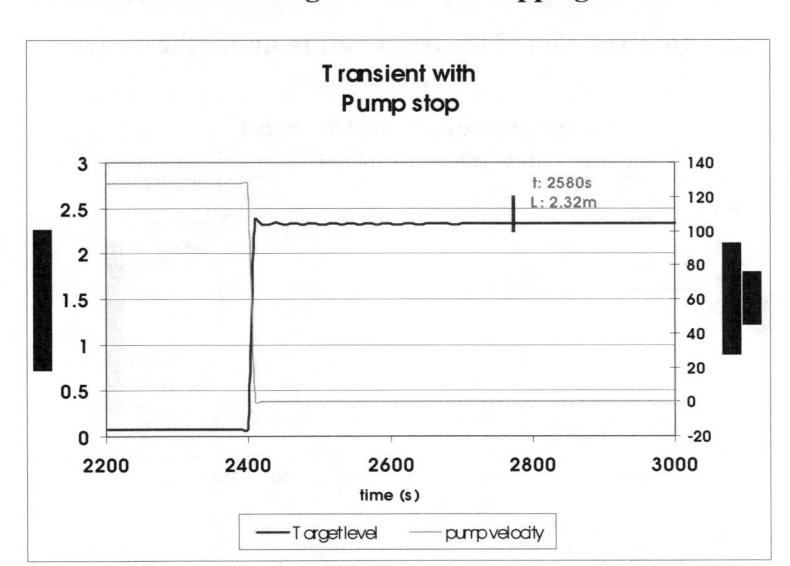

Figure 14. Evolution of the target level after stopping the mechanical pump

Steady state with power

When the steady state without power is reached, the flow rate, the target free surface level and the pressure are in agreement with to the design specifications. A power of 1.45 MW is then introduced into the structures around the target in 1 second.

The power effect is an increase of the LBE temperature under the target generating energy discontinuities at the target free surface. These discontinuities are not tolerated by the code when the pressure above the free surface is 0.1 Pa. The code stops by "Iteration no convergence".

For that reason the pressure above the target free surface was increased to 1 000 Pa. Figure 15 shows that the target free surface level is stable at 3 000 sec. when the power is injected.

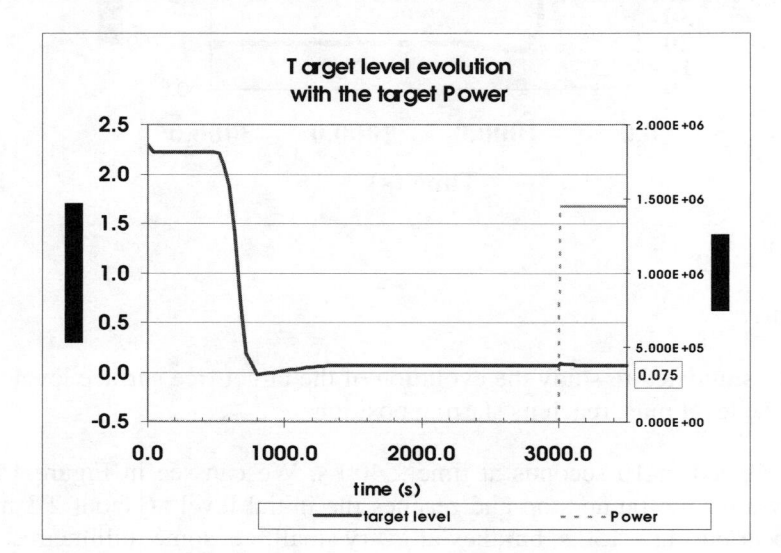

Figure 15. Evolution of the target free surface level with the power

Figure 16 shows the LBE target temperature evolution with the power. When the power is injected, the temperature increases to 689 K.

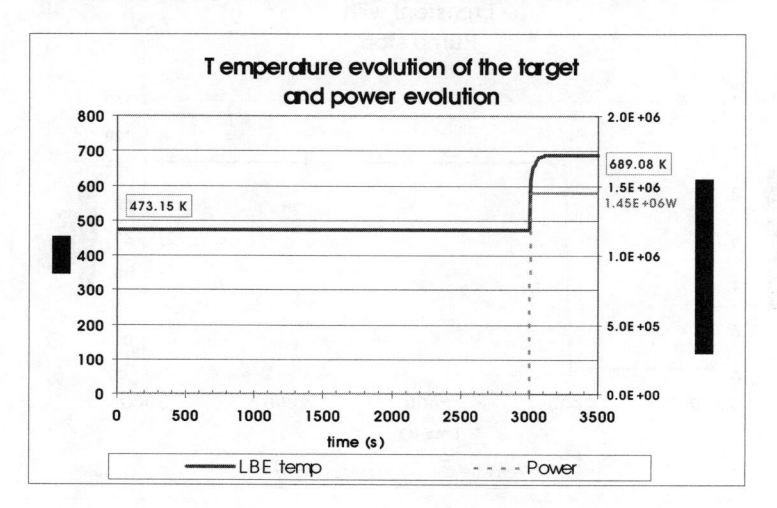

Figure 16. Evolution of the temperature under the target

The temperature increase is 216 K, which is too high with respect to the design value (100 K), but the flow rate in the loop is a little too low (99 kg/s instead of 104kg/s) and the surface of heat exchange between the LBE of the re-circulation zone under the free surface and the LBE coming from the gap is not yet evaluated. By improving and refining the model, it would be possible to substantially decrease the discrepancies with the design values.

Conclusion

The work is still in progress. Some problems were solved: pressure above the target free surface, evolution of the free surface level after a pump stop and stabilisation of the level when power is injected. When no power is released in the target, a stable steady state is obtained with a pressure as low as 0.1 Pa above the free surfaces. However, as soon as power is added, large energy discontinuities appear that are not tolerated by the code. Setting a pressure of 1 000 Pa above the free surfaces constitutes a good compromise between the model and the design requirements and allows reaching a stable steady state.

With the present model, it is not possible to simulate all the accidental situations, especially those involving small changes of the target free surface level.

At the present stage, a pump stopping or a power shutdown or power increase can be simulated to give useful indications for example on the free surface target level behaviour and on the temperature variations.

Improvement of the heat transfer between the re-circulation zone under the target free surface and the loop will be performed.

The control of the target free surface level is acceptable with a time-dependent junction which adds or removes a small quantity of LBE to adjust the free surface level in normal operation. With the action of a PID, stability will be improved. The possibility to model the MHD pump is still under study.

REFERENCES

[1] Dierckx, Marc, *1"100_PS_KVT_MD_StRep_Draft-2-spallation-target_1.0.0.doc*, SCK•CEN, Mol, Belgium.

[2] Idel'cik, I.E., "Memento des Pertes de Charge" (1999).

[3] *RELAP5/MOD3 Code Manual*, Volume II: "User's Guide and Input Requirements", Idaho National Engineering Laboratory, Idaho Falls.

POLONIUM BEHAVIOR IN EUTECTIC LEAD-BISMUTH ALLOY

J. Neuhausen, F. von Rohr, S. Horn, S. Lüthi, D. Schumann
Paul Scherrer Institut, Villigen, Switzerland

Abstract

It is planned to use liquid eutectic lead-bismuth alloy (LBE) as the target material in future high-power spallation targets for neutron production. These types of targets will be an important component of accelerator-driven systems (ADS) for the transmutation of long-lived nuclear waste. One of the most significant problems when using LBE as a target material is the production of significant amounts of the highly radiotoxic ^{210}Po by neutron capture on ^{209}Bi. A thorough knowledge of its behaviour within the target system is a prerequisite for the safe operation of the target system and its disposal. The behaviour of polonium in solidified LBE may be of interest in the context of target handling after irradiation and final disposal. Knowledge about the release behaviour of polonium in vacuum is important for the design of windowless target systems. We report here preliminary results of our investigations on the behaviour of polonium in solidified LBE as well as its release from the eutectic under vacuum conditions.

Introduction

Liquid eutectic lead-bismuth alloy (LBE) is planned to be used as target material in future high-power spallation targets for neutron production. In 2006, PSI performed the first test experiment to demonstrate the feasibility of such a target system in the MW range, MEGAPIE. One of the most significant problems when using LBE as target material is the production of significant amounts of the highly radiotoxic ^{210}Po by neutron capture on ^{209}Bi. According to Monte Carlo simulations ^{210}Po is expected to be produced in gram amounts in such target systems [1]. In future accelerator-driven systems (ADS) for the transmutation of long-lived nuclear waste the amount of polonium produced will be even orders of magnitudes larger due to prolonged irradiation times, higher neutron fluxes and increased system size. Therefore, a thorough knowledge of the behaviour of polonium within the target system is a prerequisite for the safe operation of the target system and its disposal. Previous studies performed in our laboratory dealt with the evaporation of polonium from molten LBE [2] under ambient pressure, using an Ar/H$_2$ atmosphere. For high-power targets of ADS, the current window materials cannot withstand the high-energy deposition. Therefore, windowless target designs are considered. In such a system, the liquid target material will be exposed to the vacuum of the proton beam line. Vacuum conditions will have a drastic influence on the evaporation behaviour of volatile radionuclides. However, investigations on the release behaviour of volatiles from liquid metals under vacuum are scarce. In case of a liquid eutectic lead-bismuth alloy (LBE) target, as considered e.g. within the EUROTRANS project, ^{210}Po is the most relevant nuclide due to its high radiotoxicity. There are a few literature data on Po release under vacuum [3], but for a reliable assessment of Po release in such a system, more detailed studies are necessary.

Furthermore, the behaviour of polonium in solidified LBE may be of interest in the context of target handling after irradiation and final disposal. Literature indicates that the distribution of polonium in irradiated LBE samples becomes inhomogeneous after some time [4]. In this work, we study the distribution of ^{210}Po in LBE samples irradiated in SINQ at PSI using the NAA rabbit system. Additionally, we show how a homogeneous distribution of polonium in LBE can be achieved. The homogenised samples are then used to study polonium release in vacuum.

Experimental

Polonium containing LBE samples were produced by irradiation of pieces of LBE in SINQ at PSI using the NAA rabbit system. Typical samples had a mass of ≈8 g and were irradiated for 1 h without Cd shielding. The samples investigated here were previously used to investigate the thermal release of Hg and Tl nuclides and afterwards stored in the laboratory for about three years at ambient temperature in closed PE bags. Those samples that were not heated above 600°C in the release studies mentioned above still contained enough ^{210}Po to yield acceptable count rates in α spectroscopy. Therefore, they were considered to be useful to study the thermal release behaviour of Po from LBE under vacuum. However, for release experiments, it had to be ensured that Po be homogeneously distributed within the samples. Therefore, the distribution of ^{210}Po within these samples was studied.

The distribution of ^{210}Po in the samples was determined by successively etching-off a certain amount of the sample surface and analysing its ^{210}Po content. For this purpose, a sample of typically ≈1 g LBE containing carrier-free amounts of ^{210}Po was immersed in 7M HNO$_3$ for a certain time. Afterwards, the weight loss was measured and the ^{210}Po concentration of the dissolved LBE was analysed using Liquid Scintillation Counting (LSC) with the following procedure: the leaching solution was evaporated to dryness and dissolved in 5 ml 1M HNO$_3$. An amount of 1.25 ml of this solution was mixed with 14 ml Aquasafe 300 and measured in the LSC spectrometer (Packard Tri-Carb 2200CA) for 5 h. Standard solutions containing varying known amounts of ^{210}Po and LBE matrix were used for calibration.

For the use of such samples for Po release studies a method for homogenisation of the Po distribution was devised. Homogeneity can be achieved by heating the LBE samples at 500°C in a H$_2$ atmosphere under mild agitation. Samples treated in this manner no longer show a surface enrichment of ^{210}Po immediately after solidification.

Release experiments were performed in the apparatus depicted in Figure 1. Before the experiment, a small part of the sample was cut off. The remaining part was put in a quartz boat lined with a quartz fleece, which was placed inside the quartz reaction tube. Additionally, a gold foil was placed in the tube to absorb the evaporated Po. The tube was evacuated using a turbo molecular pump. Typical pressures achieved were in the order of 1×10^{-5} mbar. A furnace was heated to the desired temperature. The reaction tube was then inserted in the furnace and kept there for 1 h. Afterwards, the tube was removed from the furnace, cooled to room temperature and vented. A small part of the LBE sample was cut off. The parts that were cut off from the sample before and after heating were analysed by LSC to determine the release of Po.

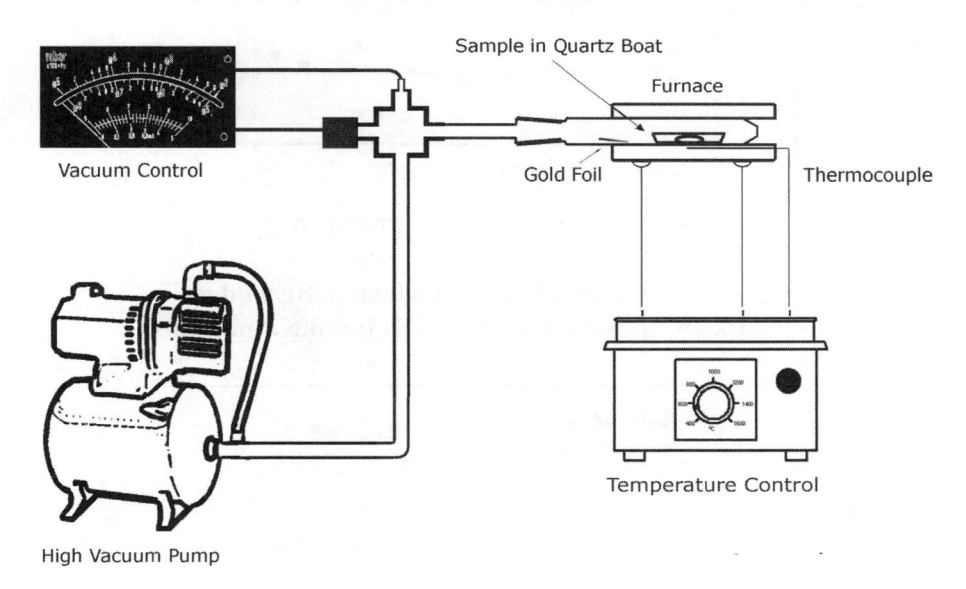

Figure 1. Set-up for release experiments in vacuum

Results

Figure 2 shows the variation of the ^{210}Po concentration in a typical sample as a function of mass remaining in the successive dissolution process. The diagram clearly shows that ^{210}Po is enriched in the surface region, while in the inner part of the sample its concentration remains more or less constant. The thickness of the enriched layer is estimated to be roughly in the order of 20 μm, based on the weight loss. The driving force for the segregation and its kinetics are a topic of further studies.

Figure 3 shows a comparison of the first results of release experiments performed in vacuum and those obtained in an Ar/7%-H$_2$ atmosphere under ambient pressure [2]. The fractional release of Po is plotted as a function of temperature. It is evident that the evaporation of Po is significantly enhanced under vacuum, as expected. Apart from the missing points in temperature, further release experiments are planned, including long-term studies at temperatures close to the onset of Po release and investigations under different pressures, with the final goal of quantification of Po release rate. Already from the first results it can be concluded that in a windowless spallation target suitable measures have to be taken to catch the highly radiotoxic polonium isotopes.

Figure 2. Specific activity of LBE fractions successively dissolved from a ^{210}Po-containing LBE sample as a function of mass

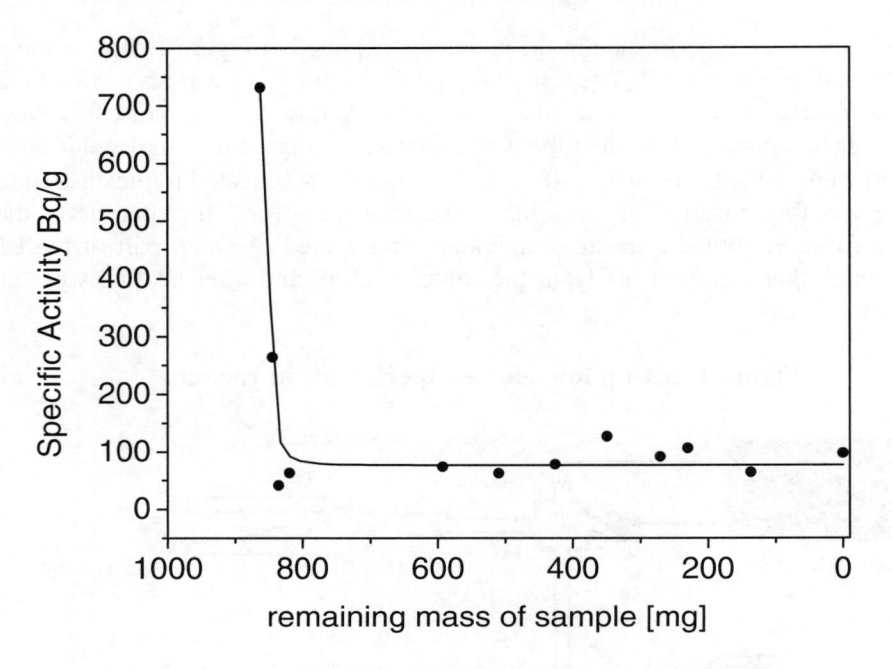

Figure 3. Fractional release of Po from LBE under vacuum ($\approx 1 \times 10^{-5}$ mbar) and Ar/7%-H$_2$ (ambient pressure)

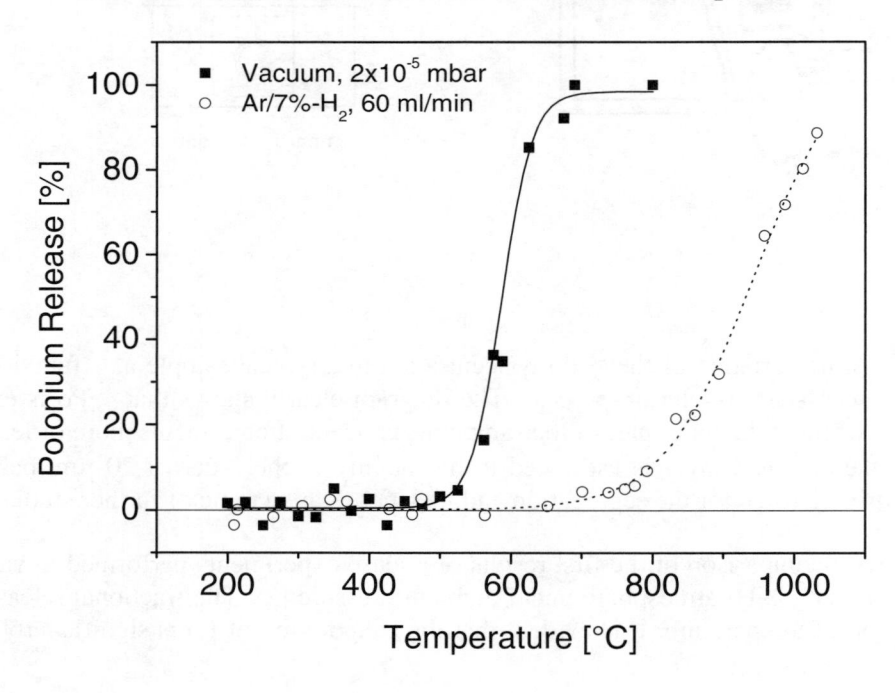

REFERENCES

[1] Zanini, L., *Summary Report for MEGAPIE R&D Task Group X9: Neutronic and Nuclear Assessment*, PSI Report Nr. 05-12, Paul Scherrer Institut, Switzerland (2005).

[2] Neuhausen, J., U. Köster, B. Eichler, *Radiochim. Acta*, 92, 917 (2004).

[3] Tupper, R.B., B. Minushkin, F.E. Peters, Z.L. Kardos, *Proc. Int. Conf. on Fast Reactors and Related Fuel Cycles*, Kyoto, Japan (1991).

[4] Miura, T., T. Obara, H. Sekimoto, *Appl. Radiat. Isot.*, 61, 1307 (2004).

A BRIEF REVIEW OF THERMODYNAMIC PROPERTIES AND EQUATION OF STATE OF HEAVY LIQUID METALS

Vitaly Sobolev
Belgian Nuclear Research Centre SCK•CEN

Abstract

Heavy liquid metals (HLM), such as mercury (Hg), lead (Pb), bismuth (Bi) and lead-bismuth eutectic (Pb-Bi), have been studied in several countries as potential materials for intensive spallation neutron sources, due to a high neutron yield per incident proton and heat removal capacity. However, experimental data on their thermal properties in the temperature range of interest are rare. This renders rather difficult the design calculations and the prognosis of the normal and abnormal behaviour of the spallation targets. In this case, physically based correlations and an equation of state (EOS), developed on the basis of the fundamental thermodynamics and limited experimental data, can play a very important role. Recently a compilation with the recommendations for HLM properties was prepared by the WPFC (OECD) Expert Group on Lead-bismuth Eutectic Technology. A brief review of this compilation is proposed in the present article. A set of correlations for temperature dependence of the main thermodynamic properties of Hg, Bi, Pb and Pb-Bi(e) and EOS based on a modified Redlich-Kwong equation are presented.

Introduction

Spallation neutron sources are considered an important tool in many fields of research and for development of new techniques where neutrons are used. Since the mid-1990s, the interest and the active studies concerning heavy liquid metals (HLM) – Pb-Bi(e), Bi, Pb, and Hg – have been growing, due to the perspective of their application in the intensive spallation neutron sources and in fast reactors of the new generation (subcritical and critical). Heavy metals for spallation targets should preferably have the following properties: a high neutron yield per proton in unit volume, radiation stability, low chemical activity, low melting temperature, high boiling temperature, low saturated vapour pressure, large heat capacity, small viscosity and large thermal expansion. A detailed knowledge of the HLM thermodynamic properties is a necessary step for design of a spallation source and modelling of its behaviour under representative normal and abnormal conditions. Main properties of the HLM were mainly measured at the normal atmospheric pressure and at relatively low temperatures in many laboratories. The available information has recently been collected by the WPFC (OECD) Expert Group on Lead-bismuth Eutectic Technology in the Handbook on HLMC Properties; a Version 0 will be issued in 2007 [1]. For prognosis of the missing thermodynamic properties at high temperatures and pressures, work is under way to develop a relevant equation of state (EOS) based on the available experimental data and proved physical models.

This article gives a brief review of the compilation results on the main thermodynamic properties of the molten Hg, Pb, Bi and Pb-Bi(e) (i.e. density, thermal expansion, heat capacity, compressibility, surface tension and saturated vapour pressure) and on the development of EOS for these HLM.

Temperature range

A temperature range of the normal operation of an HLM is determined by its melting and boiling temperatures. Under accidental conditions it can even reach the critical point region.

The melting temperatures of the chemically pure Hg, Bi, Pb and Pb-Bi(e) were measured with a high precision at normal atmospheric pressure. However, for the technically pure metals, the uncertainty can be a few tenths of degree of Celsius [2]. The melting temperature increases when pressure increases (e.g. for Pb, the rate of this increase is 7.92×10^{-8} K Pa^{-1} in pressure range of 15-200 MPa [3]).

The boiling temperatures of these HLM were measured with a lower precision than their melting temperatures. The uncertainty ranges from 10 to 20°C for Pb, Bi, Pb-Bi(e) and less than 1°C for Hg. The boiling temperature increases (with the decreasing rate) when temperature increases. The most probable values of the normal melting and boiling temperatures together with the latent enthalpies of melting and boiling of Hg, Bi, Pb and Pb-Bi(e), recommended in [2,3,5], and the operation temperature ranges of these HLM are presented in Table 1.

Table 1. Characteristic temperatures and temperature ranges of Hg, Bi, Pb and Pb-Bi(e)

	T_{M0} (K)	ΔH_M (kJ·mol^{-1})	T_{B0} (K)	ΔH_B (kJ·mol^{-1})	$T_{B0} - T_{M0}$ (K)
Hg	234.2±0.1	2.3±0.2	629.7±0.5	59.1±1	396
Bi	544.4±0.3	11.0±0.4	1 806±50	181±4	1 264
Pb	600.6±0.1	4.9±0.2	2 016±10	177.8±0.4	1 415
Pb-Bi(e)	397.7±0.6	8.1±0.1	1 943±20	178.0±0.4	1 845

A very large uncertainty still exists in the critical temperatures, pressures and densities of the HM of interest. These parameters are very important for the EOS development and the extension of the properties recommendations to higher temperatures and pressures. They were measured with satisfactory precision only for Hg. Many results were obtained for Pb but with a large dispersion. A few data sets are available for Bi. Only theoretical estimations were found in the open literature for the critical parameters of Pb-Bi(e). Mean values of the measured and calculated critical parameters collected in [6] are presented Table 2 below; a deviation from each mean value is given in the brackets for the calculated parameters.

Table 2. Critical parameters of Hg, Bi, Pb and Pb-Bi(e)

		T_c (K)	p_c (MPa)	ρ_c (kg m^{-3})	$Z_c \equiv \dfrac{p_c \cdot M}{RT_c \cdot \rho_c}$
Hg	Measured	1 747±50	162±5	5357±700	0.419±0.040
	Calculated	1 656 (7%)	143(10%)	4 885 (4%)	0.426 (21%)
Pb	Measured	5 400±400	250±30	3 200±300	0.361±0.050
	Calculated	4 808 (14%)	118 (36%)	2 468 (19%)	0.247 (59%)
Bi	Measured	3 780	132	3 010	0.292
	Calculated	4 457 (15%)	119 (25%)	2688 (34%)	0.249 (84%)
Pb-Bi(e)	Measured	–	–	–	–
	Calculated	4 830 (?)	166 (47%?)	2 170 (??)	0.396 (>50%?)

Properties and recommendations

Almost all main thermodynamic properties of the HLM (such as density, thermal expansion, heat capacity, enthalpy, surface tension, sound velocity and compressibility) are measured with a satisfactory precision in the region of melting temperature. An exception is the saturated vapour pressure, which is well measured for Hg, but not so well for other three heavy liquid metals: Pb, Bi and, especially, for Pb-Bi(e) at low temperatures. The recommended values of all above mentioned parameters, calculated for the considered HLM at the normal atmospheric pressure and at the melting temperatures using the correlations given in [2,7] and hereafter, are presented in Table 3.

Table 3. Recommended values of main thermodynamic parameters of
Hg, Bi, Pb and Pb-Bi(e) at normal atmospheric pressure and melting temperatures

Parameter	Hg	Pb	Bi	Pb-Bi (e)
M_a (g mol^{-1})	200.59	207.20	208.98	208.18
ρ (kg m^{-3})	13 691	10 673	10 062	10 570
α_{vol} (10^{-5} K^{-1})	17.7	22.7	24.1	23.0
u_{sound} (m s^{-1})	1 480	1 774	1 677	1 773
C_p (J mol^{-1} K^{-1})	28.3	30.6	30.5	31.1
σ (10^{-3} N m^{-1})	498.1	451.1	382.0	410.8
p_s (Pa)	3.4·10^{-4}	5.7·10^{-7}	1.4·10^{-8}	3·10^{-15}

Density and thermal expansion

The temperature and pressure dependence of density allow directly constructing a thermal EOS. At the normal atmospheric pressure, density was measured better than other properties for all four HLM. For Hg, Bi and Pb the experimental data are available from the normal melting point up to the normal boiling point. For Pb-Bi(e), the upper temperature limit is about 1 300 K.

With uncertainty of 0.7-0.8%, the temperature dependence of the density can be described by the linear correlation:

$$\rho(T, p_0) = \rho_{m0} - A_\rho (T - T_{m0}) \quad (1)$$

A volumetric coefficient of thermal expansion (CTE) is determined from the density as follows:

$$\alpha_p(T, p_0) \equiv -\frac{1}{\rho(T, p_0)}\left(\frac{\partial \rho(T, p_0)}{\partial T}\right)_p = \frac{A_\rho}{\rho_{m0} - A_\rho (T - T_{m0})} \quad (2)$$

Correlations (1) and (2), with the recommended in Table 3 density values at the normal melting temperature and with the coefficients A_ρ given in Table 4, are illustrated in Figure 2. Density decreases with temperature due increasing interatomic distances caused by thermal expansion. In its turn CTE increases with temperature due reduction of the interatomic forces with the distance.

Table 4. Coefficients of the correlations (1), (3), (6) and (7)

	$A_{\rho 0}$ (kg m^{-3} K^{-1})	A_{u0} (m s^{-1} K^{-1})	$A_{\sigma 0}$ (N m^{-1} K^{-1})	A_s (Pa)	ΔH_e (kJ mol^{-1})
Hg	2.428	0.200	0.200	$1.06 \cdot 10^{10}$	60.5
Bi	1.2208	0.797	0.080	$2.47 \cdot 10^{10}$	190.0
Pb	1.2795	0.277	0.113	$6.57 \cdot 10^{9}$	185.0
Pb-Bi(e)	1.3236	0.193	0.066	$1.11 \cdot 10^{10}$	187.5

Figure 2. Density (a) and CTE (b) of liquid Hg, Bi, Pb and Pb-Bi(e) ($p \approx p_{atm}$)

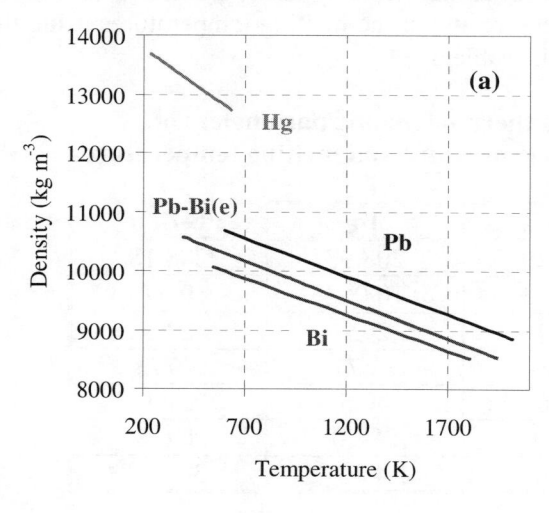

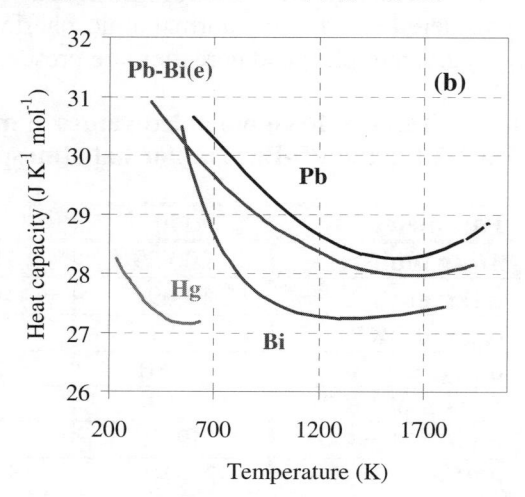

Compressibility and sound velocity

The velocity of propagation of sound waves in the HLM was measured at the normal atmospheric pressure and mainly close to the melting temperature. In this limited temperature region it decreases with temperature almost linear:

$$u_{sound}(T, p_0) = u_{m0} - A_u(T - T_{m0}) \tag{3}$$

The recommended values of the sound velocity at the normal melting temperature (u_{m0}) are given in Table 3 and the coefficients A_u are given in Table 4. The experimental results at higher temperatures were found only in a few publications: for Hg and Pb the measurements were performed in the temperature region from T_M to T_B; for Bi and Pb-Bi(e) up to 603 K [2]. At high temperatures the sound velocity in lead strongly deviates from the linear temperature dependence [see Figure 3(a)] and can be described with the following parabolic correlation [8]:

$$u_{sound(Pb)}(T, p_0) = 1951.75 - 0.3423 \cdot T + 7.635 \times 10^{-5} \cdot T^2 \tag{4}$$

The adiabatic bulk modulus B_S (or the adiabatic compressibility K_S) can be deduced from the results of measurement of the sound velocity and density:

$$B_S \equiv \frac{1}{K_S} = \frac{\rho}{(\partial \rho / \partial p)_S} = \rho \cdot u_{sound}^2 \tag{5}$$

The results of calculation of B_S are presented in Figure 3(b) below.

Figure 3. Ultrasound velocity (a) and adiabatic bulk modulus (b) of liquid Hg, Bi, Pb and Pb-Bi(e) ($p \approx p_{atm}$)

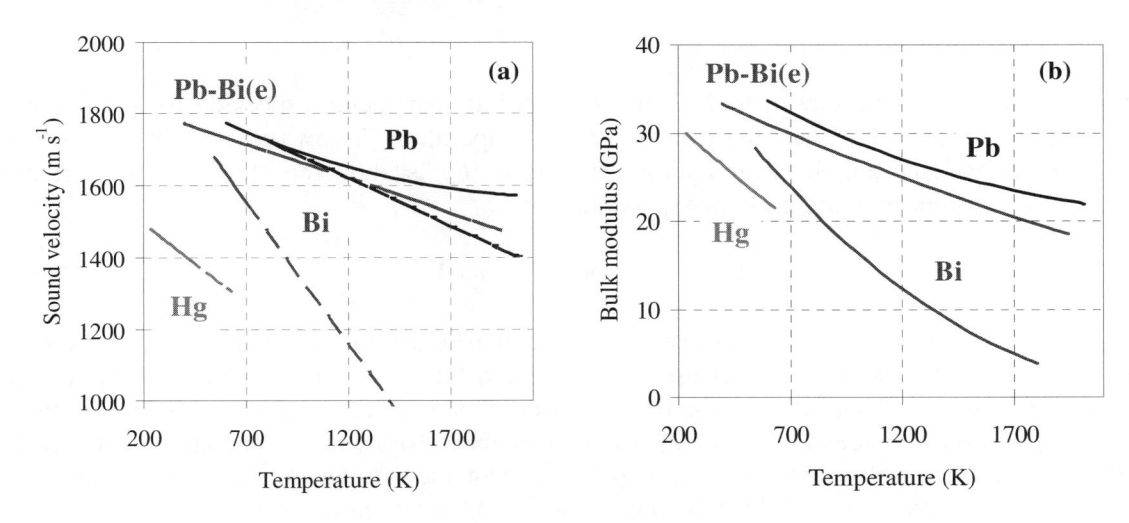

Surface tension

A surface tension is a measure of the cohesive energy of atoms and correlates with the latent heat of evaporation. The surface tension of normal liquids should decrease with temperature linearly and becomes zero at the critical temperature (Eötvös' law). So the linear correlation can be recommended:

$$\sigma(T, p_0) = \sigma_{m0} - A_\sigma(T - T_{m0}) \tag{6}$$

The recommended coefficients of correlation (6) are given in Tables 3 and 4, and the calculated surface tension is presented in Figure 4.

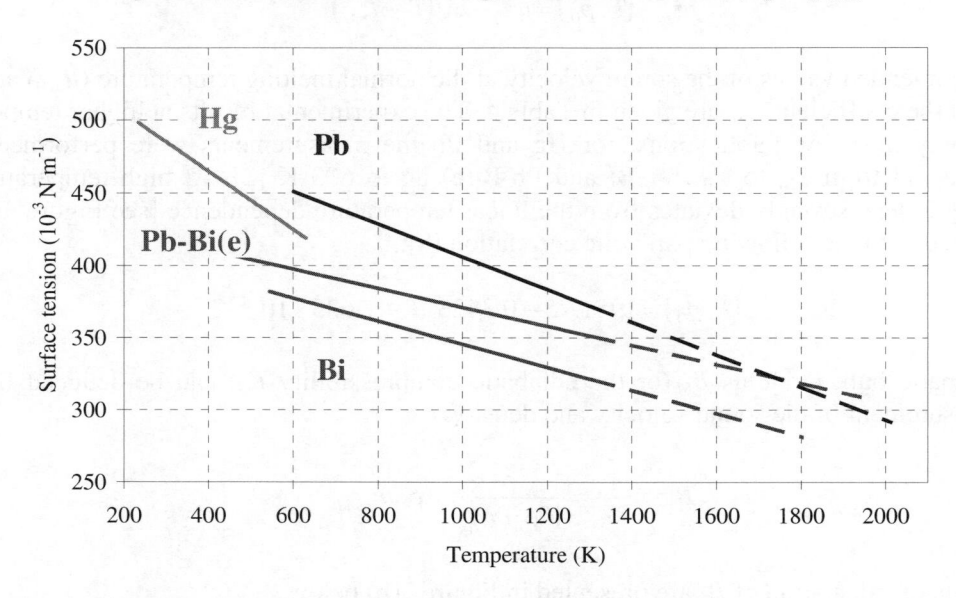

Figure 4. Surface tension of Hg, Bi, Pb and Pb-Bi(e) ($p \approx p_{atm}$)

A dispersion of ±(3-5)% exists between the surface tension values given by different sources at $T < 1\,300$ K. At higher temperatures, the scatter of the experimental results increases significantly.

Saturated vapour pressure

The saturated vapour pressure of HLM is a very important parameter, especially for windowless spallation targets. It is directly related to the latent heat of evaporation. At low pressures, where vapour behaves as a perfect gas and the evaporation enthalpy (ΔH_E) is about constant, the temperature dependence of the vapour pressure is about exponential:

$$p_s(T) = A_s \cdot \exp(-\Delta H_E / RT) \tag{7}$$

$R = 8.314$ J·mol^{-1}·K^{-1} is the universal gas constant. The coefficients of correlation (7) are given in Table 4. Figure 5, where the vapour pressure is plotted as a function of temperature, illustrates the correlation (7) for Hg, Bi, Pb and Pb-Bi(e) in the temperature range of T_{M0} to T_{B0}. It should be noticed that for Hg, Pb and Bi the uncertainty of this correlation is about 10% when temperature is 50-100°C higher than the normal melting temperature and lower than the normal boiling temperature. The uncertainty increases rapidly beyond this interval, at lower and higher temperatures.

For Pb-Bi(e) the uncertainty of this correlation is about 10% at temperatures higher than 1 500 K, however it becomes more than 50% when temperature decreases below 1 000 K. At lower temperatures the saturated pressure is very low to be measured correctly. The values of ΔH_E presented in Table 4 are in satisfactory agreement with ΔH_B presented in Table 3.

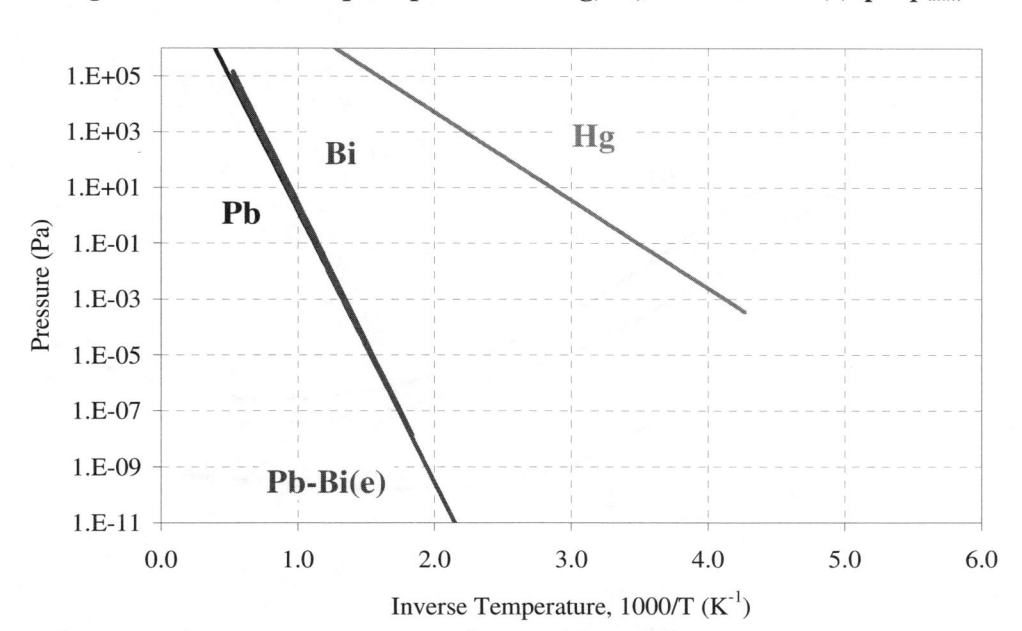

Figure 5. Saturated vapour pressure of Hg, Bi, Pb and Pb-Bi(e) ($p \leq p_{atm}$)

Heat capacity

It is a very difficult task to measure the heat capacity of HLM. The heat capacity is more or less well measured only in the region of the melting temperatures at the normal pressure. At higher temperatures, the available experimental data often give contradictory results. The theoretical calculation of heat capacity of liquids is restricted by the extreme complexity of the motion of atoms in liquids. They indicate that the heat capacity should decrease with temperature after melting, and then passes through a large plateau and begins increasing. Usually, an empirical correlation is used (deduced from the "standard" thermodynamic polynomial applied for enthalpy):

$$C_p(T, p_0) = \left(\frac{\partial H}{\partial T}\right)_{p0} = a + b \cdot T + c \cdot T^2 + d \cdot T^{-2} \tag{8}$$

The recommended values for the coefficients of correlation (8) are given in Table 5 and the calculated heat capacities are presented in Figure 6. Uncertainty is about 5-7% up to $T = 1\,200\text{-}1\,400$ K and increases at higher temperatures were no experimental data were found in open literature.

Table 5. Recommended coefficients of correlation (8)

	a ($J\,K^{-1}\,mol^{-1}$)	b ($J\,K^{-2}\,mol^{-1}$)	c ($J\,K^{-3}\,mol^{-1}$)	d ($J\,K^{2}\,mol^{-1}$)
Hg	30.38	$-1.146 \cdot 10^{-2}$	$1.016 \cdot 10^{-5}$	0
Bi	36.50	$-1.020 \cdot 10^{-2}$	$3.2 \cdot 10^{-6}$	$-3.158 \cdot 10^{5}$
Pb	24.70	$1.240 \cdot 10^{-3}$	0	$1.5011 \cdot 10^{6}$
Pb-Bi(e)	34.30	$-8.20 \cdot 10^{-3}$	$2.6 \cdot 10^{-6}$	$-9.5 \cdot 10^{4}$

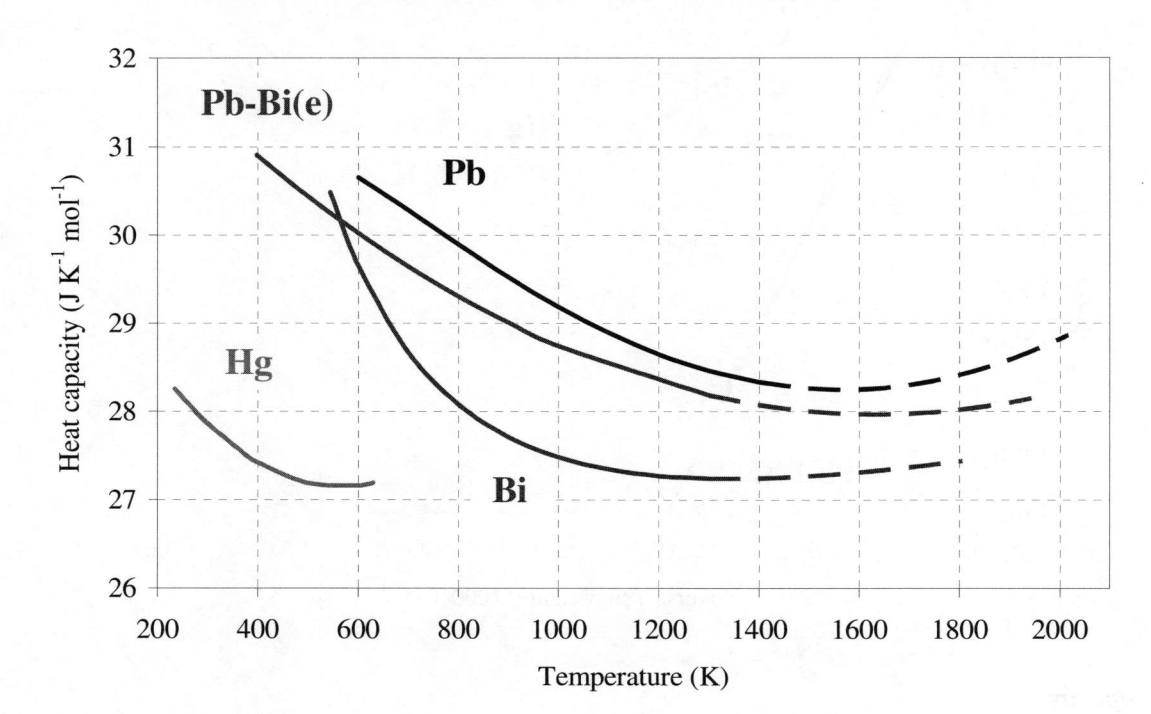

Figure 6. Isobaric heat capacity of Hg, Bi, Pb and Pb-Bi(e) ($p \approx p_{atm}$)

Equation of state

In recent years, considerable progress has been achieved in the development of empirical, semi-empirical and mechanistic EOS for fluids. Simple thermal EOS is frequently used for HM, which binds the main TD variables' pressure, temperature and volume (or density): $F(p,T,\rho) = 0$. The simplest approach to construct the thermal EOS is to use available information on temperature and pressure dependence of density:

$$\rho(p,T) = \int_{T_0}^{T} \left(\frac{\partial \rho}{\partial T}\right)_P dT + \int_{p_0}^{P} \left(\frac{\partial \rho}{\partial p}\right)_T dp \qquad (9)$$

The temperature dependence was described in one of the precedent sections. The effect of pressure can be estimated using the information on the HLM sound velocity, thermal expansion and heat capacity and the following thermodynamic relationship:

$$\left(\frac{\partial \rho}{\partial p}\right)_T = \left(\frac{1}{u_{sound}^2(T)} + \frac{M \cdot T \cdot \alpha_p^2(T)}{C_p(T)}\right) \qquad (10)$$

The results of calculation of the coefficient (10) for Hg, Bi, Pb and Pb-Bi(e) are presented in Figure 7 and they show that the correction factor is rather small. The density correction does not exceed 0.1% per 1 MPa in the region of the normal operation of the HLM.

Figure 7. Pressure correction factor for density of Hg, Bi, Pb and Pb-Bi(e) ($p \approx p_{atm}$)

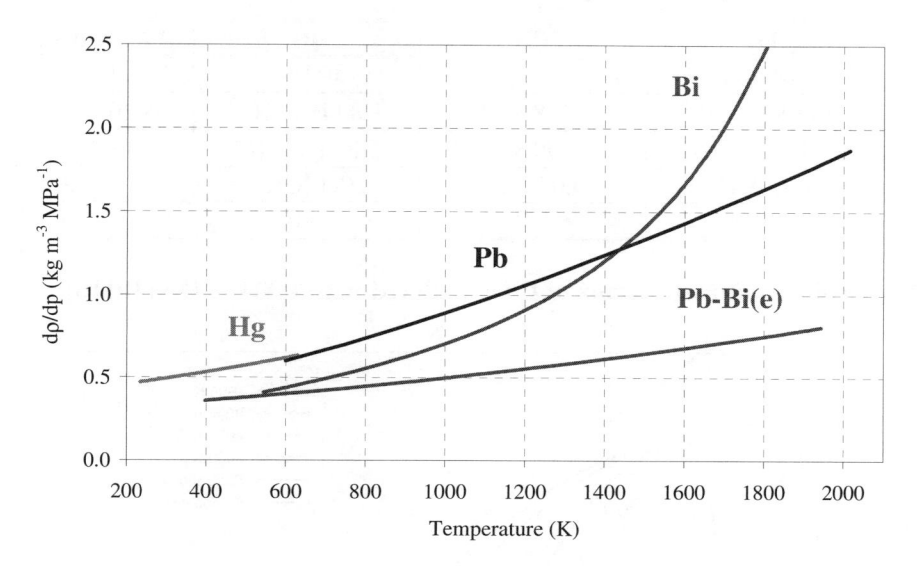

However, the estimation performed above is only valid for liquid phase and under the condition that thermodynamic coefficients do not depend on pressure. In order to take into account the pressure dependence of the HLM thermodynamic parameters, different thermal EOS were developed. In the design and industrial applications, a few semi-empirical equations of the Van der Waals family are widely used, in particular those based on the EOS proposed by Redlich and Kwong [9] (RK EOS):

$$p(\upsilon, T) = \frac{R \cdot T}{M \cdot (\upsilon - b)} - \frac{a}{\upsilon \cdot (\upsilon + b) \cdot T^{0.5}} \qquad (11)$$

where $\upsilon = \rho^{-1}$ is specific volume. In order to improve the prognosis capacity of this two-parameter equation of state, Soave [10] introduced a more general temperature dependence $a(T)$ in the attractive term of RK EOS. Later, Morita and Fischer [11] provided more flexibility for this EOS by introducing a third parameter c (in the place of b) in the attractive term and a dimer molar fraction x_d in the repulsive term (in order to consider a vapour with dimer and monomer molecules):

$$p = \frac{RT}{M \cdot (1 + x_d) \cdot (\upsilon - b)} - \frac{a(T)}{\upsilon(\upsilon + c)} \qquad (12)$$

$a(T)$ is presented as a power function of temperature:

$$a(T) = a_c \cdot \left(\frac{T}{T_c}\right)^n \text{ at } T \leq T_c \text{ and } a(T) = a_c + \left.\frac{da}{dT}\right|_{T_c} (T - T_c) \text{ at } T > T_c \qquad (13)$$

Parameters a_c, b and c can be found from the estimated values of the critical parameters T_c, p_c, ρ_c; x_d is determined using the "best estimate" values of the saturated vapour pressure.

Recently, Eqs. (12) and (13) were applied to Pb-Bi(e) [12] and then to Bi, Pb and Hg [6]. The coefficients of MRK EOS extracted from [6] are presented in Table 6. Figure 8 shows (as an example) the calculated vapour isotherms of Pb on a $p - \upsilon$ diagram.

Table 6. Coefficients of MRK EOS (10) [6]

	Hg	Pb	Bi	Pb-Bi(e)
a_c	14.377	203.24	49.064	159.33
b	7.4988×10^{-5}	3.3918×10^{-5}	1.8448×10^{-4}	6.2682×10^{-5}
c	-3.1436×10^{-5}	7.5099×10^{-4}	-5.8218×10^{-5}	8.1187×10^{-4}
n	-0.028413	0.048956	0.14618	0.37836
y_2	0.0?	0.0	$\neq 0$	$\neq 0$

Figure 8. Vapour isotherms of Pb calculated with RMK EOS (10) [6]

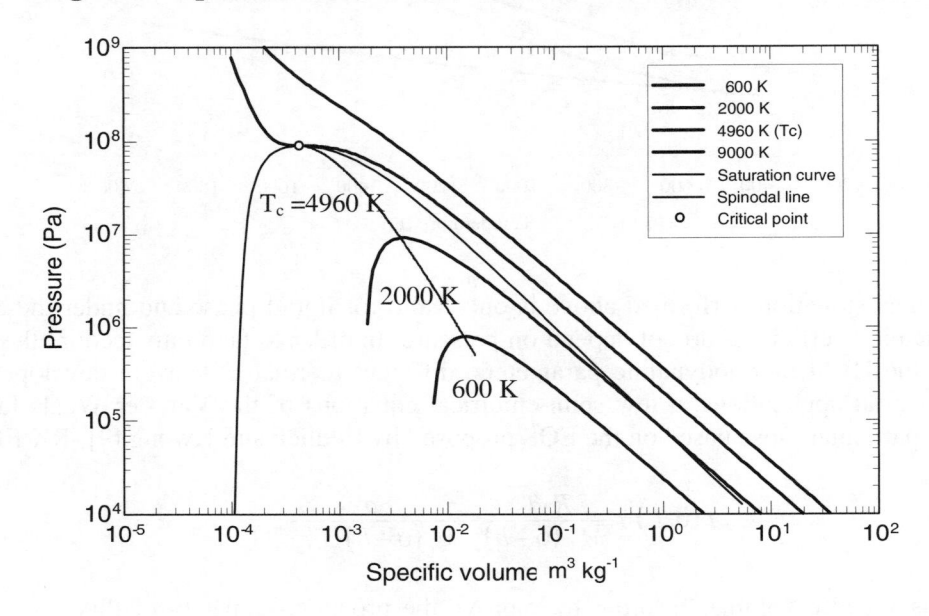

Conclusions

For liquid Hg, Pb Bi and Pb-Bi(e), the experimental data are available for most thermodynamic parameters of interest in the temperature region of the normal operation of nuclear installations, but only at the atmospheric pressure. Moreover, some of these parameters (heat capacity, saturated vapour pressure, sound velocity, critical point) have not yet been determined with sufficient accuracy.

The critical parameters of Hg, Pb, Bi and Pb-Bi given in the literature were analysed and compared aiming to develop more reliable recommendations. The modified four-parametric Redlich-Kwong EOS can, in principle, be applied for Hg, Pb, Bi and Pb-Bi. However, supplementary experimental data are needed to fit better the constants.

Co-ordinated international and national R&D programmes (*experimental and theoretical*) are needed to complete and extend the OECD database on HLM thermal properties.

Acknowledgements

This work was supported by funds of the MYRRHA project (SCK•CEN) and of the EURATOM FP6 ELSY project. Recommendations for thermodynamic properties of Pb, Bi and Pb-Bi(e) were mainly extracted from Chapter 2 of the OECD Handbook prepared with the support of many members of WPFC EG on HMLC Technology.

REFERENCES

[1] HLMC Handbook, C. Fazio (Ed.), OECD, Paris (2007).

[2] Sobolev, V., G. Benamati, *Chapter 2: Thermophysical and Electric Properties,* in HLMC Handbook, OECD, Paris (2007).

[3] Hofmann, W., *Lead and Lead Alloys*, Springer-Verlag, Berlin, New York (1970).

[4] Iida, T., R.I.L. Guthrie, *The Physical Properties of Liquid Metals*, Clarendon Press. Oxford, UK (1988).

[5] *Smithells Metals Reference Book* (8th edition), W.F. Gale, T.C. Totemeier (Eds.), Elsevier, Amsterdam (2004).

[6] Morita, K., V. Sobolev, M. Flad, *J. Nucl. Mater.*, 362, 227 (2007).

[7] Sobolev, V., *J. Nucl. Mater.*, 362, 235 (2007).

[8] Mustafin, G.M., G.F. Shaikhiev, *Russ. J. Phys. Chem.*, 45, 422 (1983).

[9] Redlich, O., J.N.S. Kwong, *Chem. Rev.*, 44, 233 (1949).

[10] Soave, G., *Chem. Eng. Science*, 27, 1197 (1972).

[11] Morita, K., E.A. Fischer, *Nucl. Eng. Design*, 183, 177 (1998).

[12] Morita, K., W. Maschek, M. Flad, Y. Tobita, H. Yamano, *J. Nucl. Sci. Tech.*, 43, 526 (2006).

DESIGN AND R&D SUPPORT OF THE XT-ADS SPALLATION TARGET

P. Schuurmans[1], A. Guertin[2], J.M. Buhour[2], A. Cadiou[2], M. Dierckx[1], J. Heyse[1], K. Rosseel[1], K. Van Tichelen[1], R. Stieglitz[3], D. Coors[4], L. Mansani[5], F. Roelofs[6], H. Aït Abderrahim[1]

[1]SCK•CEN, Boeretang 200, B-2400 Mol, Belgium
[2]Subatech, UMR 6457, École des Mines de Nantes – IN2P3/CNRS – Université de Nantes
4 rue A. Kastler, BP 20722; F-44307 Nantes Cedex 3, France
[3]Forschungszentrum Karlsruhe, P.O. Box 3640, D-76021 Karlsruhe, Germany
[4]Areva N, P.O. Box 3220, D-91050 Erlangen, Germany
[5]Ansaldo Nucleare, Corso F.M. Perrone, 25, Genoa, I-16161 Italy
[6]NRG Petten, Westerduinweg 3, P.O. Box 25, NL-1755 ZG Petten, The Netherlands
Tel.: +32 14 33 22 93·• Fax: +32 14 32 15 29 • Eml: paul.schuurmans@sckcen.be

Abstract

The experimental accelerator-driven system XT-ADS is being developed within the framework of the European FP6 project EUROTRANS. The device will serve a twofold purpose. Firstly it will act as an ADS concept demonstrator and secondly it will serve as a flexible fast neutron spectrum experimental irradiation tool for materials, fuel materials and radioactive isotopes studies. Because of the functional similarity between the XT-ADS and the MYRRHA concept that was developed earlier at SCK•CEN in Mol, Belgium, the design file of the latter was chosen as a starting point for the development of the XT-ADS. Within an ADS system the spallation target unit is the link between the accelerator and the subcritical core as here the neutrons that are fed into the core are produced. The design of the spallation target must meet the boundary conditions set by the different concepts in the XT-ADS with respect to the primary system, fuel handling, proton beam properties, etc. Because these boundary conditions were changed with respect to the MYRRHA case and in order to improve the spallation target design, several alterations as compared to MYRRHA were investigated. These affect the thermal-hydraulics of the target loop, the vacuum interface, the LBE handling and several active components that are housed in the target shell. In this paper, the altered design concept, the proposed technical solutions and the related R&D effort are discussed.

Introduction

High-level nuclear waste is the most important issue weighing on the acceptability of nuclear fission as a long-term energy source. Transmutation is the only method to fundamentally reduce the radiotoxic burden of the waste in a more efficient way than natural decay. In this respect the technology of accelerator-driven systems is an important track to be investigated. An ADS basically consists of a subcritical core to which neutrons from an accelerator-driven spallation neutron source are fed. Because the operation of the "reactor" system is no longer based on reaching criticality in the core but rather on the intensity of the neutron source, operation with any kind of fissile material is in principle possible. This feature makes an ADS particularly suited for waste transmutation.

The European project EUROTRANS is dedicated to the advancement of nuclear transmutation techniques [1]. The project aims to reach two main goals: to realise the conceptual design of the lead alloy European Transmutation Demonstrator (ETD) and to perform a detailed design of the small-scale experimental XT-ADS aiming at the short-term realisation of the system. The XT-ADS will serve a twofold function. Firstly, it will be an ADS concept demonstrator and component test bench for the industrial level nuclear waste transmuter ETD. Secondly, the XT-ADS will be designed as a flexible experimental irradiation device for fuel, materials and radioactive isotopes studies for present and future nuclear energy concepts. Because of its function as an experimental irradiation device, the XT-ADS subcritical core will need to be designed in a very compact geometry to achieve high flux levels ($\phi_{Tot} = \sim 3.10^{15}$ n/cm^2.s) [2] within a reasonable small core (~0.5 m^3). Evidently, the spallation target design must match the requirements determined by the general concept of the XT-ADS. It should produce a sufficient amount of neutrons to feed the subcritical core at its specific k_{eff} value (≈ 0.95). For this purpose the spallation target must accept the appropriate high-power proton beam that is currently set at 2.5 at 600 MeV. At the end of the cycle the required beam current may rise to 4 mA to compensate for fuel burn-up. Because the thermal energy of about 1-1.5 MW that is deposited by the proton beam requires forced convection cooling, liquid lead-bismuth eutectic (LBE) is chosen as target material. LBE is likewise the coolant of the main vessel although both circuits are separated. In addition, the target must fit into the space that is available in the subcritical core. The target space is created by removing three of the hexagonal fuel assemblies in the centre of the core. Furthermore, the design of the target should not hamper the fundamental role of XT-ADS as a flexible high-intensity experimental irradiation device. Finally, although a replacement of the spallation target within the envisaged lifetime of the XT-ADS is unavoidable this operation should not be required too often. Thus, the spallation target unit should be able to survive operation within the ADS system for a sufficient amount of time.

Spallation target design concepts

Due to the functional similarity between the XT-ADS and MYRRHA that was developed earlier at SCK•CEN [3], the design of the latter was chosen as a starting point for the development of the XT-ADS target loop. The design of the XT-ADS target is driven by the functional and spatial constraints given above. Because of the very high proton beam densities of about 150 µA/cm^2, the spallation target is designed without a window between the liquid metal target area and the vacuum of the beam line. It may be noted that the focus of the EUROTRANS project on a windowless design is complementary to the work that was carried out in the FP5 programme PDS-XADS and the MEGAPIE initiative in which a window concept for a high-power spallation target was studied.

The spallation zone is formed by free flowing LBE shaped by a concentric feeding nozzle thus creating a confluent flow (Figure 1). The compact core of the XT-ADS only allows a single passage of the LBE target material in one direction from top to bottom. The feeder line passes above the core and

Figure 1. Schematic drawing of the vertical confluent flow concept

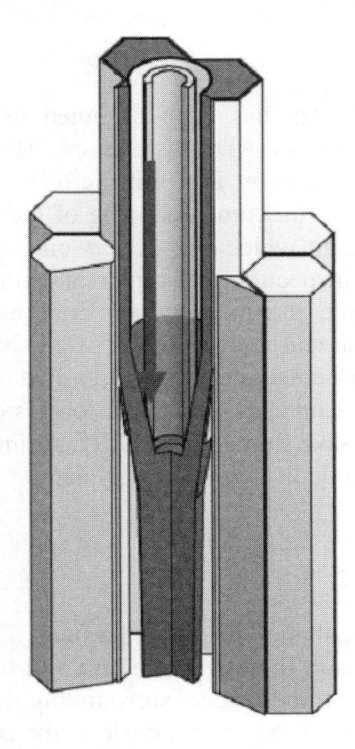

the return line passes underneath it, thus interlinking the core. In this configuration the spallation target loop has an off-axis housing for all active components. The off-axis design of the spallation loop leaves the top and bottom of the subcritical core accessible for fuel manipulations and the installation of irradiation experiments. In addition, the main part of the spallation loop is moved away from the high radiation zone which is beneficial for its lifetime.

The common vacuum of the target zone and the proton beam line puts requirements on the vacuum system. Firstly, the vacuum pressure directly above the spallation target should be below the 10^{-5}-10^{-6} Pa range to guarantee compatibility with the vacuum of the proton beam line and to avoid plasma formation caused by the interaction between residual gas above the target and the proton beam. This pressure condition implies that the outgassing of the spallation target material must be limited and that care should be taken in the design of the vacuum system to ensure sufficient vacuum conductance and pumping capacity. The second essential function of the vacuum system is the confinement of volatile radioactive spallation products. Due to the spallation interaction of the proton beam with the target material, radioactive elements with a high vapour pressure (e.g. Hg isotopes) are produced. These products are likely to emanate from the free surface of the target and should be confined, either within the spallation loop or in the vacuum system. For this purpose the latter is equipped with a closed back end composed of sorption and getter pumps. In order to minimise emanation of spallation products into the proton beam line, a large second free LBE surface is foreseen in the servicing vessel, directly underneath the main vacuum pumps.

The heat deposited in the spallation target by the 600-MeV proton beam is about 975 kW for the present target configuration and 2.5 mA current [4]. This heat is removed by pumping the liquid target material through a heat exchanger that is situated at the bottom of the off-axis part of the spallation loop. The secondary side of the heat exchanger is cooled by the main vessel coolant. The flow rate of the LBE in the spallation loop is about 10-20 l/s. The lower limit of this value is determined by the maximum temperature allowed in the target. In order to limit LBE evaporation and corrosion of structural materials in the target loop, this temperature is set at about 430°C. Because the spallation target material is cooled against the main vessel coolant, the lowest achievable temperature during normal operation is 330°C.

The XT-ADS spallation target system has been designed to be compatible with the remote handling scheme envisaged for the entire XT-ADS installation. The spallation loop can be removed from the main vessel after unloading of the core. The prior unloading of the core is to avoid criticality issues, for general safety and to allow *in situ* commissioning of the target unit. In addition, all active elements are placed in a separate sub-unit which allows servicing of these parts without removal of full the spallation loop. Maintenance, inspection and repair of the spallation unit are foreseen to be performed in the XT-ADS hall outside the main vessel pool under cover of a protective inert atmosphere. This includes disconnection and reengagement of all service jumpers, replacement of the embrittled loop parts close to the target zone and removal and re-installation of the interior column with all active parts. Before and after maintenance the LBE loop is drained and later refilled into and from a special container. This allows to save storage of the LBE during maintenance and simultaneously permits conditioning of the material in a dedicated off-line system

Spallation loop layout and operation

Figure 2 shows the interior of the spallation loop together with a schematic layout. All components are indicated. The liquid metal is fed from the off-centre unit and traverses to the central axis of the subcritical core. It descends through the three feeders surrounding the beam transport line. The target surface is formed at the confluence point of the target nozzle in the centre of the subcritical core. Here the proton beam impinges from the top. The LBE subsequently returns from the beam impact zone through the central tube, the lower U-bend and the heat exchanger to the pumps in the off-centre unit.

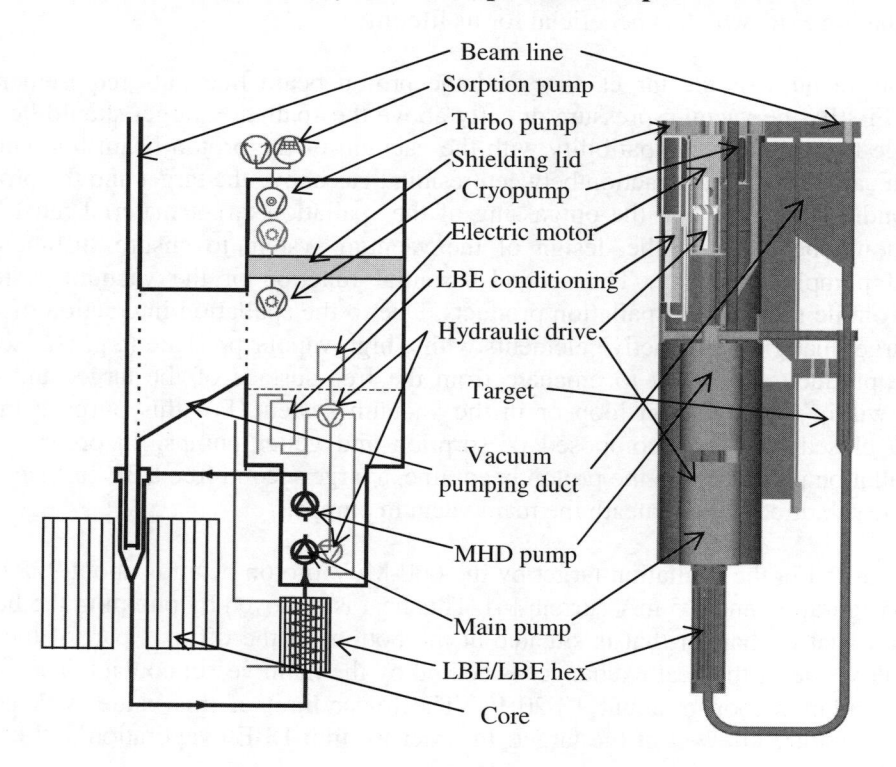

Figure 2. Schematic layout of the spallation loop and its interiors

For proper operation of the liquid target, the formation of the target free surface and a firm control over the size of the recirculation zone that is formed in the centre of the target free surface is essential. The size of the recirculation zone can be determined indirectly by its height. The latter is measured by a LIDAR positioned in line with the beam.

Within the XT-ADS spallation target design work of the EUROTRANS project, a specific effort is put on the development of the target nozzle. In the MYRRHA Draft 2 [3] that serves as the input for the XT-ADS, the target nozzle was designed to have a straight LBE flow without detachment of the LBE from the nozzle walls. However, real size water flow experiments performed at the Université Catholique de Louvain (UCL) and LBE flow experiments that were done at the KALLA laboratory of the Forschungszentrum Karlsruhe (FzK) have shown that a more stable flow is achieved if some detachment of the LBE flow is allowed. Also, introducing a mild swirl has a stabilising effect. These results are now used as input for further development of the target nozzle. Firstly, the feasibility of a target nozzle that explicitly forces flow detachment is studied. In order to achieve detachment at the right position and nowhere else, a rig in the nozzle wall is created (Figure 3). In addition, the nozzle shape and the in- and outflow cross-sections must be optimised. For this purpose, several nozzle proposals are looked into with CFD calculations, followed by experimental tests using water flow experiments at UCL. A similar strategy is being followed for the investigation of the influence of swirl on the LBE flow. Here, CFD calculations indicate that in the current geometry, 5% swirl, corresponding to a 1/20 ratio of the tangential relative to the axial LBE velocities, stabilises the position of the recirculation zone whereas at 10% swirl the onset of the creation of a hollow vortex in the centre of the nozzle is observed [5]. In water flow experiments similar results are obtained albeit that already at 5% swirl vortex formation is found. In the near future, water flow experiments with a target nozzle introducing less swirl will be carried out.

Figure 3. Target nozzle with a detached flow profile

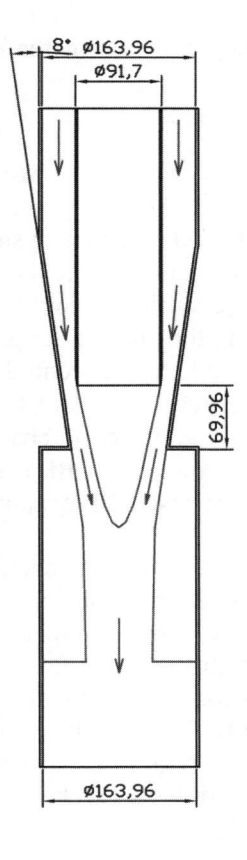

LBE pumps

The present design of the spallation loop is equipped with two active pumps: the main pump in the return line and an auxiliary pump in the feeder line. The pumping system is laid out to provide a total pressure boost of 4.4 bar, to drive the LBE in the spallation loop at a flow-rate of 10-20 l/s.

For the main pump, the reference design consists of a mechanical pump powered by an indirect hydraulic drive to avoid the use of a long shaft. In this configuration, a canned electric motor drives a pump that transmits its power to the hydraulic drive/hydraulic pump situated directly above the outlet of the heat exchanger. The hydraulic transmission fluid is taken to be the same as the LBE that it is circulating in the spallation loop (although at higher pressure). For the design of the pump-drive tandem, common impeller technology or a screw spindle technology are considered. In the latter case, the fluid smoothly follows the spindle motion without being subject to the accelerating or decelerating phases of impeller pumps. Such a smooth flow is assumed to be less bothered by corrosion and cavitation problems. Although the hydraulic drive pump is at present the reference design, the long shaft option has not been ruled out. Here however, the proper design and testing of the shaft bearings is the most crucial point. The use of an annular linear induction or magneto-hydrodynamic (MHD) pump is also not excluded, provided an efficient design can be produced which fits into the available space (diameter ~800 mm, length ~1 000 mm).

For the pump in the feeder line, the reference design is an MHD type pump. The basic concept is similar to that used for the MEGAPIE Project at PSI (CH). The MEGAPIE pump is constructed at IPUL (University of Latvia). The MHD pump envisaged for the XT-ADS differs from this pump in certain respects in order to obtain the pressure gain of 1.2 bars as required. By better matching the LBE velocity to the mean magnetic field velocity, the slip ratio can be decreased, resulting in almost doubled pump efficiency. Furthermore, the pump length is increased to 1.2 m, including an integrated annular magnetic flow meter and the number of poles is increased accordingly. Taking into account the MEGAPIE model findings, the efficiency could be increased even more by grading the magnetic core flux over its length.

Vacuum system

The absence of a window implies that a vacuum system must be provided that maintains sufficiently low pressures above the target to accommodate the vacuum of the accelerator and to avoid plasma formation. In addition, it must ensure sufficient confinement of volatile radioactive spallation products that may emanate from the target. In the present design the low conductance of the beam line is overcome by connecting the spallation unit to the central beam line via a duct. An integrated vacuum system provides a vacuum over the free surfaces of less than 10^{-5} Pa. At the back end of the system all radioactive volatile emanations are collected in absorption pumps from where they can be batch-wise removed. Because of the target vacuum, the LBE flow is optimised to keep the temperature of the free surface low in order to prevent excessive LBE evaporation into the beam line.

During target nozzle experiments, it was observed that the pressure in the target module was reduced to two orders of magnitude below the minimum pressure reachable by the vacuum pump installed at that time. This indicates that the fast flowing LBE in the spallation target zone has a significant vacuum pumping effect. Although the effect has not been duly quantified in the present target geometry, it does open possibilities for a simplification of the vacuum system design. If the vacuum pumping effect of the LBE is sufficient to reduce the vacuum pressure in the target zone during normal operation to below the level required, the central duct can be abandoned. In this scenario, the confinement of the spallation products can be improved as the vacuum chamber above the free surface in the main vessel has no contact with the beam line, thus preventing migration of volatile radioactive spallation products that were emanated from this free surface to the beam line. Separation also allows operating the vacuum vessel at higher pressures of the order of 1-100 Pa which would greatly simplify the vacuum system without changing the LBE flow behaviour significantly. In addition, the higher operating pressure allows to envisage a room temperature vapour trap in the vessel that would immobilise the bulk of the volatile spallation products, so that the load on the absorption pumps is reduced.

LBE conditioning

Conditioning of the LBE eutectic in the spallation loop is required for two main reasons: to inhibit corrosion and to prevent conglomeration of insoluble impurities (e.g. PbO) that may lead to a blocking of the flow. These two requirements lead to the necessity to control the oxygen content in the LBE target material to the level of ca 1.10^{-6} wt.%. For this a hydrogen and water vapour gas treatment system is foreseen in the design of the MYRRHA ADS to reduce the amount of oxygen in the LBE when required. However, since the spallation unit is a vacuum system the treatment is only possible during maintenance times. During operation, the generally reducing nature of the spallation products and the hydrogen from the proton beam needs to be counteracted. For this purpose a dedicated conditioning unit is foreseen. Its active component is a heated basket with PbO pebbles housed in an insulated vessel. The oxygen exchange rate between the pebbles and the bypass LBE flow is governed by controlling the temperature of the basket. In addition, magnetic filtering is foreseen at the entrance of the MHD pump to extract magnetic corrosion products (mainly Fe and Ni compounds) that could otherwise clog the MHD pump. Finally, at the top of the second free surface filtering/skimming is envisaged to remove floating debris.

Summary

The design of the XT-ADS spallation target is performed within the European integrated project EUROTRANS that started in April 2005. At the current status of the spallation target design process, the boundary conditions for the spallation target loop with respect to the XT-ADS performance requirements and the design of the subcritical core and primary system have been established. In this paper, the next steps are outlined that will concentrate on further development of the spallation target nozzle, the vacuum and spallation product confinement system and the pumping system.

This work is accomplished in the frame of the FP6 Contract FI6W-CT-2005-516520.

REFERENCES

[1] Aït Abderrahim, H., *et al.*, *Proceedings of the 9th Information Exchange Meeting on Actinide and Fission Product Partitioning and Transmutation*, NÎmes, France, 25-29 September 2006 (433-441).

[2] Van den Eynde, G., *et al.*, *Proceedings of the 9th Information Exchange Meeting on Actinide and Fission Product Partitioning and Transmutation*, NÎmes, France, 25-29 September 2006 (561-567).

[3] Aït Abderrahim, H., *et al.*, *MYRRHA Pre-Design File – Draft-2*, SCK•CEN Report R-4234, June 2005.

[4] Malambu, E., Th. Aoust, *Compared Design Parameters of a 50 MWth MYRRHA Core: 350 MeV Versus 600 MeV Proton Beam*, SCK•CEN, Boeretang 200, 2400 Mol, Belgium, internal report RF&M/EM/em.34.B043200/85/MYRRHA-Design/05-39.

[5] Roelofs, Ferry, Arné Siccama, "Status of Free Surface Modelling", presented at the *DEMETRA* technical meeting, FZR, Germany, June 2006.

STUDY OF MEASUREMENT METHOD OF HIGH-ENERGY NEUTRONS FOR ADS

Kazuaki Abe, Tomohiko Iwasaki, Satoshi Gunji
Tohoku University, Japan

Cheol Ho Pyeon, Hiroshi Shiga, Masamitsu Aiba, Hiroshi Yashima
Kyoto University, Japan

Toshiya Sanai
High Energy Accelerator Research Organization, Japan

Abstract

The system conditions of an accelerator-driven subcritical system (ADS) must be closely monitored to ensure safe operation. The energy spectrum of neutrons in the ADS is largely varied by changing the conditions of the accelerator and the target, as well as the core. Therefore, by measuring the energy spectrum of those neutrons, the system conditions can be monitored. Bismuth activation was proposed for measuring the high-energy neutrons and the experimental verification for the Bi activation method was performed with a high-energy accelerator. The experiment to establish a measurement system for thermal and fast regions was also performed. Based on those experiments, we have designed a measurement system of the whole energy spectrum of neutrons existing in the ADS.

Introduction

The ADS is a hybrid system, which consists of a GeV-order proton accelerator, a spallation target, and a subcritical reactor. High-energy neutrons with an energy range of up to several hundred MeV are generated in the ADS. For example, Figure 1 shows the typical spectrum in the ADS proposed by Japan Atomic Energy Agency, which has a 1.5-GeV proton accelerator, a Pb-Bi target and a nitride fuel core. The spectrum of neutrons distributing in the wide energy range from thermal to high energy is formed. The spectrum is largely varied by changing the conditions of the accelerator, the target and the core. This fact inversely means that, by measuring the energy spectrum of those neutrons, the system conditions of ADS can be monitored.

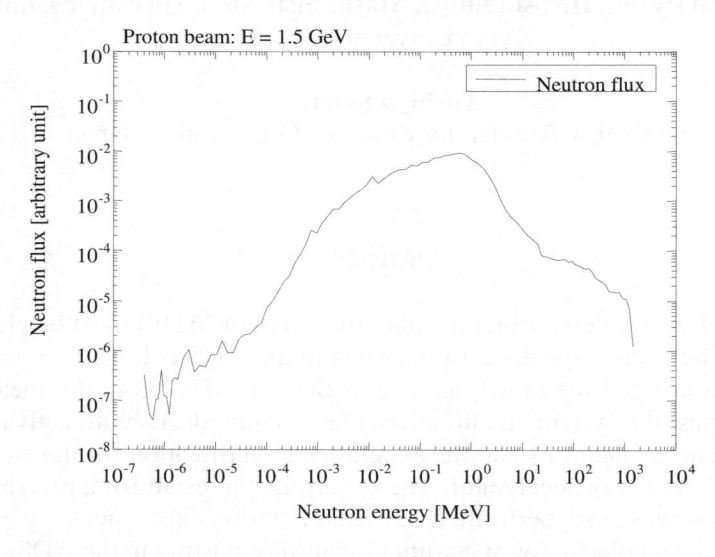

Figure 1. Typical spectrum in the ADS

At this research stage of ADS, there is no measurement method for the neutron energy spectrum including those high-energy neutrons. Therefore, this study aims: 1) to develop a measurement method for the energy spectrum of high-energy neutrons existing in the ADS; 2) to establish a measurement system of the whole energy spectrum inside the ADS core.

As a detector to measure neutron spectrum, a scintillation detector is one of the candidates as it can directly acquire the neutron energy spectrum. However, the scintillation detector is not useable in the ADS core since the ADS field has both very high neutron flux and very large γ contamination. A fission counter can be employed to measure the neutrons in the ADS core, but it cannot provide the neutron spectrum. Therefore, we selected an activation detector since it can measure the neutron spectrum with high-energy resolution and it is useable in the strong γ field.

Development of measurement method of high-energy neutron spectrum

Selection of foil

We propose bismuth for measuring the high-energy neutrons. Bi continuously occurs the (n,Xn) reactions successively from 10 to 100 MeV with the same energy interval of 8 MeV. Figure 2 shows the cross-sections of the Bi (n,Xn) with the threshold energies from 20 to 100 MeV. The activated nuclides from those reactions have the suitable half-lives and the γ-ray energies for γ-counting. Bi will be suitable to obtain the energy spectrum of high-energy neutrons.

Figure 2. (n,Xn) threshold reaction cross-section of Bi

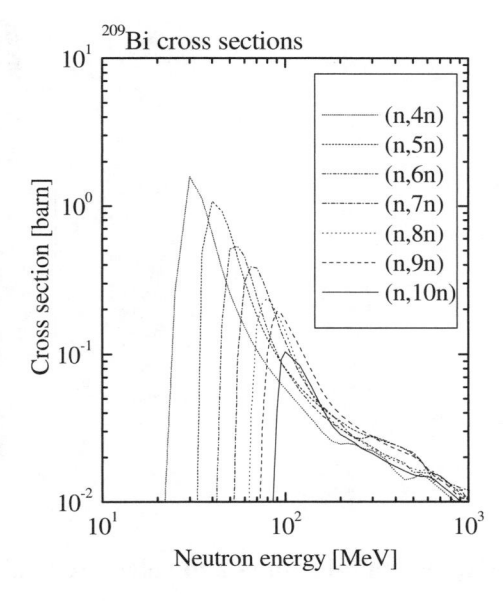

^{209}Bi cross sections

- (n,4n)
- (n,5n)
- (n,6n)
- (n,7n)
- (n,8n)
- (n,9n)
- (n,10n)

Verification experiment

The experimental verification of the Bi activation was made by using a Fixed Field Alternating Gradient (FFAG) accelerator at High Energy Accelerator Research Organization (KEK).

The FFAG accelerator is an accelerator, which has the combined advantages of a cyclotron and a synchrotron. The FFAG used in the experiment is the accelerator with a beam energy of 150 MeV. However, unfortunately in this experiment, the proton energy was not stable and the beam did not converge well. For this reason, the beam energy could not be well defined although it was supposed between 70 and 80 MeV.

Figure 3 shows the experimental set-up. The sample consists of Bi as activation detector and W as a spallation target into which the proton beam was injected. The thickness of W was 9 mm, which is enough to stop the protons inside the W targets. The Bi sample was placed behind the W target. The thickness of Bi is 3 mm. The diameter of W and Bi was the same diameter of 5 cm. That thickness and the diameter of Bi and W were decided by the MCNPX calculation.

The sample was irradiated for three hours and was measured using a Ge detector. Reactions observed are as follows: ^{209}Bi(n,4n)^{206}Bi with a threshold energy of 22 MeV and ^{209}Bi(n,8n)^{202}Bi with threshold energy of 54 MeV.

Comparison with experimental and calculated results

From the γ counting data, we derived the neutron energy spectrum by the unfolding method. Figure 4 shows the experimental result using the unfolding method and the calculated result using the MCNPX code. The calculated values are two spectra for the proton energy of 70 and 80 MeV as shown in the figure. The reaction rates are the relative values normalised by the reaction rate of the (n,6n) reaction.

Figure 3. Experimental set-up

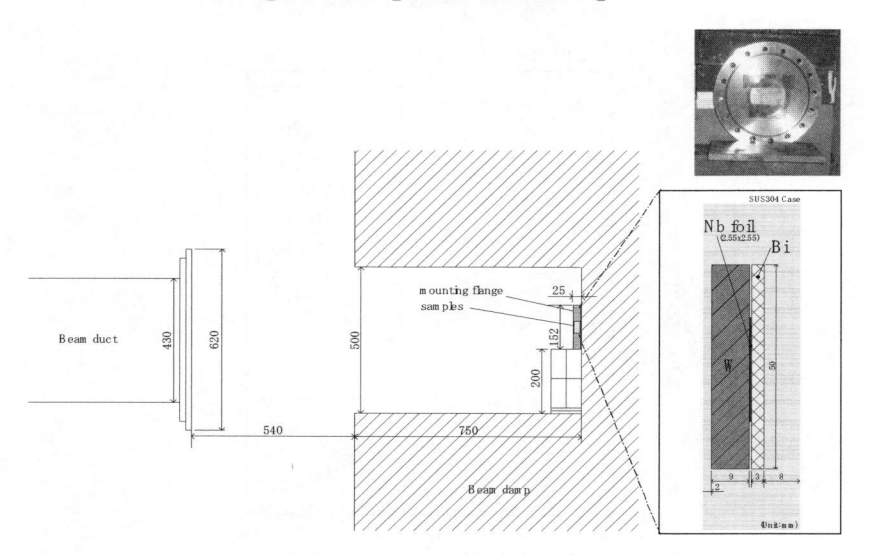

Figure 4. Experimental results and calculated values

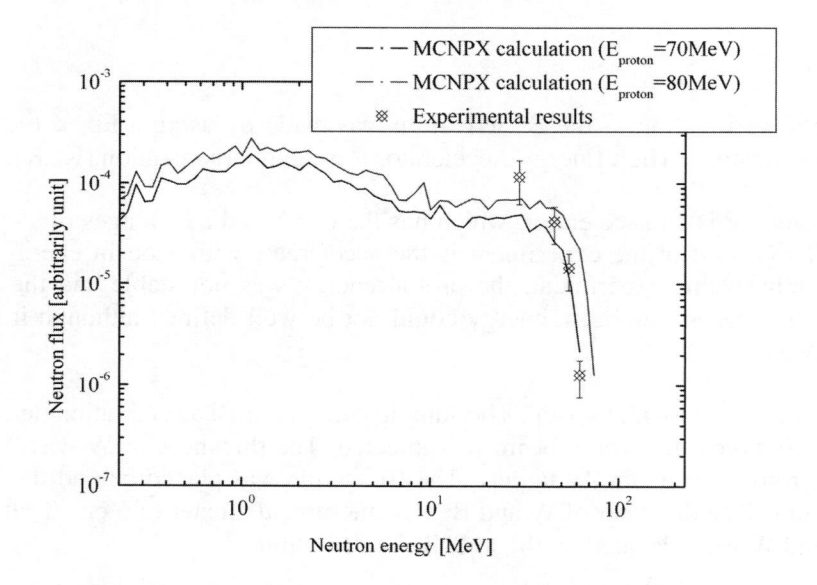

It is found from the figure that the experimental result is between the calculated values of 70 and 80 MeV. This result confirms that the Bi activation is available for measuring the high-energy neutron spectrum. More accurate experiments should be undertaken to verify this Bi method.

Establishment of measurement system of whole energy neutron spectrum

Verification experiment for thermal and fast neutrons region

To measure the whole energy spectrum, it is required to measure the thermal and the fast neutrons. In this study, the experiment was performed to establish the measurement method of the thermal and the fast neutrons using the Kyoto University Critical Assembly (KUCA). In the KUCA, a subcritical core driven with 14-MeV neutrons was mocked-up and the activation experiment was made by using the convectional foils such as Au and In.

Although the experiments were performed for the thermal and the fast neutron spectrum, the description presented here was limited to the experiment for the fast neutron higher than 0.1 MeV. Figure 5 shows the experimental layout of the KUCA experiment. The accelerator is the conventional Cockcroft-Walton accelerator with an energy of 300 KeV, which produces the neutrons via the D-T reaction. The subcriticality for the ADS core was about 1.2%Δk/k. In, Fe, Al and Nb were adopted for measuring the fast neutrons with an energy threshold interval of 5 MeV. The foils were set at the centre of the core and near the neutron target. Each parameter of the foils employed is shown in Table 1. For measuring the thermal neutron flux, ^{197}Au was used. The irradiation time of all the foils was five hours and the γ-rays were detected by using a Ge detector.

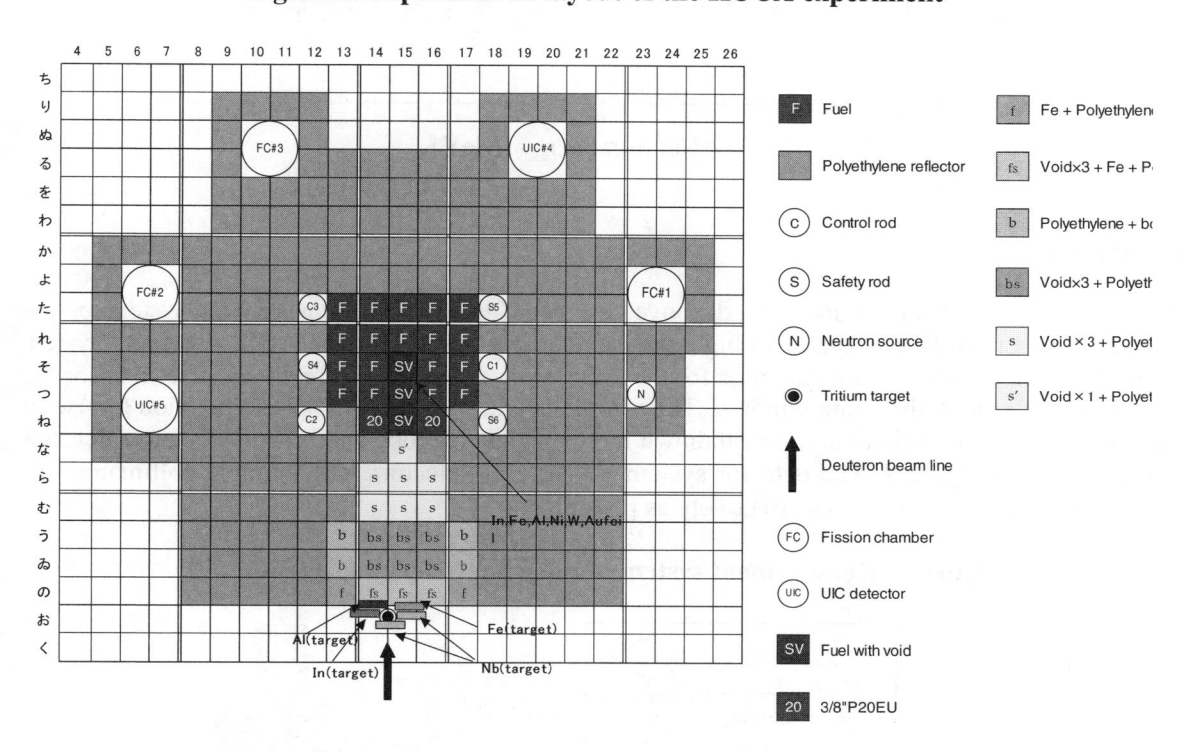

Figure 5. Experimental layout of the KUCA experiment

Table 1. Parameter of used foils

	Threshold [MeV]	Mass [g]	Volume [mm³]	Half-life
^{115}In(n,n')	0.5	42.5	45 × 45 × 3	4.49 h
^{56}Fe(n,p)	4.9	77	45 × 45 × 3	2.58 h
^{27}Al(n,α)	4.9	27.3	45 × 45 × 3	15.3 h
^{93}Nb(n,2n)	9	34.6	45 × 45 × 3	244 h
^{197}Au(n,γ)		8.01	20 × 20 × 1	2.70 d

The reaction rates for all foils were unfolded by SAND-II code using the cross-section of JENDL Activation Cross-section File96. The derived spectrum is shown in Figure 6 with the neutron spectrum calculated using the MCNP code at the centre of the core.

It is found that the experimental result agrees with the calculated one, thus establishing the measurement method for the neutron spectrum in the thermal and fast regions.

Figure 6. Derived spectrum

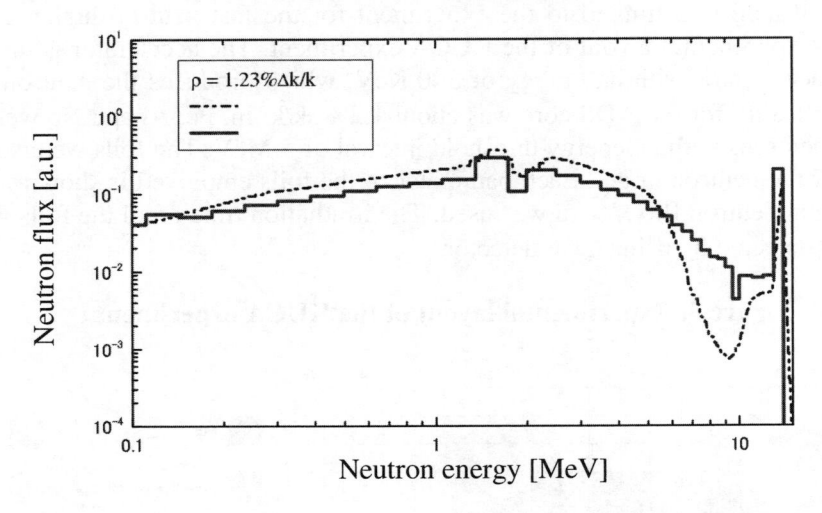

Monitoring system

Based on the experiments mentioned above, we have designed a measurement system of the whole energy spectrum from thermal to high-energy neutrons existing in the ADS. A typical system design is shown in Figure 7. This system is for the ADS proposed by JAEA and it aims to monitor the neutron spectrum around the beam window. The sample is placed in an appropriate position by use of a pneumatic tube. After irradiation, it is withdrawn by the pneumatic tube and is set inside a Ge detector system as shown in Figure 7. The detector system has several detectors with suitable collimators for measuring the γ-rays as soon and as effectively as possible.

Figure 7. Measurement system of the whole energy spectrum

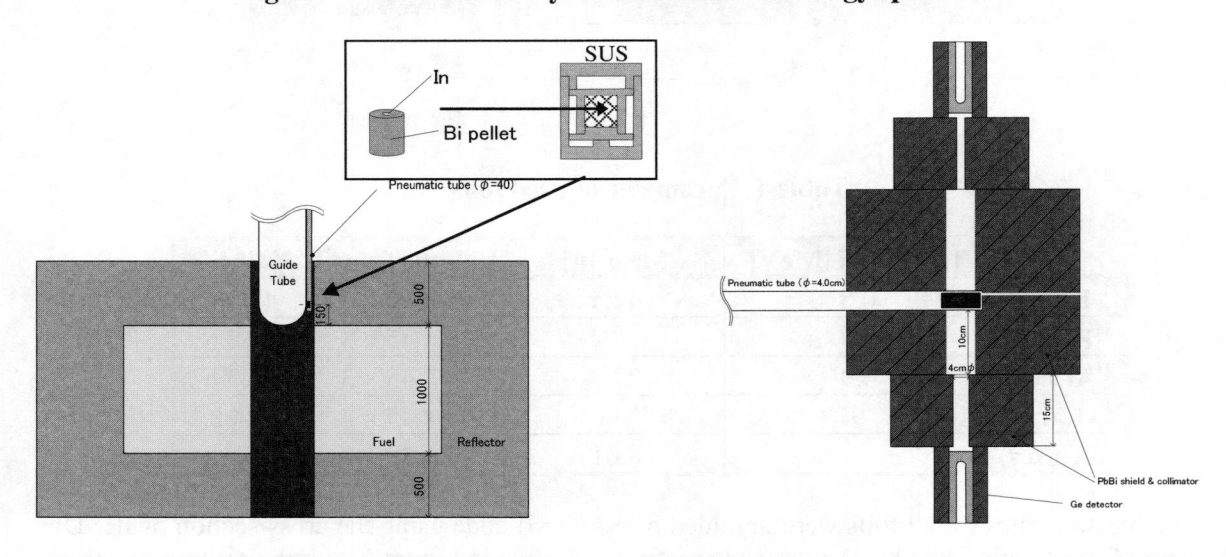

The sample is also shown in Figure 7. The sample presented here consists of Bi and In. In is for measuring the thermal neutrons by the In(n,γ) reaction and the fast neutrons by the In(n,n′) reaction. Other foils for measuring the spectrum more precisely will be added if necessary. The sample is supposed to be encapsulated as shown in Figure 7. The shapes and masses of the activation detectors were decided by MCNPX calculation as follows:

	Mass	Shape	Size
Bi	12.25 g	Pellet	6.32 mm × 10 mm (diameter × height)
In	0.18275 g	Flat	5.0 mm × 5.0 mm × 1.0 mm (length × width × thickness)

Assuming the operation conditions of the ADS proposed by JAEA, this system was analysed using MCNPX and ENDF/B-IV. Table 2 shows the results when the irradiation time is set to three hours and three types of combinations of the waiting time and the detection time are arranged as follows.

i) Waiting time: 30 minutes, detection time: 3 hours.

ii) Waiting time: 6 hours, detection time: 1 day.

iii) Waiting time: 12 hours, detection time: 30 minutes.

Table 2. Activity of Bi and Ge detector by MCNPX

Reaction	Threshold	Waiting time	Detection time	Peak counts
Bi(n,10n)	70.89 MeV	30 min.	3 hrs.	105 448
Bi(n,9n)	61.69 MeV	30 min.	3 hrs.	61 883
Bi(n,8n)	54.24 MeV	30 min.	3 hrs.	571 995
Bi(n,7n)	45.37 MeV	6 hrs.	1 day	76 424
Bi(n,6n)	38.13 MeV	30 min.	3 hrs.	126 098
Bi(n,5n)	29.62 MeV	6 hrs.	1 day	773
Bi(n,4n)	22.55 MeV	6 hrs.	1 day	12 909
In(n,n)	Fast	12 hrs.	30 min.	130 843
In(n,γ)	Thermal	12 hrs.	30 min.	99 403

It is found that enough counts are obtained for every reaction of Bi(n,6n) to Bi(n,10n) and In(n,γ) and In n,n'). It is confirmed that the spectrum for the whole energy region inside the ADS can be measured by the present system.

Conclusion

In this study, bismuth activation was proposed for measuring the high-energy neutrons and the experimental verification for the Bi activation method was performed using a high-energy accelerator. In addition, the experiment to establish for the thermal and the fast regions was performed by a use of subcritical reactor and an accelerator. Based on the experiments mentioned above, we designed a measurement system of the whole energy spectrum existing in the ADS.

As the results, we have developed the measurement method of the neutron spectrum for the high-energy neutrons and established the measurement system of the whole energy spectrum from thermal to high-energy neutrons existing in the ADS.

REFERENCES

[1] Gungi, Satoshi, *Study on the Monitoring System of Accelerator Driven Subcritical System Operation by In-core Neutron Measurements*, Tohoku University (2006).

[2] Hirano, Yoshiyuki, *Preliminary Study on ADSR by Using FFAG Accelerator in KUCA*, Kyoto University (2004).

[3] Hirano, Yoshiyuki, *Preliminary Study on ADSR by Using FFAG Accelerator in KUCA*, Kyoto University (2004).

[4] Takizuka, T., K. Tujimotto, T. Sasa, *et al.*, "Design Study of Lead-bismuth Cooled ADS Dedicated to Nuclear Waste Transmutation", *Progress in Nuclear Energy* (2002).

[5] *MCNPX User's Manual, Version 2.4.0.*, LA-CP-02-408, Los Alamos National Laboratory (2002).

[6] *MCNP – A General Monte Carlo N-particle Transport Code, Version 4C*, Judith F. Briesmeister (Ed.), LA-13709-M, Los Alamos National Laboratory (2000).

[7] Cross, B., *SAND-II/CATAPE* (Memo, June 1975).

[8] Berg, S., W.N. McElory, "A Computer-automated Iterative Method for Neutron Flux Spectra Determination by Foil Activation", *Vol. II: SAND II (Spectrum Analysis by Neutron Detectors II) and Associated Codes*, AFWL-TR-67-41, Vol. II (September 1967).

APPLICATION OF THE INCL+ABLA REACTION MODEL TO THE STUDY OF THE EVOLUTION OF SPALLATION TARGETS

Th. Aoust, J. Wagemans
SCK•CEN, Advanced Nuclear Systems Institute, Boeretang 200, B-2400 Mol, Belgium

J. Cugnon
University of Liège, AGO Department, Allée du 6 Août 17, Bât. B5, B-4000 Liège, Belgium

A. Boudard, S. Leray, Y. Yariv
CEA Saclay, DAPNIA/SPhN, F-91191 Gif-sur-Yvette Cedex, France

Abstract

The design of spallation targets of accelerator-driven systems requires the use of reliable spallation reaction models embedded inside general transport codes. Nuclear reactions above ~150 MeV are often reasonably well described by a Monte Carlo implementation of an intranuclear cascade (INC) model followed by a de-excitation model. The Liège intranuclear cascade model (INCL4) coupled to the ABLA evaporation-fission model has been shown to give a fairly good description of a large amount of experimental measurements for proton-induced spallation reactions on thin targets in the 200 MeV-2 GeV range, without free parameters. This model was recently improved in order to extend its capabilities and to cure some remaining shortcomings. These modifications bear on nuclear mean field, some aspects of pion dynamics, light charged clusters emission and low-energy extensions. In this paper we will compare both the modified and the standard versions of the INCL model with respect to their ability to reproduce thin target experimental measurements. After, we will tentatively validate the improved INCL model for particle emission and for residue production in thick targets. In order to estimate the residue production, MCNPX has been updated and coupled to the ORIGEN evolution code, through the use of the ALEPH system developed at SCK•CEN. The constraints on the reaction models brought by the thick target data, especially concerning radiotoxicity, will be compared with those provided by the similar comparison with thin target data.

Introduction

The projects of accelerator-driven systems for the transmutation of nuclear waste have increased the interest in spallation reactions as primary neutron sources. The design of such facilities requires the use of reliable spallation reaction models, embedded inside general transport codes, to accurately evaluate emitted particle distributions, radiation damages and radiotoxicity of the target.

A spallation reaction is currently described by a first brief phase of intranuclear cascade process governed by nucleon-nucleon collisions, leading to an excited nucleus after ejection of a few energetic particles, followed by a second longer phase corresponding to the decay by evaporation with a possible competition with fission.

The intranuclear cascade model of Liège INCL4 [1] coupled to the ABLA evaporation-fission model [2] has been shown to give a fairly good description of a large amount of experimental measurements for proton-induced spallation reactions on thin targets in the 200 MeV-2 GeV range [1]. These last years this model was improved to extend its capabilities and to cure some remaining shortcomings [3]. The INCL4 model and these recent extensions will be shortly presented in the next section.

After comparing the standard and the modified versions concerning their predictions for particle emission and residue production in thin targets, validating the modified model, we will investigate whether these extensions have an influence on particle emission and on the residue production inside a thick target. This validation will allow studying the impact of spallation model improvements on particle transport inside media.

The INCL4 model

The standard version of the Liège INC model is described in detail in Ref. [1]. It is important to recall that the particle-nucleus collision is described by a Monte Carlo simulation. Initially the position and momentum of all target nucleons are randomly distributed in the nuclear volume and in a Fermi sphere, respectively. In the INCL4 version, a smooth initial density distribution is introduced, in concordance with electron scattering data. NN⇔NN, NN⇔NΔ and Δ⇔πN collisions are based on realistic parameterised cross-sections and are subjected to a consistent Pauli blocking. The cascade is stopped according to a criterion based on the time evolution of some physical quantities. The code accommodates nucleons and light composites as incident particles.

This leads to a parameter-free code in the 200 MeV-2 GeV range with its own absolute normalisation (the computed total reaction cross-section is correctly reproduced). Neutron and proton energy spectra are well reproduced for a vast set of target nuclei and incident energies (see Figures 1 and 2). Concerning residue production, the rates are on average well reproduced, except for the isotopes containing more protons than the target nucleus, which are overestimated (see Figures 3 and 4) and for the rare earths which are underestimated (not shown here).

The recent extensions of the INCL42 model consist of:

- the introduction of an isospin and energy-dependent baryon mean field [4];

- an improvement of the pion dynamics: new cross-sections and introduction of the pion mean field [5];

- several refinements allowing the extension to lower energy [6-8];

- the implementation of d, t, ^{3}He and ^{4}He production during the intranuclear cascade step [9];

- extension to higher energy, up to 20 GeV [10].

Since the last two extensions are still in progress, they will not be considered in our studies below. Despite the first reasonably good results obtained for incident energies as low as a few tens of MeV, among the low-energy extensions, only the refinement of the Pauli blocking for the first collision will be used in our studies. Some of these low-energy ingredients can not be easily taken into account at the same time as the energy dependence of the nucleon potential.

Thin target results

The modification of the baryon mean field leads to an improvement (reduction) of the multiplicity of neutrons per cascade. The excitation energy of the nucleus after the INC step and the multiplicity of neutrons per evaporation are increased. Figure 1 gives the double-differential cross-section for neutron production induced by 800-MeV protons on ^{208}Pb. Globally, the effects of our modifications are rather small: a slight decrease of the intensity of the quasi-elastic peak and a shift towards lower energy are observed, the production of intermediate-energy neutrons is slightly reduced and the emission of evaporated neutrons is increased. Compared to the experimental measurements, our modified model improves the capabilities of INCL. Since the quasi-elastic peak, and thus events involving few emitted particles, is improved by our modifications, we also expect an improvement in the production of residues close to the target isotope (see below).

Figure 1. Double-differential cross-section of production of neutrons induced by 800-MeV protons on ^{208}Pb

A zoom of the quasi-elastic peak is given at the top of the figure. The results obtained with the standard and the modified models are given by the dotted line and the continuous line, respectively. Experimental data come from Ref. [11].

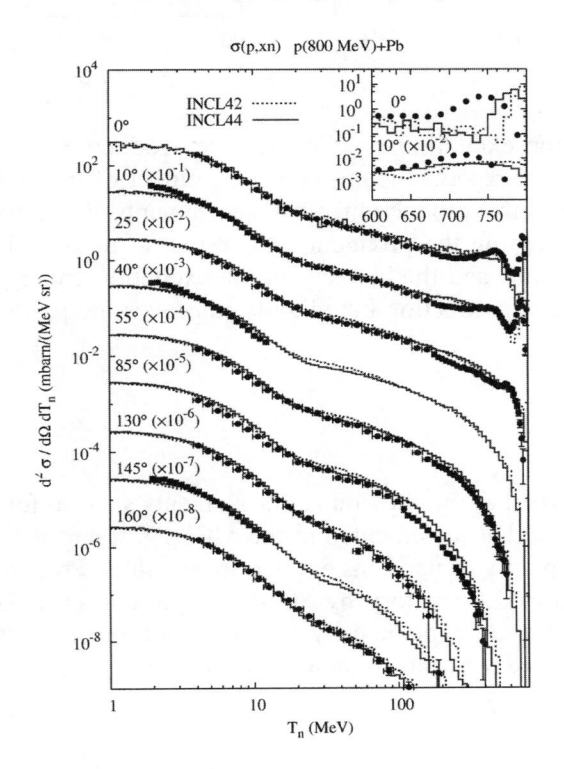

Figure 2 gives the double-differential cross-section for π^- production induced by 730-MeV protons on ^{208}Pb. This figure shows that the production of pions is considerably improved by our modifications. This will also have an impact on the production of some residues: above 100 MeV, the only way to produce ^{210}Po from ^{209}Bi is the (p,π^0) channel. The globally good agreement seems to show that the most important characteristics of the pion and Δ dynamics are taken into account in the modified INCL model.

Figure 2. Double-differential cross-section for production of negative pions induced by 730-MeV protons on ^{208}Pb

Same convention as in Figure 1. Experimental data are taken from Ref. [12].

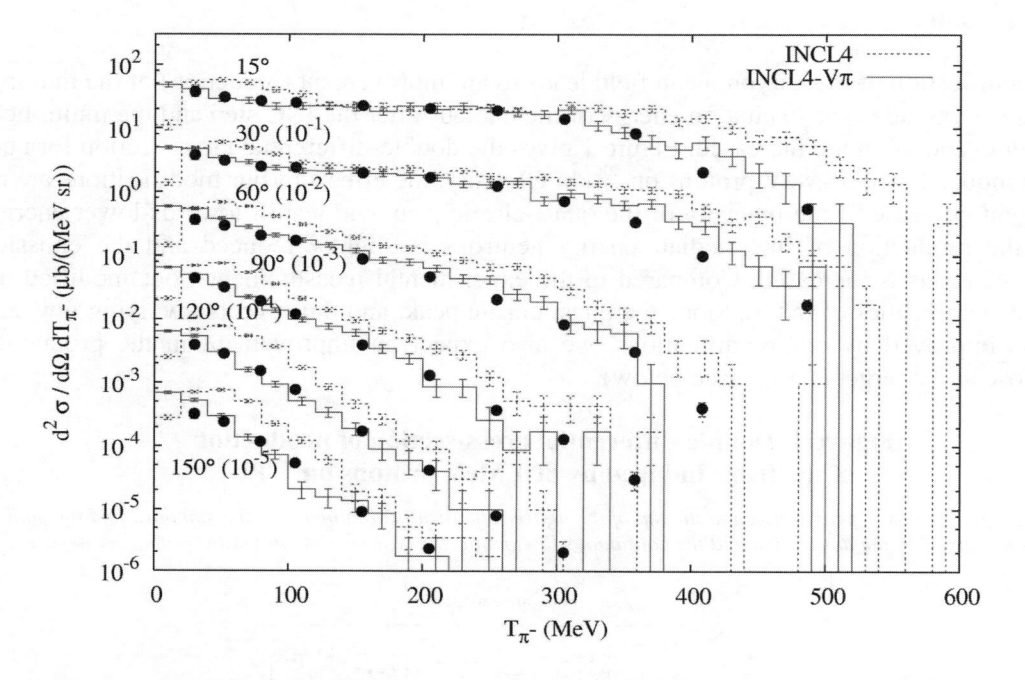

Figures 3 and 4 give the excitation function for the production of ^{210}Po and ^{209}Po induced by protons on a ^{209}Bi nucleus. As expected from the improvement of the estimation of the quasi-elastic peak and of the pion sector, the production of these elements is improved by our modifications. It should be noted that concerning the production of residues, the overall good agreement observed with the standard model remains and the increase of the excitation energy after the cascade step does not degrade the fission residue production but slightly improves the production of rare-earth isotopes (not shown here, see Ref. [5]).

Thick target results

We have investigated whether the previous improvements survive for a thick target. To this aim, we have updated the INCL model implemented in the MCNPX transport code [16]. Figure 5 displays the double-differential multiplicity of neutrons emitted from a thick Pb target (60 cm length and 10 cm radius) irradiated at the Saturne laboratory by 800-MeV protons [17]. Globally, our modifications bring smaller effects. This can be explained by compensation effects: less primary neutrons but of higher energy will produce more neutrons in secondary collisions.

Figure 3. Excitation of the cross-section for ^{210}Po production induced by protons on ^{209}Bi

Same convention as in Figure 1. Experimental data are taken from Ref. [13].

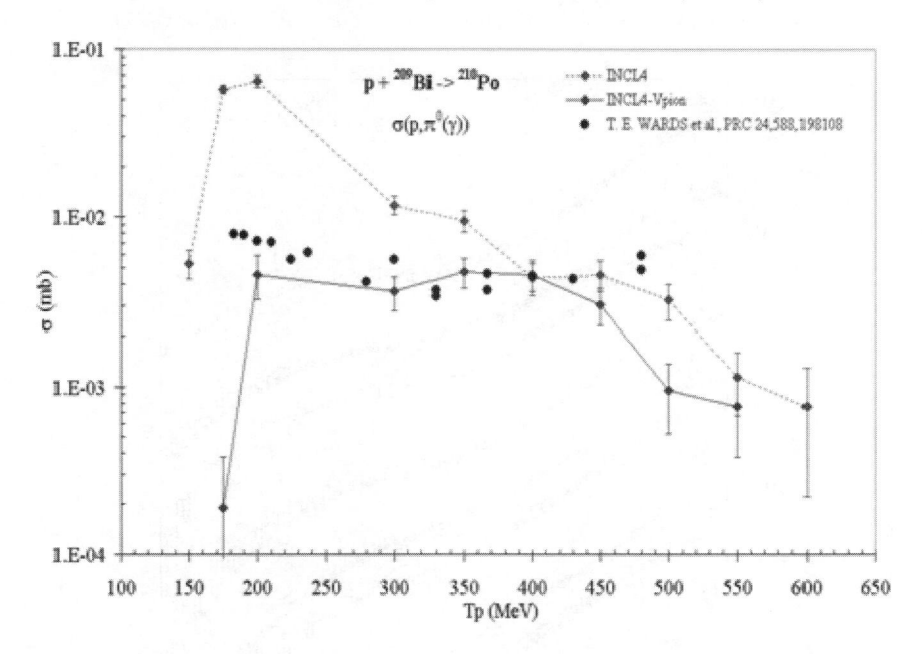

Figure 4. Excitation function of the cross-section for ^{210}Po production induced by protons on ^{209}Bi

Same convention as in Figure 1. Experimental data are taken from Refs. [14,15].

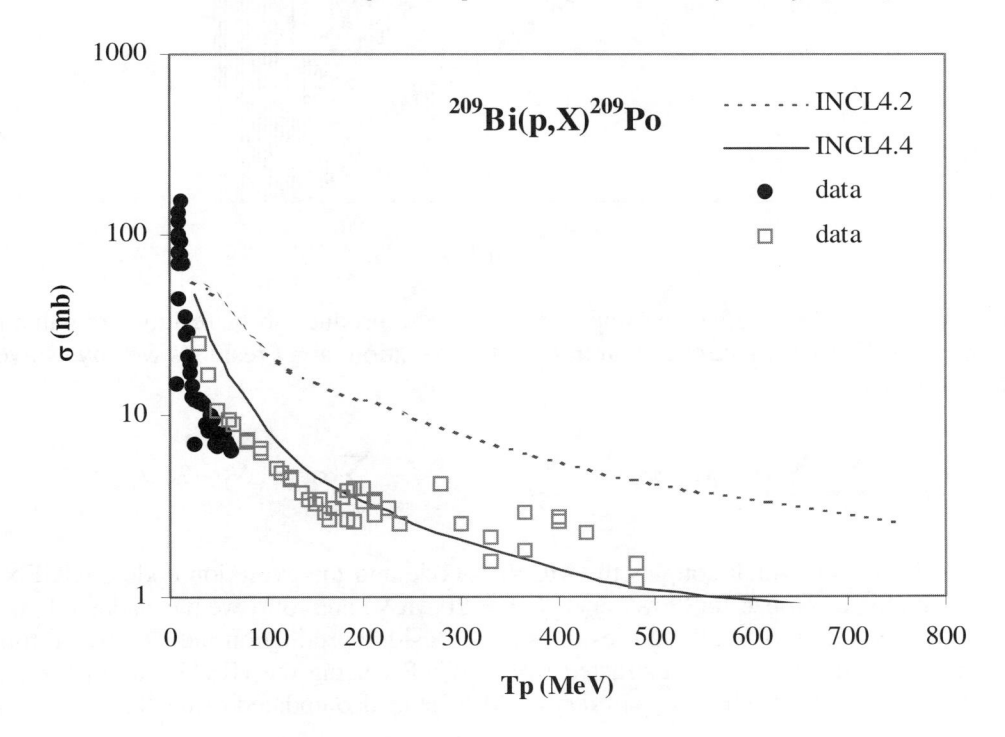

Figure 5. Double-differential multiplicity of neutrons induced by protons of 800 MeV on a thick Pb target (see text for detail)

Same convention as in Figure 1. Experimental values are taken from Ref. [17].

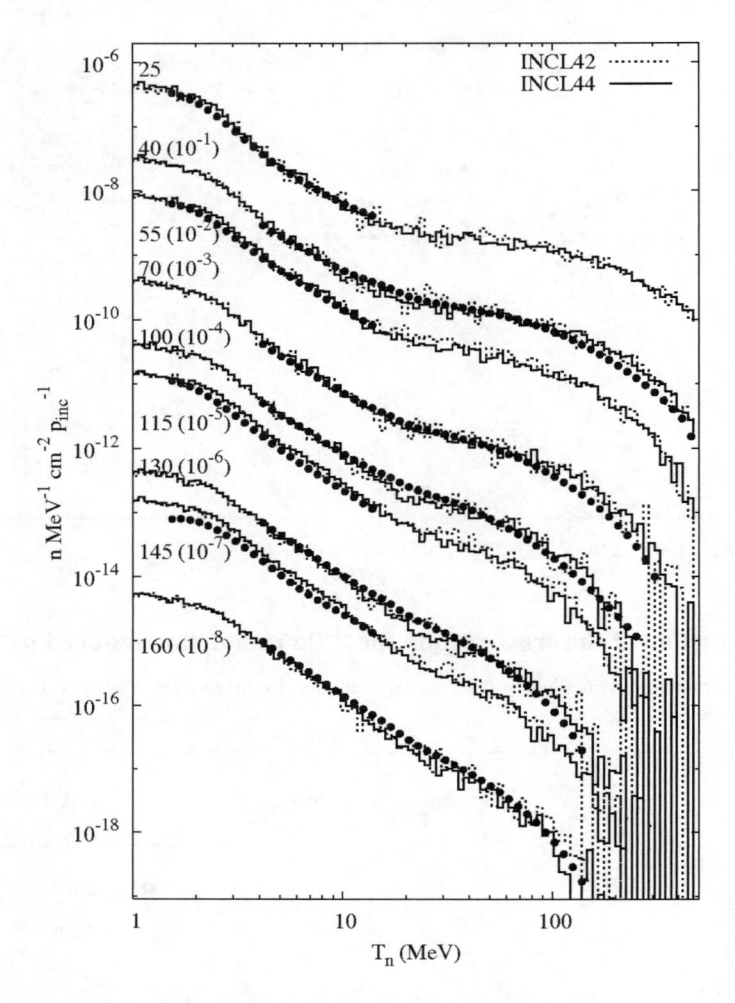

We will now investigate whether the improvement of the production of residues remains in the thick target case. In order to estimate the evolution of the spallation target residues we have solved the multi-particle Bateman equations:

$$\frac{dN_i}{dt} = \sum_{j \neq i} \sum_{k=n,p,\pi} \alpha_{j,k,i} \sigma_{j,k} \phi_k N_j + \sum_{j \neq i} \beta_{j,i} \lambda_j N_j - \sum_{k=n,p,\pi} \sigma_{i,k} \phi_k N_i - \lambda_i N_i$$

using the ALEPH [18] code which couples the MCNPX code and the evolution code ORIGEN [19]. As ALEPH was initially developed for "low" energy (E < 20 MeV) neutrons, we have adapted this code to accommodate the evolution of spallation residues [20]. Residue production induced by protons and higher energy neutrons' interaction are extracted from MCNPX using the HISTP output file and the HTAPE3X post-processing tool. The decay library of ALPEH is also updated using NUBASE97 [21].

The experimental measurements used to validate the production of residues inside a thick target are taken from Ref. [22]. In this experiment, a stack of alternating Pb and Bi disks was shortly irradiated by 590-MeV protons at PSI and the residues were measured by γ-spectrometry. In Figure 6

Figure 6. Isotopic density of At, Po, Bi and Pb produced in the second disk (^{209}Bi)

Experimental data are taken from Ref. [24]. Figure adapted from Ref. [25].

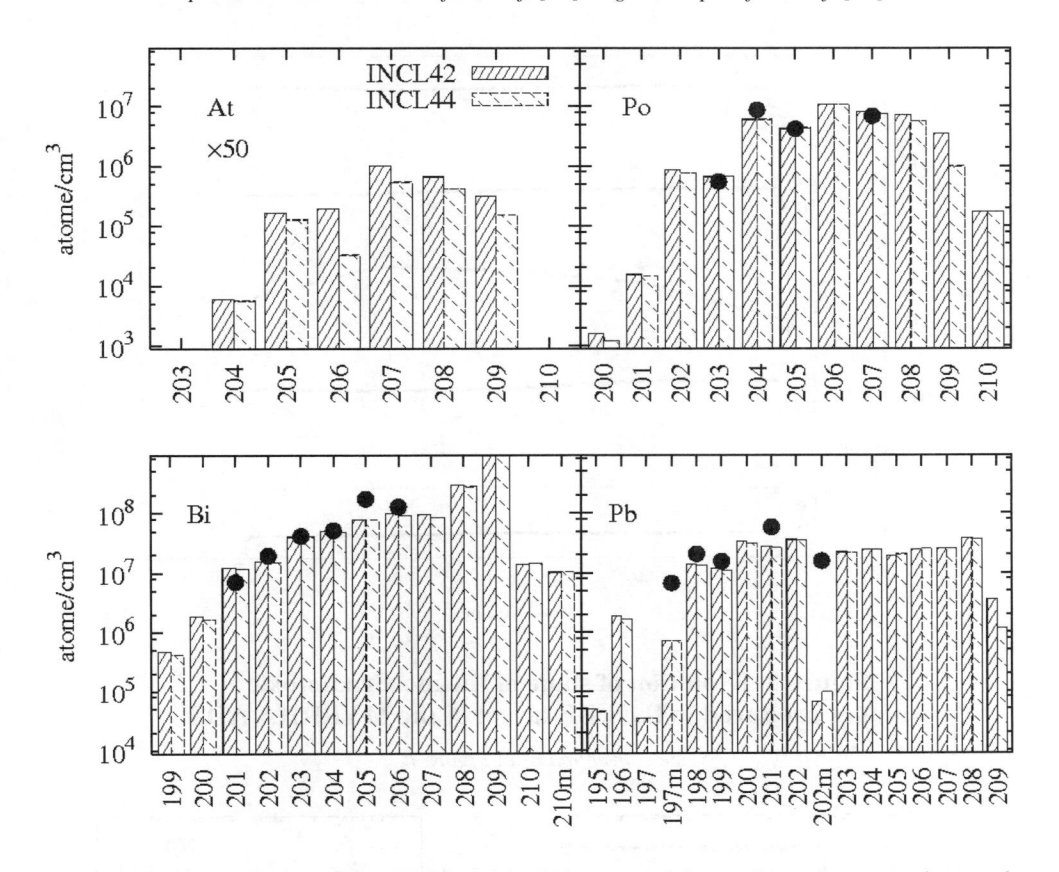

the theoretical estimates obtained in the first Bi disk are confronted to the experimental results of Ref. [23]. Both the standard and the modified INCL models reproduce the experimental results. The production of isomeric nuclei is strongly underestimated. This is due to the use of old data for the residual γ de-excitation (PHTLIB data file) provided with MCNPX version 26a [24]. As foreseen from thin target analyses, ^{208}Po and ^{209}Po are reduced with the same magnitude. The production of ^{210}Po is not influenced by our modifications and more ^{210}Po are produced inside a thick target than a thin one. In a thick target, the (n,γ) reaction is of course the dominant source of production of ^{210}Po.

Only the first eleven disks were analysed in Ref. [23]. The calculated production of the most highly radiotoxic isotopes is given in Figure 7 for the Bi disks, which contain most of the radiotoxic isotopes [25]. Except for the production of ^{210}Po, which is fed by the low-energy neutron capture reactions, the production of the other isotopes remains quite constant inside the successive disks and the reduction of Po isotopes is basically constant inside these disks.

Figure 8 displays the calculated evolution of some of the most radiotoxic isotopes produced in the LBE target of the MYRRHA ADS [26]. The irradiation consists of 1.88 mA and 600-MeV protons during 270 EFPDs. After this irradiation, the target decays during 1 000 years. We can draw the same conclusion as for the PSI target except that here the production rates of ^{210m}Bi and of ^{210}Po are still larger due to the contribution of the neutrons coming from the surrounding subcritical core which enhances the (n,γ) reaction rate on ^{209}Bi.

Figure 7. Activity (in Bq/cm^3) of ^{204}Bi and ^{194}Hg induced in the Pb disks

Same convention as in Figure 1. Figure adapted from Ref. [25].

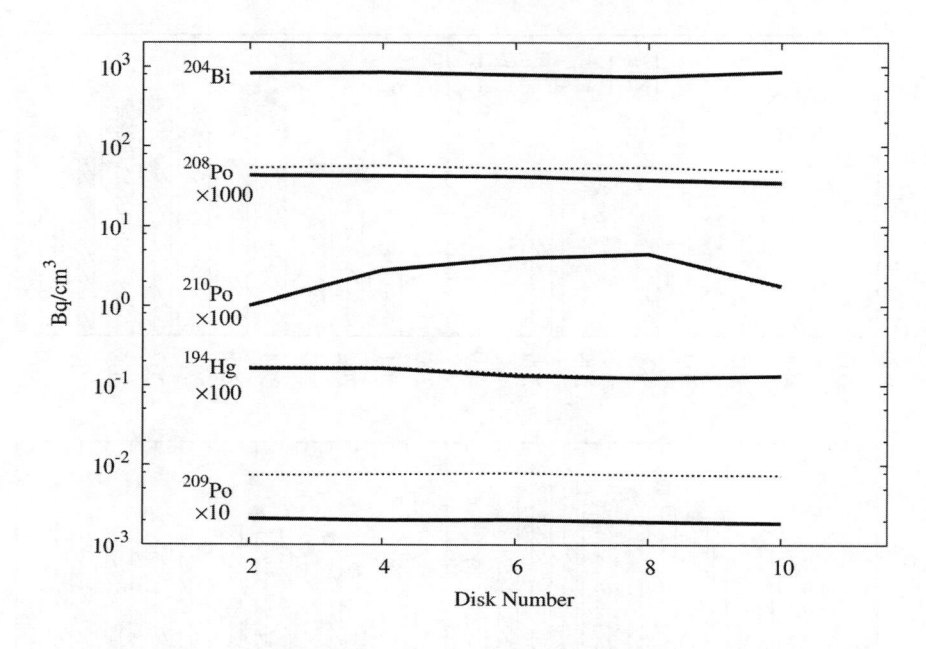

Figure 8. Evolution of some of the most radiotoxic isotopes produced in the LBE target of the MYRRHA ADS

Same convention as in Figure 1

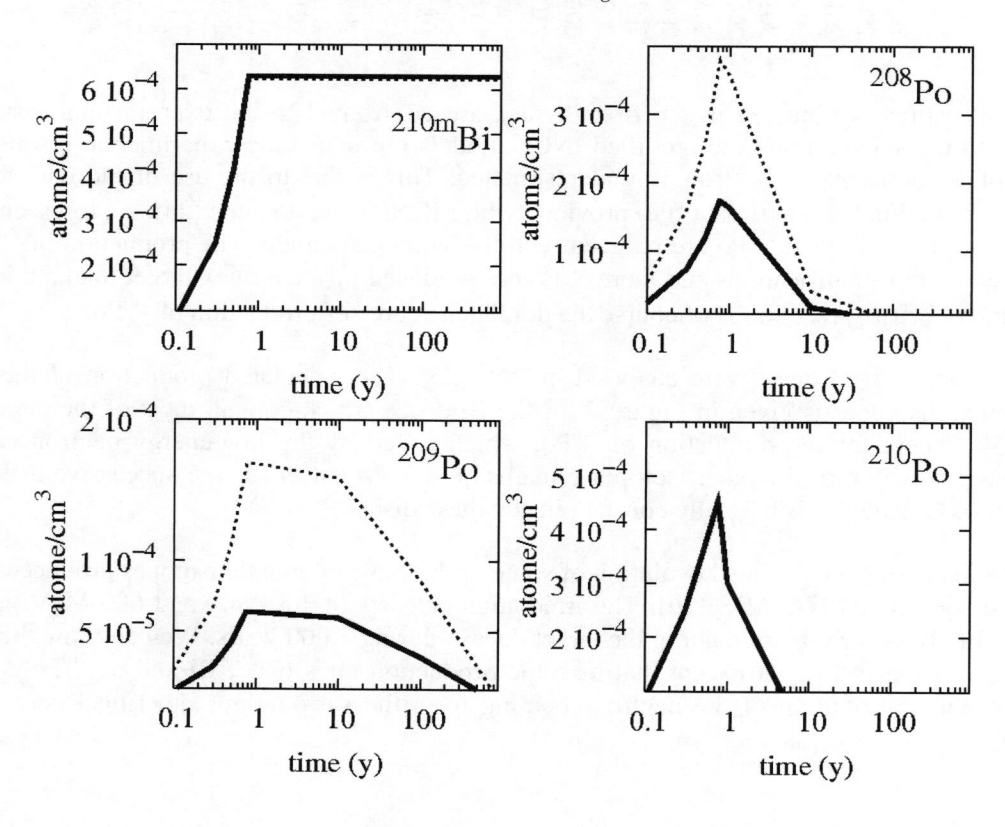

Conclusion

Extensions of the INCL model allowed a better description of events involving the emission of few particles and, consequently, the production of polonium residues induced by protons in thin ^{209}Bi targets. The predicted ^{210}Po and of ^{209}Po production cross-sections are considerably reduced and are now very close to the experimental measurements.

To assess the influence of spallation models' improvements on thick spallation target design, the INCL model implemented in the MCNPX code was updated. If the impact of our modifications on the emission of nucleons is reduced by compensation effects inside thick targets, for the production of residues, the variation observed is of the same magnitude for thin and thick targets.

The effect of our improvements for the production of some radiotoxic isotopes induced in thick targets were first performed for the stack of alternating Pb and Bi disks bombarded with 590-MeV protons at PSI. The evolution of this spallation target was performed by adapting the ALEPH code. For the few experimental measurements, the predictions of the standard and the modified INCL models remain close to each other and give satisfactory results. The use of old data for the residual γ de-excitation in MCNPX26a gives a strong underestimate of the production of isomeric states. As observed from our thin target analysis, the production of ^{209}Po and ^{208}Po is strongly reduced. Comparing the effects of our modifications in thin and thick targets, the observed variations due to our modifications remain similar. The production of ^{210}Po and of ^{210m}Bi in the bismuth disks are not influenced by our new implementations, since these isotopes are mainly produced by (n,γ) reactions, followed, for the ^{210}Po, by β$^-$ decay.

In the case of the LBE target of the MYRRHA ADS, the same conclusion as for the PSI target can be drawn, except for of ^{210}Po and of ^{210m}Bi for which the production rates are larger due to the presence of the surrounding multiplying media.

REFERENCES

[1] Boudard, A., *et al.*, *Phys. Rev. C*, 66, 044615 (2002).

[2] Benlliure, J., *et al.*, *Nucl. Phys. A*, 628, 458 (1998).

[3] Aoust, Th., *et al.*, *Nucl. Instr. Method A*, 562, 810 (2006).

[4] Aoust, Th., J. Cugnon, *Eur. Phys. J. A*, 21, 79 (2004).

[5] Aoust, Th., J. Cugnon, *Phys. Rev. C*, 74, 064607 (2006).

[6] Cugnon, J., P. Henrotte, *Eur. Phys. J. A*, 16, 393 (2003).

[7] Boudard, A., *et al.*, *Eur. Phys. J. A*, forthcoming.

[8] Yariv, Y., et al., "Intra-nuclear Cascade Models at Low Energy?", *International Conference on Nuclear Data for Science and Technology (ND2007)*, Nice, France, to be published (2007).

[9] Boudard, A., et al., *Nucl. Phys. A*, 740, 95 (2004).

[10] Cugnon, J., S. Pedoux., "Extension of the INCL4 Model for Spallation Reactions Up to 20 GeV", *International Conference on Nuclear Data for Science and Technology (ND2007)*, Nice, France, to be published (2007).

[11] Leray, S., et al., *Phys. Rev. C*, 65, 044621 (2002).

[12] Cochran, D.R.F., et al., *Phys. Rev. D*, 6, 3085 (1972).

[13] Wards, T.E., et al., *Phys. Rev. C*, 24, 588 (1981).

[14] D'auria, J.M., et al., *Phys. Rev. C*, 30, 236 (1984).

[15] Miyano, K., et al., *Journal of the Physical Society of Japan*, 45, 1071 (1978).

[16] *MCNPX User's Manual*, Los Alamos Report LA-CP-02-408.

[17] Ménard, S., "Production de neutrons en cible épaisse par un faisceau de protons de haute énergie", PhD Thesis, Institute of Nuclear Physics Orsay, University of Paris South, 6 Jan. 1988.

[18] Haech, W., B. Verboomen, *Nucl. Sci. Eng.*, 156, 1 (2007).

[19] Croff, A.G., *Nuclear Technology*, 62 (1983).

[20] Aoust, Th., J. Cugnon, W. Haeck, "Actinide Transmutation by Spallation", *Proceedings of the 12th International Conference on Emerging Nuclear Energy Systems (ICENES05)*, Brussels, Belgium, B. Verboomen, et al. (Eds.), ISBN 90-7697-110-2, August 2005.

[21] Audi, G., et al., *Phys. A*, 624, 1 (1997).

[22] van der Meer, K., et al., *Nucl. Instr. Method A*, 217, 202 (2004).

[23] Dams, C., "Analyse van gamma-spectrometriegegevens van door 590 MeV protonen bestraalde Pb- en Bi-schijven", XIOS Hogeschool Limbourg (2005).

[24] Hendricks, J.S., et al., *MCNPX, Version 26b*, Los Alamos Report, LA-UR-06-3248 (2006).

[25] Aoust, Th., J. Cugnon, J. Wagemans, "Production of Radiotoxic Isotopes in Proton-induced Spallation Reactions on Thick LBE Targets. Recent Extensions of the INCL Model and Experimental Validation", *International Conference on Nuclear Data for Science and Technology (ND2007)*, Nice, France, to be published (2007).

[26] Aït Abderrahim, H., et al., *MYRRHA Pre-design File – Draft 2*, SCK•CEN Report R-4234 (2005).

ENGINEERING DESIGN OF THE EURISOL MULTI-MW SPALLATION TARGET

Adonai Herrera-Martínez[1], Yacine Kadi[1], Morteza Ashrafi-Nik[2],
Karel Samec[2], Janis Freibergs[3], Ernests Platacis[3]
On behalf of the EURISOL DS Task 2
[1]European Laboratory for Nuclear Research (CERN), Geneva, Switzerland
[2]Paul Scherrer Institute (PSI), Villingen, Switzerland
[3]Institute of Physics at the University of Latvia (IPUL), Riga, Latvia

Abstract

The *Eur*opean *I*sotope *S*eparation *O*n-*l*ine Radioactive Ion Beam project (EURISOL) is set to design the "next-generation" European *I*sotope *S*eparation *O*n-*l*ine (ISOL) *R*adioactive *I*on *B*eam (RIB) facility. It will extend and amplify current research on nuclear physics, nuclear astrophysics and fundamental interactions beyond the year 2010.

In EURISOL, four target stations are foreseen, three direct targets of approximately 100 kW of beam power and one multi-MW target assembly, all driven by a high-power particle accelerator. In this high-power target station, high-intensity RIBs of neutron-rich isotopes will be obtained by inducing fission in several actinide targets surrounding a liquid metal spallation neutron source.

This article summarises the work carried out within Task 2 of the EURISOL Design Study, with special attention to the coupled neutronics of the mercury proton-to-neutron converter and the fission targets. The overall performance of the facility, which will sustain fast neutron fluxes of the order of 10^{14} n/cm^2/s, is evaluated, together with the production of radionuclides in the actinide targets, showing that the targeted 10^{15} fissions/s can be achieved.

Some of the greatest challenges in the design of high-power spallation sources are the high-power densities, entailing large structural stresses, and the heat removal, requiring detailed thermo-hydraulic calculations. The use of a thin martensitic steel beam window and a well-controlled mercury flow has been shown to reduce the von Misses stress in the former below the 200 MPa limit, with reasonable maximum flow rates of ~6 m/s.

Alternatively, a windowless target configuration has been proposed, based on a liquid mercury transverse film. With this design, higher power densities and fission rates may be achieved, avoiding the technical difficulties related to the beam window. Experimentally, several tests have been performed at IPUL (Riga, Latvia) in order to study the stability of the liquid metal flow and validate the mercury loop design.

Introduction

The EURISOL DS project [1] aims at a design study of the "next-generation" *Eu*ropean *I*sotope *Se*paration *On-l*ine (ISOL) *R*adioactive *I*on *B*eam (RIB) facility, which will extend and amplify, beyond the year 2010, the work presently being carried out at the first generation RIB facilities in Europe and other parts of the world, in the fields of nuclear physics, nuclear astrophysics and fundamental interactions.

The scientific case for high-intensity RIBs using the ISOL method includes: (a) the study of atomic nuclei under extreme and so-far unexplored conditions of composition (i.e. as a function of number of protons and neutrons, or the so-called isospin), rotational angular velocity (or spin), density and temperature; (b) the investigation of the nucleosynthesis of heavy elements in the universe, an important part of nuclear astrophysics; (c) study the properties of the fundamental interactions governing the universe, and in particular of the violation of some of their symmetries; (d) potential applications of RIBs in solid-state physics and in nuclear medicine. These cases require a next-generation infrastructure such as the proposed EURISOL facility, with intensities several orders of magnitude higher than those presently available or planned, allowing the study of hitherto completely unexplored regions of the Chart of the Nuclei.

The main components of the proposed facility are: a driver accelerator, a target/ion source assembly and a mass selection system [2]. As shown in Figure 1, the proposed ISOL facility would use: (a) three 100-kW proton beams on a thick solid target to produce RIBs directly; (b) a liquid metal 1-5 MW proton-to-neutron converter, similar to intense spallation neutron sources such as ESS [3], SINQ [4] and SNS [5], to generate high neutron fluxes, which would then produce RIBs by fission in secondary actinide targets. An alternative windowless liquid mercury-jet "converter" target to generate the neutrons was also proposed for this multi-MW target station [2].

Figure 1. EURISOL DS schematic layout, presenting the three direct targets and the multi-MW target station

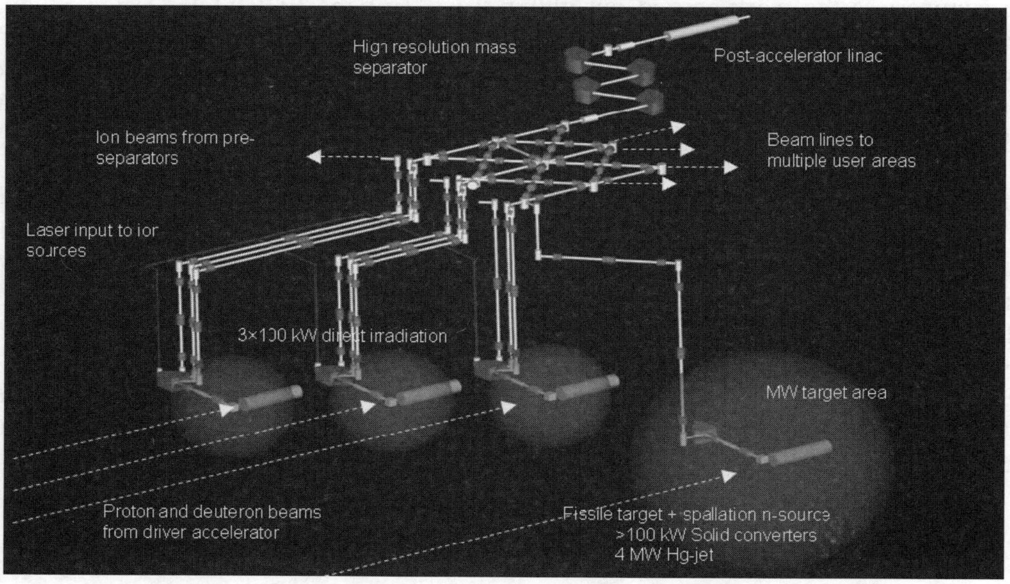

Since the purpose of the facility is to produce certain radioisotopes, maximising the yield of such isotopes (e.g. ^{74}Ni, ^{81}Ga, ^{90}Kr or ^{132}Sn) is the main objective in the design. For the proton-to-neutron converter this implies increasing the neutron yield and reducing the parasitic absorptions in the spallation target. The compactness and efficiency of the assembly is mandatory in order to minimise the total inventory of material in the facility and attain the specified neutron flux and fission density. Moreover, to increase the fission rate in a non-enriched target, the neutron energy spectrum should lie in the fast region, since fission cross-sections for non-fissile isotopes are higher at these energies. This harder neutron spectrum may be achieved by decreasing the moderation of the spallation neutrons in the target.

Finally, minimising the power densities is a requirement to allow for the evacuation of the heat from the converter, in particular from the liquid mercury target and the beam window interface. This is one of the most complex issues when dealing with high-power spallation targets. Consequently, a sensitivity analysis, covering a broad range of parameters, was performed [6] in order to propose some alternatives for the design.

Neutronic design of the multi-MW target

Following the results from the aforementioned study, performed using the Monte Carlo particle transport code FLUKA [7], a baseline design was defined [8,9]. In order to maximise the neutron production, favour a fast neutron spectrum and confine the charged particles inside the assembly, a 8-cm radius, 40-cm long mercury proton-to-neutron converter was proposed, surrounded by fission targets and, possibly, by a neutron reflector (Figure 2). For the latter, beryllium oxide (BeO) was proposed due to the high albedo of this material and to produce ^{6}He (via n,α reactions in ^{9}Be) for neutrino physics (β-beams) [10].

Figure 2. Schematic representations of the baseline design, where several components of the facility have been integrated

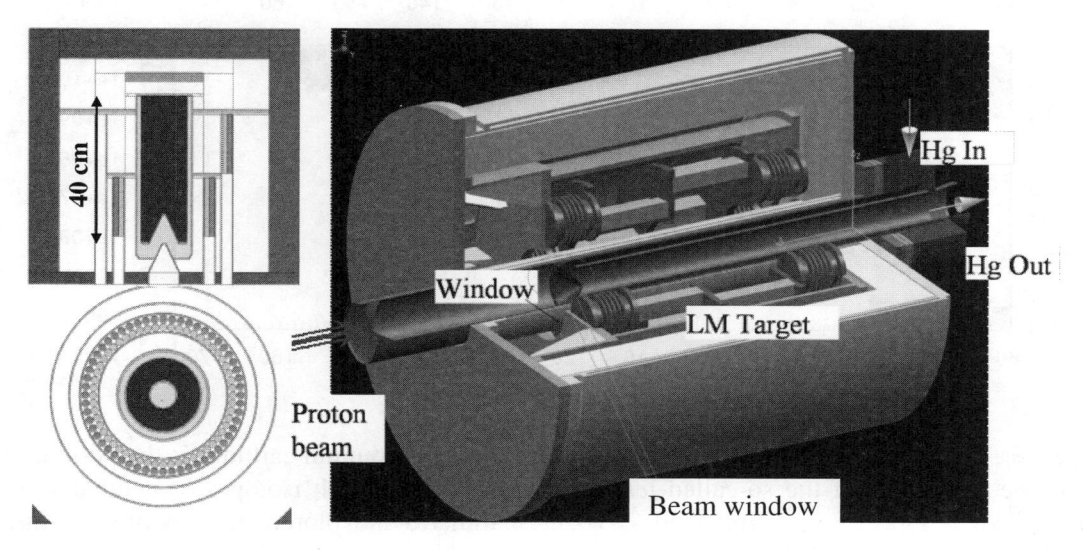

The neutron flux distribution in the baseline design is rather isotropic a few cm away from the centre of maximum production (from 0 to 10 cm from the impact point), as elaborated in [8]. The flux in the fission target is $\sim 10^{14}$ n/cm^2/s per MW of beam, similar to those of conventional nuclear reactors. As elaborated below, these flux levels are more than sufficient to produce the aimed $\sim 10^{15}$ fissions per second [8] with reasonable fission target volumes (5 litres) and using an acceptable beam power.

Multiple calculations were carried out to assess the performance of fission target materials for the baseline design, taking advantage of the new developments implemented in FLUKA [11]. The use of natural uranium carbide was analysed and compared to thorium oxide, for the same target densities, the latter producing one order of magnitude lower fission rates.

The detailed isotopic distribution of the fission fragments may be observed in Figure 3, allowing the prediction of RIB intensities for specific isotopes. These distributions show the nature of the isotopes produced by fission: these lie on the unstable, neutron-rich area of the chart of nuclides (β^- emitters), ranging from manganese to terbium. The use of depleted uranium carbide or thorium oxide entails a reduction in the production of asymmetric fission fragments ($32 < Z < 42$ and $50 < Z < 58$); thus, the presence of ^{235}U is advantageous for the production of elements such as krypton or tin, major references in the physics case for EURISOL [2].

Figure 3. Fission fragment distribution (nuclei/cm^3/s per MW of beam) as a function of atomic number (Z) and mass number (A), for two actinide targets. Stable isotopes are represented by black squares.

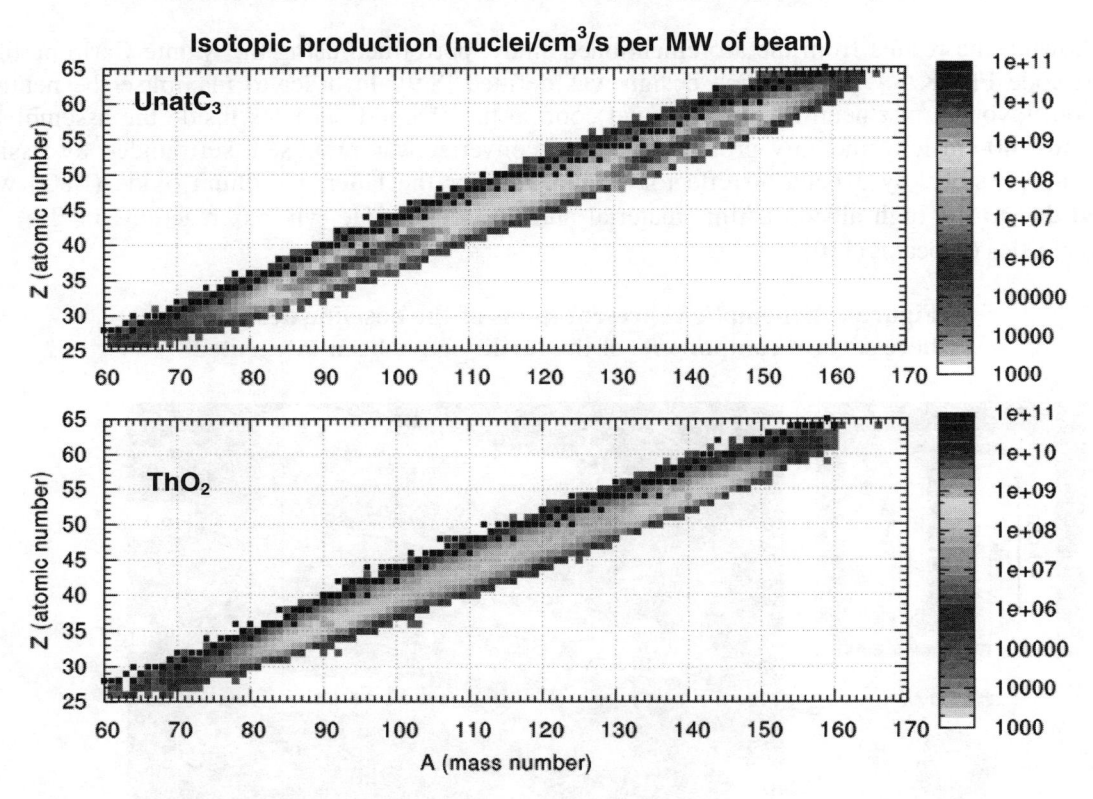

Moreover, a relevant benefit of the large fission densities in uranium carbides is the possibility to investigate the lower end of the so-called *terra incognita*, neutron-rich isotopes of neodymium and above (e.g. ^{157}Nd, ^{159}Pm, ^{162}Sm, ^{163}Eu, ^{166}Gd, ^{167}Tb, etc.), hitherto unexplored. A study of the neutronic design of the facility and its RIB production potential was published in [12].

An estimation of the actinide inventory produced after 3 000 hours of operation is presented in Ref. [13]. Between 40 and 70 g of ^{239}Pu would be produced in a diluted manner within 30 kg of fissile material surrounded by a neutron reflector and by up to 10 m of concrete shielding. The shielding of the assembly was designed to maintain the effective contact dose below 1 µSv/h at its outer surface.

Thermo-hydraulics and beam window design

A key parameter in the design of the experiment is the power distribution, since it will determine the maximum beam intensity that the system may withstand, which in turn is correlated with the fission rates. As elaborated in [8], the energy deposition peaks at ~2 cm after the interaction point, reaching 1.9 kW/cm^3 per MW of beam, and decreases rapidly, along the beam axis. The beam window is enduring lower heat deposition (~900 W/cm^3/MW of beam). These power densities demand an innovative liquid mercury flow design and a careful choice of beam window material.

An iterative design process was necessary to limit the large thermal stresses in the beam window, which were higher than the 200 MPa limit for martensitic steel T91 (below 2 dpa of radiation damage), and the temperature gradient. Finally, these stresses were reduced to ~150 MPa for a beam of 4 MW [14]. Nevertheless, a wider beam profile was considered, 25 mm standard deviation (σ) for the Gaussian beam distribution instead of 15 mm, to further reduce stresses in the window. Figure 4 shows a comparison of power densities for both cases. There is a 2.6 reduction factor brought about by the beam enlargement, and a decrease in temperature gradients, proven by the more homogenous power distribution for a 25 mm σ beam.

Figure 4. Power density distribution for different beam widths, along the beam axis and around the window

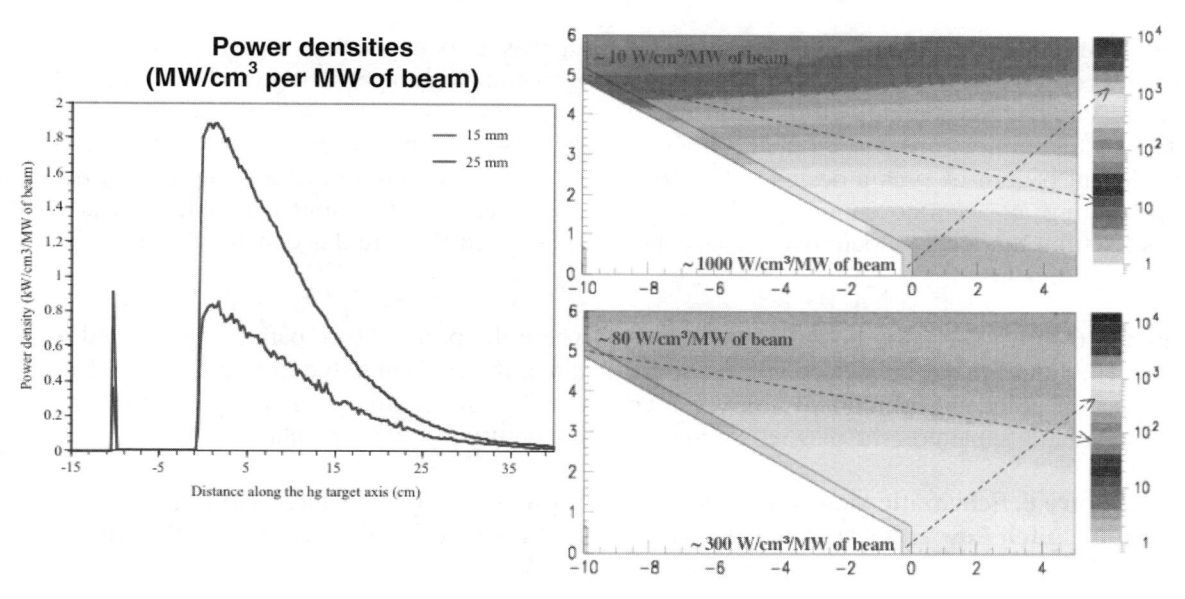

Once the beam window was optimised, the liquid mercury flow inside the target container was recalculated to minimise pressure losses while ensuring adequate cooling and preventing boiling and cavitation in the back-swept surfaces. Several design changes were performed to improve the flow, such as the use of annular blades along the beam window to accelerate the flow, increase the local cooling and reduce the pressure drop at the 180° turn. Holes through the guide tube were also foreseen to avoid recirculation.

With this design and a bulk pressure of 7.5 bar, the maximum temperature in the beam window is ~200°C and the maximum von Misses stress ~135 MPa. Concerning the mercury flow, the peak temperature is 180°C (at the beam axis, 2 cm away from the interaction point) and 6 m/s is the maximum velocity (in the channels formed by the flow guides and the walls, at the 180° turn).

Figure 5 represents the temperature distributions within the structural materials (a), and in the flowing mercury (b). Note the sharp temperature gradient in the beam window, main source of difficulties in the design of this element.

Figure 5. Temperature distribution in the structures of the container (a), and within the mercury flow (b)

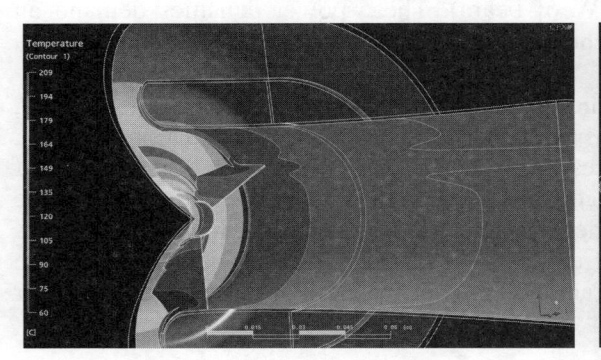

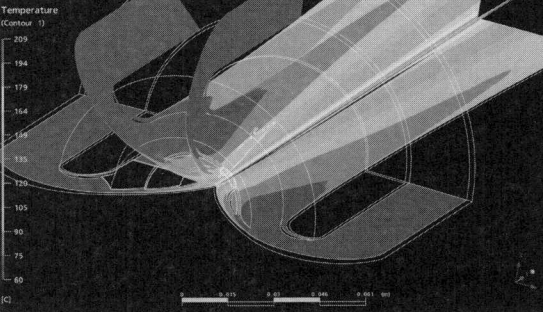

Alternative transverse mercury film spallation target

An alternative and innovative windowless design was also developed to avoid the technical difficulties related to the beam window, also presenting several advantages in terms of neutronics. The transverse windowless mercury film would fall by gravitation, interacting with the proton beam to produce spallation neutrons and efficiently removing the heat deposited at reasonable flow rates. The most relevant benefit of such a design is the brief exposition of the liquid metal to the proton beam, thus permitting an accurate control of the temperature increase in the molten metal. This is also achieved setting the local velocity by varying the pitch between flow guides depending on the beam cooling requirements.

Figure 6 shows the basic layout of such design, where the proton beam path is represented by a red line. The liquid metal flows through the upper tube and the fins guide the falling mercury. Below the interaction point, the mercury is recovered and driven to the auxiliary circuit some 10 m away, where the volatile separator, the magnetic pump and the mercury reservoir are placed.

Figure 6. Schematic view of the liquid metal transverse flow target, including the variable pitch flow guide segments to regulate the flow rate for different power densities

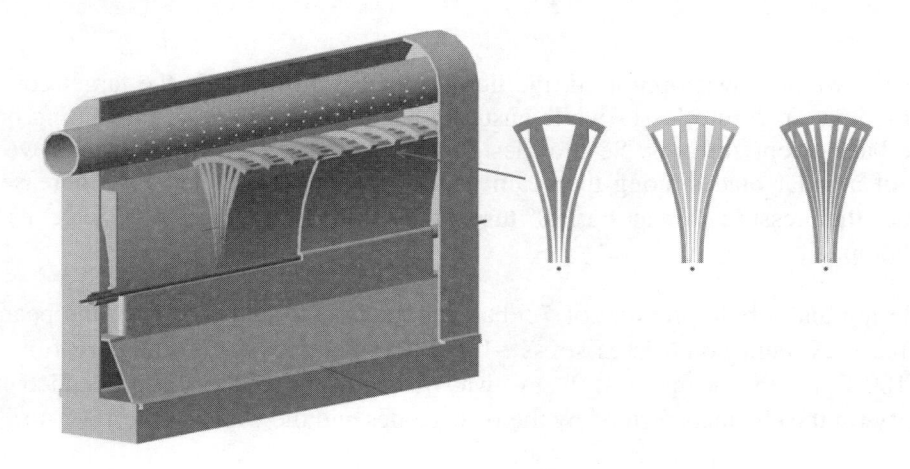

The technical simplicity of the system, in particular of the beam-target interaction as a free surface, facilitates its operation for extended periods of time by eliminating the need to exchange targets every few months due to beam window radiation damage and aging [5]. Moreover, the reduced thickness of the film produces a harder neutron spectrum and permits the positioning of flat actinide fission targets closer to the interaction point. This fact increases the fission density rates and reduces the higher actinide production, by favouring fission reactions rather than capture. The film is decoupled in two regions, a central one (~1 cm thick), receiving the direct impact of the beam and flowing at greater speed, and an external one (~1.5 cm thick on each side), confining the former, to reduce the high-energy escapes and maximise the production of spallation neutrons.

The proton beam is mostly contained within the proton-to-neutron converter, as opposed to the mercury jet design previously proposed in [2]. The neutron flux in the fission targets reaches 2×10^{14} n/cm^2/s per MW of beam and the proton and neutron distributions are similar to that of the mercury jet option [2,9]. High fission densities (~5×10^{11} fissions/cm^3/s per MW of beam, Figure 7) are achieved, allowing for the aimed RIB yields with reduced fission target volumes (1-5 litres). The neutron balance shows that the neutron-producing region extents to ~40 cm along the beam axis.

Figure 7. Neutronic parameters for the transverse film configuration, showing high and homogeneous fission densities in the UC$_3$ target and the power density distribution

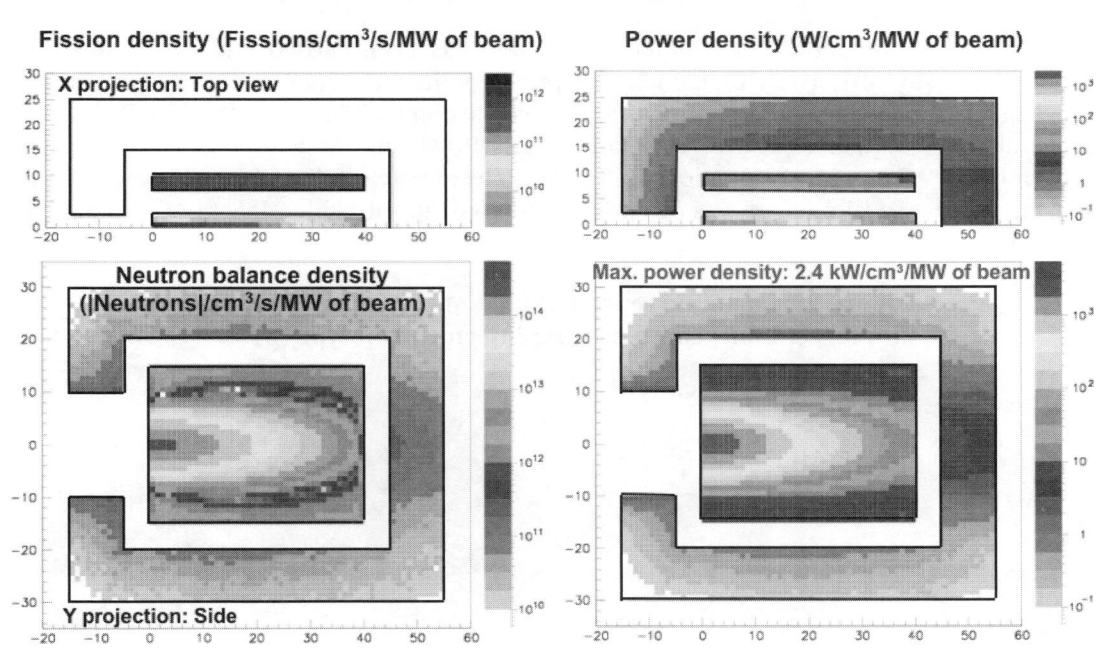

The maximum power density for a wide beam (σ_x ~4 mm, σ_y ~30 mm) is 2.4 KW/cm^3 per MW of beam. Even for more concentrated proton beams, i.e. $\sigma_{x,y}$ ~2 mm and 20 KW/cm^3 per MW of beam, this design would remove the energy deposited with moderate flow rates (~4 m/s), keeping temperatures well below the boiling point of mercury (ΔT ~100 K) [15].

In terms of RIB production, there is a clear advantage in using this design, particularly for the symmetric neutron-rich fission fragments. In fact, the transverse film design equals, and for some isotopes exceeds, the mercury jet estimated performance, as presented in Figure 8. As an example, since krypton and tin are two of the most relevant elements in the EURISOL DS physics case [2], this system could produce up to 5×10^{13} isotopes/s of ^{90}Kr and ^{132}Sn for a 4-MW proton beam and a total UC$_3$ target volume of 5 litres.

Figure 8. Fission fragment distribution (a), and Sn isotopic distribution (b), in UC$_3$ targets for different three different target designs (nuclei/cm^3/s per MW of beam)

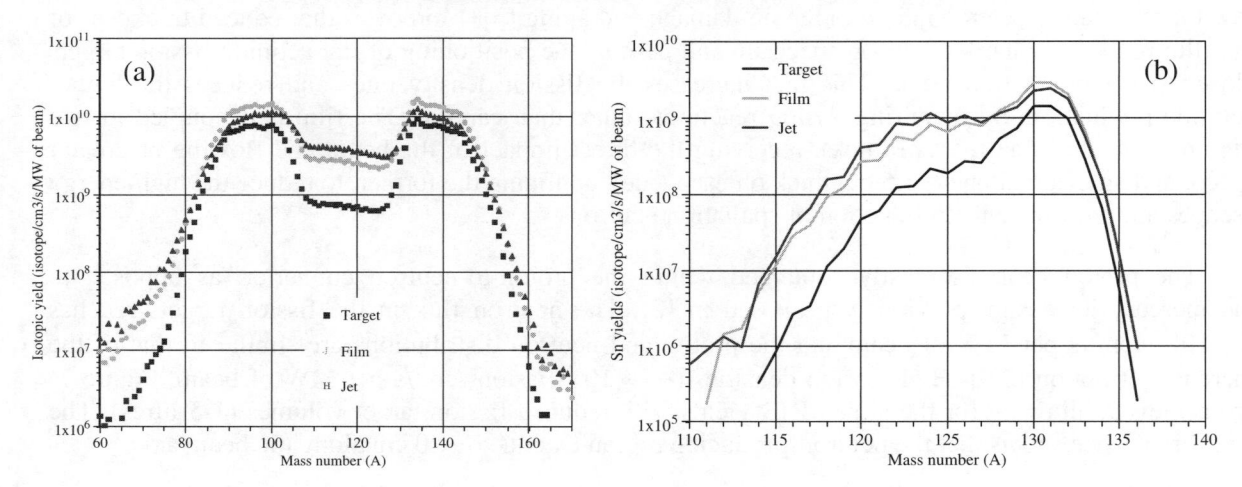

Liquid metal loop experimental tests

The experimental validation of the liquid metal loop has been carried out by the Institute of Physics at the University of Latvia (IPUL), where a prototype mercury loop and transverse film have been studied. The film behaviour and flow stability seem compatible with the EURISOL design requirements, although further tests involving larger mass flows and the addition of heat deposition should be performed. As previously mentioned, in order to test the feasibility of the proposed design, a mock-up of the transverse film target was developed and constructed, with a length of 100 mm and a width of 10 mm (Figure 9).

Figure 9. Schematic representation of the transverse film target honeycomb (a), and injector prototype (b)

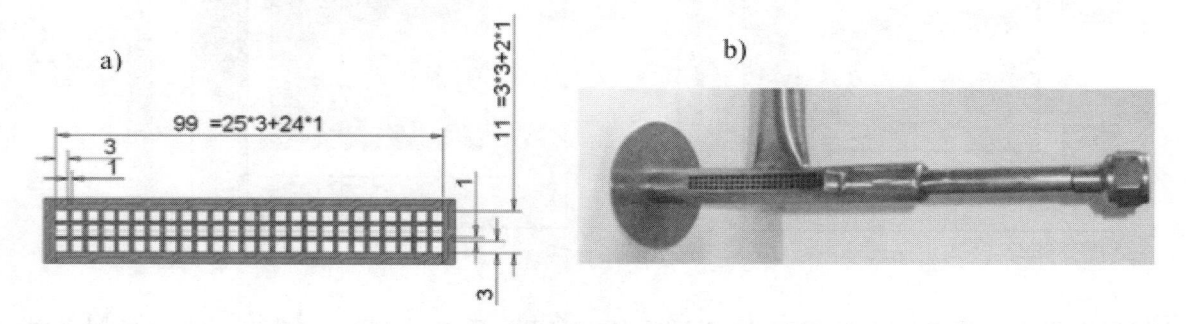

The model was installed in an adapted indium gallium tin loop, presented in Figure 10. The film observed was rather stable, with a few droplets detaching from the main body [15]. Nevertheless, further studies shall be carried out, improving the visualisation of the film and increasing the flow rates and dimensions to progressively approach the real configuration.

Conclusions

The technical feasibility of such an innovative target assembly has been confirmed by Monte Carlo and finite element calculations as well as by experimental tests. The high-energy neutron-induced

Figure 10. View of the liquid metal experimental loop (a), including the magnetic pump (to the left, in blue). Front view of the liquid metal transverse film (b).

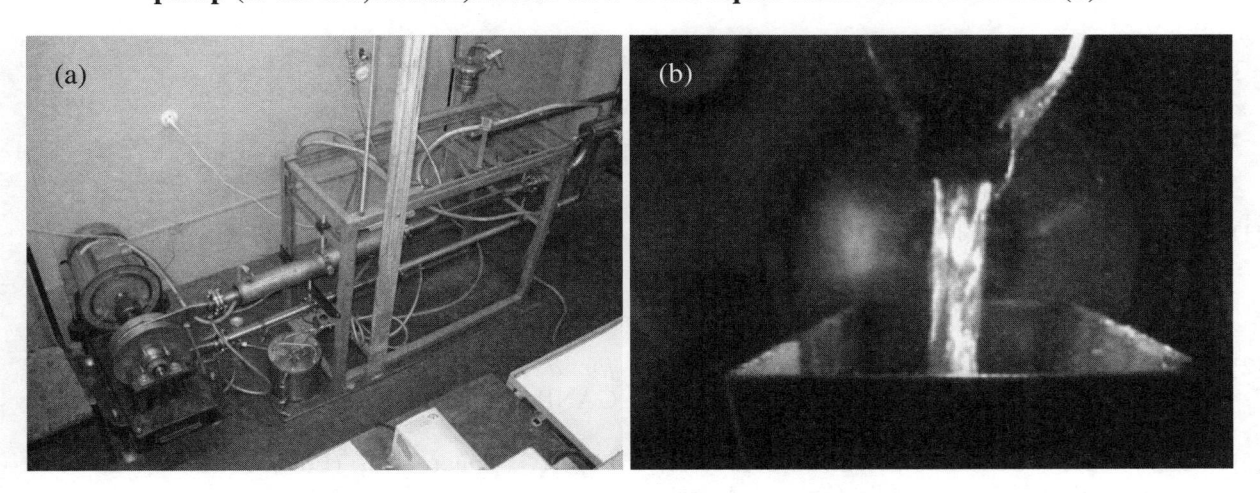

fission densities aimed for can be achieved with the proposed multi-MW target baseline design, by using moderate proton beam intensities and reasonable fission target volumes, independently of the actinide composition. A 1-GeV proton beam on a compact mercury proton-to-neutron converter seems favourable to obtain fluxes above 10^{14} n/cm^2/s/MW of beam, producing more than 10^{11} fissions/cm^3/s per MW of beam, and intense RIBs.

The beam window and mercury flow design have been carefully addressed, producing a configuration which will keep stresses below rupture limits under radiation in the window and avoid cavitation within the liquid metal. Maximum flow rates of ~6 m/s under 7.5 bars of pressure seem acceptable, removing the 2.4 MW of beam power deposited within the system and maintaining stresses below 150 MPa in the beam window.

Nevertheless, an alternative windowless spallation neutron target design has been produced based on the transverse film concept, technically simple and showing improved neutronic performances. This innovative design, experimentally tested, would allow for higher-power densities, avoiding the need to replace the beam window every few months.

Due to the intense neutron flux within the assembly, there are obvious potential synergies between EURISOL and other nuclear physics activities. For example, a neutron escape line could be foreseen for neutrino physics, time-of flight measurements and other neutron applications (e.g. material science), without hindering the performance of the ISOL facility.

Acknowledgements

We acknowledge the financial support of the European Commission under the 6th Framework Programme "Research Infrastructure Action – Structuring the European Research Area" EURISOL DS Project Contract No. 515768 RIDS. The EC is not liable for any use that may be made of the information contained herein.

REFERENCES

[1] *EURISOL DS; European Isotope Separation On-line Radioactive Ion Beam Facility Design Study*, EC – FP6 Research Infrastructure Action – Structuring the European Research Area, Project Contract No. 515768 RIDS.

[2] *The EURISOL Report: A Feasibility Study for a European Isotope Separation On-line Radioactive Ion Beam Facility*, J. Cornell (Ed.), GANIL, France (2003).

[3] *ESS: A Next Generation Neutron Source for Europe*, G. Bauer, *et al.* (Eds.), ISBN 090 237 6 500 and 090 237 6 659, ESS Council, March 1997.

[4] Bauer, G., *et al.*, *Target Options for SINQ – A Neutronic Assessment*, TM-36-97-04, Paul Scherrer Institute, Villingen PSI, Switzerland (1998).

[5] The SNS Collaboration, *National Spallation Neutron Source Conceptual Design Report – Volume 1*, NSNS/CDR-2/V1, Oak Ridge National Laboratory, United States, May 1997.

[6] Herrera-Martínez, A., Y. Kadi, *EURISOL-DS Multi-MW Target: Preliminary Study of the Liquid Metal Proton-to-neutron Converter*, CERN-AB-Note-2006-013 and EURISOL-DS/TASK2/TN-2005-01.

[7] Fassò, A., A. Ferrari, J. Ranft, P.R. Sala, *FLUKA: A Multi-particle Transport Code*, CERN-2005-10 (2005), INFN/TC_05/11, SLAC-R-773.

[8] Herrera-Martínez, A., Y. Kadi, *EURISOL-DS Multi-MW Target: Neutronic Calculations for the Baseline Configuration*, CERN-AB-Note-2006-035 and EURISOL-DS/TASK2/TN-2006-05.

[9] Herrera-Martínez, A., Y. Kadi, *EURISOL-DS Multi-MW Target: Comparative Neutronic Performance of the Baseline Configuration vs. the Hg-jet Option*, CERN-AB-Note-2006-037 and EURISOL-DS/TASK2/TN-2006-07.

[10] M. Lindroos, *The Technical Challenges of Beta-beams*, EURISOL-DS/TASK12/CONF-05-01.

[11] Ferrari, A., *et al.*, "Recent Developments in the FLUKA Nuclear Reaction Models", *Proc. of the 11th International Conference on Nuclear Reaction Mechanisms*, Varenna, Italy, 12-16 June 2006.

[12] Herrera-Martínez, A., Y. Kadi, *Radioactive Ion Beam Production by Fast-neutron-induced Fission on Actinide Targets at EURISOL*, Proceedings of Science (POS), International Workshop on Fast Neutron Detectors and Applications (FNDA 2006), Cape Town, South Africa, 3-6 April 2006 and EURISOL-DS/TASK2/CONF-2006-08.

[13]   Felcini, M., A. Herrera-Martínez, Y. Kadi, T. Otto, L. Tecchio, *Design of the Multi-MW Target Assembly: Radiation and Safety Issues*, Proc. of the 8th International Meeting on Shielding Aspects of Accelerators, Targets and Irradiation Facilities (SATIF-8), Pohang, Republic of Korea, 22-24 May 2006.

[14]   Ashrafi-Nik, M., *Thermo-hydraulic Optimisation of the EURISOL DS Target*, TM-34-06-0, Paul Scherrer Institute, Villingen PSI, Switzerland (2006).

[15]   Freibergs, J., E. Platacis, Engineering Design and Construction of a Function Hg-loop, EURISOL DS Task 2 Internal Note, Institute of Physics at the University of Latvia (IPUL) (2007).

DELAYED NEUTRON YIELDS AND TIME SPECTRA FROM 1-GeV PROTONS INTERACTING WITH ^{nat}Fe, ^{nat}Pb AND ^{209}Bi TARGETS

Danas Ridikas[1], Anatoli Barzakh[3], Valentin Blideanu[1], Jean-Christophe David[1], Diane Doré[1], Xavier Ledoux[2], Fedor Moroz[3], Vladimir Panteleev[3], Arturas Plukis[4], Rita Plukiene[4], Aurelien Prévost[1], Oleg Shcherbakov[3], Alexander Vorobyev[3]

[1] CEA Saclay, DSM/DAPNIA, F-91191 Gif-sur-Yvette, France
[2] CEA/DIF, DPTA/SPN, F-91680 Bruyères-le-Châtel, France
[3] Petersburg Nuclear Physics Institute, 188350 Gatchina, Leningrad District, Russia
[4] Institute of Physics, Savanoriu pr. 231, 02300 Vilnius, Lithuania

Abstract

This paper presents the experimental results on the DN yields and time spectra from 1-GeV protons interacting with the ^{nat}Pb, ^{209}Bi and ^{nat}Fe targets of variable thicknesses from 5 cm to 55 cm. Both absolute yields and time constants are reported. In parallel, the MCNPX and PHITS codes were used to predict the DN precursors and construct the theoretical DN tables. Different model parameters are examined and show significant dependence on the choice of the intra-nuclear cascade and fission evaporation models used. Finally, the above experiment allowed the determination of the production cross-sections of a number of DN precursors such as ^{87}Br, ^{88}Br and ^{17}N. These results permit the examination of two different reaction mechanisms, namely fission and fragmentation, implemented in high-energy transport codes. These data and modelling are of great importance for the new generation spallation neutron sources based on liquid metal technologies.

Introduction

Increased proton beam power results in target thermal-hydraulic issues leading to new target designs, very often based on liquid metals such as Hg, Pb, Pb-Bi. Indeed, present spallation neutron sources aiming at beam power of 1 MW or higher [MEGAPIE (PSI), SNS (ORNL), JSNS (JAEA), ESS and EURISOL (both Europe)] all focus on liquid metal targets. Due to high-energy fission-spallation reactions a significant amount of the delayed neutron (DN) precursor activity can be accumulated in the target fluid. The transit time from the front of a liquid metal target into areas where DNs may be important can be as short as a few seconds, i.e. shorter than one half-life of many DN precursors. Therefore, it is necessary to evaluate the DN neutron flux as a function of time and determine if DNs may contribute significantly to the dose rates, including the activation of surrounding structure materials. These issues were addressed in detail in an earlier work [1], including the estimation of the DN flux and corresponding time spectra in the case of the MEGAPIE spallation target at PSI (Switzerland) [2]. In Ref. [1] it was also demonstrated that the final estimates of DNs were very much model-dependent within the MCNPX code [3]. Here we will only summarise some of these findings.

The main goal of this work is to report on the measurements of DN yields from high-energy fission-spallation reactions on different targets of variable thicknesses. These measurements were realised in December 2005 and April 2006 at PNPI/Gatchina (Russia). Here we present our results from the analysis of the DN yields and time spectra from 1-GeV protons interacting with thick ^{nat}Fe, ^{nat}Pb and ^{209}Bi targets. Corresponding calculations with MCNPX [2] and PHITS [4] codes were also performed leading to some recommendations related to the use of these tools. We note that the data obtained with thin and thick ^{nat}Pb targets were already partially reported in [5]. Finally, the obtained results are applied for the case of a "typical" high-power spallation to illustrate the importance of the DNs in liquid target technologies.

The case of the MEGAPIE

Liquid Pb-Bi eutectics (LBE) loop in the case of the MEGAPIE spallation target, as in most of the high-power spallation targets based on liquid metal technologies, extends much further compared to the primary proton interaction zone. As it is presented in Figure 1, the activated LBE reaches as high as 400 cm, where it enters the heat exchanger, and from where it returns to its initial position. It takes ~20 s for the entire ~82 litres of Pb-Bi to make a "round trip" at a flow rate of ~4 litres/s. Indeed, a big part of the DN precursors, created in the interaction region via high-energy fission-spallation, will not have enough time to decay completely even during the "round trip" of the circulating liquid metal. The main concern is the neutron flux level due to the DNs at the very top position of the heat exchanger.

To estimate the DN flux we employed the multi-particle transport code MCNPX [3] combined with the material evolution program CINDER'90 [6], as was detailed in [1]. The DN data (emission probabilities and decay constants) were based on the ENDF/B-VI evaluations [7]. For the MEGAPIE target characteristics we used the design values, i.e. a 575-MeV proton beam with 1.75 mA intensity, interacting with the liquid LBE target. The 3-D geometry of the target has been modelled in detail by taking into account all materials used in the design. The estimation of the DN parameters for MEGAPIE was performed in steps according to the following procedure: 1) calculation of independent fission fragment and spallation product distributions with MCNPX; 2) calculation of cumulative fission fragment and spallation product yields with CINDER'90; 3) identification of all known DN precursors and construction of the six-group DN tables.

Figure 1. Middle: a schematic view of the entire MEGAPIE target (5.3 m long).
Left: zoom of the lowest part of the target-proton-PbBi interaction zone.
Right: zoom of the target at the level of the heat exchanger (~400 cm above the target widow).

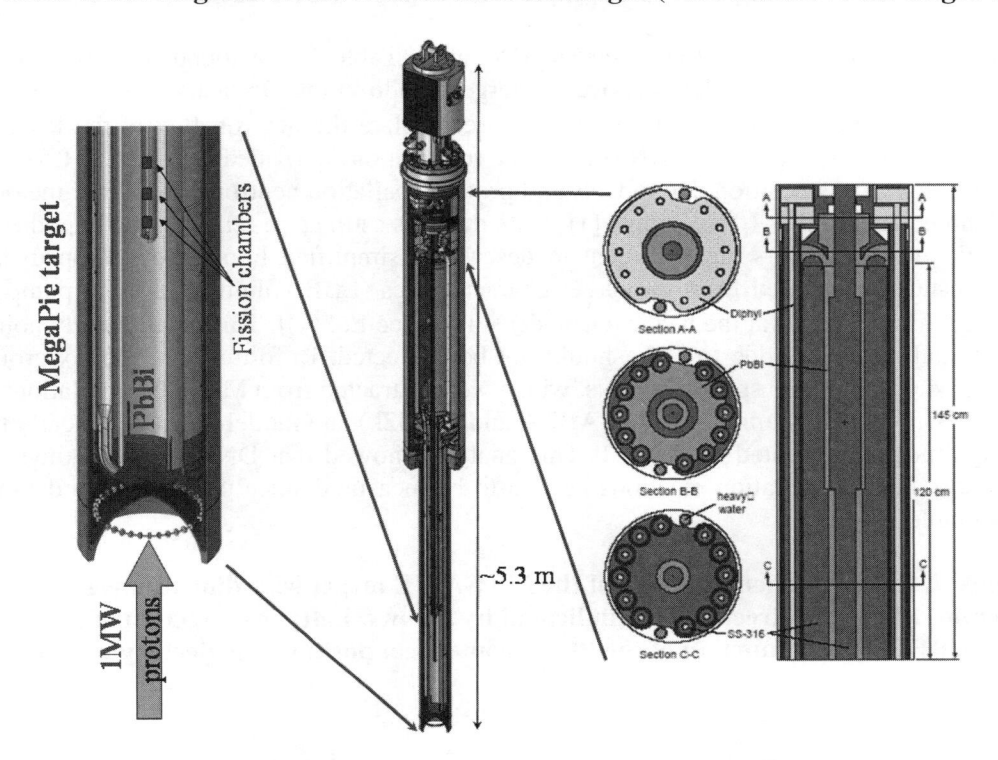

Table 1. DN parameters (six-group representation) in the case of the MEGAPIE target using two different model calculations. Note that DN yields are normalised for 10^6 incident protons.

Group	INCL4-ABLA model		CEM2k model	
	$T_{1/2}$, s	a_i, n/p × 10^6	$T_{1/2}$, s	a_i, n/p × 10^6
1	55.49	0.87	55.60	6.78
2	16.29	0.89	16.35	15.25
3	4.99	0.44	4.66	23.58
4	1.90	1.19	1.63	174.24
5	0.52	0.21	0.45	129.95
6	0.20	0.00	0.11	233.52
Total (averaged)	(18.703)	3.59	(1.903)	583.35

After having built the DN table (Table 1) we developed a generalised geometrical model to estimate the DN activity densities at any position x of the MEGAPIE target loop as presented in Figure 2. It can be noted in this figure that the LBE cross-section changes over the loop, meaning that the transit time of a given LBE volume depends on its position. In particular, the activation time of an LBE volume under irradiation (in the so-called spallation region) is very short (~0.5 s) compared to the total circulation time (~20 s). Within this model the DN activity at position x can be expressed as:

$$a(x) = \sum_{i=1}^{n} a_i(x) = \sum_{i=1}^{n} a_i^a \frac{1-\exp(-\lambda_i \tau_a)}{1-\exp(-\lambda_i T)} \exp(-\lambda_i \tau_d(x)) \tag{1}$$

where τ_a is the activation time of the Pb-Bi under irradiation, T represents the total circulation period of the LBE, i.e. duration of the "round trip", τ_d is the transit (decay) time to reach the point x, λ_i are the decay constants of the DN precursor i while a_i express the density of DNs due to the precursor i.

Using the above equation and the six-group DN table (Table 1) we found that at the very top position of the LBE loop (400 cm level above the target window) the DN activity is of the order of $2 \cdot 10^5$ n/(s cm^3). This intermediate result permitted to recalculate the neutron flux at the level of the heat exchanger inserting the volumetric DN source as a function of x provided in Figure 2. It was found that the neutron flux at this position due to DNs and prompt spallation neutrons is of the same order of magnitude, both equal to a few 10^6 n/(s cm^2) [1]. Note that this estimation relies on the hypothesis that three averaged time parameters are sufficient to describe a simplified liquid metal loop dynamics. These time constants, estimated from the target characteristics (LBE volume and main pump debit) are: $\tau_a \sim 0.5$ s, $T \sim 20$ s and τ_d (at the heat exchanger) ~ 10 s [see Eq. (1)]. This result clearly points out that activation and dose rates due to DNs should not be neglected. Unfortunately, the six-group DN parameters, i.e. yields and time spectra of DNs, which were extracted from MCNPX simulations based on different physics models (namely INCL4+ABLA and CEM2k), are model-dependent nearly by two orders of magnitude as presented in Table 1. This analysis showed that DN yields and time spectra from high energy fission-spallation reactions needed to be measured since up to now no data of this type were available.

Figure 2. Right: a schematic view of the MEGAPIE target with dimensions given in mm. LBE flow directions are indicated by arrows. Left: cross-section of the liquid LBE loop as a function of the PbBi geometrical position – trajectory x in cm.

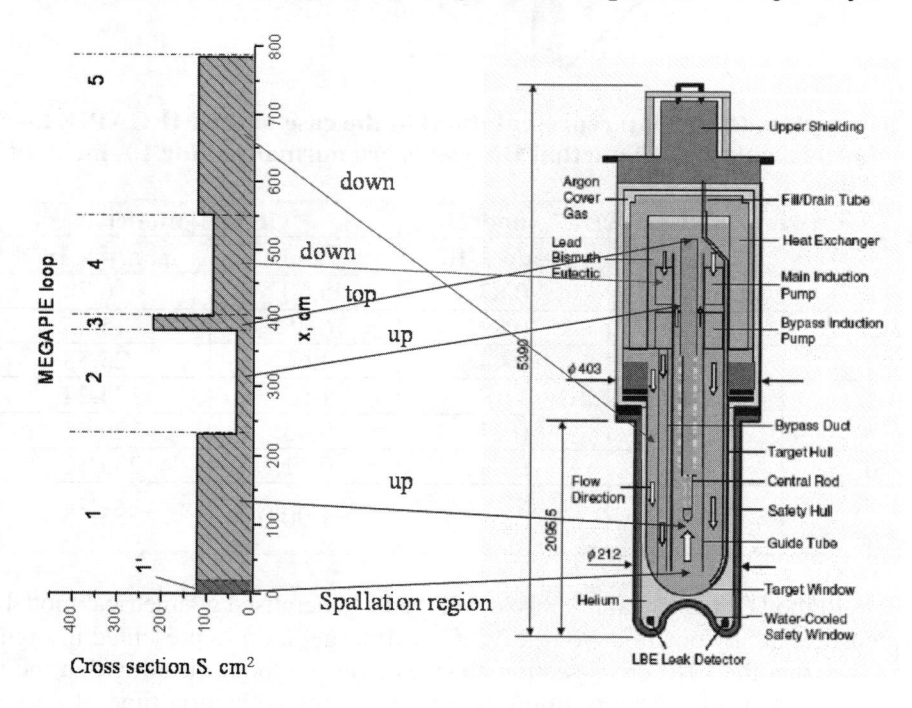

The experiment at PNPI/Gatchina

A schematic view of the experiment performed at PNPI/Gatchina (Russia) is shown in Figure 3. Protons of 1 GeV from accelerator impinged on the ^{nat}Pb, ^{209}Bi and ^{nat}Fe targets of various thicknesses (0.5, 5, 10, 20, 40 and 55 cm). The emitted DNs were detected with the optimised ^{3}He detector

Figure 3. A photo of the experimental set-up with a cylindrical target (in the centre) and the ^{3}He counter (in the back)

following specific irradiation periods after the beam is switched off. Long (300 s), intermediate (20 s) and short irradiation (350 μs – a single pulse) times were used to optimise the extraction of different time parameters of DN groups. Each individual irradiation-decay cycle was recorded on line and summed off line afterwards to accumulate the statistics.

The ^{3}He counter was surrounded by 5-cm thick polyethylene (CH$_2$) and coated by 1-mm ^{nat}Cd foils in order to increase the neutron detection efficiency in terms of neutron moderation and to avoid the background due to the thermal neutrons correspondingly. The detector optimisation was performed using Monte Carlo simulations with MCNPX [2]. Finally, the detector was calibrated with a standard ^{252}Cf neutron source, and its efficiency was reproduced to within 9% by MCNPX. The proton beam intensity was monitored with relative errors of around 10% by activation of ^{27}Al foils (see Figure 3) and the γ-spectroscopy off-line from the ^{7}Be, ^{22}Na and ^{24}Na activity. The diameter of the beam spot and its alignment was obtained using the photo films as presented in Figure 3.

Some measurements without target and with iron or brick concrete targets were also performed to characterise the active background contribution due to the environment of the experimental area. During these test measurements without any target and with non-fissile targets we observed only two dominating decay periods of $T_{1/2}$ ~ 4.2 s and ~0.18 s, which we attributed to the reaction products ^{17}N and ^{9}Li with $T_{1/2}$ = 4.173 s and $T_{1/2}$ = 0.178 s, respectively (see Figure 4). Indeed, these two DN precursors can be produced in the reactions from 1-GeV protons on different targets such as Al, Cu, Fe or any other materials present in the experimental hall or accelerator structures [8]. It is important to note that no longer half-lives were observed during these test irradiations, as shown in Figure 4. Once the irradiations with ^{nat}Pb or Bi targets started, we noticed that the DN decay curves also extended to the longer half-lives than ~4 s [9], which are characteristic for neutron-rich fission products only.

Analysis of the DN decay curves for all ^{nat}Pb and Bi target thicknesses and with identical irradiation-decay periods of 300 s were performed by fitting data with exponential sums using the TMinuit class implemented in the ROOT toolkit [10]. The following expression was used to obtain the $\{a_i;\lambda_i\}$ values:

$$DN(t) = \sum_i a_i \exp(-\lambda_i t)(1-\exp(-\lambda_i T_{irr})) + C \quad DN(t) = \sum_i a_i \exp(-\lambda_i t)(1-\exp(-\lambda_i T_{irr})) + C \quad (2)$$

with $T^i_{1/2} = \ln 2/\lambda_i$, C – a constant representing the background, and the irradiation time T_{irr} = 300 s. Systematic errors due to the estimation of the C values have been taken into account.

Figure 4. The accumulated DN decay curve from p (1 GeV) + Fe (55 cm) (on the bottom); individual contributions from different precursors are indicated separately

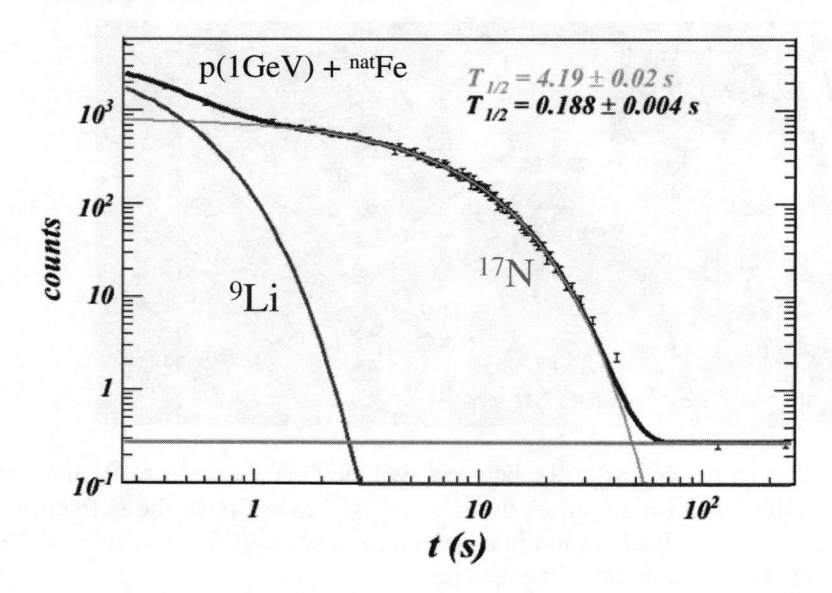

In contrast to the conventional six-group approach to reproduce the DN decay curves from neutron-induced fission on actinides [7], four terms of the above expression were sufficient for our purpose. Half-lives for these four terms correspond to the previously identified spallation products, ^{9}Li and ^{17}N, in addition to the two fission products ^{87}Br and ^{88}Br (with $T_{1/2}$ = 55.6 s and $T_{1/2}$ = 16.29 s, respectively). As a representative example, the accumulated data for the ^{209}Bi target (55 cm thick), are presented together with the fits of exponential sums in Figure 5. Similar quality fits using the same four terms in the exponential sum, i.e. with four half-lives fixed to those of ^{9}Li, ^{17}N, ^{87}Br and ^{88}Br, were obtained for all target thicknesses and both ^{nat}Pb and ^{209}Bi targets [9] (not shown here).

Figure 5. The accumulated DN decay curve from p (1 GeV) + ^{209}Bi (55 cm); individual contributions from different precursors are indicated separately

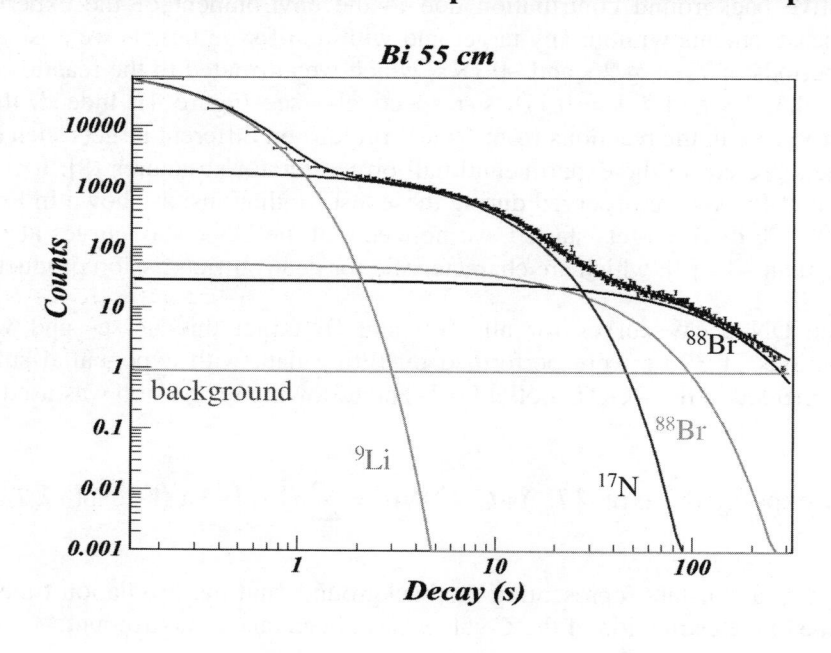

Discussion

Some conclusions can already be drawn just looking at the fitted DN decay curves for the ^{nat}Pb and ^{209}Bi targets (e.g. for ^{209}Bi in Figure 5). First, the DN contribution from light mass products such as ^{9}Li and ^{17}N is clearly more important than the contribution from fission products. In addition, this "unusual" DN emission dominates the DN decay curve for Pb and Bi targets up to the cooling time of 10-20 s (see Figure 5). For longer decay times, say 50-100 s, the long-lived DN precursors, being "usual" fission products such as ^{88}Br and ^{87}Br, remain the only contributors to the DN activity. Note that this finding is very important in the case of high-power liquid metal targets, where the liquid metal makes the "round trip" typically in a period of 10-30 s (e.g. the MEGAPIE loop makes a turn in ~20 s [1]).

Due to the efficiency determination of the ^{3}He counter and to the proton beam intensity monitoring, it was possible to extract the DN yields in absolute values. In brief, once the $DN(t)$ is parameterised during the fitting procedure [see Figure 5 and Eq. (2)] and when t approaches zero, one obtains:

$$DN(t \to 0) = \sum_i a_i + C \qquad (3)$$

Here the following expression can be employed to calculate the individual DN yields:

$$P_n^i Y^i = a_i / \left(\varepsilon_{He-3} I_p \Delta t_{ch} N_{cycles} \right) \qquad (4)$$

with P_n^i being the DN emission probability, Y^i the individual DN precursor yield (atoms per incident proton), ε_{He-3} the total efficiency of the ^{3}He counter (events per emitted source neutron), I_p the proton beam intensity (protons per second), Δt_{ch} the channel width (seconds) and N_{cycles} a number of accumulated irradiation-decay cycles.

Figure 6 presents the obtained DN precursor production yields Y^i for i = ^{17}N, ^{87}Br and ^{88}Br as a function of target thickness. Note that in addition to the uncertainties in detector efficiency, proton beam intensity monitoring and statistical errors, we took into account systematic errors due to the fitting procedure and background contribution. We add that the extraction of ^{9}Li yields was impossible due to its short half life, which is comparable with the experimental channel width Δt_{ch} = 200 ms.

Finally, in order to illustrate the importance of our results, let us assume that we deal with a stopping liquid Pb-Bi target interacting with 1 GeV protons at 1 MW beam power. Figure 7 shows the resulting absolute DN intensity as a function of cooling time after 5 min of irradiation period. In brief, more than ~10^{13} n/s will be created at the very short decay times, while more than ~ 10^{11} n/s will remain even after ~20 s of cooling.

Conclusions

The experimental data on measured DN yields and time spectra from 1-GeV protons interacting with thick ^{nat}Fe, ^{nat}Pb and ^{209}Bi targets are presented in this work. For all these targets the emission of DNs is dominated by light reaction products such as ^{9}Li and ^{17}N during the decay time from 0 to ~20-30 s. In the case of fissile targets after the longer decay time the fission fragments as ^{88}Br and ^{87}Br are the major contributors. The DN yield production per incident proton is increasing with the target thickness and the majority of the DNs from 1 GeV protons are produced in the first half of the stopping target.

The above experimental observations are reproduced with the PHITS transport code, which is able to predict the production of the DN precursors within a factor of 2 or better. These new data are of great importance for the new generation high-power spallation targets based on liquid metal technology as well as for further development of high-energy spallation-fission fragmentation models.

Finally, we add that in the framework of the Neutronic and Nuclear Assessment Task Group of the MEGAPIE experiment the DN flux at the top of the liquid PbBi target was recently measured. A preliminary comparison between the experimental DN decay curve at MEGAPIE and its interpretation using the DN parameters extracted from this work is in a good agreement [11]. This example illustrates that, thanks to this new data, the DN fluxes can be evaluated even in complex spallation target scenarios.

Figure 6. The DN precursor yields Y^i (atoms per proton) as a function of target thickness for Pb. The experimental data (filled symbols) are compared with the PHITS predictions (open symbols). The lines are to guide the eye.

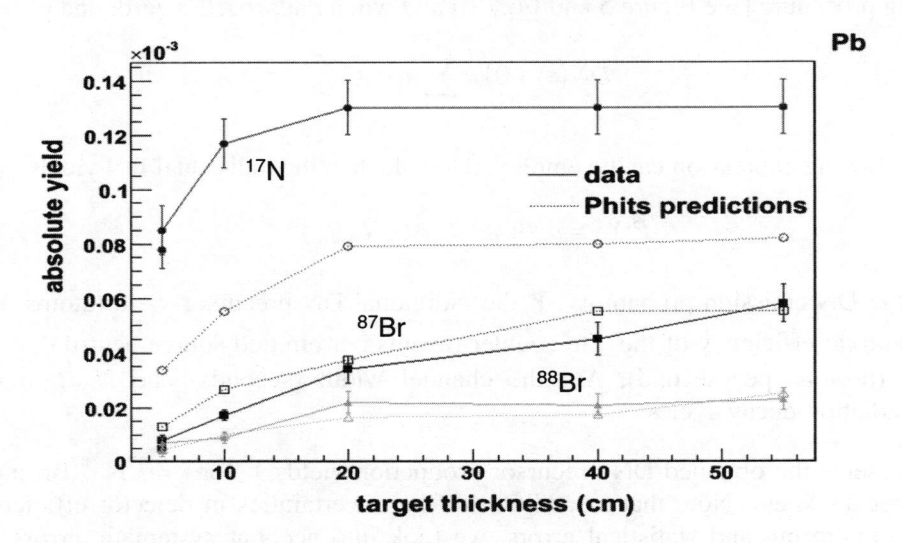

Figure 7. The resulting DN activity in (n/s) as a function of cooling time after the irradiation of a stopping PbBi target with 1-GeV protons at 1 MW beam power for 5 min.

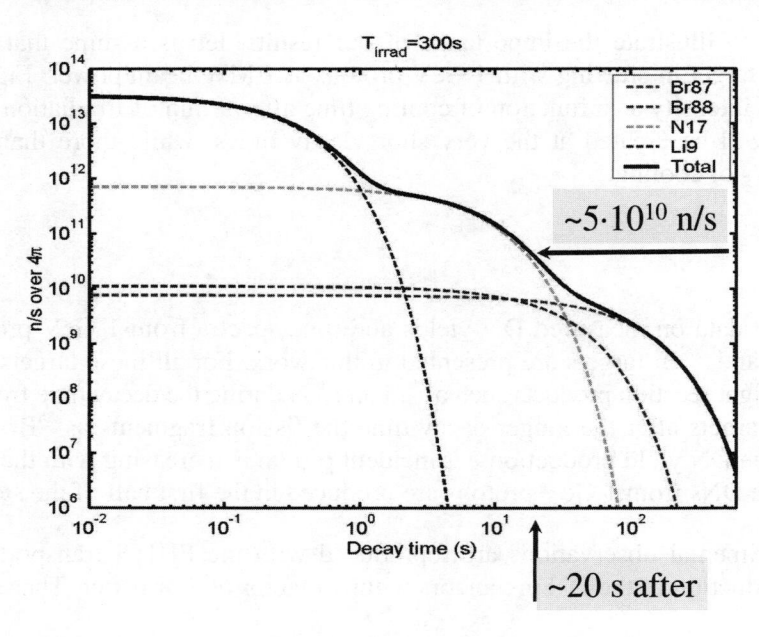

REFERENCES

[1] Ridikas, D., *et al.*, "Delayed Neutrons from High Energy Fission-spallation Reactions", *Proc. of the 3rd International Workshop on Nuclear Fission2005*, 11-14 May 2005, CEA Cadarache, France; published in AIP Conf. Proc. 798, 277 (2005).

[2] Bauer, G.S., M. Salvatores, G. Heusener, "MEGAPIE, a 1 MW Pilot Experiment for a Liquid Metal Spallation Target", *Proc. of the 15th Int. Conference ICANS-XV*, J. Suzuki, S. Itoh, (Eds.) JAERI, Tsukuba, Japan (2001).

[3] Pelowitz, D.B., "MCNPX^{TM} User's Manual – Version 2.5.0", LA-CP-05-0369, LANL, USA, April 2005.

[4] Iwase, H., *et al.*, "Development of heavy ion transport Monte Carlo code", Nucl. Instr. &. Meth. B 183 (2001) 374.

[5] Ridikas, D., *et al.*, "Measurement of Delayed Neutron Yields and Time Spectra from 1 GeV Protons Interacting with Thick ^{nat}Pb Targets", *European Physical Journal A*, 32, 1-4 (2007).

[6] Wilson, W.B., T.R. England, *A Manual for CINDER'90 Version C00D and Associated Codes and Data*, LA-UR-00-Draft, April 2001.

[7] Wilson, W.B., T.R. England, "Delayed Neutron Study Using ENDF/B-VI Basic Nuclear Data", *Progress in Nuclear Energy*, Vol. 41, 71 (2002).

[8] Dostrovsky, I., *et al.*, "Cross Sections for the Production of ^{9}Li, ^{16}C, and ^{17}N in Irradiations with GeV-energy Protons", *Phys. Rev.*, 139, 1513 (1965).

[9] Ridikas, D., *et al.*, "Relative Delayed Neutron Yields and Time Spectra from 1 GeV Protons Interacting with Thick ^{nat}Pb Targets", *Proc. of Int. Conference PHYSOR 2006*, Vancouver, Canada, 10-14 September 2006, ISBN: 0-89448-697-7, B104, 9 pgs; ANS (2006).

[10] Brun, R., F. Rademaker, "ROOT – An Object Oriented Data Analysis Framework", *Nucl. Instr. & Meth. A*, 389, 81 (1997).

[11] Panebianco, S., *et al.*, "Delayed Neutron Measurement at the MEGAPIE Target", *Proceedings of the Int. Conf. on Nuclear Data 2007*, Nice, France, 22-27 April 2007.

SESSION IV

Subcritical System Design and ADS Simulations

Chairs: P. Baeten, S. Monti

PRE-CONCEPTUAL DESIGN OF A HELIUM-COOLED ADS: He-EFIT

P. Richard[1], G. Rimpault[1], J.F. Pignatel[1], B. Giraud[2], M. Schikorr[3], R. Stainsby[4]
[1]CEA/DEN/DER Cadarache, F-13108 Saint-Paul-lez-Durance, France
[2]AREVA NP, [3]FzK, [4]AMEC-NNC

Abstract

Within the EUROTRANS integrated project, the domain DESIGN has the task to provide the pre-design of a European Facility for Industrial Transmutation (EFIT) able to demonstrate the feasibility aspect of the nuclear waste transmutation/burning in an ADS at industrial scale. The helium-cooled EFIT is one of the solutions studied for this purpose. The present paper summarises the work carried out during the first two years of the project. The approach followed to design the plant is presented first. The plant characteristics (proton beam characteristics, core power, inlet/outlet temperatures, etc.) are then discussed and the He-EFIT core is described as well as the spallation module design.

Introduction

Within the EUROTRANS Integrated Project (partially funded by the European Commission) in FP6 [1], the domain DESIGN has the task to provide the pre-design of a European Transmutation Demonstrator (ETD) able to demonstrate the feasibility aspect of the nuclear waste transmutation/burning in an ADS at industrial scale. This is done within the frame of the DOMAIN1 (DESIGN) of the EUROTRANS project. Three different concepts of accelerator-driven systems (ADS) are studied: a Pb-cooled European Facility for Industrial Transmutation (Pb-EFIT); a back-up solution, based on a He-cooled ADS (He-EFIT); and as an intermediate step towards this industrial-scale prototype, an eXperimental Transmuter based on the ADS concept (XT-ADS) able to demonstrate the feasibility of both the ADS concepts.

Only the He-EFIT concept is considered here. This is a nuclear power plant project intended to transmute minor actinides produced by nuclear industry. The plant has a power of approximately 400 MWth. As for all the ADS plants, it consists of three main components: accelerator, spallation module and subcritical core.

The present paper summarises the work carried out during the first two-year stage of the project. Emphasis is given to the spallation module and the He-EFIT core designs. The design presented here constitutes the first partial set-up of the plant, which will be subject to optimisation during the second part of the project.

The EFIT missions

The general objectives assigned to IP EUROTRANS are focused on the evaluation of the industrial practicability of transmutation of high-level nuclear waste in an ADS, along with the development of the basic knowledge and technologies needed. In this context, the XT-ADS being more oriented to the technology demonstration, the EFIT missions are logically oriented toward industrial transmutation.

Thus, the first objective assigned to the plant is to transmute Am and Cm. As stated in D1.1 of the EUROTRANS project [2], the optimisation of an ADS in order to optimise its transmutation capability has its limits since each actinide burnt is generating around 200 MeV. Hence, there is a relation between the installed power and the optimum transmutation capability. This limit value is around 45 Kg/TWhth.

The second objective is related to plant safety: as with any nuclear facility the reactor must operate safely; but the He-EFIT safety analysis must also properly address ADT as well as gas-cooled plant peculiarities (MA-containing fuels, subcritical level, absorber/safety rods, He low heat capacity, etc.).

Finally, the third important objective assigned to the plant is to produce electricity. One has to take into account that the accelerator is an important electricity consumer, it is suitable to use the power produced by the plant to supply the accelerator with electricity.

Design approach

General approach

First of all, it worth mentioning that He-cooled EFIT will take as far as possible benefits from other programmes dealing with gas-cooled reactors and particularly from the GCFR studies [3]. Moreover, it will be largely extrapolated from the gas-cooled XADS design [4].

Beyond the above-mentioned missions, it was agreed that the feasibility of the spallation target module is one the key issues for ADT feasibility. The target design should be investigated first and the design of the other main components should then be defined accordingly. Thus, the proposed scheme for the design studies can be summarised as follows:

1. Define the target size. This task takes as far as possible benefits from the outcome of the PDS-XADS project (technological feasibility).

2. Define the proton beam intensity compatible with the spallation module design for a maximum proton energy of 800 MeV (hypothesis accounted for the maximum proton energy).

3. Define the reactor power and the K_{eff}. At this first stage of the design, assumptions on the potential reactivity insertions and burn-up swing are made *a priori* as a basis to define the K_{eff} such that the plant remains subcritical under all conditions – normal operating conditions as well as incidental/accidental situations. These assumptions must of course be finally verified on the basis of the real plant design.

4. Design the core taking into account the core design constraints (fuel composition, cladding composition, pin and sub-assembly geometry, etc.): definition of the fuel (pin), sub-assembly and whole core dimensions.

5. Define the approach for DHR and design the DHR main components (blowers, HX, etc.).

6. Design the primary system the balance of plant and containment, including main services (fuel handling, radiological protections, etc.).

This approach is obviously an iterative process and, if necessary, the design options considered in the first steps of the studies can be revisited according to the outcome of latter stages. Nevertheless, to start this iterative process, a first set of design parameters/characteristics was agreed upon during the first meetings devoted to He-EFIT.

The status of the He-EFIT design is currently as follows: the design of the spallation module was preliminarily addressed. The results of this study led to a He-cooled solid target with a cold window. The core power has been fixed to 400 MWth and K_{eff} to 0.97. For stage 5, the approach for the DHR has been defined. The detailed design of the DHR loops still remains to be carried out. The design of the following stage (6) will be performed in the upcoming future.

The current He-EFIT core design is shown in Figure 1. The main design parameters are summarised in Table 1.

Design constraints and criteria

Associated with the approach described above, a certain number of main objectives and/or criteria for the design were defined for the He-EFIT. They are as follows:

- an efficiency for the elimination of minor actinides close to 42 kg/TWhth;

- the small variation of K_{eff} versus burn-up rate;

- a rather flat distribution of power: $f_{rad} < 1.4$, $f_{ax} < 1.3$;

- a core pressure drop less than 1 bar;
- a gas pressure in the fuel plenum limited to 50 bar at the end of fuel life;
- an average helium velocity limited to 100 m/s.

At this stage, it is important to note that the definition of the reactor power and K_{eff} is a compromise; if, as in the gas-cooled XADS plant design, no control rods are considered for the He-EFIT core design, the core must be designed with a relatively large margin of subcriticality (smaller K_{eff}) as the operational reactivity variations (transients, burn-up, etc.) cannot be compensated by any control rod insertion. Conversely, with control rods, the operational reactivity variations can partly be accommodated by the control rods. On the basis of the rationale as adopted within the PDS-XADS project, no control rods are considered in the He-EFIT design.

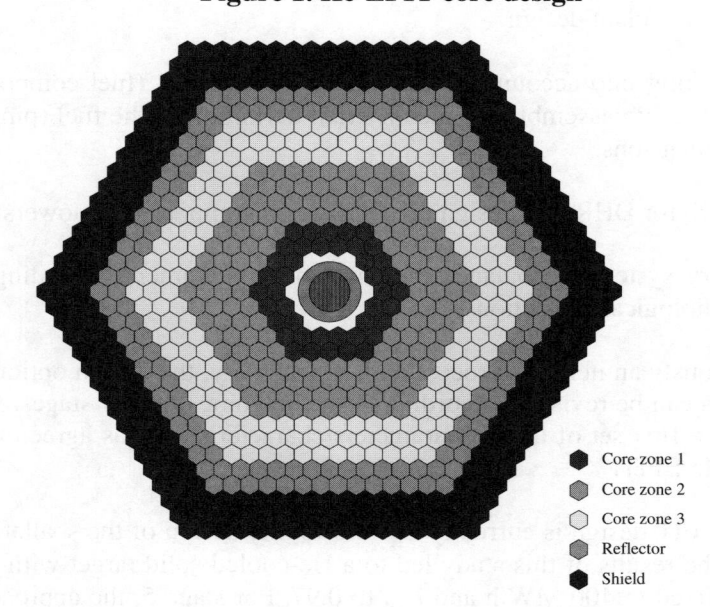

Figure 1. He-EFIT core design

Table 1. Characteritics of the three-zone He-EFIT core

EFIT type	He-EFIT
Thermal power (MW_{th}) – volumic power	**400-55 MW/m³**
Coolant type	He
Coolant pressure (bar)	70
Inlet temperature (°C)	350
Outlet temperature (°C)	550
Plant efficiency	43.3% – indirect $S-CO_2$ cycle with re-compression
Cladding type – thickness	SiC/SiCf, thickness = 1 mm
Spallation module	Solid W He-cooled
Accelerator	Linac, E = 800 MeV, I = 18-22 mA

Table 1. Characteritics of the three-zone He-EFIT core (cont.)

Zone	1 (inner)	2 (intermediate)	3 (outer)
Matrix type	MgO	MgO	MgO
Matrix fraction	50%	50%	50%
Core geometry			
Number of fissile sub-assemblies	42	156	180
Fissile height (H)	1.25 m	1.25 m	1.25 m
Wrapper width over outer flats	137 mm	137 mm	137 mm
Wrapper thickness	4.0 mm	4.0 mm	4.0 mm
Inter-wrapper gap	5.0 mm	5.0 mm	5.0 mm
Coolant volumic fraction (%)	**49.80%**	**44.75%**	**37.19%**
Fuel + matrix volumic fraction (%)	**11.16%**	**21.70%**	**34.98%**
Structure volume fraction (%)	**39.05%**	**33.56%**	**27.83%**
Sub-assembly bundle geometry			
Number of pin rows per S/A – pin per S/A	12 – 469	8 – 217	5 – 91
Pin outer diameter	4.38 mm	6.87 mm	11.57 mm
Pitch	5.89 mm	8.60 mm	13.20 mm
Pellet/cladding gap	0.040 mm	0.081 mm	0.160 mm
Pellet diameter	2.30 mm	4.71 mm	9.25 mm
Core thermal-hydraulics			
Average coolant speed in the fissile core	30 m/s	35 m/s	45 m/s
Maximum pin linear power	33 W/cm	72 W/cm	171 W/cm
Total core pressure drop	**0.84 bar**	**0.74 bar**	**1.01 bar**
Hottest channel thermics			
Max. cladding temperature in the hottest channel	**686°C**	**700°C**	**711°C**
Max. fuel temperature in the hottest channel	**792°C**	**950°C**	**1 310°C**

Plant description

Main plant characteristics/general characteristics

On the basis of the above-mentioned objectives several characteristics of He-EFIT have been estimated and bounded. They concern the thermal power of the facility, the proton beam characteristics, and the choice of fuel.

Ranges for the general characteristics of the plant were agreed upon at the very beginning of the project. Those values were optimised during the first stage of the project. Their justification is detailed further on.

Definition of the plant power

In the studies carried out within the PDS-XADS project [4], it was demonstrated that a lower power unit with about 50-80 MW(th) would not preclude the envisaged demonstrator since the limited amount of TRU loading will not yield any significant transmutation. Thus, it is important to emphasise that for units of a modular system, a power range of about 300-600 MW(th) seems to be appropriate to meet these requirements. Such an ADS system could be loaded with about 5-12 tonnes of TRUs, and a transmutation rate of about 80-140 kg per year could be obtained.

Definition of the beam characteristics

The linac concept of accelerator proposed within the PDS-XADS is maintained for the EFIT design. The choice of a similar accelerator type for EFIT than for the XT-ADS brings some credibility to the overall plans. Because of the relative large size envisaged for the core and correspondingly for the spallation module, the source efficiency will decrease significantly and all the various means available to compensate for this effect will be required. A proton beam energy not lower than 600 MeV and up to 800 MeV should be adopted, as its realisation is relatively consequential and simple to implement.

As for the required accelerator current, it can be indicatively assessed by a generic parameterisation, that for a power level of P_{max} = 400 MW and a core load with U-free type fuel and with a multiplication factor $k_0 = k_{max} = 0.97$, the current needed for a 600-MeV proton energy beam and a source efficiency of 0.75 is about 16 mA. With a core multiplication factor as low as $k_0 = 0.96$, which might be the lower limit during the overall cycle, the current would rise up to about 22 mA.

Finally, it is useful to point out that a core is designed to accept a limited number of beam SCRAMs (five per year). In the SCRAM definition, only beam trips longer than 1 sec. are considered.

Choice of fuel

The choice for the advanced fuel is one of the most demanding parts of the transmutation feasibility demonstration. Oxide fuels are considered the most promising fuel materials but nitride and metal fuels are also being studied. A preliminary recommendation for a uranium-free fuel candidate is necessary for the development of the EFIT. This preliminary selection has mainly been made on the basis of the work carried out in the FUTURE project of FP5, with important data coming from the FUTURIX-FTA experiment, now part of AFTRA [5].

A number of fuel candidates have been assessed with regard to their fabrication/reprocessing, high-temperature stability/margin to failure and ultimate behaviour under accident conditions, etc. As a result of this selection process, detailed studies were performed on two composite fuel types:

- $(Pu,Am)O_{2-x}$ – ^{Enr}Mo (CERMET)

- $(Pu,Am)O_{2-x}$ – MgO (CERCER), where the ^{Enr}Mo of the CerMet fuel is ~93% enriched in ^{92}Mo.

Fabrication of the two preferred candidate fuels has been successfully accomplished in ATALANTE and in the MA lab within the FUTURIX-FTA project.

Given the significantly lower transmutation efficiency observed with CERMET fuel, and the higher manufacturing cost due to ^{92}Mo enrichment, the study of CERCER fuel is given priority as concerns the EFIT.

The most important properties for both CERCER fuels (as proposed by the AFTRA domain) to start the optimisation design process are presented in Table 2.

The PDS-XADS studies have shown that due to the rather low heat capacity of helium, a cladding material having an elevated operating temperature was recommended [4]. For this reason, a SiC/SiCf cladding (material similar to the one considered for the GCFR) is chosen. Table 2 also presents the maximum temperatures considered in the design for the fuel pellets as well as for the cladding in the different operating conditions.

Table 2. U-free fuel characteristics to be considered in EFIT studies

Characteristics to be considered	CERCER with MgO matrix
Minimum matrix volume fraction	50%
Fuel theoretical density – filling density	11.46 g/cm^3 – 95%
Matrix theoretical density	3.58 g/cm^3
Melting (or dissociating) temperature	1 930 K
T_{fuel_max}	1 380°C (DBC cat. I)
T_{clad_max}	1 200°C (DBC cat. I)

Main plant characteristics/detailed characteristics

Definition of inlet/outlet temperatures – power conversion cycle

The core inlet/outlet temperatures have been optimised in order to maximise the power conversion efficiency. Basic scoping studies aimed at developing a supercritical carbon dioxide (S-CO$_2$) gas turbine power conversion unit for the helium-cooled version of the EFIT (He-EFIT) accelerator-driven system (ADS) were carried out. Several conversion cycle options have been studied for the reference core coolant outlet temperature of 550°C and for a proposed higher temperature of 600°C. Cycle maximum pressures of 200 bar and 250 bar and core temperature increases of 150°C and 200°C were considered. Better performance was achieved for cycles based on the higher maximum cycle pressure. This was further improved by increasing the core temperature rise from 150°C to 200°C, either by lowering the core inlet temperature or by increasing the core outlet temperature, with the latter giving the best performance of the five options. On this basis, T_{in}/T_{out} have been set to 350°C/550°C. These temperatures allow the 40% target to be exceeded easily (calculated efficiency of 43.3%).

Spallation module – solid target design

The spallation module design is presented in Figure 2. Figure 3 presents the detail of the rod bundle. This target is made of tungsten. The central part geometry corresponds to a pin bundle made of 46 staggered rows in the axial direction (see Figure 3) whereas its peripheral part is a tungsten ring in the thickness of which longitudinal channels are arranged allowing for helium circulation. Helium is circulating downwards in the central part and upwards in the peripheral part. In the peripheral part, one can consider that 50% of the ring is tungsten and 50% is helium. The purpose of these upper and lower peripheral parts is to maximise the efficiency of the target in terms of neutron production. It was seen in previous calculations that without these devices a significant fraction of high energy particles were escaping the target. The height of the target is of 690 mm. Target first rods are located 65 mm under the core mid-plane. The proton beam impacts the target from above. The main characteristics of the proton beam are summarised in Table 3 hereafter.

Subcritical core design

General core characteristics

The subcritical core is constituted by hexagonal sub-assemblies (SA), distributed in a hexagonal scheme in concentric rows (see Figure 1). At the centre of the core, 19 SA are removed in order to accommodate sufficient place for the spallation module. Each SA is embedded in a hexagonal wrapper

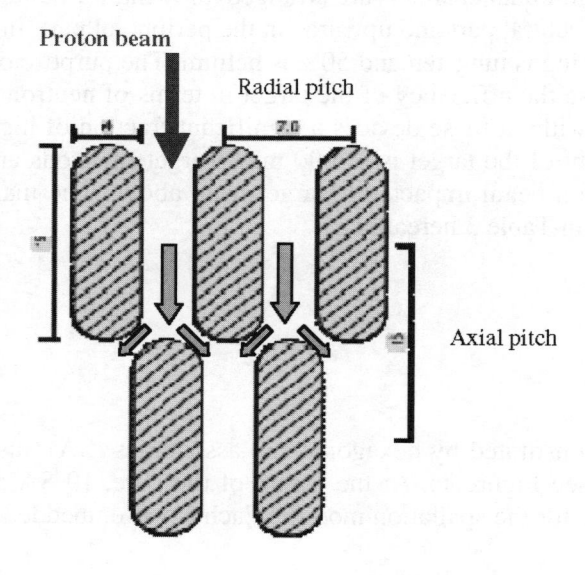

Figure 2. Spallation module design

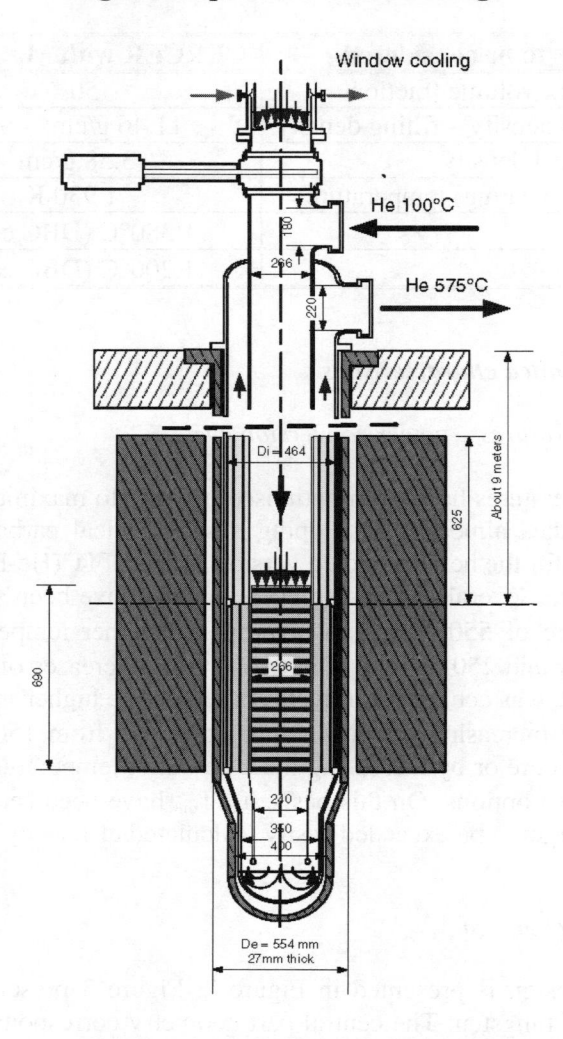

Figure 3. Rod bundle characteristics

Table 3. Spallation module design parameters

Parameters	Value	Remarks
Proton beam diameter (cm)	22.6	Calculated for an average flux < 50 mA/cm^2
Vacuum tube diameter (cm)	26.6	About 2 cm margin
He inlet pressure (bar)	14	
Inlet temperature in the target (°C)	100	Objective : minimise the outlet temperature
Max. helium outlet temperature (°C)	575	Creep still moderate for stainless steel 316 LN
Helium flow target (kg/s)	4.62	
External diameter of thimble (cm)	55.4	

which ensures its sealing and is separated from the other SAs by a gap which allows for thermal expansion as well as handling of the fuel. Each SA includes a certain number of vertical and parallel heating fuel pins separated by helium channels which extract the power produced in the fuel.

The core and the cooling systems of the plant are dimensioned from a thermal-hydraulic point of view in nominal operating conditions as well as at residual power. The nominal regime corresponds to the operation of the plant at full power. The residual power is the power produced by the core following the reactor shutdown.

The DHR strategy has been preliminarily assessed and proposed. The detailed design of the DHR components will benefit as much as possible from the development of gas-cooled fast breeder reactors.

Thermal-aeraulics – thermics – geometric features

In order to model the thermal and aeraulical phenomena involved and to allow the optimisation of an ADS reactor, the COPERNIC tool (CEA internal tool) was used. It consists of a worksheet (Excel) associated with a library of functions programmed in Visual BASIC. The phenomena brought into play in the hotttest channel are more particularly analysed as this channel sees the strongest temperatures of the core.

An important design criterion consists of limiting the core pressure drop to about 1 bar in order to ensure the removal of the residual power by natural convection in case of accidental transients such as the loss of coolant flow.

As far as thermics are concerned, work consists in determining the temperatures in various locations of the fuel pin in the hottest channel: at the centre of the fuel (Tic), at the external wall of the pellet (Tce), at the inner and outer part of the cladding (Tig and Teg). In this calculation, axial conduction is neglected. The fuel and cladding maximum temperatures are then compared to the acceptable values.

Coherently with the neutronic calculations (see the following section) and after several iterations between thermal-aeraulics and neutronics, a core with three zones was proposed with the pin diameter being different in the three zones.

Preliminary neutronic assessment

To perform the neutronic design of the of the He-EFIT core, the nuclear library used was ERALIB1 based on integral experiments specifically starting from the nuclear evaluations of the data JEF2.2 and one was based on the procedure MECONG of the system code ERANOS [6].

Initially, with the standard version of the procedure, a single zone core is defined on the basis of RZ geometry. In this single zone core, the content Pu FIMA [Pu (Pu+AM)] is defined to ensure the core subcriticality of at least 3 000 pcm ($K_{eff} < 0.97$). Then, using this Pu content, potential reactivity insertions are calculated. A modified version of MECONG was developed to answer specificities of the ADS of calculation with a source. In this version, the calculation of flux is carried out with source ("inhomogeneous" flow) and this makes it possible to obtain a K_s. Using K_{eff} and K_s, one can calculate the importance of the source, φ^*, which defines the effectiveness with which the neutrons of the source are amplified by the subcritical core: $\varphi^* = (1 - 1/K_{eff})/(1 - 1/K_s)$.

It is then possible to determine the intensity of the current of protons necessary to operate the plant power at 400 MWth.

In the second step, ERANOS procedures lead to define a core with three different radial core zones with the primary aim of radially flattening the power. The presence of three radial zones (of different compositions which impact at the same time neutronics and thermal-aeraulics) however tends to move fissile fuel away from the spallation module and thus to reduce the importance of the source neutrons.

The Pu vector used is that resulting from a MOX fuel irradiated in PWRs at 45 GWj/t and cooled for 30 years [2]. Burn-up calculations are then carried out for a frequency of 1, for a residence time of 2 770 EFPD. Then, a management of fuel with a frequency of 5 for which each cycle would last 554 days was calculated. The maximum burn-up rate is then 136 MWd/t and the average burn-up rate is 81 MWd/t for its average value. The maximum calculated damage is 55 dpa.

The reactivity variations during this burn-up cycle with a frequency of 1 (2 770 days) is shown in Figure 4.

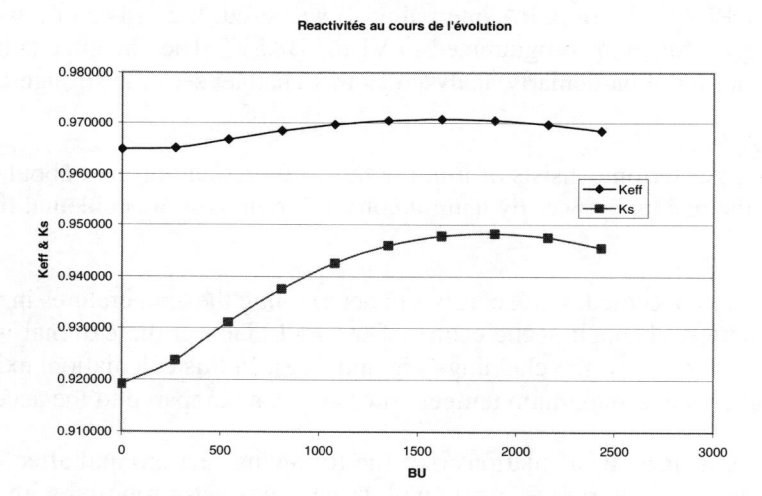

Figure 4. Reactivity evolution versus burn-up

The variation of reactivity (K_{eff}) during the cycle varies in a range of 650 pcm approximately. The management of fuel is carried out with a frequency of 1 for reasons of simplicity in this phase of the preliminary design. The maximum subcriticality during the operation, $K_{eff} = 0.965$, yields a source importance of 0.414 and a current of 22 mA for a proton beam of 800 MeV taking into account an appropriate K_s at this time in the burn-up. Fuel assembly management is certainly possible. It would allow an optimisation of the current intensity as well as flattening of the power profile. This would make it possible that the current remain below an acceptable maximum current of 20 mA as planned in the design of the target.

The axial and radial power distributions are then calculated to check the values of hot spots used in the thermal-aeraulical dimensioning of the assemblies. They are presented in Figures 5 and 6. The axial and radial peaking factors are 1.35 and 1.4, respectively. The transmutation rate calculated is 42 Kg/TWhth with the distribution shown in Table 4.

Figure 5. Axial form factor (BOL)

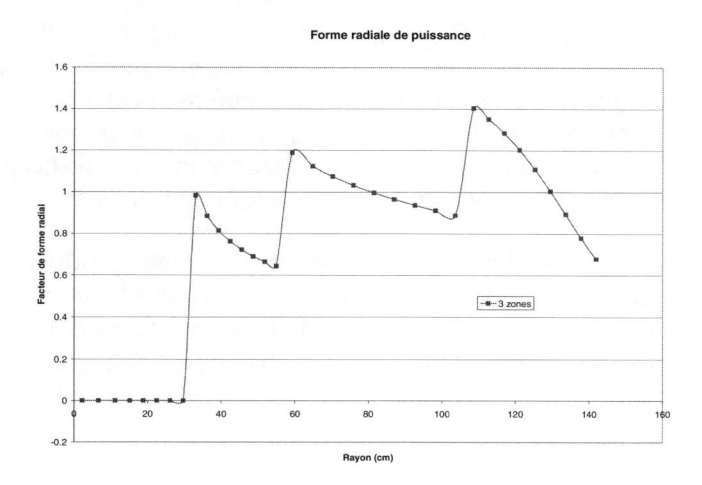

Figure 6. Radial form factor (BOL)

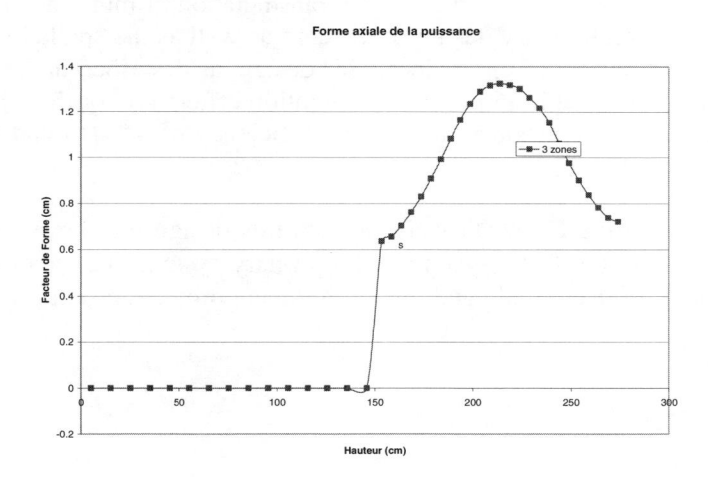

Table 4. Transmutation rate between BOL and EOL (2 790 FPD, 400 MW$_{th}$)

Element	Kg	Kg/TWhth
U	27.42	1.02
Np	-30.32	-1.13
Pu	176.47	6.59
Am	-1 463.36	-54.64
Cm	178.90	6.68
Total	**-1 110.89**	**-41.48**

Decay heat removal

In the present analysis two principal different approaches for extracting the decay heat in the He-EFIT plant have been compared: the first one is the CEA GFR600 approach proposed in 2004. It makes use of passive systems during the first 24 hours of an accident. The second approach is the SCS systems as proposed by AREVA within the PDS-XADS project.

Despite the attractiveness of the guard containment option studied for the GFR600, there are some specificities in an ADS reactor which must be taken into account. In the He-EFIT case, the guard containment adds a lot of complexity, and notably at the penetration of the proton beam towards the spallation module: the double envelope doubles the numbers of tight penetrations to be built. That also imposes frequent openings of the second envelope for the maintenance of this beam, as well as for the spallation target which is subject to frequent maintenance and/or replacement.

For He-EFIT, the PDS-XADS solution seems to be the best strategy of decay heat removal due to the same issues associated with proton beam spallation modules. As a consequence, the guard containment option will not be considered for the He-EFIT design. On this basis, it is proposed to design a DHR system making use of blowers/compressors having operating conditions ranging from 1 to 70 bar ("SCS-like" design).

Conclusion

The helium-cooled EFIT is an ADS devoted to the transmutation of minor actinides studied within the EUROTRANS project. A preliminary design of the core as well as the spallation module has been established during the first two years of the project. This design, as described in the present paper, fits with the plant objectives as specified in terms of transmutation efficiency for the elimination of minor actinides (42 kg/TWh$_{th}$), electricity production (with an efficiency of 44.4%) and the design criteria (temperature limits, core pressure drop, etc.).

During the second half of the EUROTRANS project, this design will be further optimised, and the safety features (particularly the DHR systems), the primary system, the balance of plant and the containment including main services (fuel handling, radiological protections, etc.) will be integrated in the plant design.

REFERENCES

[1] IP_EUROTRANS DM1_DESIGN, 6th Framework Programme – EU Contract FI6-CT-2004-516520.

[2] Rimpault, G., *Définition des missions détaillées de l'XT-ADS refroidi au Pb-Bi et de EFIT refroidi au Pb et de son option de repli refroidie au gaz*, Delivrable D1.1, EUROTRANS.

[3] Garnier, J.C., "Status of GFR Pre-conceptual Design Study", *ICAPP'06*, 4-8 June 2006.

[4] Giraud, B., *Synthesis Report of the PDS-XADS Project*, Deliverable 86, PDS-XADS Project.

[5] Thetford, R., V. Sobolev, *Recommended Properties of Fuel, Cladding and Coolant for EFIT*, Deliverable 3.1.4, Draft 0.2, o f EUROTRANS.

[6] Rimpault, G., *et al.*, "The ERANOS Code and Data System for Fast Reactor Neutronic Analyses", *PHYSOR2002*, Seoul, Korea, 7-10 October 2002.

A-BAQUS, A MULTI-ENTRY GRAPH ASSISTING THE NEUTRONIC DESIGN OF AN ADS. CASE STUDY: EFIT

Carlo Artioli
ENEA, Italy

Abstract

In a complex system such as an ADS reactor, a large number of parameters influence each other in a manner that is not self-evident, and in many cases the linkages are indirect and difficult to monitor. Even with reference only to the neutronic field (as in this case), the multiple natures of the indirect correlations increase the complexity of the design task. Moreover, on completion of certain parametric calculations, misunderstandings may arise in the identification of the cause-and-effect relations between pairs of parameters, misleading the designer. In order to facilitate a complete vision at a glance of most of the neutronic parameters involved in the design and their interaction, a multi-entry graph is proposed: A-BAQUS. Although making no claim to refinement, it displays at a glance the results of a change in any parameter and/or how they have to vary in order to achieve a stated goal. Based on an *a priori* definition of the required subcriticality, the kind of fuel, the coolant and its permitted velocity, i.e. the maximum linear power rating and the required coolant volume fraction, few parametric calculations will make it possible to draw curves characteristic of different fuel enrichments. Starting from a "status point" it will be possible to read in the A-BAQUS the related main neutronic parameters.

This paper will explain how to construct and use the A-BAQUS for the neutronic design of the core of an EFIT reactor, where paramount importance is given to performance in terms of MA transmutation and the required proton current.

In the top-right quarter two axes show the performance (MA burning capability and the conjugated Pu breeding) and the BU reactivity swing as merely a rough function of the enrichment. The bottom-right quarter indicates the power of the core (and its geometrical dimension) as a function of the content of matrix in the fuel. The bottom-left quarter shows the corresponding proton current and its range during the cycle, as a function of the core power and the enrichment and selected. The top-left quarter reflects necessarily the other quarters and shows the relationship between the enrichment (and BU reactivity swing) and the proton current.

Introduction

In the neutronic design of an ADS there are a number of parameters that have to be kept under control simultaneously. Their mutual interrelations sometimes could suggest misleading relationships among them. In fact, aside some self-evident cause-effect links, there are other ones rather indirect and not easy to monitor in the same time.

For this purpose a graph is proposed, in such a way as to allow the full vision of the main parameters, at a glance, and to see immediately what happens changing one or more of those, or even what has to be done to obtain the goal of the design. This kind of graph has to be considered as a "logical" guide, providing only rough results to be refined with suitable calculations. For this reason the axes are here presented not fully quoted; nevertheless the graph, if built up with care, can provided rather accurate results.

This graph, which can be re-elaborate according to individual feeling and needs, is presented here along the main considerations that have ruled the Pb-EFIT neutronic design.

Figure 1. Main parameters to keep under control simultaneously in ADS neutronic design

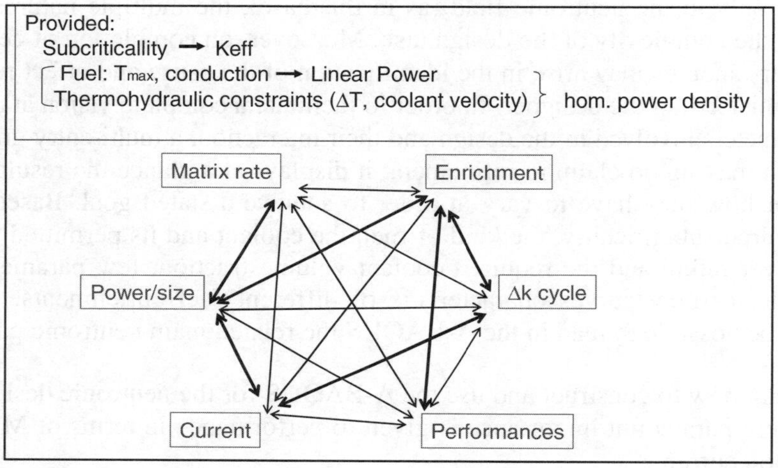

Provided:
Subcriticallity → Keff
Fuel: T_{max}, conduction → Linear Power
Thermohydraulic constraints (ΔT, coolant velocity) } hom. power density

Matrix rate — Enrichment
Power/size — Δk cycle
Current — Performances

Fuel enrichment

Pb-EFIT is a lead-cooled ADS, devoted to destroy (fission) the minor actinides (MA). It is fuelled with Pu and MA oxides dispersed in an inert MgO matrix. Since the fissions occur both on the Pu and MA atoms, and transmutations chains exchange Pu isotopes into MA ones and *vice versa*, some important parameters are directly related to the ratio between Pu and MA, i.e. the enrichment Pu/(Pu+MA). So in the first quarter, top right, on the ordinates there will be, along the enrichment itself (e), the performances in term of burning and the BU swing (ΔK/cycle). The total burning capability must be about 42 kg/TWh$_{th}$, as corresponding to the fission energy yield of 200 MeV/fission. It means that the "performances" ordinate can be duplicated in Pu balance rate and MA balance rate, being their overall -42 kg/TWh$_{th}$ everywhere (emphasis is given to the pair "-42/0").

Of course the scale of the different axes (BU swing and kg/TWh$_{th}$) are deformed as compared to the enrichment one.

Figure 2

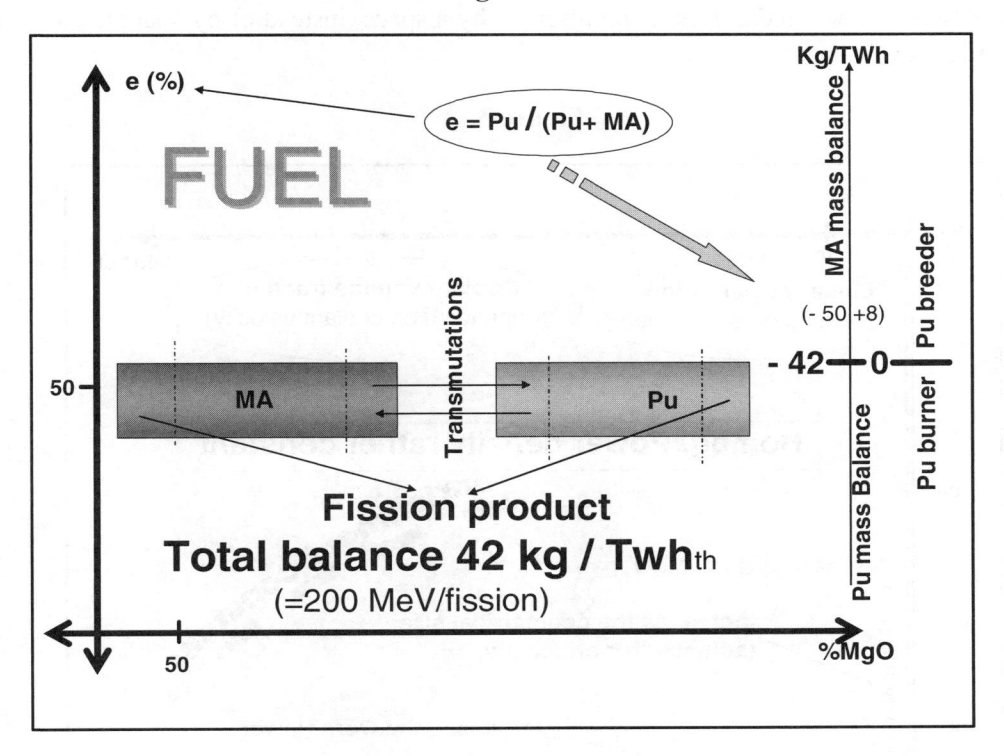

Matrix rate

The fuel is dispersed in an inert MgO matrix, whose minimum rate in the pellet must be 50%, for heat conduction reason. Nevertheless the matrix rate can be modulated to different higher rates, so varying the Pu+MA oxides volumetric fraction in different core regions, for flattening radially both neutronic flux and power density.

As the linear power rate is determined by the thermal properties of the fuel (melting T and conduction) and the coolant volume fraction is defined by the thermal-hydraulic requirements, the homogenised power density can be assumed rather constant, no matter the core size and enrichment. This means that along the increasing of the matrix rate (abscissa), there is an increase of the core volume and therefore of the core power. This leads to the establishment of a relationship between matrix rate (abscissa) and core power through core dimension (two ordinate axes).

Proton current

The relationship between core power and proton current (for a stated proton energy, 800 MeV in this case) should be a self-evident linear one; in first approximation approach (stated the subcriticality level).

But it should be pointed out that the linearity is spoiled by the different source efficiencies related to different spallation module sizes. The overall effect is that increasing the core dimensions (or power), the spallation module has to be increased with a loss of efficiency and this drag an even larger increase of the current.

The BU swing changes the k_{eff} in the cycle, so requiring a variable current. Therefore the relationship between power and current is represented by a space, instead of by a single curve, except the case where Δk cycle is zero.

Figure 3

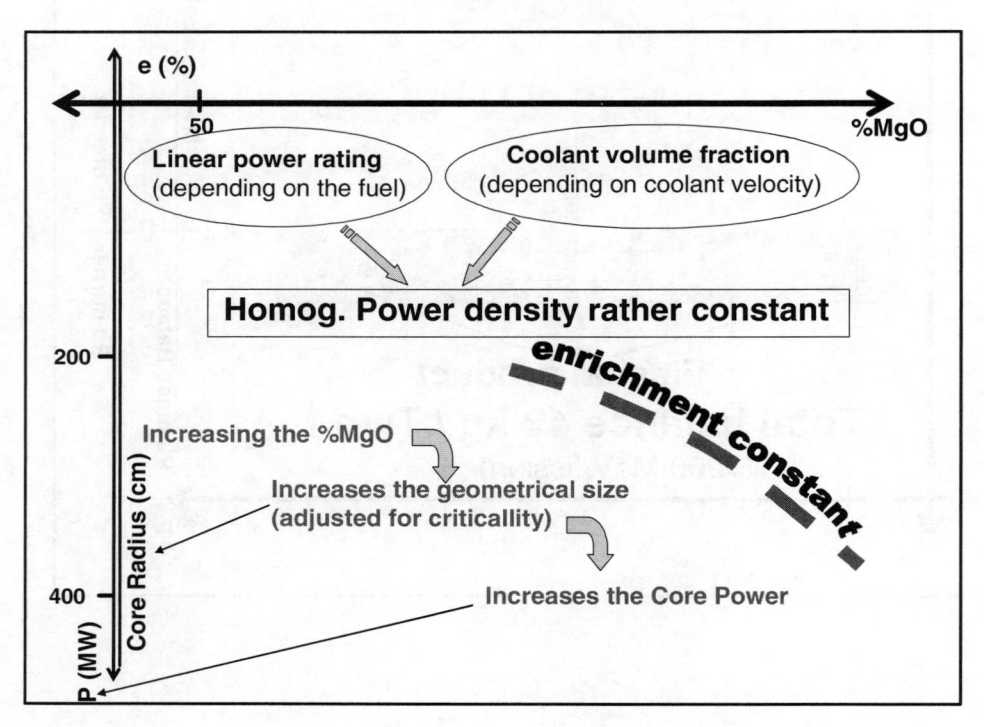

Figure 4

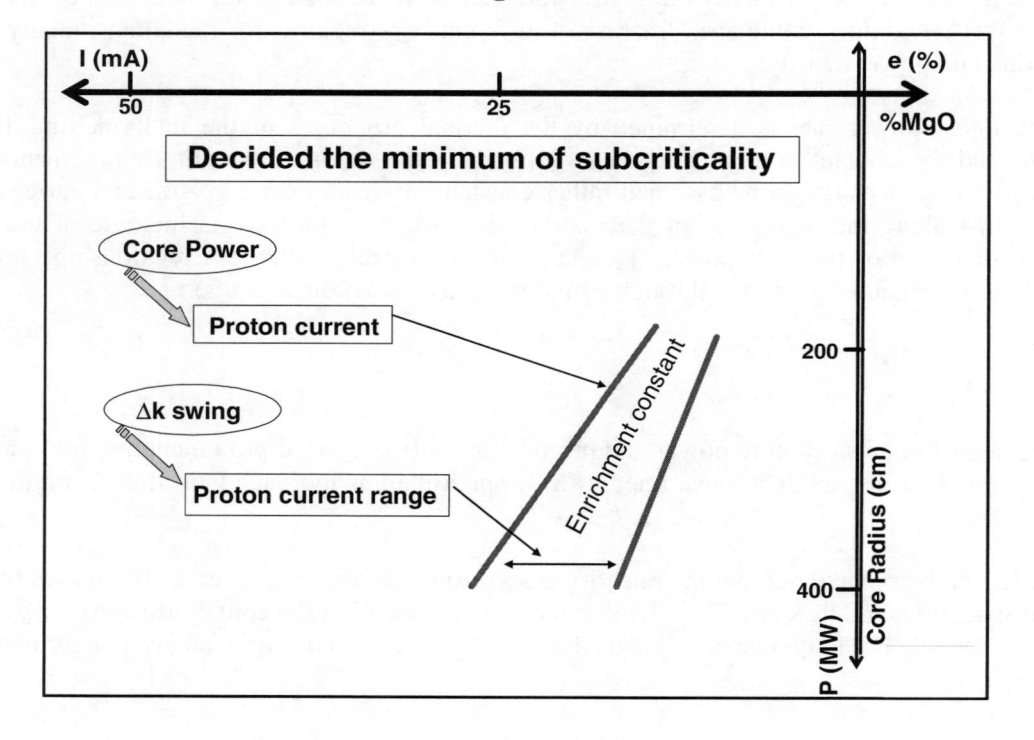

Different approaches to the neutronic design

To show how the graph can collect and manage the main neutronic parameters and how it can be used, three different approaches that have been studied in the EFIT design are here presented. For each of those there will be on the graph the characteristic curves that aid to monitor the parameters and to select the best actions to be taken.

"Δk zero" approach

This approach is based on the main consideration that in an ADS it is important to keep the range of proton current (and its value) as low as possible. This requires a BU swing of nearly zero. To this goal the right enrichment has been calculate resulting in e = 50%. The corresponding burning performances are 36 and 6 kg/TWh$_{th}$, respectively, for MA and Pu net fissions. According to different average rates of matrix, aimed to the radial flattening, three scenarios have been calculated: three core sizes, three corresponding core powers (210, 245 and 275 MW) and the three related proton currents (5, 6 and 6.5 mA), all being constant during the cycle. In the quarters, characteristic curves labelled e = 50% represent all the possible cores belonging to this approach. Changing whatever parameter (matrix rate, core power, proton current) the other ones are directly shown; keeping the enrichment (50%), the BU swing (0) and the burning performances (36/6).

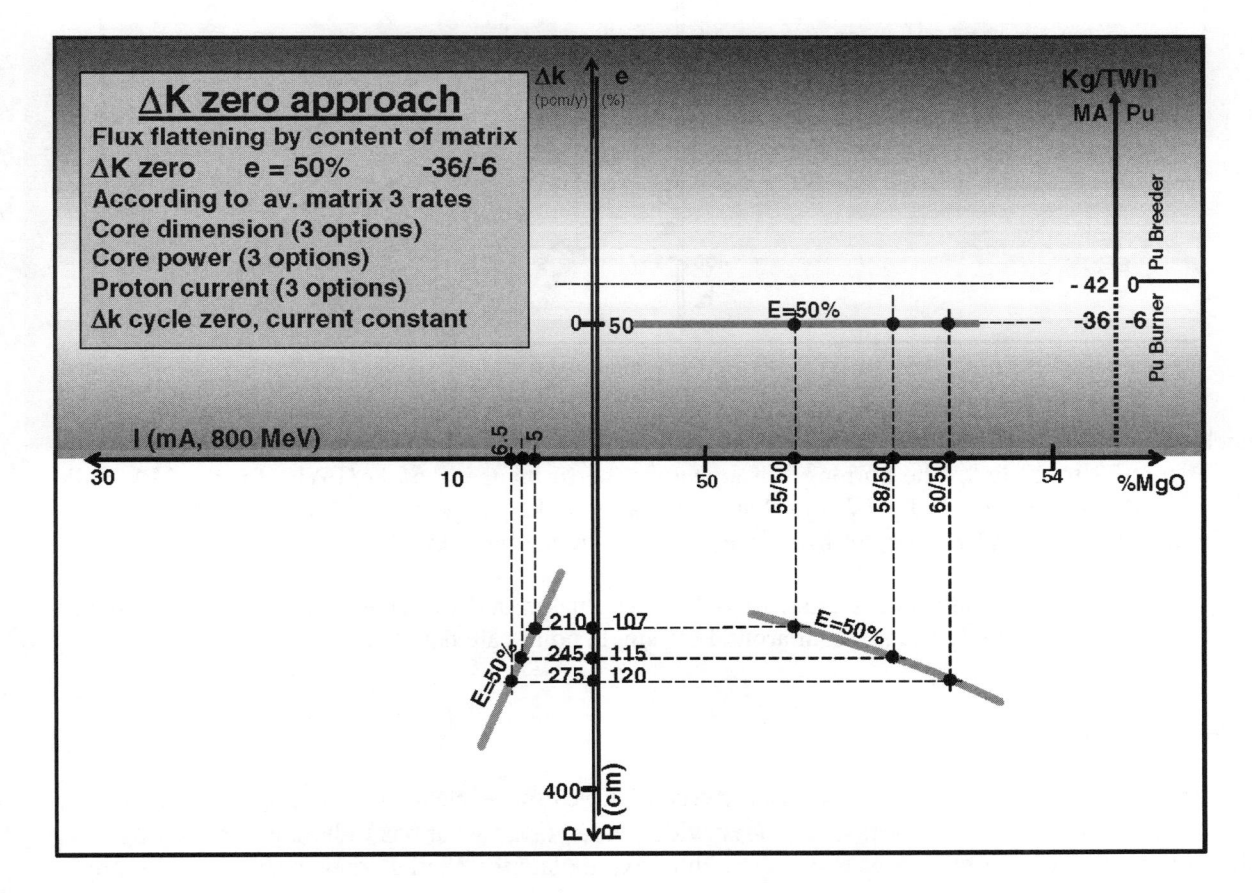

Figure 5. "Δk zero" approach

"400 MW" approach

In this scenario the starting point is the core power, 400 MW in this case. The matrix rate has been kept at its minimum 50%, while the radial flattening has been obtained by different pin sizes in two core regions.

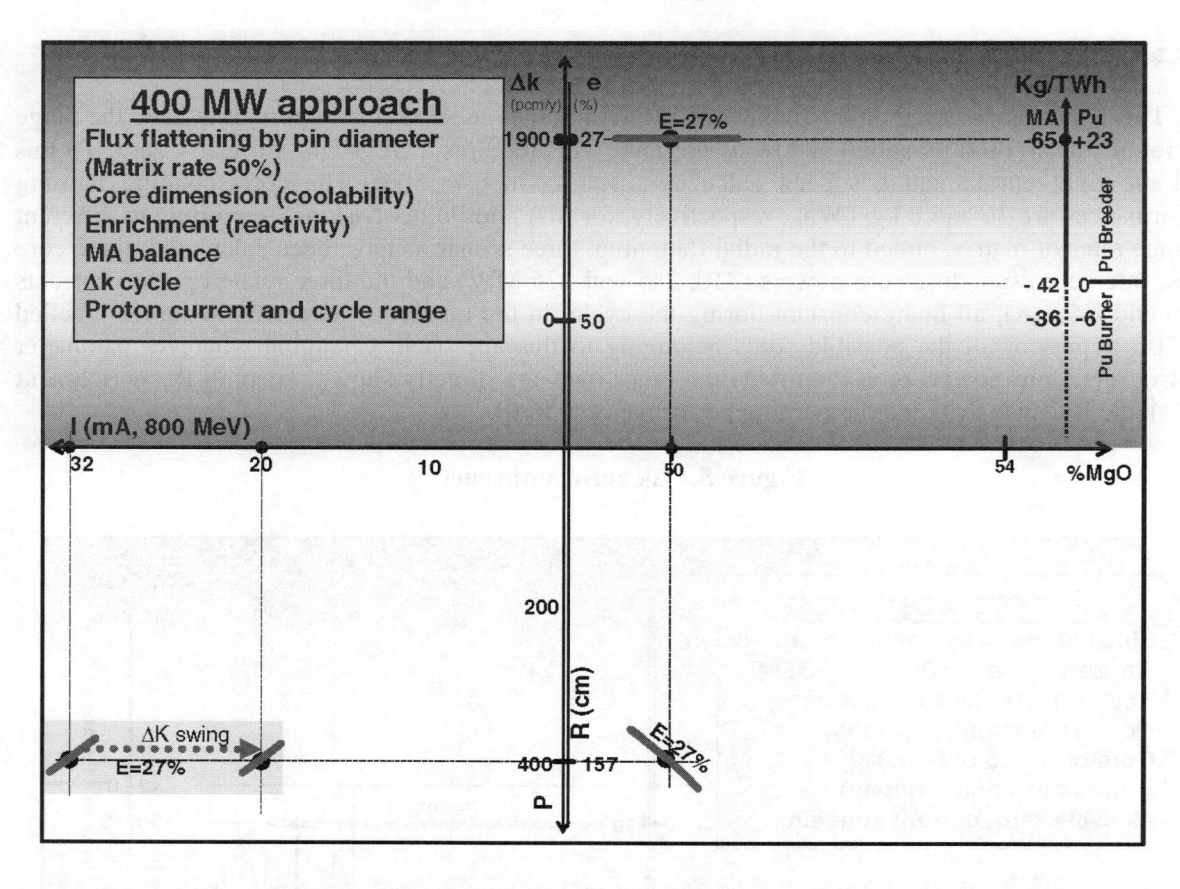

Figure 6. "400 MW" approach

[Figure: "400 MW approach" — Flux flattening by pin diameter (Matrix rate 50%); Core dimension (coolability); Enrichment (reactivity); MA balance; Δk cycle; Proton current and cycle range. Axes show Δk (pcm/y) 1900, e(%) 27, E=27%, Kg/TWh MA/Pu −65/+23, Pu Breeder, −42/0, −36/−6, Pu Burner, %MgO 50, 54; I (mA, 800 MeV) 32, 20, 10; R (cm) 200; P 400, 157; ΔK swing E=27%.]

The resulting core dimension and fuel volume fraction has required an enrichment of 27% to reach the expected reactivity (0.97). The associated BU swing was 1 900 pcm/year and the current range from 32 to 20 mA. The burning (disappearing) performance is 65 kg/TWh$_{th}$ for the MA while there is a Pu breeding of 23 kg/TWh$_{th}$. That means a mere 42 kg/TWh$_{th}$ of MA is actually fissioned, while the remaining 23 kg/TWh$_{th}$ of MA have been transmuted into new Pu.

For this approach only one scenario has been evaluated and the characteristic curves, labelled e = 27%, have been simply roughed out around the single point calculated.

"42-0" approach

The analysis of the graph and its characteristic curves made clear the situation and preliminary optimisation could be done. For instance, it is evident at a glance what has to be done for having a Δk zero core with 400 MW core power. It is enough to extrapolate the characteristic curves (see Figure 7) to see that it would require an average matrix rate of about 55% and the proton current would be about 10 mA. Current and burning performances (36/6) would be kept constant, as well as the enrichment.

In the same way it easy to see in the first quarter, on the right ordinate (kg/TWh$_{th}$), how the space is divided in two parts: above the -42/0 quote where the core acts as a Pu breeder (Pu increases) and below where it acts as a Pu burner (Pu decreases) at some extension.

These considerations lead to the so-called "42-0" approach, meaning that all the fissions are devoted to the MA (-42 kg/TWh$_{th}$) while the Pu net balance is kept nearly zero.

By a rough interpolation of the characteristic curves already available, can be divined that such a core, tuned on about 400 MW as reference, will have: enrichment about 45%, BU swing of some hundreds pcm/year, proton current ranging by some mA around 15 mA. All of this suggests that the core in question is rather viable.

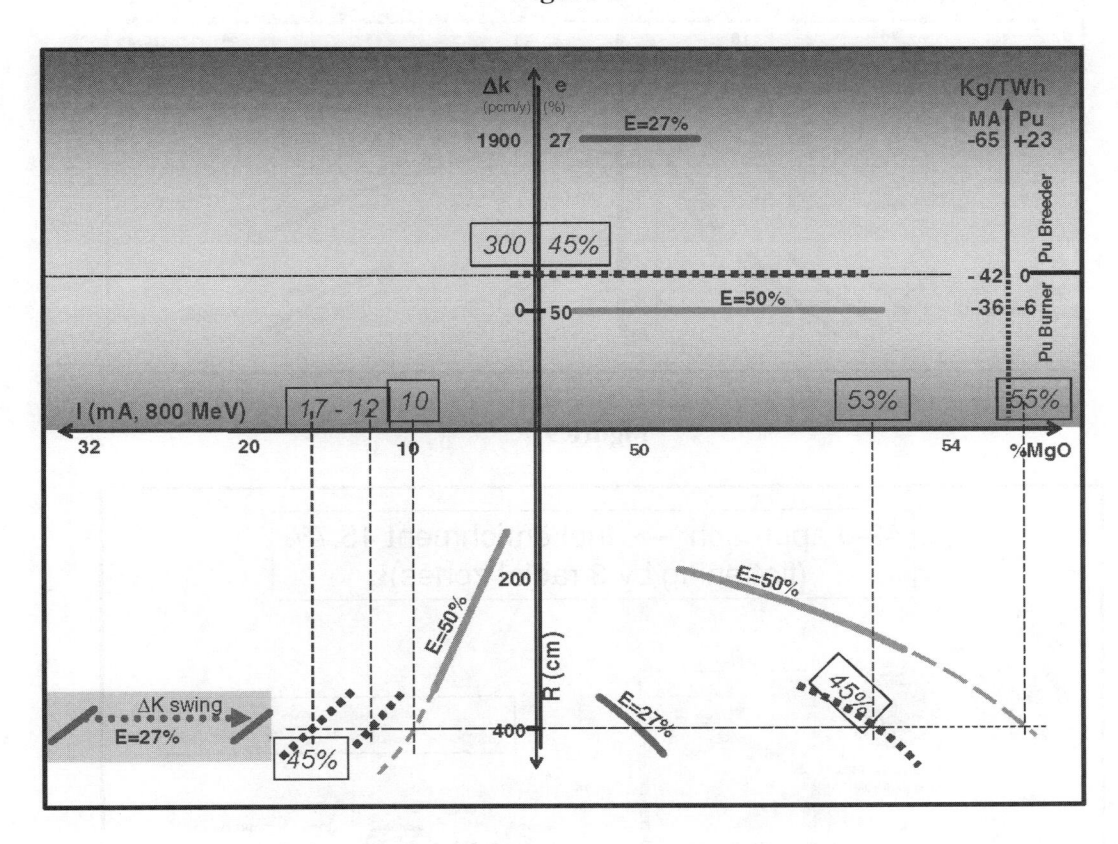

Figure 7

Some devoted calculations allowed to tracks the characteristic curves, labelled with e = 45.7%, for this final approach. Optimisation considerations, aiming to core size, have to be done within this approach and therefore laying on these last curves (Figure 8).

The remaining figures show the sketches of the subassemblies and the core, as actually resulted and some relevant characterisations.

Figure 8

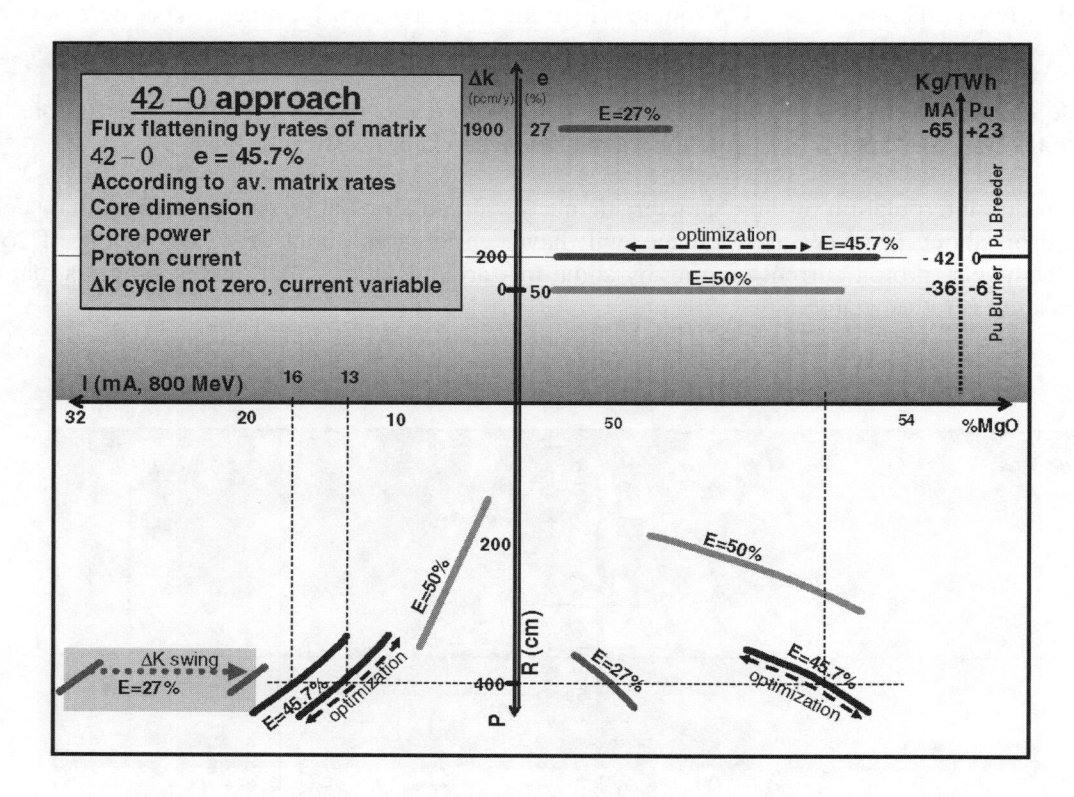

Figure 9

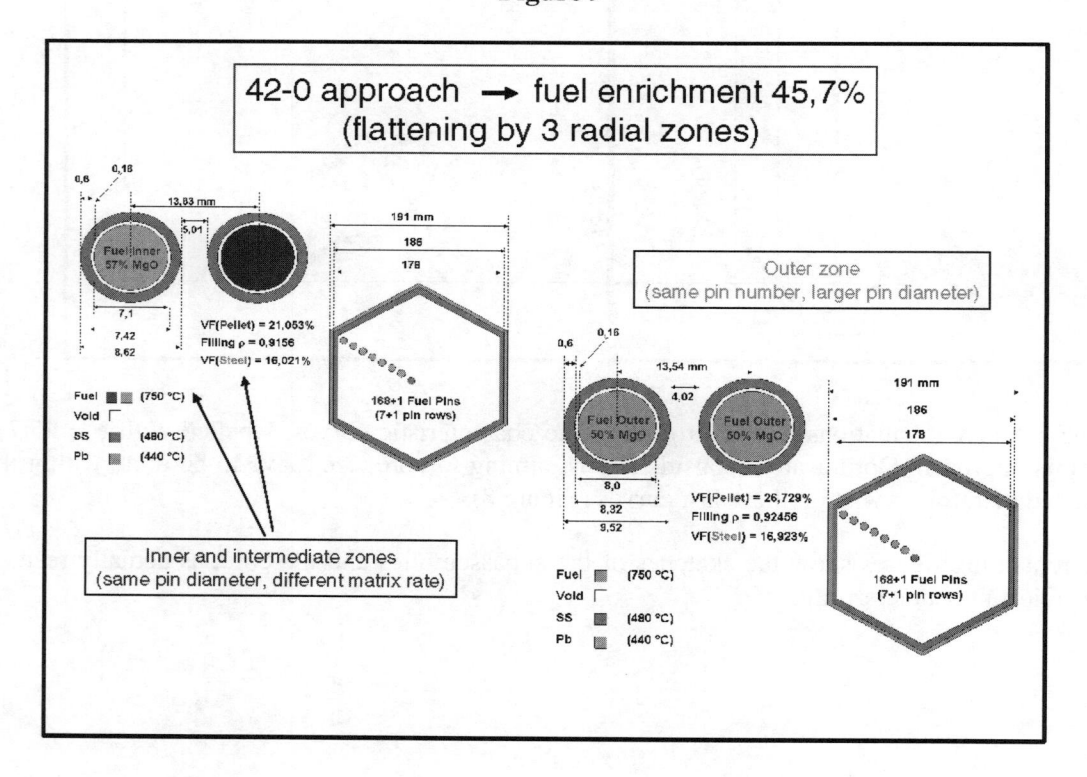

Figure 10

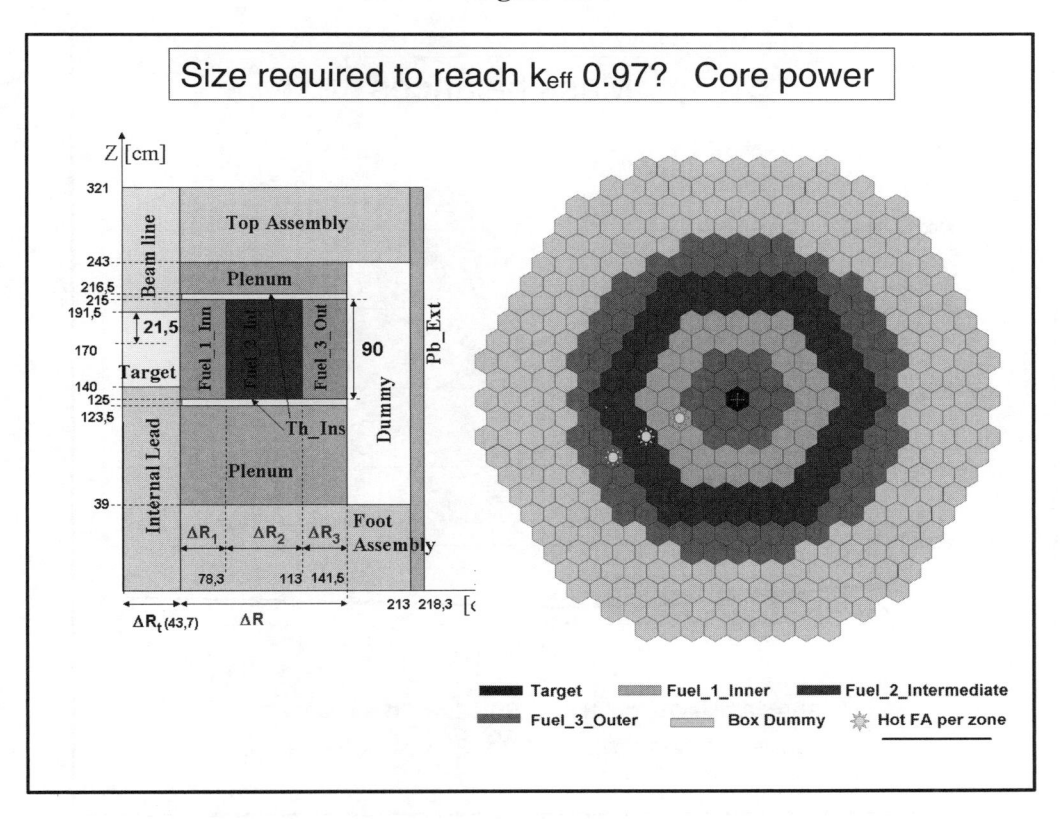

Figure 11

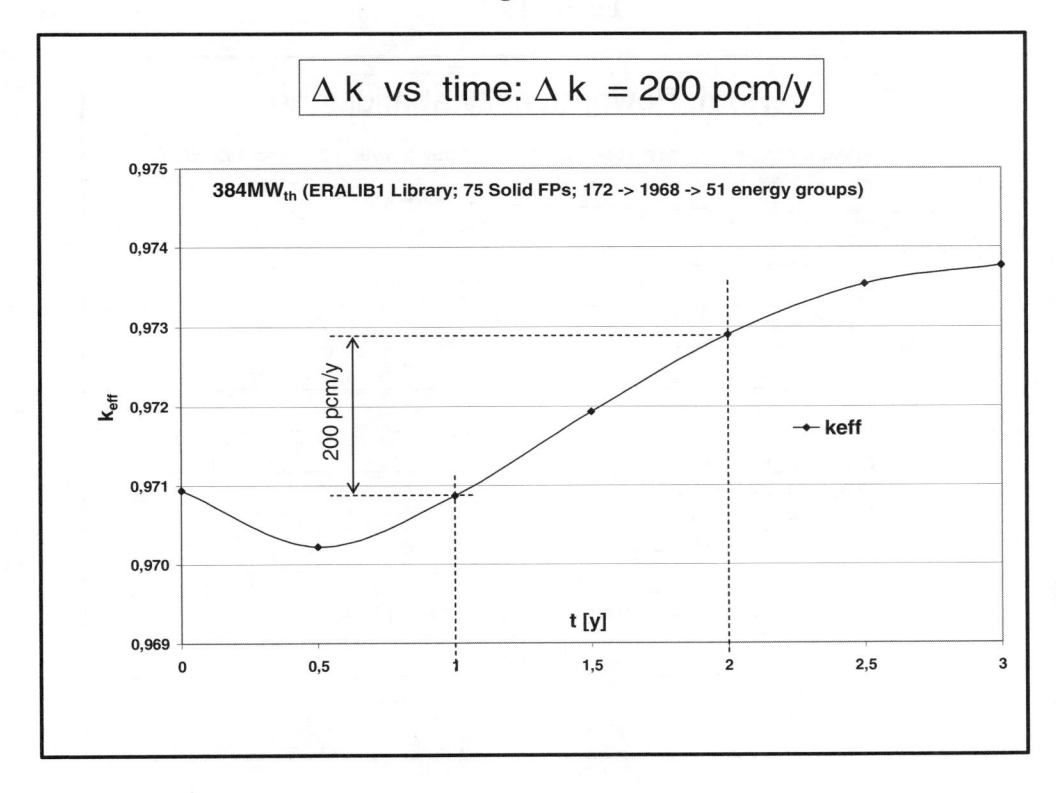

Figure 12

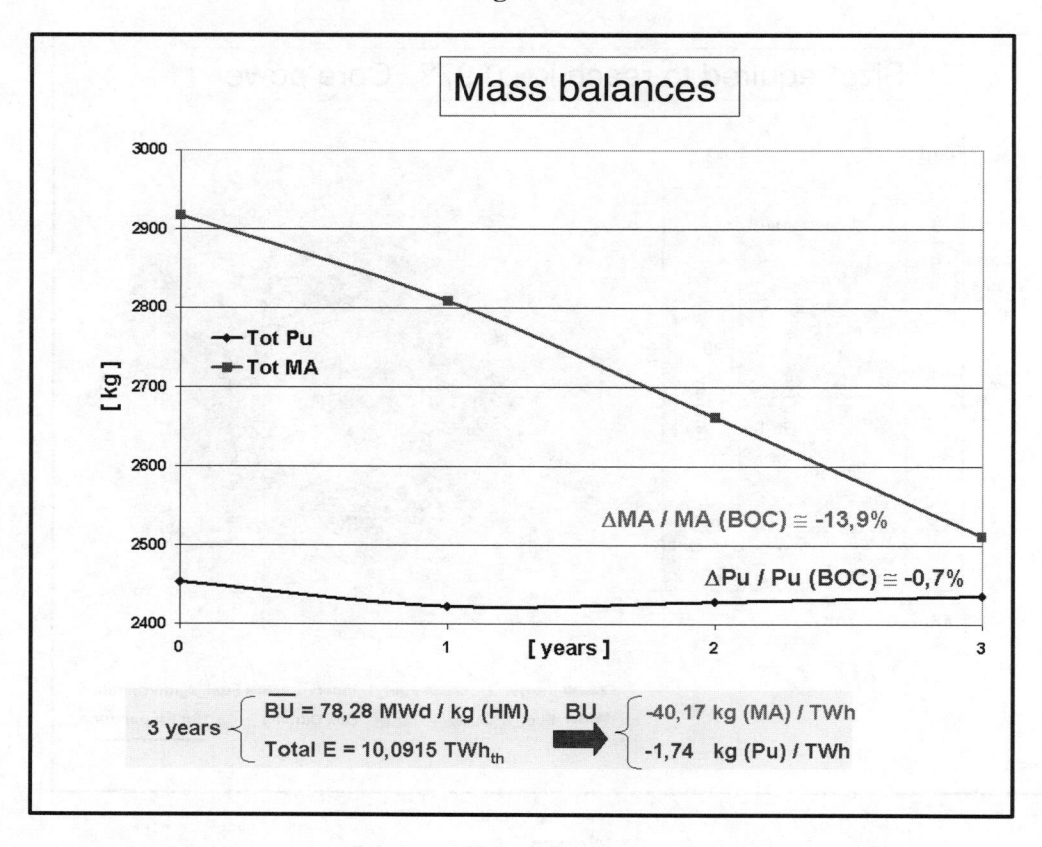

Mass balances

- Tot Pu
- Tot MA

$\Delta MA / MA\ (BOC) \cong -13{,}9\%$

$\Delta Pu / Pu\ (BOC) \cong -0{,}7\%$

3 years:
- $\overline{BU} = 78{,}28$ MWd / kg (HM)
- Total E = 10,0915 TWh$_{th}$

BU →
- -40,17 kg (MA) / TWh
- -1,74 kg (Pu) / TWh

Figure 13

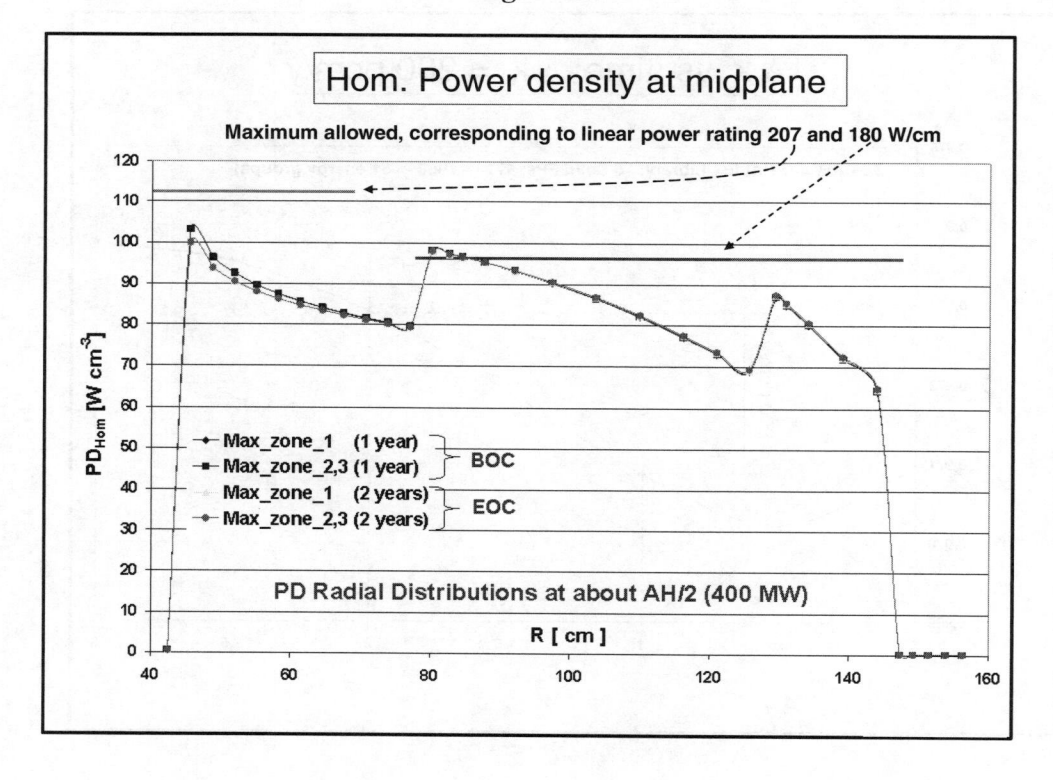

Hom. Power density at midplane

Maximum allowed, corresponding to linear power rating 207 and 180 W/cm

- Max_zone_1 (1 year) ⎫ BOC
- Max_zone_2,3 (1 year) ⎭
- Max_zone_1 (2 years) ⎫ EOC
- Max_zone_2,3 (2 years) ⎭

PD Radial Distributions at about AH/2 (400 MW)

PD_{Hom} [W cm^{-3}]

R [cm]

REFERENCES

[1] *EUROTRANS Specialist Meeting on Pb-EFIT Core Design*, CEA Cadarache, 13-14 June 2006.

[2] Sarotto, M., C. Artioli, V. Peluso, *Preliminary Neutronic Analysis of the Three-zones EFIT-MgO/Pb Core*, Tech. Rep. ENEA FPN P815 004 rev01, December 2006.

COMPARATIVE ANALYSIS OF NEUTRON SOURCES PRODUCED BY LOW-ENERGY ELECTRONS AND DEUTERONS FOR DRIVING SUBCRITICAL ASSEMBLIES

Dmitry Naberezhnev, Yousry Gohar, Henry Belch, Jose Duo
Argonne National Laboratory
9700 S. Cass Avenue, Argonne, IL 60439, USA

Igor Bolshinsky
Idaho National Laboratory
P.O. Box 2528, Idaho Falls, Idaho 83403, USA

Abstract

A conceptual design of an accelerator-driven subcritical assembly has been developed using the existing accelerators at Kharkov Institute of Physics and Technology (KIPT) in Ukraine. This paper examines two different external neutron source options for driving the subcritical assembly. The neutron source is produced by low-energy particles: either electrons with energies below 200 MeV or deuterons with energies below 100 MeV. Comparative analysis of these two options is presented and discussed.

Introduction

Argonne National Laboratory (ANL) in collaboration with Kharkov Institute of Physics and Technology (KIPT) of Ukraine is developing a conceptual design of an accelerator-driven subcritical assembly [1,2]. The main functions of this assembly are medical isotope production, neutron experiments and subcritical physics experiments. Two different options are considered for generating the required neutron source. In the first option, neutrons are produced from photonuclear reactions and the photons are generated from the interaction of electrons with a high atomic number target material, uranium or tungsten. The second option generates the neutrons from the interaction of deuterons with beryllium target. For both options, the particle beam power is 100 KW. The main focus is to maximise the neutron production from the available beam power.

Over the course of this design activity, several studies were performed to define the system's parameters and performance. The neutron source performance was characterised in terms of: (1) neutron source strength; (2) neutron spatial and energy distributions; (3) energy deposition in the target material; (4) target geometrical configurations; (5) beam radius relative to the target radius. The paper presents the results of these studies.

Neutron yields

Electron accelerator option

High atomic number target materials are used to produce neutrons from the electron interactions. In addition, high melting point, high thermal conductivity, chemical inertness, high radiation damage resistance and low neutron absorption cross-section are the desirable properties for the target materials. Previous physics studies [3] and our analyses show that uranium, tungsten, lead, bismuth and tantalum materials produce the highest neutron yield per incident electron. Based on our analyses, we also concluded that the use of the uranium and tungsten materials results in highest neutron yield at any electron energies in the range 50-200 MeV. All the calculations were performed with the Monte Carlo MCNPX computer code [4]. The photon and neutron spectra produced in different materials do not vary significantly from one material to another. Based on the operating experience of different accelerator facilities around the world, and the highest generated neutron yields, uranium and tungsten materials are selected for further consideration. The neutron yields per electron for both materials are shown in Figure 1. The neutron yield steadily increases as the electron energy increases. However, the neutron source production per second reaches a saturation value as the electron beam energy increases above 150 MeV when the neutron yield is normalised to the available beam power as shown in Figure 2. Therefore, electrons with energy above 150 MeV produce the maximum neutron yield for the available beam power.

The neutron spectra from tungsten and uranium target materials were analysed to quantify the high energy component as function of the electron energy. For both materials, the results of Figures 3 and 4 show that the neutron spectra are centred about 1 MeV, similar to a fission spectrum. The high energy component of the spectrum is not sensitive to the electron energy and its magnitude is very small.

Deuteron accelerator option

The second option to generate the neutron source for driving the subcritical assembly is to use a deuteron beam with a beryllium target material. The neutron yield from the (d,xn) reactions is much greater than the yield from the photonuclear reactions produced with the electron beam as illustrated in Figure 5, and consequently the external source intensities are high as shown in Figure 6. However, the creation of 100 KW deuteron beam with deuteron energy in the range of 20-100 MeV requires high deuteron current, which is difficult to generate and very expensive to produce. Only lower deuteron energies around 20 MeV are possible. As discussed further, when a reasonable current ~0.5 mA and low-energy deuteron beam of ~20 MeV is used, the resulting neutron intensity is inferior to the neutron intensities obtained with electron beam. The resulting neutron spectra for different deuteron energies are shown in Figure 7. The obtained neutron spectrum is much harder than the fission spectrum and the neutron spectrum generated using electron accelerators.

Targets and beam parameters optimisation

Electron accelerator option

In order to choose appropriate beam energy, the performance of uranium and tungsten target materials has been analysed as a function of the electron energy for the 100 KW beam power. The spatial energy deposition density per incident electron in tungsten is shown in Figure 8 as a function of the target length for different electron energies. The peak value of the energy deposition occurs a few millimetres away from the electron beam window. This peak value increases and shifts further away from the electron beam window as the electron energy increases. The spatial energy deposition normalised to 2 KW/cm^2 beam power density on the beam window, corresponding to the total beam power of 100 kW, is plotted in Figure 9. The thermal-hydraulics analyses require the use of thin layers of the target material to remove the deposited energy without generating high thermal stresses. The layers are separated by the water coolant channels. A large number of coolant channels slows down and absorbs electrons without generating neutrons and reduces the target performance. To avoid this situation either a low beam power density (less than 2 KW/cm^2) must be used or an electron beam with energies above 150 MeV must be employed. Reducing the beam power density increases the target cross-section area, which decreases the number of source neutrons delivered to the subcritical assembly. To avoid this scenario, the electron beam with energy above 150 MeV is used. A similar conclusion can be drawn for the uranium target when Figures 10 and 11 are analysed.

The next step in the analysis defines the target length to maximise the neutron yield. The choice of the target length is important since an extra target length acts as a neutron absorber and decreases the neutron yield. Figure 12 and Figure 13 illustrate the dependence of the neutron yield on the target length for tungsten and uranium, correspondingly. For the tungsten target, the neutron yield decreases when the target length exceeds ~6.9 cm. The neutron yield behaviour for the uranium target is different; no decrease in the yield is noted. This is due to the extra fission neutrons that are produced by the photonuclear neutrons at the bottom of the target. In the first part of the target length, most of the neutrons are generated from the electron and neutron interactions. As the electron beam vanishes inside the target material, the fission reactions caused by the bottom neutron fraction slightly increase the neutron yield. This slight increase was quantified and it is limited to less than 10% of the total yield. Therefore, about 8 cm target length is utilised for the uranium target to get some of these extra fission interactions to occur in the fuel material of the subcritical assembly.

The next geometrical parameter to consider is the target radius relative to the beam radius. If the two radii are very close, a portion of the source electrons might leak out of the target material without producing neutrons. To study this situation, the neutron yields were defined as function of the difference between the two radii. The results of Figure 14 and 15 of this analysis show that for the tungsten target material, the electron beam radius should be ~0.5 cm smaller than the target radius. For the uranium target, this difference should be ~1 cm. In addition, the choice of the target radius is controlled by the subcritical assembly geometry, the nuclear fuel geometry of the assembly and the thermal-hydraulics requirements.

The spatial distribution of the generated neutrons is analysed to maximise their utilisation in the subcritical assembly. In these analyses, a stacked target disk design is employed including the water coolant channels. The neutron fraction that leaves the target along the beam axis has the potential of disappearing without utilisation. A large fraction of the neutrons leaving from the electron beam window (top fraction) exits the system through the vacuum beam tube or interacts with the beam tube structure. A fraction of these neutrons reaches the subcritical assembly by leaking from the beam tube. Also, the neutrons leaving the target in the beam direction (bottom fraction) interact with the water coolant or the bottom reflector, which reduce their chances for reaching the subcritical assembly. Table 1 shows the distribution of the neutron fractions from the target assembly as a function of the electron energy. These results confirm that the use of high-energy electrons in the range of 150-200 MeV is beneficial for improving the system performance.

Table 1. Target parameters normalised to 100 kW beam power

Target material	Electron energy, MeV	Target radius (Beam radius), cm	Neutron yield (% of total)				Intensity, n/s
			Top	Bottom	Side	Total	
W	100	3.84145 (3.34145)	0.0084 (27.85)	0.0012 (4.08)	0.0206 (68.07)	0.0302 (100)	1.89E+14
W	200	3.84145 (3.34145)	0.0150 (24.46)	0.0029 (4.80)	0.0434 (70.74)	0.0613 (100)	1.91E+14
U	100	3.77145 (3.27145)	0.0131 (21.39)	0.0010 (2.34)	0.0385 (76.27)	0.0526 (100)	3.28E+14
U	200	3.77145 (3.27145)	0.0228 (24.89)	0.0025 (1.94)	0.0814 (73.18)	0.1067 (100)	3.33E+14

Deuteron accelerator option

The low-energy deuterons do not penetrate deeply in the target beryllium material. Only 2 to 3 mm of the beryllium material is enough to stop a low-energy (20 MeV) deuteron beam as shown in Figures 16 and 17. This fact is also supported by the dependence of the neutron yields on the target length shown in Figures 18 and 19. At the same time, the energy deposition in the first few millimetres of the target material is extremely high. As mentioned before, it is difficult and expensive to obtain 100-kW deuteron beam with high-energy deuterons. Thus, it is possible to consider a beam with 20-MeV deuterons and 0.5-mA current. In this case, the energy deposition in the first few millimetres of the beryllium material is extremely high. This situation calls for a special target design that allows for a large cooling surface. For example, a conical layer of beryllium material is considered. To assess the losses in neutron yield produced by low-energy deuterons, Figures 18 and 19 demonstrate the difference in neutron yields obtained with low-energy (23 MeV) and high-energy (100 MeV) deuterons. The results show that that the neutron yield with low-energy deuterons is significantly lower than that obtained with an electron beam option.

Comparative analysis between deuteron and electron options

Two main options for generating external neutron source were analysed and discussed. One of these options uses an electron accelerator with high atomic number materials (U and W), another option is based on the Be(d,xn) reactions. It was shown that with fixed beam power of 100 kW and particles energy of about 100 MeV, the deuteron accelerator option provides the maximum neutron strength. However, a deuteron accelerator with such characteristics is technologically challenging to operate in steady-state mode and expensive relative to an electron accelerator. With lower deuteron energies (~20 MeV), a very high current must be provided to achieve the 100-kW deuteron beam power. This results in a costly and inefficient option. On the other hand, an electron accelerator provides reasonable neutron strengths and does not require high accelerator current. The choice of the target for the electron accelerator option is between tungsten and uranium. Table 2 compares the three different options for the external neutron source. For a reasonable beam current and 100-kW beam power, the electron accelerator option is possible and its performance is superior to the deuteron accelerator option. The uranium target is preferred relative to the tungsten target because of the higher neutron source intensity.

Table 2. Comparison of different options for producing external neutron source

Target concept	Beam power, KW	Particle energy, MeV	Beam current, mA	Neutron source strength, n/s
Uranium with electrons	100	200	0.5	3.28×10^{14}
Tungsten with electrons	100	200	0.5	1.91×10^{14}
Beryllium with deuterons	100	100	1.0	1.90×10^{15}
	100	23	4.35	4.03×10^{14}
	11.5	23	0.5	4.63×10^{13}

Conclusion

The conceptual design activity of a driven subcritical assembly facility analysed the use of electron and deuteron accelerators for neutron generation. The electron and deuteron energies are less than 200 and 100 MeV, respectively. Physics studies have been carried out to investigate the target design choices and the accelerator beam parameters for the best neutron source performance. Physics analyses with the MCNPX code were focused on the neutron source intensity, the neutron spectrum, the spatial neutron generation, and the spatial energy deposition in the target assembly as a function of the beam parameters, the target materials, and the target design details. The main focus is to maximise the neutron production from the available beam power. The results show that an electron accelerator with electron energy in the range of 150-200 MeV is the preferred option. The uranium target material with electron beam produces the highest neutron yield. The selected neutron source produces 3.3×10^{14} n/s from the uranium target operating with the 100 KW electron beam power.

Figure 1. Neutron yield per electron from uranium and tungsten targets as a function of the electron energy

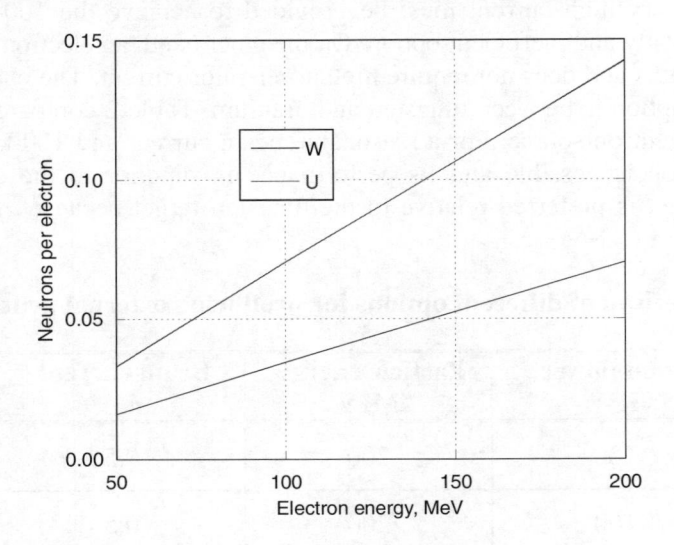

Figure 2. Neutron source strength from tungsten and uranium targets as a function of the electron energy for 100 kW beam power

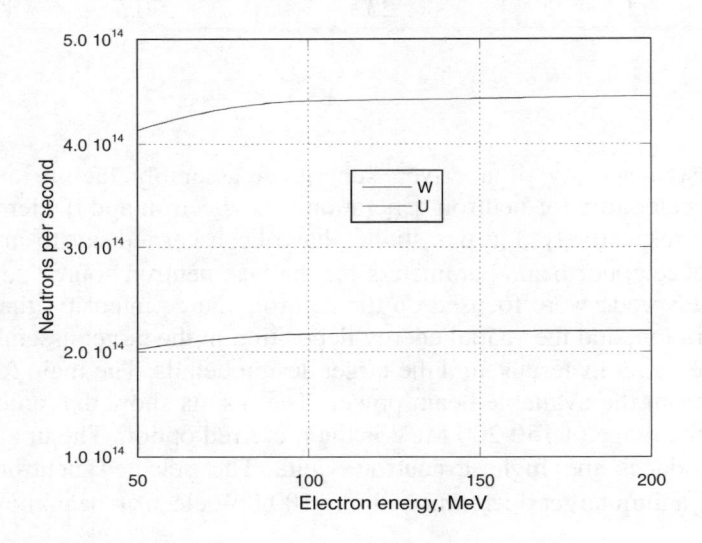

Figure 3. Neutron spectrum from tungsten target material for different electron energies

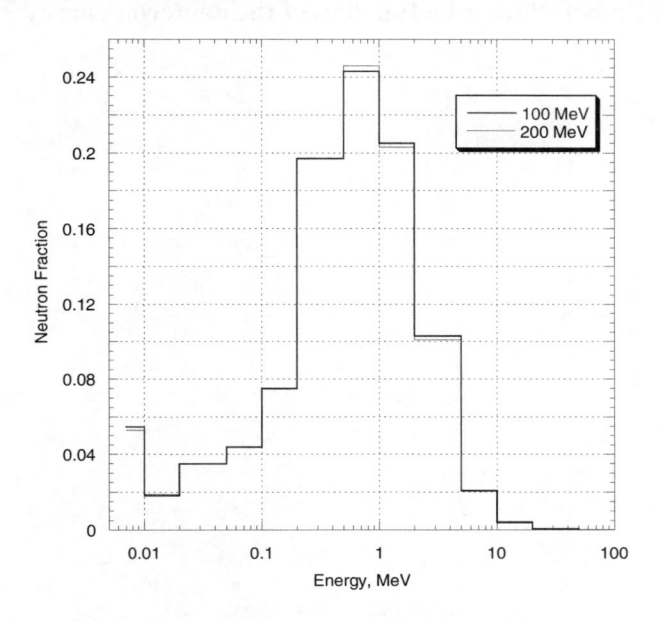

Figure 4. Neutron spectrum from uranium target material for different electron beam energies

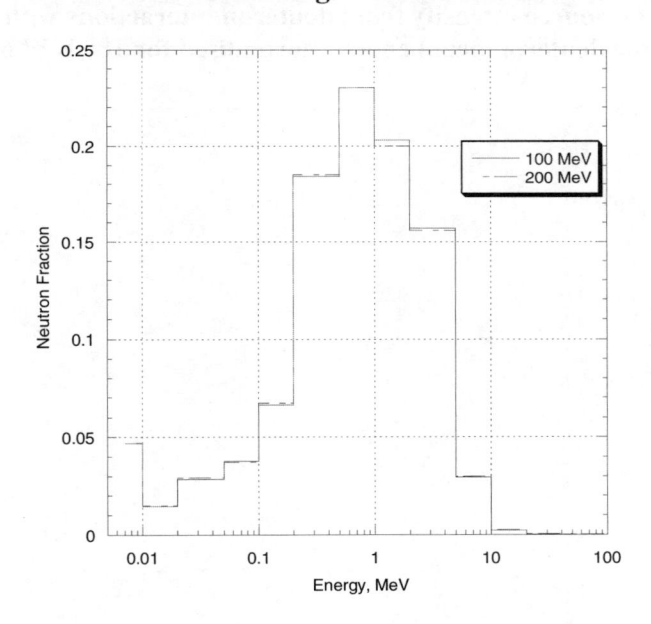

Figure 5. Neutron yields from deuteron interactions with beryllium as a function of the deuteron energy

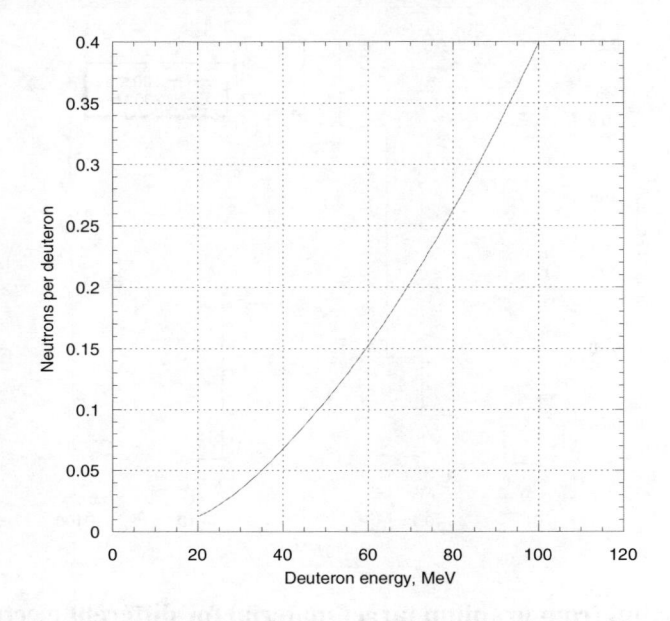

Figure 6. Neutron source intensity from deuteron interactions with beryllium as a function of the deuteron beam energy normalised for 100 KW beam power

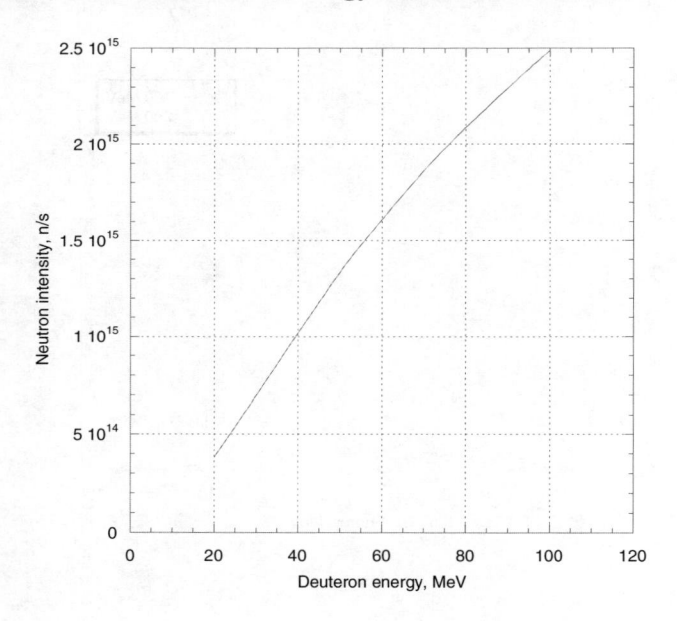

Figure 7. Neutron spectra generated from deuteron-beryllium interactions for different deuteron energies

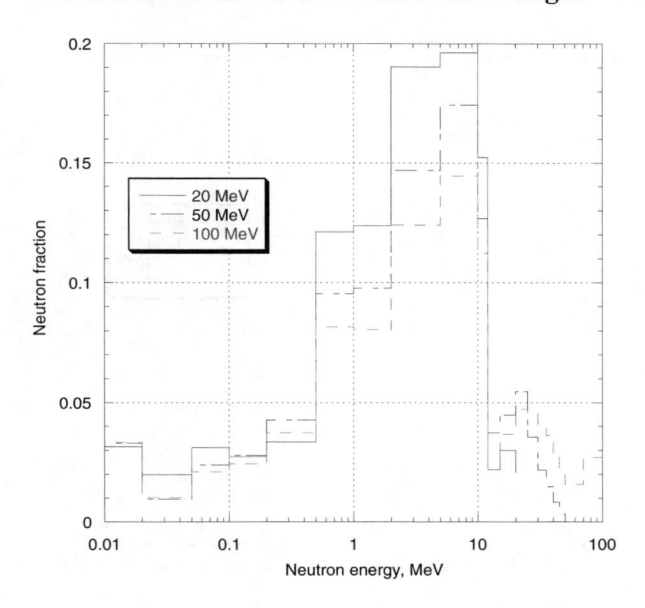

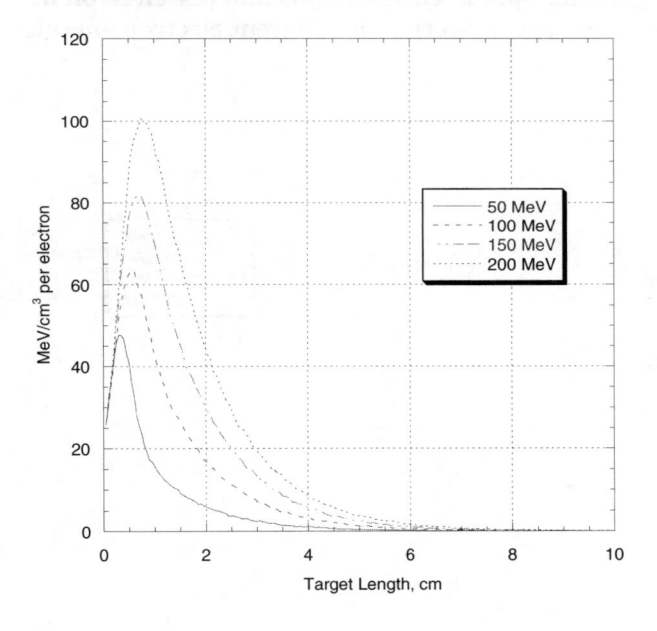

Figure 8. Spatial energy deposition per electron in the tungsten target material for different electron energies

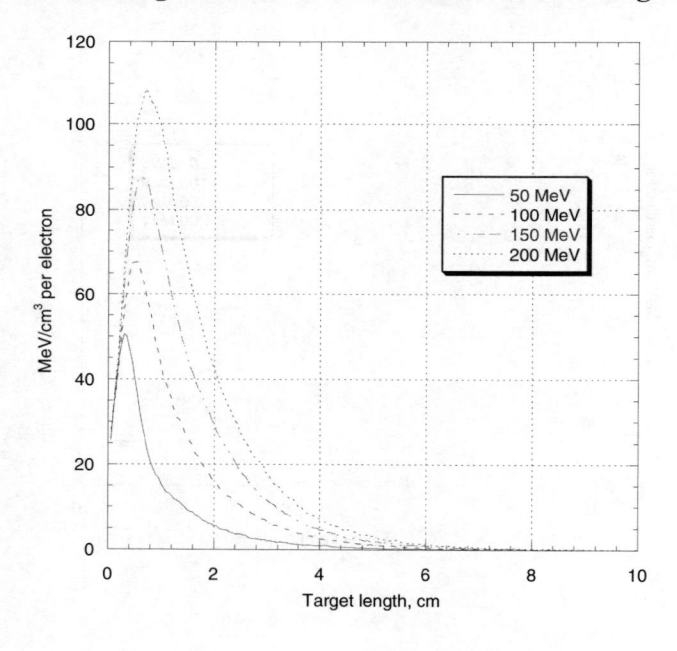

Figure 9. Spatial energy deposition in the tungsten target material normalised to 2 KW/cm^2 on the beam window for different electron energies

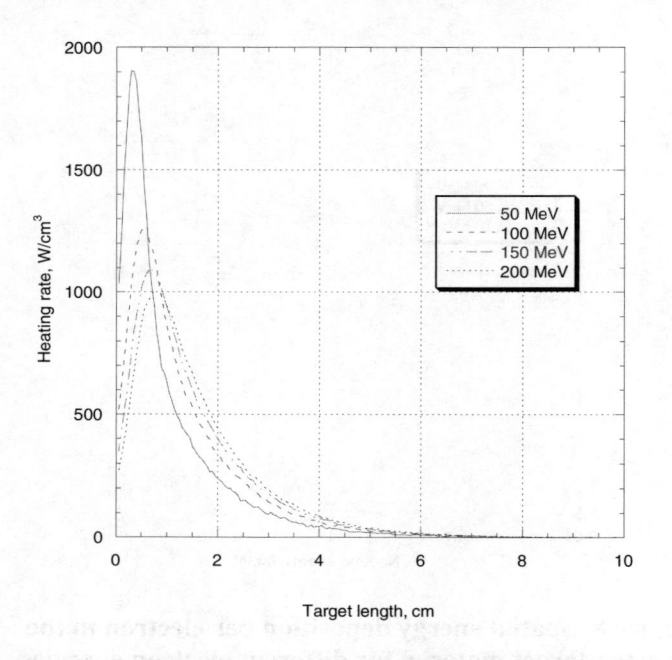

Figure 10. Spatial energy deposition per electron in uranium target material for different electron energies

Figure 11. Spatial energy deposition in uranium target material normalised to 2 KW/cm² on the beam window for different electron energies

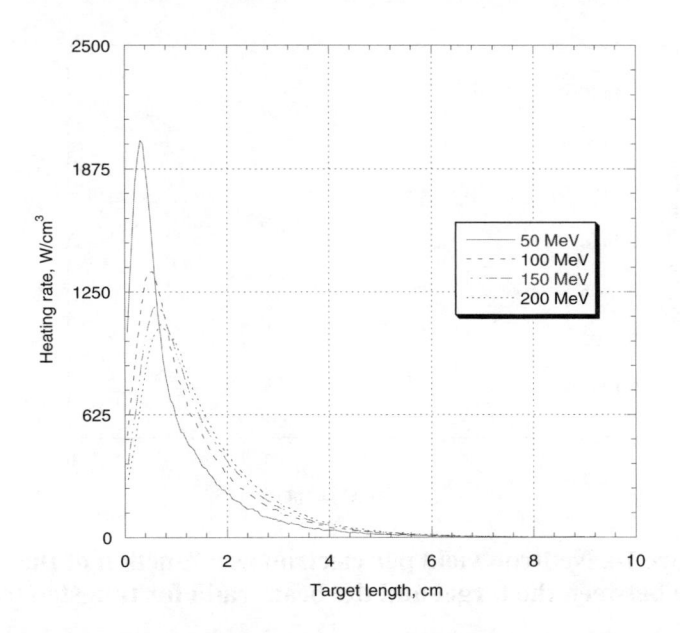

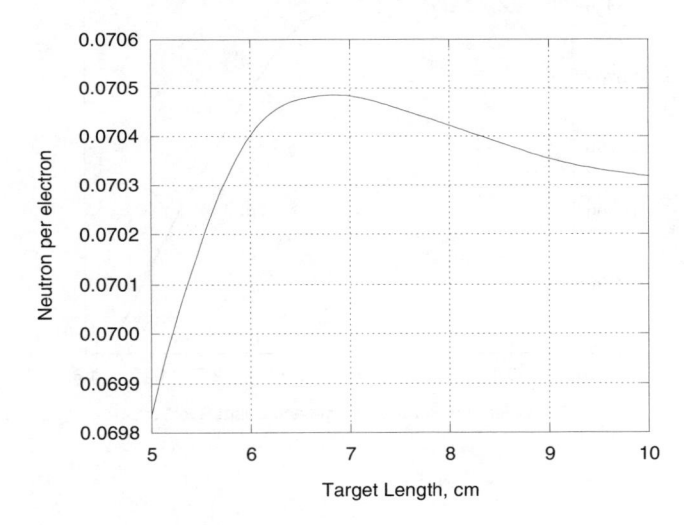

Figure 12. Number of neutrons per electron from pure tungsten material as a function of the target length for the 200-MeV electron energy

Figure 13. Number of neutrons per electron from pure uranium material as a function of the target length for the 200-MeV electron energy

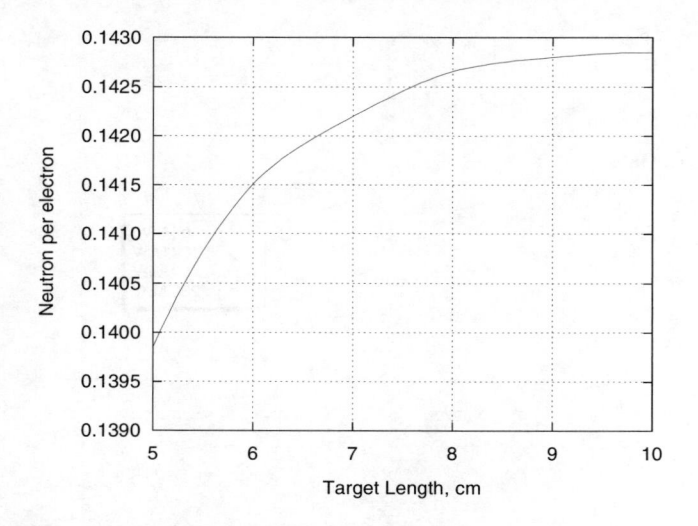

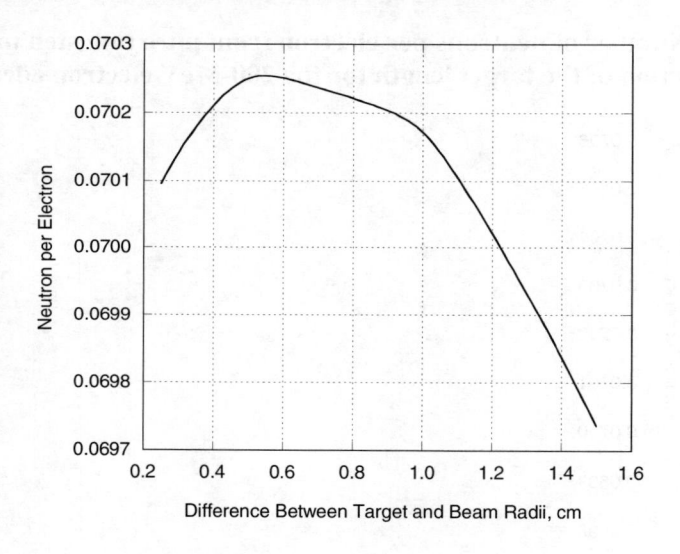

Figure 14. Neutron yield per electron as a function of the difference between the target and the beam radii for tungsten target

Figure 15. Neutron yield per electron as a function of the difference between the target and the beam radii for uranium target

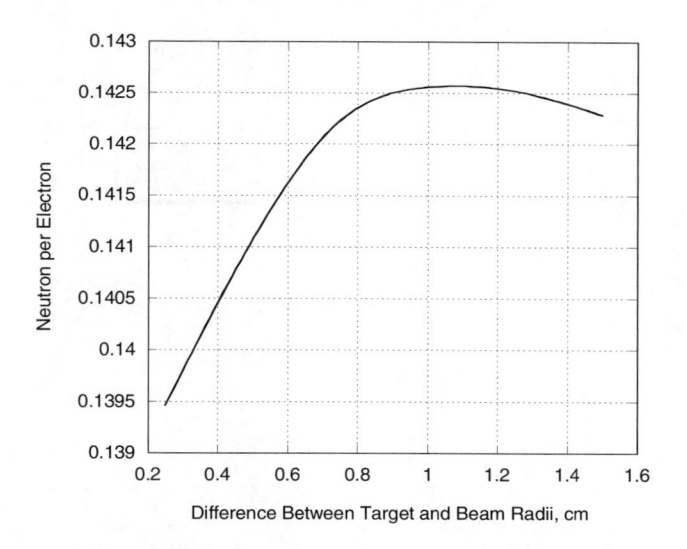

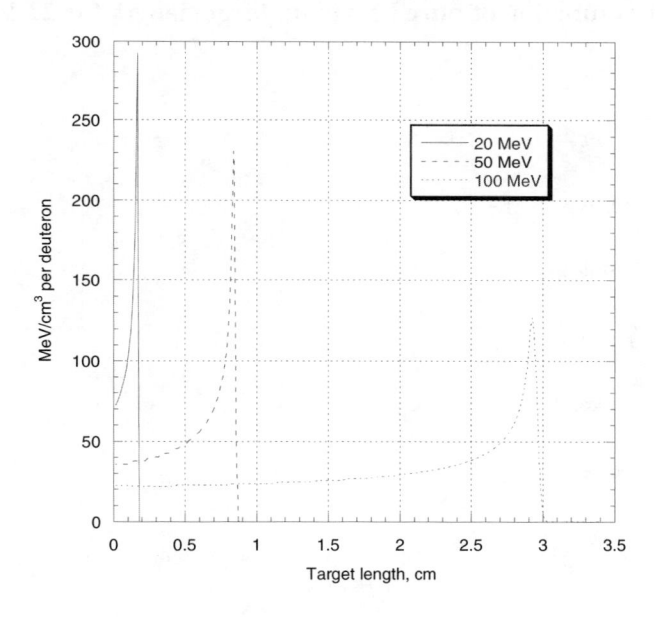

Figure 16. Spatial energy deposition per deuteron in pure beryllium target material for different deuteron energies

Figure 17. Spatial energy deposition in pure beryllium target material for different deuteron energies normalised to 2 kW/cm² on the beam window

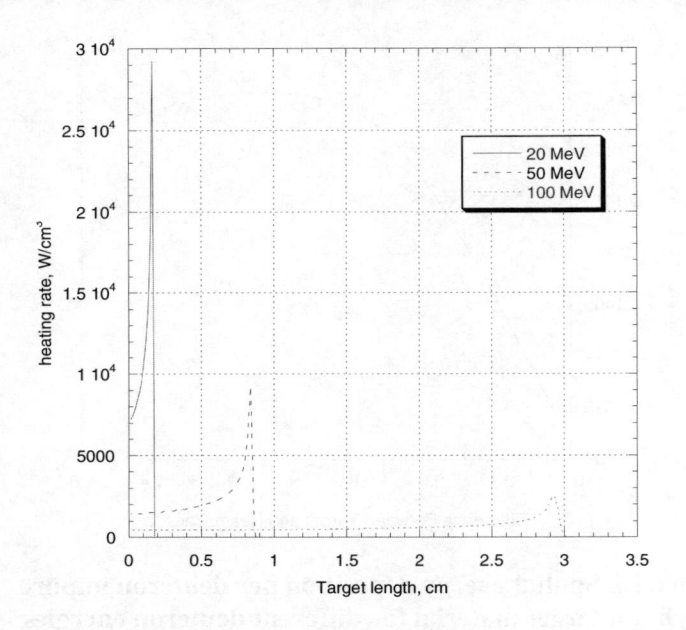

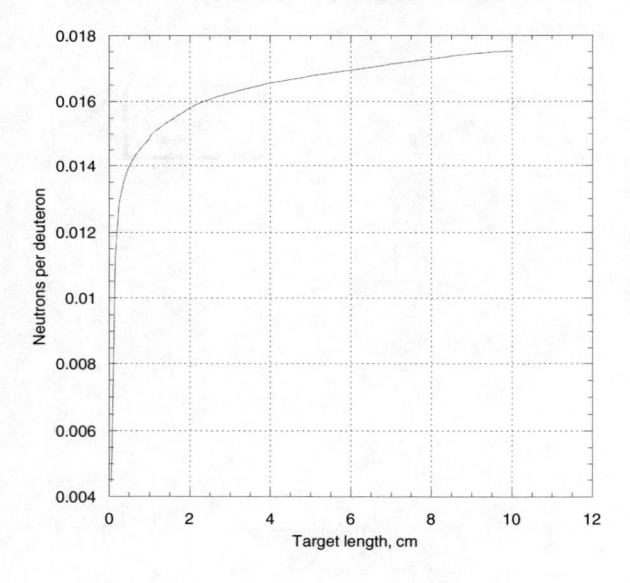

Figure 18. Neutron yield as function of pure beryllium target length for 23 MeV deuteron beam

Figure 19. Neutron yield as function of pure beryllium target length for 100-MeV deuteron beam

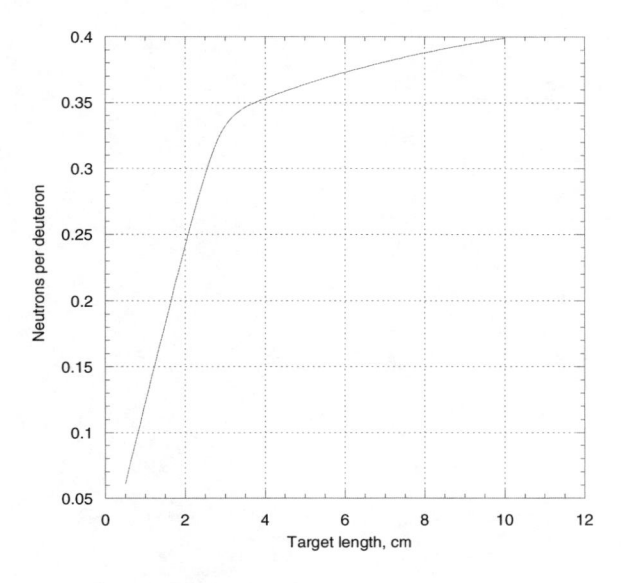

REFERENCES

[1] Gohar, Y., I. Bolshinsky, D. Naberezhnev, J. Duo, H. Belch, J. Bailey, "Accelerator-driven Subcritical Facility: Conceptual Design Development", *Nuclear Instruments & Methods in Physics Research A*, 562, pp. 870-874 (2006).

[2] Naberezhnev, D., Y. Gohar, J. Bailey, H. Belch, "Physics Analyses of an Accelerator-driven Subcritical Assembly", *Nuclear Instruments & Methods in Physics Research A*, 562, pp. 841-844 (2006).

[3] Swanson, William P., "Improved Calculation of Photoneutron Yields Released by Incident Electrons", *Health Physics*, Vol. 37, pp. 347-358, September 1979.

[4] Hendricks, John S., Gregg W. McKinney, Laurie S. Waters, *et al.*, *MCNPX, Version 2.5*, Los Alamos Report, LA-UR-04-0569, February 2004.

IMPROVEMENT OF DYNAMICS CALCULATION CODE DSE FOR ACCELERATOR-DRIVEN SYSTEM

Motomu Suzuki, Tomohiko Iwasaki, Teruyoshi Sato
Tohoku University, Aoba6-6-01-2, Aramaki, Aoba-ku, Sendai, 980-8579, Japan

Abstract

A dynamics calculation code system DSE (*D*ynamics calculation code system for *S*ubcritical system with *E*xternal neutron source) for accelerator-driven system (ADS) has been developed to research the dynamics of ADS. DSE-C deals with the diffusion code CITATION, DSE-T with the deterministic transport code THREEDANT and DSE-M with the Monte Carlo code. In this paper, we only describe DSE-M and its application. DSE-M has adopted GMVP as a Monte Carlo code. Due to GMVP, DSE-M can accurately solve both the forward and the adjoint neutron transport with exact geometrical model. The effects of the statistical error are small although the calculation time is very long. We consider that DSE-M is effective for the reference calculation. The beam variation events of an ADS were analysed using DSE-M. Due to the code's accurate calculations, it is found that the change of the beam injection position to the target has a significant effect on the core parameters such as neutron flux, core power and fuel temperature, although the changes of the beam diameter and the beam shape have a very small effect on the core parameters.

Introduction

Accelerator-driven systems (ADS) have been widely studied for the transmutation of high-level waste (HLW). The ADS is a hybrid system that consists of a high-intensity proton accelerator, a spallation target such as a lead-bismuth eutectic and a subcritical core. Accelerated protons are injected into the spallation target. Spallation neutrons are generated by a spallation reaction and drive the subcritical core. The ADS does not employ control rods and it is operated by the accelerator. Due to the subcritical core, the ADS has high transmutation efficiency, excellent inherent safety and great flexibility in core loading, etc.

Some analytical and experimental studies for an ADS have been performed; however, there are few studies concerning the dynamics of ADS, especially for inherent events which are coupled with both a reactor and an accelerator. The inherent problems include various accelerator variations such as beam variation of energy, diameter, injection position and intensity, etc. Such variations may significantly influence neutron flux, thermal power and fuel and coolant temperature and should be examined accurately. It is required for dynamics calculation to explicitly treat both the core and the accelerator. There is a code that can address both aspects: DSE.

ADS dynamics calculation code system DSE

The ADS dynamics calculation code system DSE (*D*ynamics calculation code system for *S*ubcritical system with *E*xternal neutron source) has been developed to research ADS dynamics [1,2]. Figure 1 shows the conceptual diagram of DSE, which can treat both changes of the accelerator parameters (beam intensity, energy, diameter, injection position, etc.) and the reactor parameters (coolant flow, power temperature, density, etc.). This code can calculate both static and dynamic calculations for neutronics and thermal-hydraulics. DSE calculates the neutron flux, the adjoint flux, the dynamics parameter, the external neutron source, the thermal power, the coolant temperature and the fuel temperature, etc.

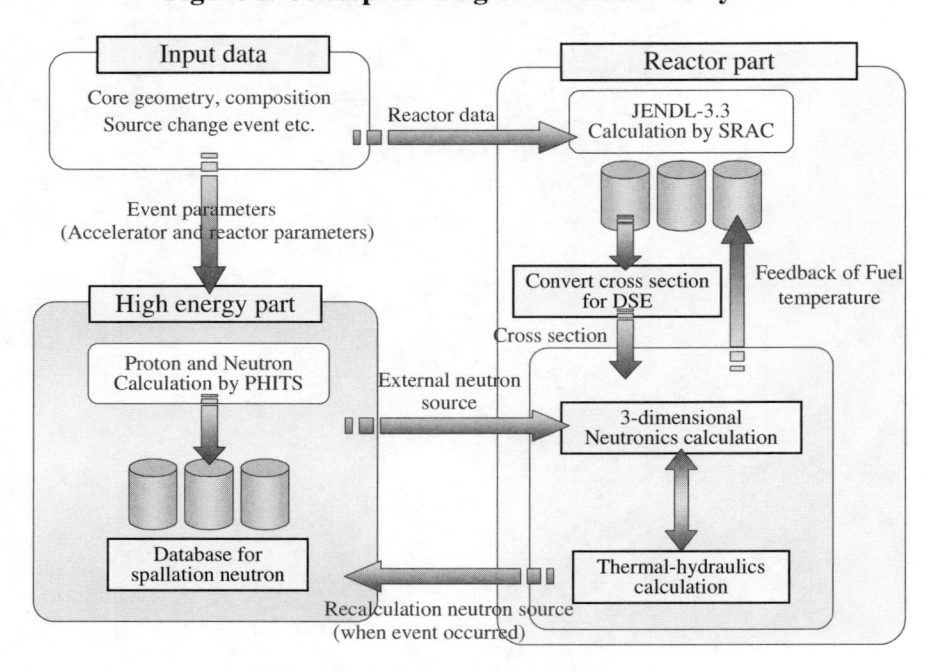

Figure 1. Conceptual diagram of DSE code system

DSE consists of two parts: (1) a high-energy part which prepares the database of the spallation neutron by using *P*article and *H*eavy-*I*on *T*ransport Code *S*ystem (PHITS); (2) a rector part which first makes the cross-section sets for neutronics calculation based on a nuclear data library such as JENDL-3.3, then calculates neutronics and thermal-hydraulics.

In the first version of DSE, the diffusion code CITATION is employed for neutronics calculation. However, the diffusion theory cannot treat angular dependence of neutrons. It is necessary to extend the neutronics calculation to solve it strictly. In the current version of DSE, the three methods have been incorporated into neutronics calculation. First is the original function DSE-C that solves the neutron diffusion equation based on CITATION. Although its calculation accuracy is insufficient, it is suitable for the calculations of lengthy dynamics or start-up, shutdown and restart operation, etc. The second function, DSE-T, carries out neutron transport calculation by using the deterministic transport SN code THREEDANT. This function is good at the calculation of general and wide range dynamic events. DSE-M is the third function that treats neutron transport by using the Monte Carlo method. DSE-M is excellent not only in terms of calculation accuracy, but also with regard to geometric flexibility, although the calculation time is huge. This function is suitable for a reference calculation.

On the other hand, the thermal-hydraulics of DSE employed multi-channel one-dimensional heat flow model (SAS1A). The fuel temperature is calculated by using the radial heat transfer equation. The coolant temperate is solved for the single-phase flow equation since LBE is not supposed to boil. Subbotin's evaluation formula is adopted for calculation of heat transfer coefficient. DSE can treat the LBE explicitly.

The verification for DSE code has been done as in Ref. [1] for the static neutronics calculation, the dynamic neutronics calculation and the thermal-hydraulics calculation. The thermal-hydraulics calculation has been verified by analysing the LBE-cooled ADS benchmark problem [3]. In this benchmark, the coolant outlet temperatures were calculated for the cases when the beam turned down at 0.01 sec. and the beam was re-entered after 1, 3, 6 and 12 sec. Figure 2 shows the calculation results of the coolant temperature. The solid lines are the results calculated by DSE and the broken line represents the average values of the many calculated results in the benchmark. The DSE results agreed with the average results. From these results, it was confirmed that the thermal-hydraulics calculation (at the same time for the neutronics calculation) was appropriate for the LBE-cooled ADS.

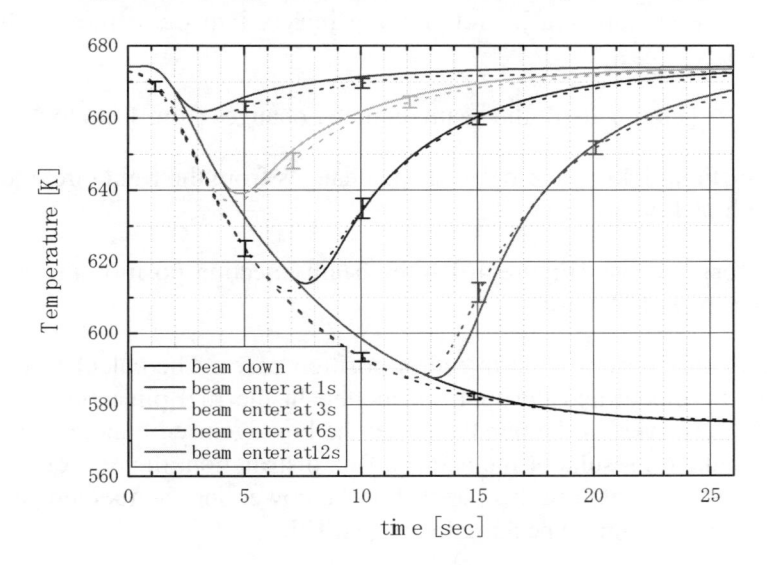

Figure 2. Comparison with benchmark problem [1]

Hereinafter, we describe the development of the DSE-M function and its application. First, the DSE-M and its development are explained, and then the application of the DSE-M is described for various beam variation events of an LBE-cooled ADS.

Development of DSE-M

DSE-M has adopted the GMVP code, which is General Purpose Monte Carlo Code for Neutron and Photon Transport Calculations based on multi-group methods developed by JAEA. [4] GMVP can solve neutron transport with the exact treatments of angular dependence and geometry description. Moreover, GMP contains the adjoint calculation function which is required to obtain the dynamic parameters. The verification of DSE-M is carried out for the static neutronics calculation and the dynamics calculation by comparing with other reference codes. The effect of the statistic error by the Monte Carlo calculation is also examined. In the static calculation, the maximum error is about 1.4% in neutron flux. The dynamics calculation error is about 1.0% in core power when the history is enough. The calculation error of the Monte Carlo to the dynamics calculation is small. However, DSE-M has a problem with calculation time. The static calculation of five million histories requires about 20 hours. The dynamics case of a 40-sec. dynamics event takes about three days. We consider that DSE-M is effective for the reference calculation.

ADS adopted in this study

Using DSE-M, we performed a highly accurate calculation for beam variation events for the ADS model proposed by JAEA [5,6]. The specification of the ADS is as follows: the proton beam energy is 2.0 GeV, the core thermal power of the system is 800 MWt, (MA+Pu)+ZrN fuel is employed, and the target and coolant material is liquid Pb-Bi eutectic. Table 1 shows the main parameters of the ADS. Figure 3 shows the conceptual drawing of the ADS.

High precision calculation for beam variation

The calculations for beam variation events were performed by DSE-M. Figure 4 shows the schematics of the beam variation events of the ADS.

(a) The reference case [Figure 4(a)]: proton beam injects into the centre of the core with the diameter of 1 cm.

(b) The diameter case [Figure 4(b)]: the beam diameter changes from 1 cm to 40 cm.

(c) The shape case [Figure 4(c)]: the beam shape changes from the right circle to the ellipse with the ellipse width of 1 cm to 13 cm.

(d) The injection position case [Figure 4(d)]: the beam injection position moves from the centre (0 cm) to the peripheral (24 cm).

In this paper, we mainly describe the injection position cases. The calculation adopted the two core models of the bundle model and the fuel pin model. In the fuel pin model, the exact pins are modelled (see Figure 5), however, in the bundle model the homogenised bundle is treated without the pin description. The calculation results of the neutron flux distribution, the reactor power and the fuel temperature distribution are presented in this paper. For the power and the fuel temperatures, we focus on the maximum values [the maximum peaking factor (MPF)].

Table 1. Main parameters of reference ADS

Proton beam energy	2.0 [GeV]
Thermal power	800 [MWt]
Core multiplication	0.95
Transmutation efficiency	250 [kgMA/year]
Core height/diameter	1.00/2.50 [m]
Target height/diameter	1.50/0.50 [m]
Core fuel element	(MA+Pu)+ZrN
Target and coolant material	Pb-Bi
Coolant temperature (in/out)	603/703 [K]
Coolant flow rate	$1.98*10^8$ [kg/h]

Figure 3. Conceptual drawing of ADS

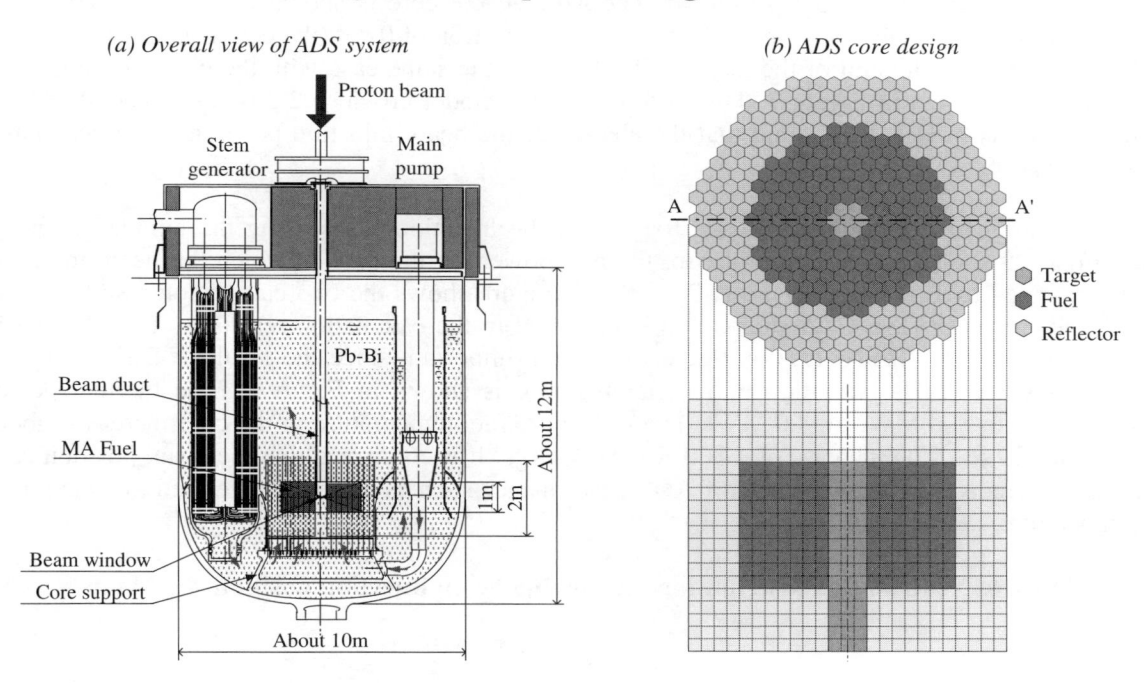

(a) Overall view of ADS system

(b) ADS core design

Figure 4. Conceptual diagram of beam variation events

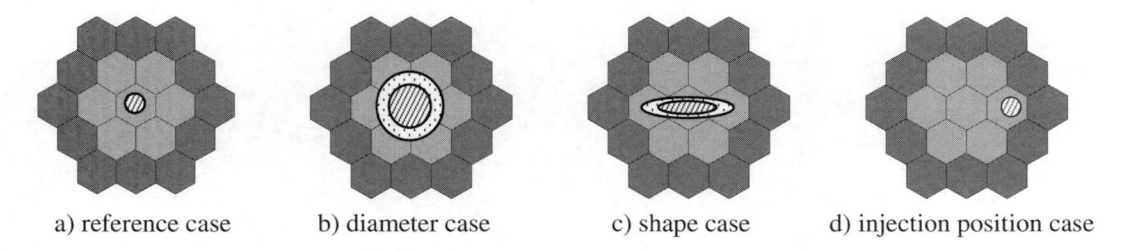

a) reference case b) diameter case c) shape case d) injection position case

Figure 5. Fuel pin model of ADS core

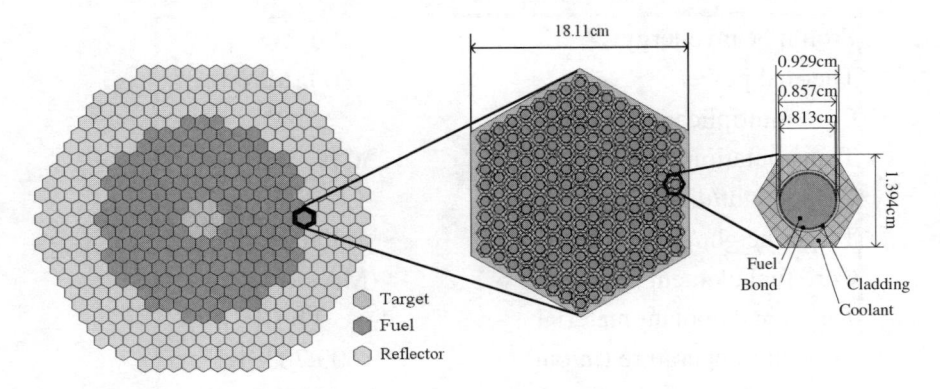

Figure 6 shows one of the results for one of the injection position cases. This figure is the neutron flux distribution along the cross-section line of A-A′ in Figure 3(b). From this figure, a large variation of the neutron flux distribution is observed. The MPF in the core is shown in Figure 7. The MPF increased from 2.5 to 4.5 by moving the beam injection position of 0 cm (the core centre of the target) to 24 cm (the most peripheral of the target). The MPF for the same case with the pin cell model (see Figure 5) is also shown in Figure 7. The MPF of the pin model provides 2.2 times larger MPF than that of the bundle model. It is found that the change of the beam injection position has a very large effect on the ADS.

The dynamics calculation was made for one of the injection cases using the bundle geometry model. Figure 8 shows the variation of core thermal power for the case in which the beam injection position changed from 0 cm to 24 cm at 20 sec. This figure shows the two calculation results, one is DSE-M and the other DSE-C. A fluctuation is observed in the calculation results by DSE-M that is caused by low history. However, this fluctuation is not important since the result of DSE-M agrees with that of DSE-C. The variation of fuel temperature is shown in Figure 9. The thermal power quickly rose from 800 MWt to 900 MWt. The maximum fuel temperature in the core increased about 1 800 K after 20 sec. The increase of 1 800 K is very large. It is confirmed that the beam position case significantly affects the core dynamics of ADS performance such as thermal power and the maximum fuel temperature.

Figure 6. Neutron flux distribution for moving beam injection position x = 24 cm

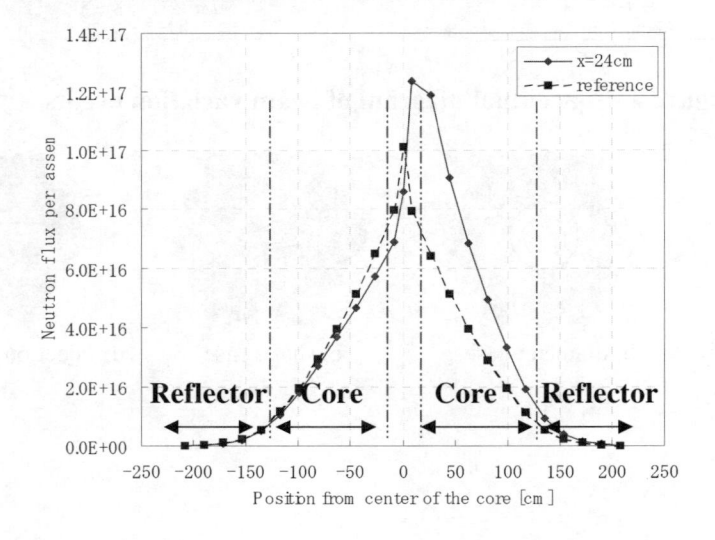

Figure 7. Calculation results for maximum peaking factor in the core

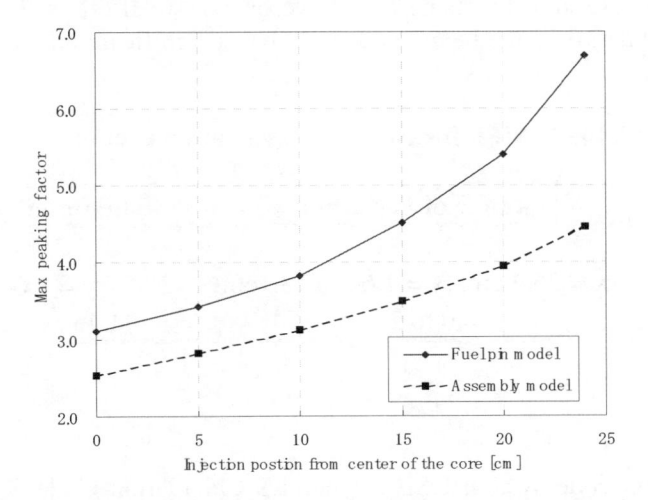

Figure 8. Calculation results for variation of thermal power

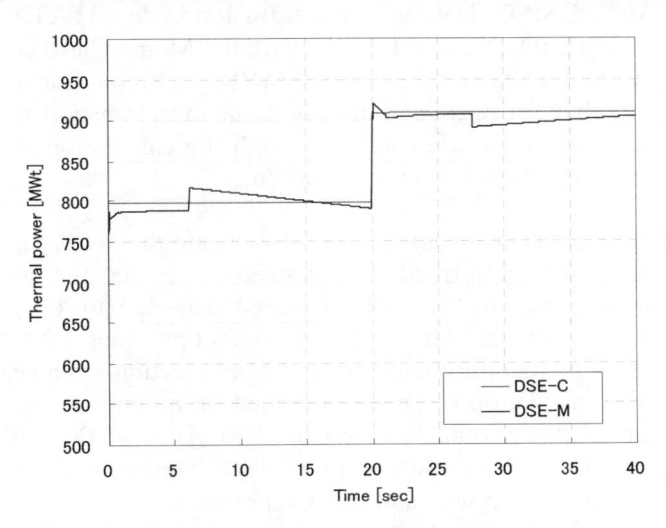

Figure 9. Calculation results for variation of fuel temperature

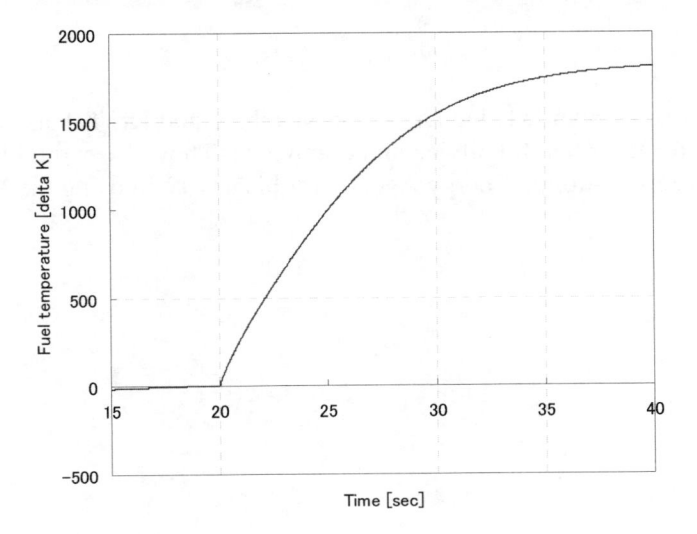

Table 2 summarises the calculation results of the MPF for all cases including the diameter and the shape cases. The diameter case and the shape case have only a small effect. From these calculation results, it can be concluded that only the beam movement has a significant effect on the core dynamics of ADS.

Table 2. MPF for each beam variation events

	Range of variation	Maximum peaking factor
Beam diameter	d = 1 to 24cm	2.70 (d = 40 cm)
Beam shape	rx = 26.2 cm, ry = 1.64 to 13.1cm	2.65 (ry = 1.64 cm, rx = 26 cm)
Beam injection position	x = 0 to 24cm	4.45 (x = 24 cm move)

Summary

A dynamics calculation code system DSE (*D*ynamics Calculation Code System for *S*ubcritical System with *E*xternal Neutron Source) for accelerator-driven systems (ADS) has been developed to research the dynamics of ADS. DSE-C deals with the diffusion code CITATION, DSE-T with the deterministic transport code THREEDANT and DSE-M with the Monte Carlo code. In this paper, we describe DSE-M and its application. DSE-M has adopted GMVP as a Monte Carlo code. Due to GMVP, DSE-M can accurately solve both the forward and the adjoint neutron transport with exact geometrical model. The effects of the statistical error are small, although the calculation time is very long. We consider that DSE-M is effective for the reference calculation.

Using DSE-M, the calculation of the beam variation (the beam diameter, the beam shape and the beam injection position) events was performed. The diameter case and the shape case have small effect. In the variation of beam position, the MPF increased from 2.5 to 4.5 by moving the beam injection position from 0 cm to 24 cm. The MPF of the pin model provides 2.2 times larger MPF than that of the bundle model. It is found that the change of the beam injection has a very large effect on the ADS performance. A dynamics calculation was also performed for moving the beam injection position from 0 cm to24 cm. The thermal power quickly rose from 800 MWt to 900 MWt and the maximum fuel temperature in the core increased about 1 800 K after 20 sec. The beam injection position case has a significant effect on neutron flux, core power and fuel temperature.

Acknowledgements

The authors would like to acknowledge Dr. Kenji Nishihara and Dr. Takanori Sugawara at Japan Atomic Energy Agency for their helpful advice and comments. They also would like to acknowledge Dr. Yasunobu Nagaya at Japan Atomic Energy Agency for his advice in using the MVP/GMVP code.

REFERENCES

[1] Sugawara, T., "Development of Dynamics Code and Study of Operation Control for Accelerator Driven System", Tohoku University Doctoral Dissertation (2005).

[2] Suzuki, M., "A Study of Operation Method by Injection Beam Control for Accelerator Driven System", Tohoku University Master's thesis (2006).

[3] D'Angelo, A., G. Bianchini, M. Carta, *Benchmark on Beam Interruptions in a Lead-bismuth Cooled and MOX Fuelled Accelerator-driven System*, NEA/SEN/NSC/WPPT(2002)6.

[4] Nagaya, Y., "General Purpose Monte Carlo Codes for Neutron and Photon Transport Calculations Based on Continuous Energy and Multigroup Methods: MVP/GMVP II", *RIST News*, No. 40, p. 9 (2005).

[5] Takizuka, T., K. Tsujimoto, T. Sasa, *et al.*, "Design Study of Lead-bismuth Cooled ADS Dedicated to Nuclear Waste Transmutation", *Progress in Nuclear Energy*, 40, 3, pp. 505 (2002).

[6] Tsujimoto, K., T. Sasa, K. Nishihara, *et al.*, "Neutronics Design for Lead-bismuth Cooled Accelerator-driven System for Transmutation of Minor Actinide", *Journal of Nuclear Science and Technology*, 29, 4, pp. 325 (2003).

CORROSION RESISTANCE OF REFRACTORY METALS AND CERAMICS IN LEAD-BISMUTH AT 700°C

Abu Khalid Rivai[1], Minoru Takahashi[2]
[1]Department of Nuclear Engineering, Tokyo Institute of Technology
[2]Research Laboratory for Nuclear Reactor, Tokyo Institute of Technology
2-12-1-N1-18, O-okayama, Meguro-ku, Tokyo 152-8550, Japan
E-mail: mtakahas@nr.titech.ac.jp

Abstract

Corrosion tests of refractory metals and ceramics were carried out employing high-temperature LBE. Oxygen concentrations in this experiment were 6.8×10^{-7} wt.% for ceramics (SiC/SiC composites) and 5×10^{-6} wt.% for refractory metals (W, Mo and Nb) and ceramics (SiC and Ti_3SiC_2). All specimens were immersed in molten LBE at 700°C in a corrosion test pot for 1 000 hours. The tungsten and molybdenum showed high corrosion resistance with no penetration of LBE into their matrices. Penetration of LBE into the matrix of the niobium was observed. The ceramic materials of SiC and Ti_3SiC_2 showed high corrosion resistance with no penetration of LBE into their matrices. The SiC/SiC composite showed LBE penetrated into the matrix due to high porosity of the material.

Introduction

Lead-alloy-cooled fast reactors (LFRs) have been considered as one of the innovative reactors that can enhance energy sustainability in the world. Chemical inertness of the lead alloys (lead and 44.5wt.%Pb-55.5wt.%Bi: LBE) with water and air in contrast with sodium makes the reactor safer and more economic than the sodium-cooled fast reactors (SFRs) through the elimination of the intermediate coolant loop. High temperature margin to boiling points makes the LFRs safer than the SFRs [1]. One of the critical issues for development of the LFRs is the compatibility of core and structural materials with high-temperature lead alloys [2]. Development of materials compatible with high-temperature lead alloys is one of the key issues for the development of the LFRs.

Research and development of materials compatible with lead alloys have been performed for critical fast reactors and the accelerator-driven systems (ADSs) [3-10]. In general, the materials which have been proposed and become candidates for the LFRs are high chromium ferrite-martensitite steels with addition of Si and Al, surface alloying-treated steels, refractory metals and ceramics. Corrosion behaviour of the high chromium steels in flowing LBE at 550°C was reported [3,4]. A stable spinel layer of chromium oxide was formed on the high chromium steels. This oxide layer was thin and compact and could protect the steel from corrosion. Aluminium alloying into the surface by surface melting using pulsed electron beam improves the corrosion oxidation behaviour with no corrosion attack in the stagnant lead at 550°C [5] and flowing LBE at 550°C [6]. Al_2O_3, $FeAl_2$ and $AlCr_2$ which was produced by the gas diffusion method exhibited corrosion resistance to liquid LBE in the stagnant LBE at 550°C, although the surface layer produced by the melt dipping method suffered from severe corrosion attack [7].

The performance of refractory metals in the lead alloy corrosion test has been reported [8,9]. Refractory metals exhibited two orders of magnitude greater corrosion resistance than the steels in the flowing LBE at 400°C [8]. In the stirred pool of LBE at 450°C, refractory metal of Mo exhibited more corrosion resistance than 12Cr steel [9]. Corrosion behaviour of ceramics has also been reported [9,10]. SiC and Si_3N_4 showed good corrosion resistance in the liquid LBE flow with negligibly small weight loss, but showed cracked surfaces in flowing LBE at 550°C [10]. SiC and Si_3N_4 exhibited good corrosion resistance with weight loss about two orders less than those of 12%Cr steels, such as HCM12A, in a stirred pool of LBE at 450°C [9]. Recently, SiC/SiC composite have been proposed as the candidate of cladding of high temperature reactors. The SiC/SiC composites showed thin surface crack layer with LBE penetration deeply in the material in the pot LBE at high temperature of 700°C [8].

Although various experimental results have been reported for refractory metals and ceramics, investigations on corrosion behaviour of those materials in lead alloy at high temperature, especially at temperatures above 650°C, are few. The experimental results of the compatibility of the materials with lead alloys at high temperature are required especially for the development of high-temperature LFRs, such as high-efficiency and/or multi-purpose reactors for hydrogen production.

Therefore, it is necessary to investigate the compatibility of various candidate materials for LFRs with high outlet temperature of the lead alloy coolant. In the present study, corrosion behaviour was investigated for refractory metals and silicon carbides types of ceramic materials in the molten LBE at temperature of 700°C under controlled oxygen concentrations.

Experimental apparatus and procedure

Experimental corrosion test

The corrosion test of the materials was conducted in a pot-type LBE test apparatus with stirred condition by continuous Ar-H_2-steam mixture gas injection. The parameters of test apparatus and test conditions are shown in Table 1. A schematic of the test apparatus is shown in Figure 1. This apparatus consists of a steam generator, a moisture sensor, a heater section of injection gas, an oxygen sensor, and a gas exhaust system. Alumina crucible was used as the pot of molten LBE with Ar cover gas, the oxygen sensor, a molybdenum lead wire, a ceramics nozzle and a thermocouple. The injection gas flows through a stainless steel tube and enters the test tank from the ceramics nozzle. Oxygen concentration was controlled by the sensor made of a sintered ceramic zirconia, i.e. magnesia-stabilised zirconia (MgO-ZrO_2) and yttria-stabilised zirconia (Y_2O_3-ZrO_2). As an internal reference electrode fluid, oxygen-saturated molten bismuth was used, where the molten Bi contained a powder of bismuth oxide Bi_2O_3 with a weight ratio of 95:5 wt.%. The electromotive force (EMF) of the oxygen sensor was detected by two molybdenum (Mo) wires inserted into the reference electrode fluid in the sensor cell and the LBE in the crucible, and measured using an electrometer.

Table 1. General condition of corrosion test apparatus

Run	No. 1	No. 2
Tested materials	SiC/SiC composites	W, Mo, Nb, SiC, Ti_3SiC_2
Oxygen concentration (wt.%)	~6.8×10^{-7}	~5×10^{-6}
Injection gas	Ar	Ar+H_2 (3%) and Ar
Type of sensor	Y_2O_3-ZrO_2	
Temperature of LBE (°C)	700	
Immersion time (hr)	1 000	

Figure 1. Schematic of corrosion test apparatus

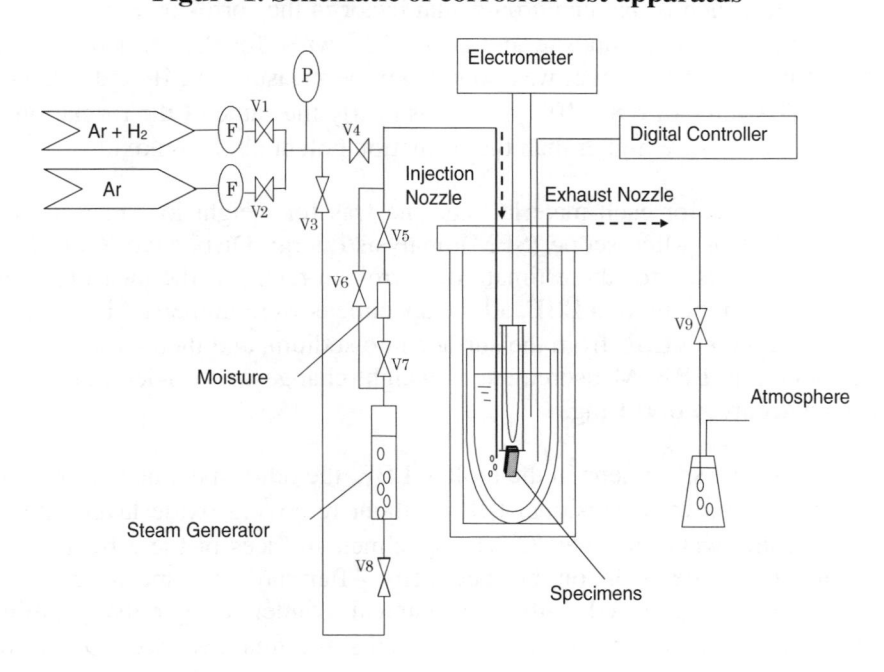

Test materials

The refractory metals tested were tungsten (W), molybdenum (Mo) and niobium (Nb). Tungsten (W) has the highest melting point of 3 410°C among metals and its density is 19 500 kg/m^3. Molybdenum has a melting point of 2 610°C and its density is 10 200 kg/m^3. Niobium has a melting point of 2 468°C and its density is 8 570 kg/m^3.

The ceramics tested were the silicon carbide, SiC (Nilaco Company, Japan), the titanium silicon carbide, Ti$_3$SiC$_2$ (3-ONE-2 LLC Company) and SiC/SiC composites (Art Kagaku Co. Ltd, Japan). The SiC specimens had compositions of 98SiC-0.2SiO$_2$-0.1Si-1.2C. The density of the SiC specimens was 3 100 kg/m^3, and the sizes of the SiC specimens for weight change analysis were $15 \times 15 \times 2.3$ mm^3.

The titanium silicon carbide, Ti$_3$SiC$_2$, is one of a new class of solids, with the general formula $M_{n+1}AX_n$, where n is 1 to 3, M is an early transition metal, A is an A group mostly IIIA and IVA element, and X is C and/or N. Its density is 4 500 kg/m^3, which is close to that of Ti. It has a strength of ~325 GPa which is roughly three times as stiff as Ti and is readily machinable with nothing more sophisticated than a manual hack saw [11]. It is stable in inert atmospheres to temperatures above 2 200°C. Mechanically, Ti$_3$SiC$_2$ is damage tolerant and resistant to thermal shock and relatively tough. A brittle-to-ductile transition is observed at 1 100°C in compressive, flexural and tensile tests. The sizes of the Ti$_3$SiC$_2$ specimens for weight change analysis were $10 \times 10 \times 2.3$ mm^3.

The specimens of SiC/SiC composite (Art Kagaku Co. Ltd.) were a type of boron nitride (BN) coating. The density of the SiC/SiC composite specimen was 1 600 kg/m^3. The density was roughly a half of the SiC specimens and much lower than the Ti$_3$SiC$_2$ specimens. The sizes of the SiC/SiC composite specimens for weight change analysis were $15 \times 15 \times 2.3$ mm^3.

Experimental procedure

Oxygen concentration, which is main environmental factor in the corrosion process, was controlled at 6.8×10^{-7} wt.% for the SiC/SiC composite and at 5×10^{-6} wt.% for the refractory metals and the other ceramics. The estimation of this level was done from the measured EMF and Gromov's oxygen solubility equation [1]. The value of 6.8×10^{-7} wt.% was nearly the same of the formation potential of iron oxide and that of 5×10^{-6} wt.% higher than the formation potential of iron oxide.

Two specimens were used for each material, i.e. one was for weight loss measurement and the other for the Scanning Electron Microscope (SEM) analysis/Energy Dispersive X-ray Microanalysis (EDX). Weight loss data was used for the estimate of corrosion rate. For the measurement of weight change, after the immersion in the molten LBE, all the specimens were immersed in a hot sodium pool at 260-280°C to dissolve adherent LBE from the surface into sodium, and then washed with ethanol to remove the sodium-containing LBE. Measurement of weight change was carried out by an electronic reading balance with an accuracy of 0.1 mg.

After the immersion of the specimens in the molten LBE, the other specimens were washed in hot glycerine at 160-180°C to remove the residual LBE without removing oxide layers. Afterwards the glycerine was removed using water at 70-80°C. The specimen surfaces of the refractory metals were coated by gold to protect the oxide layers on the specimens. Afterwards, the specimens were cut in the middle, solidified by resin and polished with a mechanical grinder using polycrystalline diamond grains. The polycrystalline diamond grains used in this polishing treatment were 9 μm, 6 μm, 3 μm and 1 μm. Then, SEM-EDX analysis was performed for the cross-sections of the specimens.

Result and discussion

Refractory metals

The SEM micrographs of the cross-sections of the refractory metals, tungsten and molybdenum, are shown in Figures 2 and 3, respectively. Neither the penetration of LBE into the tungsten and molybdenum matrices nor the former traces of a corrosion process were observed. The surfaces were smooth without any damage or cracks. There was no dissolution of tungsten and molybdenum elements into LBE. No oxide layer was observed.

Figure 2. SEM micrograph cross-section of tungsten after immersed in LBE at 700°C with oxygen concentration of 5×10^{-6} wt.% for 1 000 hours

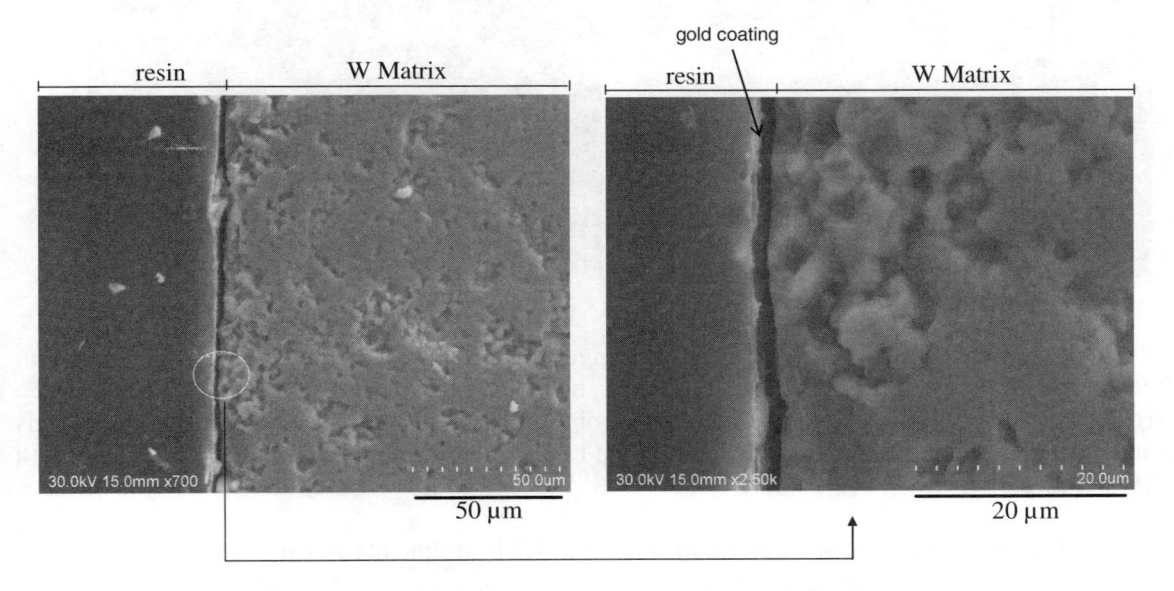

Figure 3. SEM micrograph cross-section of molybdenum after immersed in LBE at 700°C with oxygen concentration of 5×10^{-6} wt.% for 1 000 hours

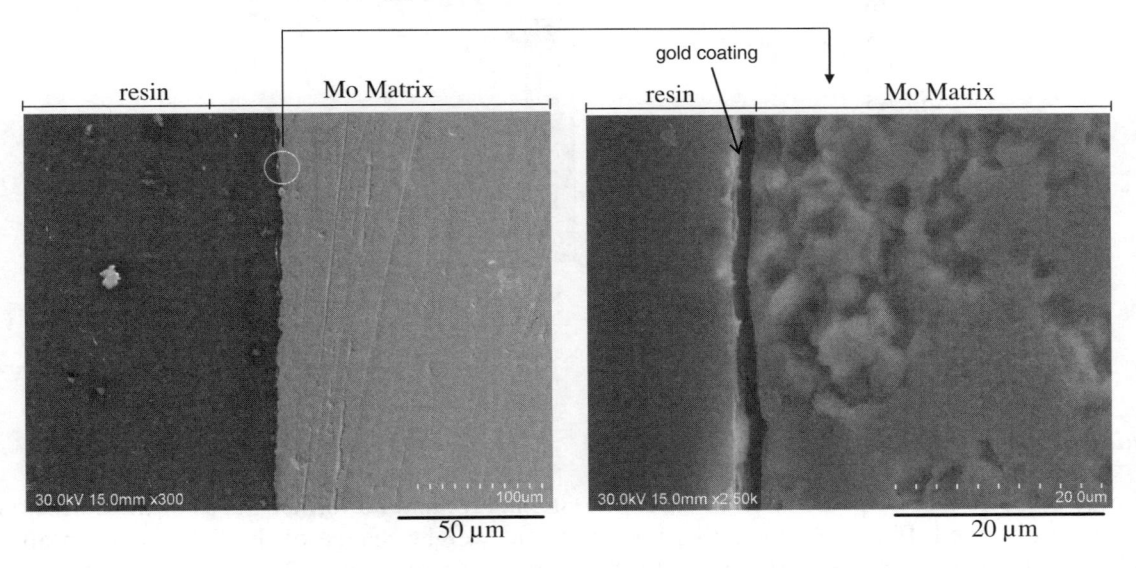

The SEM micrographs of the cross-sections of the refractory metals, niobium, are shown in Figure 4. Penetration of LBE into matrix of the niobium was observed. This penetration of LBE was not only deep but also prevailed on all parts of the surface area.

Figure 4. SEM micrograph cross-section of niobium after immersed in LBE at 700°C with oxygen concentration of 5×10^{-6} wt.% for 1 000 hours

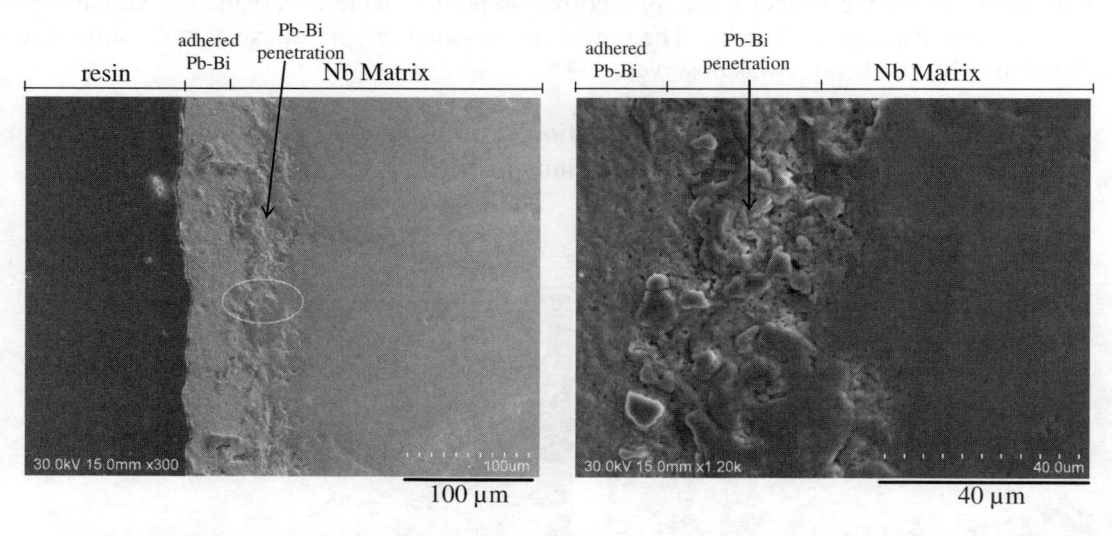

Figure 5 shows that no significant weight changes were detected for tungsten and molybdenum specimens after immersed in the LBE. SEM analysis shows that these materials have high performance of corrosion resistance. However, the weight loss of niobium was 190 g/m^2 which was not small. This result indicates that the niobium was corroded by the LBE. Therefore, niobium is not compatible with the LBE at 700°C.

Figure 5. Weight losses after immersed in LBE at 700°C for 1 000 hours

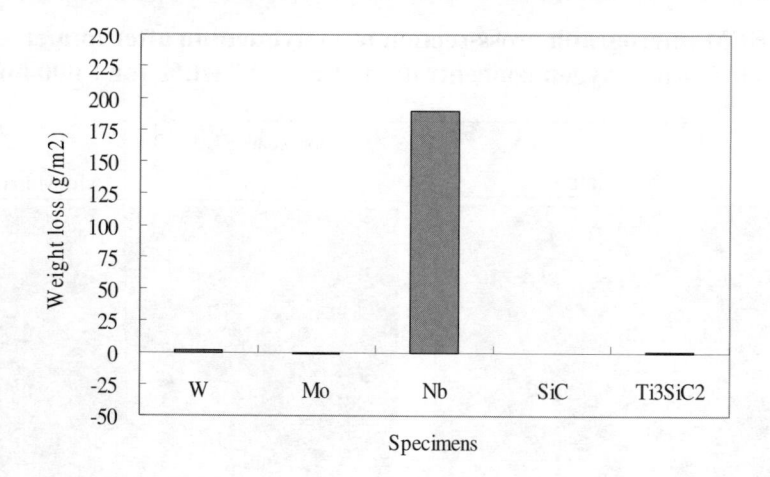

Ceramics

Figure 6 shows that no penetration of LBE into SiC matrix and no former traces of the corrosion process were observed. Figure 8 shows that no significant weight change of the SiC specimen after immersion in high temperature LBE up to 1 000 hours occurred. Figure 7 shows that neither the

Figure 6. SEM micrograph cross-sections of SiC after immersed in LBE at 700°C with oxygen concentration of 5×10^{-6} wt.% for 1 000 hours

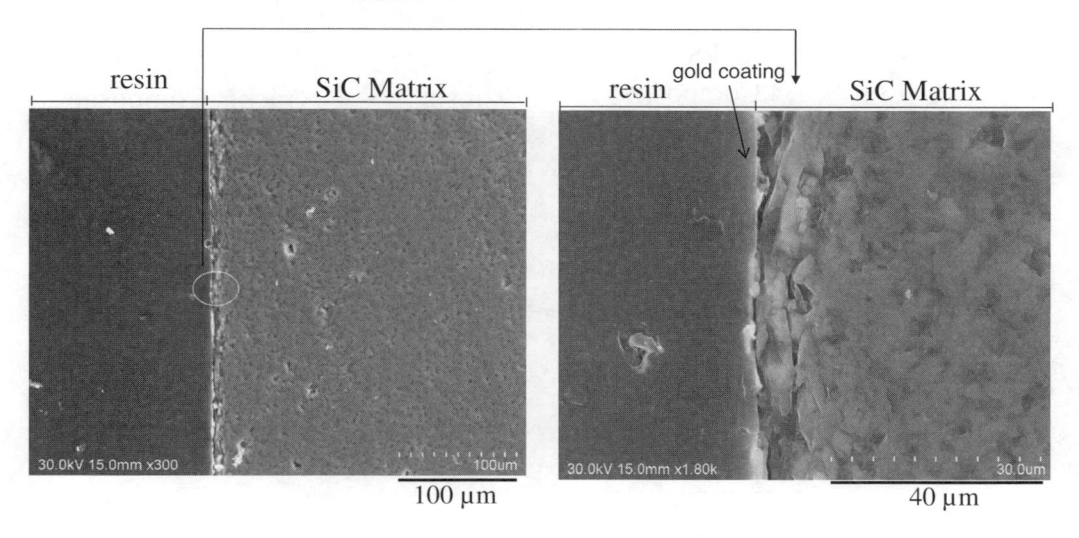

Figure 7. SEM micrograph cross-sections of Ti_3SiC_2 after immersed in LBE at 700°C with oxygen concentration of 5×10^{-6} wt.% for 1 000 hours

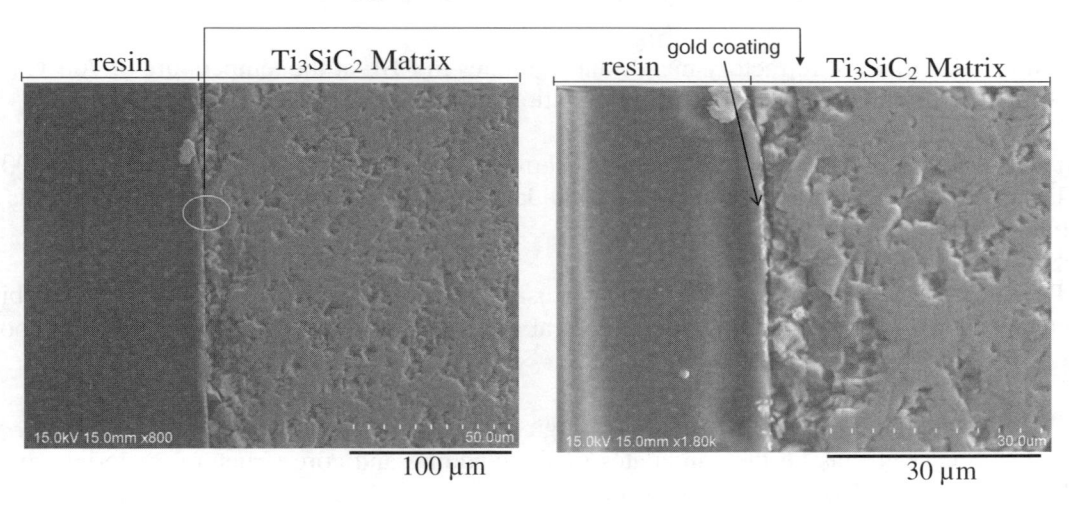

penetration of LBE into Ti_3SiC_2 matrix nor the former traces of corrosion process were observed. Figure 5 shows that no significant weight change occurred in the Ti_3SiC_2 specimen after immersion in high-temperature LBE up to 1 000 hours. Titanium in silicon carbide is effective for corrosion resistance in LBE and makes the silicon carbide easily machinable [11].

Figure 8 shows the SEM micrograph of the SiC/SiC composites after immersion in LBE at 700°C for 1 000 hours. It is found that the LBE penetrated into the pores of the SiC/SiC composites due to their high porosity. It is necessary to prevent LBE from penetration by making the SiC/SiC composites denser. It can be seen that a thin crack layer appeared. The crack layer was 20.8-33.7 μm in thickness. The crack occurred possibly because of thermal shock or thermal expansion near the surface. The crack part still remained in the specimen with neither detachment nor loss of the cracked region from the SiC/SiC composite matrix. The weight increased possibly because of the LBE penetration in the pores of the specimens.

Figure 8. SEM micrograph cross-sections of SiC/SiC composites after immersion in LBE at 700°C with oxygen concentration of 6.8×10^{-7} wt.% for 1 000 hours

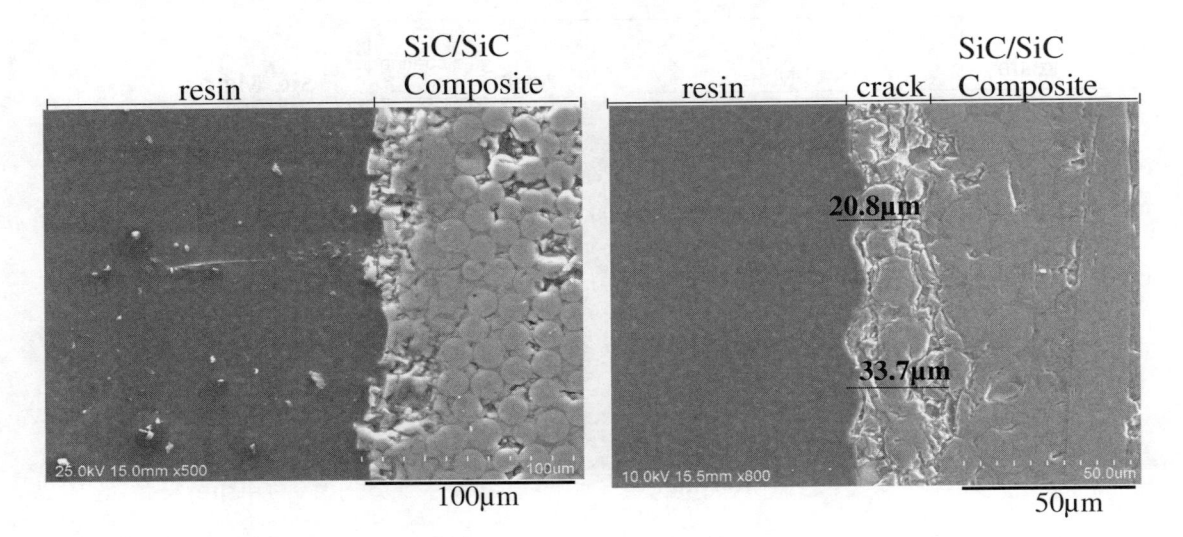

Conclusions

Corrosion behaviours of refractory metals and ceramics in LBE at the temperature of 700°C have been investigated experimentally. From the investigation and analyses, it is concluded that:

- The tungsten and molybdenum specimens showed high corrosion resistance in LBE at 700°C. Tungsten and molybdenum can be the candidates of the cladding and structural materials for high-temperature LBE-cooled fast reactors.

- The niobium specimens showed poor corrosion resistance in LBE at 700°C. The niobium cannot be used as the cladding and structural materials for the high-temperature LBE-cooled fast reactors.

- The SiC and Ti_3SiC_2 specimens showed high corrosion resistance in the LBE at 700°C. SiC and Ti_3SiC_2 can thus be the candidates of the cladding and core structural materials for the high-temperature LBE-cooled fast reactors.

- The SiC/SiC composite specimens showed that LBE penetrated into the matrix due to the high porosity of the material, and a thin crack layer appeared. It is necessary to prevent LBE penetration by making the SiC/SiC composites denser for the use of SiC/SiC composite as the cladding and core structural materials of the high-temperature LBE-cooled fast reactors.

Acknowledgement

The authors would like to express their gratitude to Mr. K. Hata of COE-INES researcher of Tokyo Institute of Technology for his assistance, to Mr. Hasegawa of Art Kagaku Co. Ltd. (Japan) for his fabrication, supply and information of the SiC/SiC composite specimens, and to Dr. T. El-Raghy of 3-ONE-2 LLC Company (USA) for his supply and information of the Ti_3SiC_2 specimens.

REFERENCES

[1] Gromov, B.F., Y.I. Orlov, P.N. Marantiv, V.A. Gulevsky, "The Problems of Technology of the Heavy Liquid Metal Coolants (Lead-bismuth, Lead)", *Proc. of HLMC98*, pp. 87-100, IPPE, Obninsk, Russia (1998).

[2] Takahashi, M., N. Sawasa, H. Sekimoto, *et al.*, "Design and Construction of Pb-Bi Corrosion Test Loop and Test Plan", *Proc. of 8th International Conference on Nuclear Engineering (ICONE 8)*, Baltimore, MD, USA, 2-6 April (2000).

[3] Kondo, M., M. Takahashi, "Corrosion Resistance of Si- and Al-rich Steels in Flowing Lead-bismuth", *Journal of Nuclear Materials*, 356, 203-212 (2006).

[4] Kondo, M., M. Takahashi, N. Sawada, K. Hata, "Corrosion of Steels in Lead-bismuth Flow", *Journal of Nuclear Science and Technology*, Vol. 43, No.2, pp. 107-116 (2006).

[5] Muller, G., G. Schumacher, F. Zimmermann, "Investigation on Oxygen Controlled Liquid Lead Corrosion of Surface Treated Steels", *Journal of Nuclear Materials*, 278, 85-95 (2000).

[6] Heinzel, A., M. Kondo, M. Takahashi, "Corrosion of Steels with Surface Treatment and Al-alloying by GESA Exposed in Lead–bismuth", *Journal of Nuclear Materials*, 350, 264-270 (2006).

[7] Kurata, Y., M. Futakawa, S. Saito, "Corrosion Behavior of Al-surface-treated Steels in Liquid Pb-Bi in a Pot", *Journal of Nuclear Materials*, 335, 501-507 (2004).

[8] Fazio, C., I. Ricapito, G. Scaddozo, G. Benamati, "Corrosion Behavior of Steels and Refractory Metals and Tensile Features of Steels Exposed to Flowing Pb-Bi in the Lecor Loop", *Journal of Nuclear Materials*, 318, 325-332 (2005).

[9] Hata, K., M. Takahashi, "Corrosion Study in Direct Cycle Pb-Bi Cooled Fast Reactor", *Proc. of GLOBAL*, Paper No. 446, Tsukuba, Japan, 9-13 October (2005).

[10] Takahashi, M., M. Kondo, "Corrosion Resistance of High Cr Steels and Ceramic Materials in Lead-bismuth Flow", *Proc. of GLOBAL*, Paper No. 425, Tsukuba, Japan, 9-13 October (2005).

[11] Barsoum, M.W., L.H. Ho-Duc, M. Radovic, T. El-Raghy, "Long Time Oxidation Study of Ti_3SiC_2, Ti_3SiC_2/SiC, and Ti_3SiC_2/TiC Composites in Air", *Journal of the Electrochemical Society*, 150 (4), B166-B175 (2003).

NEUTRONIC DESIGN OF THE XT-ADS CORE WITH IN-PILE SECTIONS

G. Van den Eynde[1], V. Sobolev[1], E. Malambu[1], D. Maes[1], D. Lamberts[1], H. Aït Abderrahim[1],
L. Mansani[2], B. Giraud[3], A. Hogenbirk[4], P. Vaz[5], Y. Romanets[5], M.C. Vincente[6],
P. Coddington[7], K. Mikityuk[7], D. Struwe[8], M. Schikorr[8], G. Rimpault[9], C. Artioli[10]

[1]SCK•CEN, Boeretang 200, B-2400 Mol, Belgium, E-mail: gvdeynde@sckcen.be
[2]Ansaldo Nucleare, [3]Areva, [4]NRG, [5]ITN, [6]CIEMAT, [7]PSI, [8]FZK, [9]CEA, [10]ENEA

Abstract

The EUROTRANS project is an integrated project in the 6th European Framework Programme in the context of partitioning and transmutation (P&T). The objective of this project is a step-wise approach to a European Transmutation Demonstration. This project aims to deliver an advanced design of a small-scale accelerator-driven system (ADS), XT-ADS, as well as the conceptual design of a European Facility for Industrial Transmutation, EFIT. The main objectives of the XT-ADS are: demonstration of the ADS concept, providing a test facility for transmutation and providing irradiation possibilities in conditions representative of EFIT operating conditions. The latter allows the use of the XT-ADS as a material testing facility as well as an aid for the fuel qualification of EFIT and for the qualification of candidate structural materials. The XT-ADS core is based on MOX fuel with relatively high Pu content and is cooled using a lead-bismuth eutectic (LBE). The combination of these two components results in a fast spectrum core. During the first 24 months of the project, the core of XT-ADS was developed starting from the MYRRHA DRAFT2 file [1]. First, a "clean core" configuration (i.e. without experimental channels and assemblies) has been set as a reference. This core is composed of 72 fuel assemblies surrounding a set of 3 positions foreseen for the spallation target module. In a second step, 8 positions were identified to house so-called in-pile sections (IPS): positions that can hold experiments and are accessible from the top of the reactor during operation (i.e. penetrations are foreseen).

The reference pin and fuel assembly design

As in MYRRHA, the XT-ADS is designed to use MOX as fuel material. Both designs start with a 30 wt.% of reactor-grade plutonium MOX. However, the composition of this MOX, i.e. the uranium and plutonium vectors, is different in these two cases. MYRRHA was designed using a reactor-grade plutonium originating from the spent fuel of "old" PWRs wherein the burn-up did not exceed 33 MWd/kg-HM, while for XT-ADS we opted for a plutonium vector coming from the reprocessing of spent PWR fuel with an initial enrichment of 4.5% in ^{235}U, a burn-up of 45 GWd/t and a cooling period of fifteen years.

The pin itself consists of a fuel column of 60 cm with on both ends a neutron reflector to increase the neutron economy, a fission gas plenum and a closing cap. The fuel pellets are of the annular type. This reduces the centreline temperature and hence gives larger margins to fuel melting limits. Figure 1 shows this reference design. The fuel assembly contains 90 of these fuel pins in a hexagonal lattice together with one "instrumentation pin" in the centre of the assembly (see Figures 2 and 3). This instrumentation pin contains no fuel but is foreseen to carry instrumentation for on-line measurements.

The major difference with the MYRRHA design is the larger fuel pin pitch. This increase was needed to reduce the pressure drop over the core. It was very clear from the beginning that the T91 cladding material would be the weakest link and hence defining the operational limits of the XT-ADS core. This reference design was the result of several iterations between neutronics calculations and thermo-hydraulic calculations with the most important feedback parameters the pressure drop over the core and the needed coolant velocity to keep the inlet and outlet temperatures as initially specified (300-400°C). Mechanical calculations showed that a wrapper wall thickness of 2 mm was sufficient. The clearance between the assemblies was fixed to 3 mm, which is double that of MYRRHA. The motivation for this came from working experience of members of the design team indicating that this was indeed needed for the easy withdrawal of an irradiated (and often deformed) fuel assembly. The final result is an assembly total width (flat-to-flat) of 93.2 mm and an assembly pitch (centre-to-centre) of 96.2 mm. During the preliminary safety analysis, a risk of early clad failure during an unprotected loss-of-flow (ULOF) accident was identified. In order to reduce this risk and allow for a larger grace time for reactor shutdown, the fission gas plenum was increased.

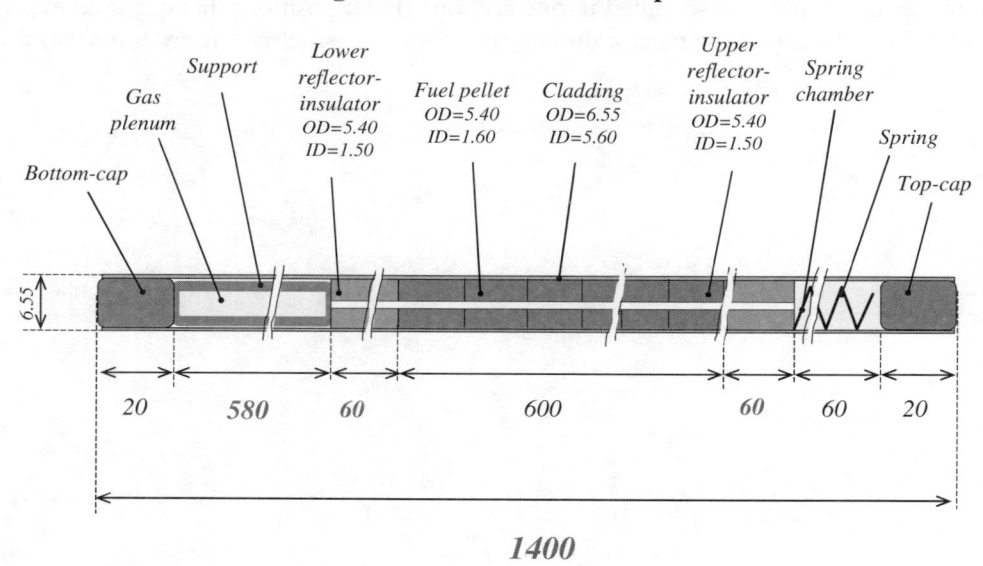

Figure 1. The XT-ADS fuel pin

Figure 2. The hexagonal fuel pin lattice

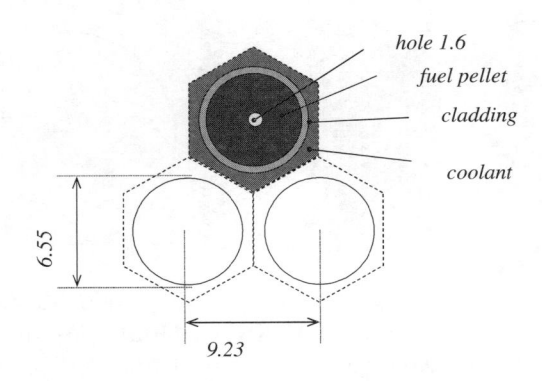

Figure 3. The fuel assembly

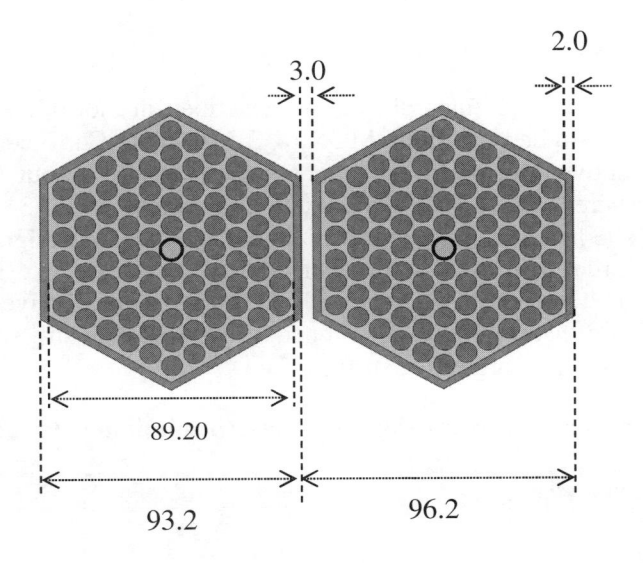

The reference core layout

As a result of the degradation of the MOX fuel and the enlargement of the fuel assemblies as well as assembly pitch (the fuel density has been reduced significantly), k_{eff} of the core dropped. Three options were open to get back to the reference value of $k_{eff} = 0.95$: 1) increase the active length; 2) use more fuel assemblies; 3) allow for a higher Pu content (more than 30 wt.%). Increasing the active length of the core has as a main consequence that the coolant sees more power, hence the outlet temperature will be higher (unless this is, of course, compensated by higher coolant velocities) and this was clearly an option we did not want to take. The second option has two major drawbacks: putting more fuel assemblies will increase the power of the system, hence increasing the dimensions of all components of the primary system and surely, room inside the vessel is limited so putting more fuel assemblies reduces the number of experiments that can be loaded. During the DM1 general meeting in June 2006, it was decided to go for the third option: the core layout was fixed to 72 fuel assemblies encircling a "gap" of 3 emptied positions to allow for the placement of the spallation target module [3,4]. Figure 4 (right-most figure) shows this core layout. During the first Task Force Meeting, it was decided to allow for 2 more rings surrounding the 72 fuel assemblies. Justification for this was twofold: increase the distance from the fuel assemblies to the core barrel to reduce the neutron damage on the latter and to allow for more flexibility in core loading.

Figure 4. Micro cell, macro cell and core cut at midplane

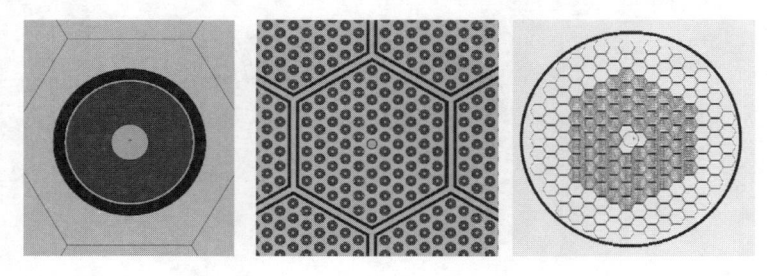

The protons impinging the target surface have an energy of 600 MeV. The choice for a higher proton energy compared to the 350 MeV of the MYRRHA Draft2 design was justified by a higher neutron yield in the first case which allowed the reduction of the needed beam current. The absence of a significant Bragg peak in the case of 600 MeV is also seen as an improvement.

The clean core neutronic results

The XT-ADS reference core as defined above and described in more detail in [5] (without the additional two rows) has been studied using the MCNPX 2.5.0 Monte Carlo code using the JEFF3.1 nuclear libraries. Figure 4 shows the micro cell, macro cell and core layout used in the MCNPX calculations. To reach the desired value of k_{eff} around 0.95, an enrichment of 31.8 wt.% in plutonium was needed in the MOX fissile material. In order to fix all parameters, the design group decided to fix the proton beam current in order to have a fuel power density of 700 W/cm^3. This results in a total power of about 57 MWth. The needed beam current is 2.39 mA. Table 1 gives the most important results for this clean core. Note that the fluxes mentioned in this table are still subject to optimisation (optimal positioning of the spallation target free surface).

Table 1. Neutronics results for the clean core (preliminary design basis)

Parameter	Unit	XT-ADS value
Proton beam energy	MeV	600
Proton beam current	mA	2.39 [†]
Proton beam deposited heat	MW	1.40
Total neutron yield per incident proton		15.3
Neutron source intensity	10^{17} n/s	2.23
Initial fuel mixture	MOX	(U-Pu)O$_2$
Initial (HM) fuel mass	kg	857
Initial Pu enrichment	wt.%	31.5
k_{eff}		0.95324
k_S		0.95711
MF = 1/(1-k_S)		23.31
Source importance φ*		1.095
Thermal power	MW	56.75 [‡]
Specific power	kW/kgHM	66.22
Peak linear power (hottest pin)	W/cm	253
Average linear power (hottest pin)	W/cm	146
Max Φ_{total} in the core near hottest pin		3.43
Max $\Phi_{>1MeV}$ in the core near hottest pin	10^{15} n/(cm^2.s)	0.56
Max $\Phi_{>0.75}$ MeV in the core near hottest pin		0.75

[†] Normalised to fuel power density of 700 W/cm^3.
[‡] 210 MeV/fission.

Figure 5. Axial core model

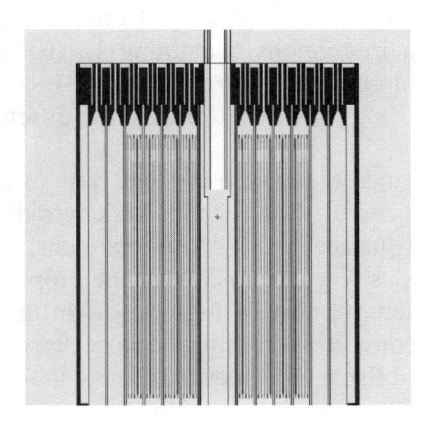

The XT-ADS core with in-pile positions

The XT-ADS machine has to perform its roll as a flexible multi-purpose irradiation machine. In order to do so, eight in-pile sections have been foreseen that are accessible from the top of the reactor to house experimental devices and allow access for measurements. As a reference for a core cross-section with positions to be used for IPS inserts a core cross-section has been selected which is shown in Figure 6. The positions for the eight IPS inserts have been chosen such that they are surrounded by six fuel assemblies. As a result one gets three positions close to the target in a high neutron flux environment and five positions closer to the core periphery in a somewhat lower neutron flux environment.

Figure 6. Proposal for a core with in-pile sections and irradiation positions

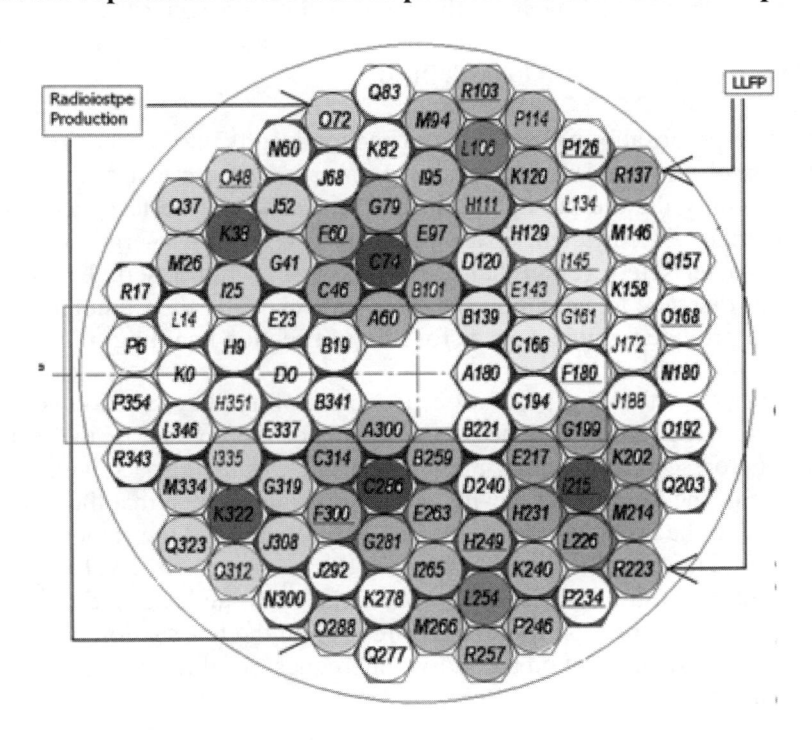

For evaluation of the impact of test sections to be inserted into IPS positions on the neutron physics characteristics of the XT-ADS core four options of different test specimen arrangements are considered: test sections for fuel pin irradiations containing 13 (P1) or 19 (P2) fuel pins, and test sections for structure materials irradiation campaigns containing 31 samples (P3), and a modified fuel assembly with a small test specimen (P4). The rational behind the different proposals is as follows.

First of all it is of utmost importance to broaden the data base of the clad and structure materials behaviour when in contact with lead and/or lead-bismuth under irradiation and to evaluate potential consequences of irradiation on a clad surface with surface treatment. For this purpose it is reasonable to start firstly with small test samples as foreseen in case of the proposal P4 with a high neutron flux level to reduce irradiation time for achievement of high doses. In these cases it is not necessary to provide the possibility of a forced convection cooling because heat production is small and radial conduction should be sufficient to establish a pre-specified temperature level within the test capsule.

In case of further testing needs for structure material it might be interesting to offer larger space for structure materials irradiation not only for a heavy metal cooling environment but for an environment with other coolants as well. For this purpose the proposal P3 defines some upper bound for irradiating a batch of samples at IPS positions of a lower neutron flux so that cooling can still be established by conduction only. However, forced convection cooling with the LBE core coolant could be considered as well as the establishment of a forced convection cooling with a different coolant as pure lead or sodium or other coolants.

The possibility of fuel pin irradiations could be provided on the basis of test section designs as indicated in the proposals P1 and/or P2 with either forced convection cooling using the LBE core coolant or establishing a separate forced convection coolant loop using either lead or sodium or even gas cooling might be possible.

The main interest at the time being is in the cross-section characteristics of the different test section proposals and their influence on volume fractions of materials within fissile core height. Details of the axial design features are still to be defined.

On the basis of a first proposal by SCK•CEN test section inserts into IPS positions contain first of all a two-fold safety vessel arrangement which allows insertion of the test section from above and decoupling partially or totally the thermo-hydraulic conditions within the test sections from the core thermal-hydraulics. To maintain sufficient space between test section inserts and the neighbouring core fuel assemblies the outer safety vessel diameter of the test section inserts has been chosen to be 76.2 mm. The minimum distance to neighbouring FA wrapper walls then amounts to a distance of 11.5 mm and thus should provide sufficient space for safe test section handling.

In case of test sections for fuel pin irradiations a separate forced convection cooling for the test specimens is to be foreseen. As a consequence a flow path for the coolant of the test specimen in a sense of a downcomer is foreseen in the outer periphery of the available space within the inner safety vessel. The cross-section of the flow in the downcomer is determined such that the cross-section of the downcomer is quite close to the cross-section of the coolant flow around the same number of fuel pins of the core fuel assembly. This approach guarantees that the maximum flow velocity in the core fuel assemblies and within the test pin arrangement becomes the same and that the pressure drop in the downcomer remains small in comparison to the pressure drop in the test section itself due to its larger hydraulic diameter. However, dependent on the individual request for different irradiation campaigns a case-to-case optimisation will become necessary.

In case of test sections for structure materials irradiation campaigns as proposed in case of P3 and P4 it might be feasible to cool the test specimen arrangement by thermal conduction to the inter-wrapper coolant flow or establishing forced convection cooling using the LBE core coolant directly. This means that there is more space available for the number of test specimens. The respective design solutions still need to be developed and adjusted on a case-by-case basis to individual requests.

For simplification of the evaluation of a first proposal of test sections for fuel pin irradiations geometrical conditions as selected for the fuel assemblies of the XT-ADS core are maintained unchanged. In future evaluations this could be changed but the respective inventories of the fuel and the structure material should not deviate from the currently specified values to a large extent. To demonstrate the intended possibilities of variations i.e. increase of the fuel inventory in the test section a second proposal has been developed, case P2.

In case of test sections for fuel pin irradiations mechanical stability of the test pin arrangement is established by providing support pins of smaller diameter at appropriate positions in the cross-sections at which spacers are fixed guaranteeing the specified geometrical positioning of the test specimens, i.e. fuel pins. Graphical illustration of the envisaged fuel pin arrangements showing the test section cross-sections are provided in Figure 7 for Case P1 (left) and for Case P2 (right), respectively.

Figure 7. Proposal for in-pile sections type P1 (left) and P2 (right)

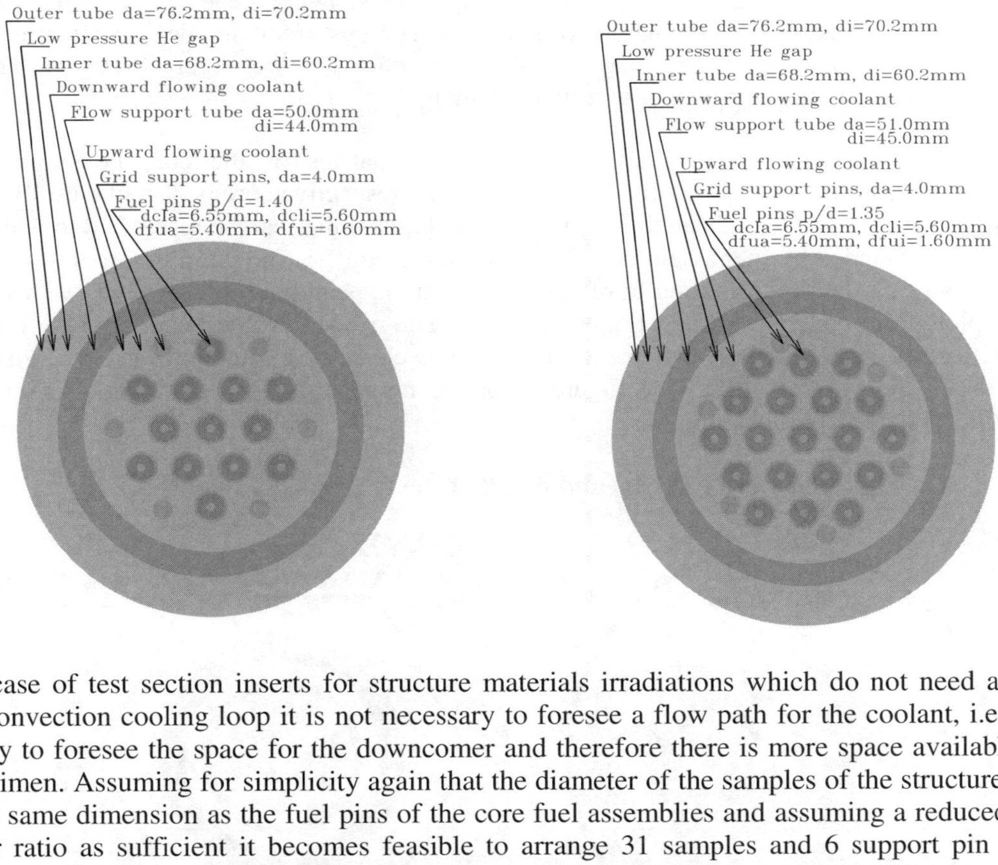

In case of test section inserts for structure materials irradiations which do not need a separate forced convection cooling loop it is not necessary to foresee a flow path for the coolant, i.e. it is not necessary to foresee the space for the downcomer and therefore there is more space available for the test specimen. Assuming for simplicity again that the diameter of the samples of the structure material have the same dimension as the fuel pins of the core fuel assemblies and assuming a reduced pitch to diameter ratio as sufficient it becomes feasible to arrange 31 samples and 6 support pin positions within the available space. To establish mechanical stability of the test section insert it is assumed that there are 6 support pins of reduced diameter at the edges of the hexagonal arrangement which provide again the possibility for fixation of spacers to guarantee appropriate positioning of the test specimen.

Figure 8. Proposal for in-pile sections type P3 (left) and P4 (right)

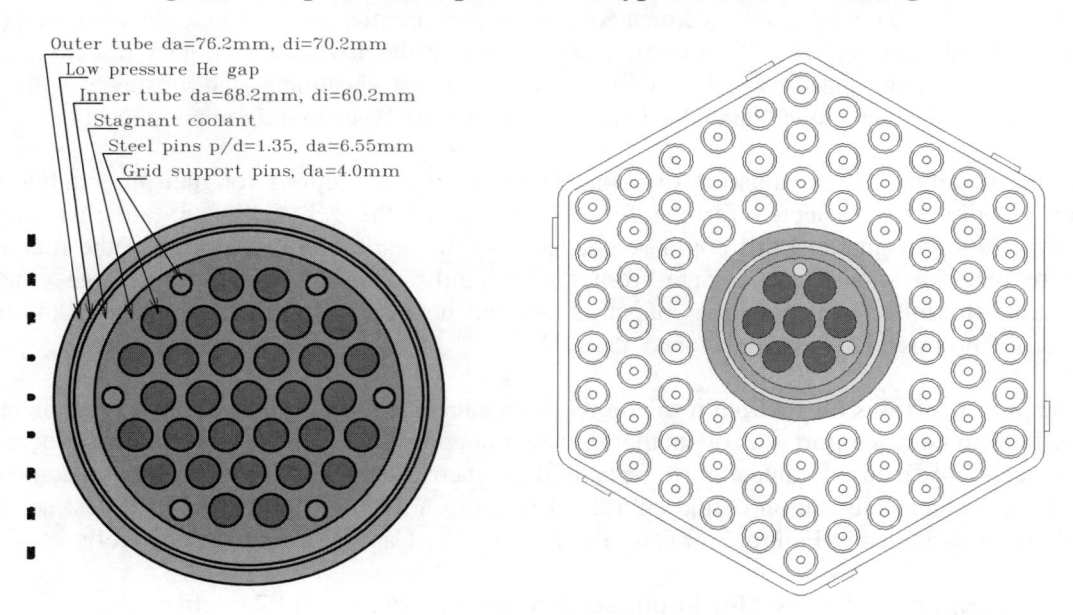

Outer tube da=76.2mm, di=70.2mm
Low pressure He gap
Inner tube da=68.2mm, di=60.2mm
Stagnant coolant
Steel pins p/d=1.35, da=6.55mm
Grid support pins, da=4.0mm

Graphical illustration of the envisaged test specimen arrangement showing the test section cross-sections is provided in Figure 8 (left). As an alternative for structure materials irradiation devices a configuration with smaller test specimens has been analysed as well in case P4, see Figure 8 (right). The test specimen is embedded in the environment of a fuel assembly.

As a starting point for the evaluation of the performances of the core housing 8 IPSs simultaneously, the case of IPS type P1 has been studied The respective core is named core P1. When inserting 8 IPSs in the positions indicated in Figure 6, 8 fuel assemblies can be reshuffled to make sure all the IPSs are surrounded by fuel assemblies. The resulting core P1 is shown in Figure 9 (left) with a detail of the IPS P1 on the right. In order to re-obtain a k_{eff} in the neighbourhood of 0.95, we needed to increase the plutonium content to 33.05 (k_{eff} = 0.95035). The needed proton beam strength does not change significantly from the 2.39 mA mentioned for the clean core situation. Table 2 gives the neutron fluxes in the different IPSs (see Figure 6 for the nomenclature of the IPSs). Again, these fluxes are still subject to optimisation.

Figure 9. Core P1 (left) and detail of the IPS type P1 (right)

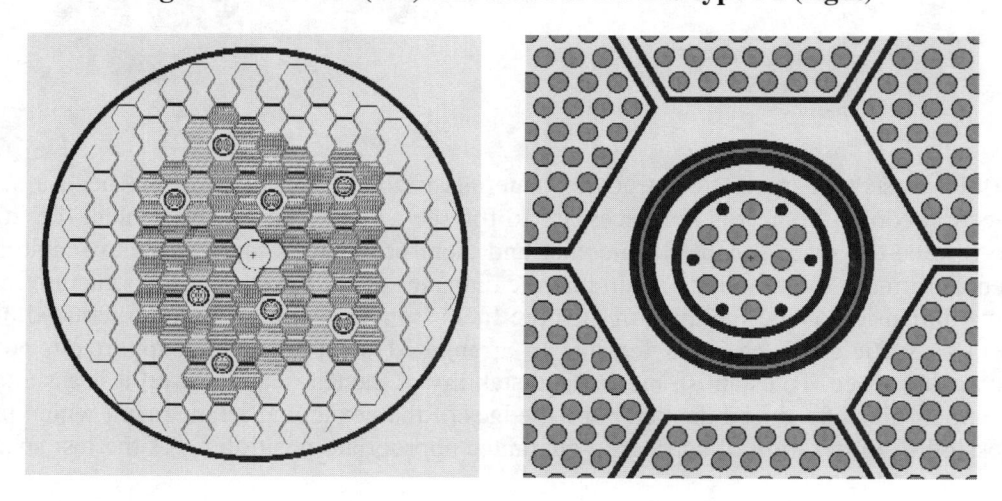

Table 2. Neutron fluxes in the central pin of the different IPSs type P1

	$\Phi > 0.75\,\text{MeV}$ (10^{15} n/cm².s)	$\Phi > 1\,\text{MeV}$ (10^{15} n/cm².s)	Φtotal (10^{15} n/cm².s)
C74, C286	0.43	0.29	2.63
K38, K322	0.28	0.19	1.82
L106, L254	0.26	0.18	1.74
I215	0.30	0.20	1.96
E143	0.43	0.30	2.46

Conclusions

The reference clean core has been fixed in the Deliverable D 1.7 of the EUROTRANS project. During the different Task Force meetings, a clean core has been proposed containing two extra rings of sub-assembly positions. A neutron physics analysis has been performed to quantify characteristics of this new core. Four types of in-pile positions have been drafted in a preliminary fashion covering fuel irradiation and material irradiation in the XT-ADS core. These four types of IPSs can be housed in eight positions strategically chosen in the XT-ADS core. One simultaneous loading of eight IPSs of the type P1 has been analysed. In order to re-obtain the reference k_{eff} value, the plutonium content of the MOX fuel had to be increased. It seems that, to be on the safe side, it would be advisable to foresee a plutonium enrichment of up to 35 wt.% in order to cover even worse (from the point of view of reactivity reduction) IPS loadings of type P3 or to compensate for burn-up effects. The clean core configuration should then be adapted (less fuel assemblies or burnable poisons) to achieve a k_{eff} around 0.95. The main drawback of increasing the Pu content in the MOX fuel is the increase in reactivity swing during the operational cycle. This issue is the subject of further analysis and a satisfactory response has to be indicated.

REFERENCES

[1] Aït Abderrahim, H., *et al.*, *MYRRHA Pre-design File – Draft 2*, SCK•CEN Report R-4234, June 2005.

[2] Rimpault, G., *et al.*, *Definition of the Detailed Missions of both the Pb-Bi Cooled XT-ADS and Pb-cooled EFIT and its Gas Back-up Option*, Deliverable D1.1 of the EUROTRANS project, June 2006.

[3] Minutes of the General DM1 meeting, 20-21 June 2006, SCK•CEN, Mol, Belgium.

[4] Schuurmans, P., *et al.*, "Design and Supporting R&D of the XT-ADS Spallation Target", *Proc. of the 9th Information Exchange Meeting on Partitioning and Transmutation*, Nîmes, France, September 2006.

[5] Van den Eynde, G., *et al.*, *Specification of the XT-ADS Core and Fuel Element*, Deliverable 1.7 of the IP EUROTRANS project, Contract No. FI6W-CT-2004-516520.

THE EVALUATION OF PRELIMINARY EXTRAPOLATION EXPERIMENTAL RESULTS OF THE CHINESE ADS SUBCRITICAL EXPERIMENTAL ASSEMBLY VENUS-1

Pu Xia, Yong-qian Shi, Zhi-xiang Zhao, Yi-guo Li, Qin-fu Zhu,
Ji-gen Li, Wei Zhang, Jian Cao, Yan-hui Quan, Huang-te Luo, Xiaofei Wu
China Institute of Atomic Energy
P.O. Box 275-75, Beijing 102413, China

Abstract

The design objectives and structure of the Chinese accelerator-driven subcritical system (ADS) subcritical assembly VENUS-1 are presented. The experiment of source multiplication method is used to determine the final loading of the Chinese ADS VENUS-1 subcritical assembly. The k_{eff} of the assembly with the same loading as extrapolation experiment have been calculated. The results of calculation and experiment are evaluated. The final loading of the Chinese ADS VENUS-1 subcritical assembly has been determined by preliminary experiments on VENUS-1.

Introduction

To develop nuclear power on a large scale, two problems must be solved. First, the technically and economically exploitable natural uranium resources are limited domestically and overseas, such that the uranium utilisation rate must be greatly increased. Second, long-lived radioactive nuclear wastes have to be disposed of to reduce their impact on the environment and public fear of nuclear power. The concept of ADS system is good to solve these two problems.

China established a research group in the 1990s, focused on the topic of the accelerator-driven system. Since 2000, a project designated "The Major State Basic Research Program (973)" was created for the energy domain. It is sponsored by the Chinese Ministry of Science and Technology (MOST), and a five-year programme of basic research for ADS physics and related technology was launched. Now, the Chinese ADS VENUS-1 subcritical assembly is completed.

The Chinese ADS VENUS-1 subcritical assembly is especially designed for subcritical reactor neutron behaviour research. Natural uranium fuel and lower-enriched uranium (LEU) fuel (3% UO_2) have been chosen. Some optimisation of the arrangement of natural fuel and LEU fuel has been done. Concerning the connection between the assembly and accelerator, the structure of the assembly has been set to a horizontal arrangement.

On 18 July 2005, the first fuel element was loaded into the VENUS-1 subcritical assembly and extrapolation experiments have been performed. The k_{eff} of assembly with the same loading as extrapolation experiment has been calculated. The results of calculation and experiment are evaluated. Final loading of the Chinese ADS VENUS-1 subcritical assembly has been determined by preliminary experiments on VENUS-1.

Design objectives

The Chinese ADS VENUS-1 subcritical assembly is especially designed for subcritical reactor neutron behaviour research. A series of experiments will be performed on this assembly. The research subjects include the following items:

- test and verify new parameters in subcritical reactor;
- test and verify ADS reactor physics and nuclear data;
- research neutron spectrum measurement technology;
- research monitor k_{eff} technology in ADS operation;
- research MA transmutation;
- other.

Design requirement

The Chinese ADS VENUS-1 subcritical assembly is a subcritical reactor. It is mainly used for research on neutron behaviour of subcritical reactor. Thus, the effective neutron multiplication factor (k_{eff}) should be less than 1.0, and the criticality of the assembly should be adjustable. The value of the effective neutron multiplication factor of China ADS VENUS-1 subcritical assembly is k_{eff} = 0.90-0.98.

The VENUS-1 subcritical assembly is also planned to be used for research on MA transmutation. The average neutron energy should be larger than fast reactors. Thus, the desired average neutron energy should be between 600-900 keV.

Moreover, the VENUS-1 subcritical assembly is planned to be used for testing and verification of the ADS reactor physics programme and nuclear data, so its structure should be simply designed.

VENUS-1 will connect with the accelerator, take the (d,t) reaction to produce extra neutrons, driven the China ADS VENUS-1 subcritical assembly running on certain power. Due to the field restrict, the direction of the VENUS-1 subcritical assembly is considered as horizontal. It should be adjustable in vertical and horizontal direction, make it easy to connect with the accelerator.

To build a new Chinese ADS VENUS-1 subcritical assembly, the fund is not enough to get new nuclear material. Currently available nuclear material is being used for the design, i.e. natural fuel and lower-enriched uranium fuel. The parameters of these fuels are list in Table 1.

Table 1. Parameters of natural fuel and LEU 3% UO_2

Item	Natural fuel	LEU fuel
Fuel meat diameter (mm)	20	6.5
Fuel density (g/cm^3)	18.6	10.5
Fuel length (mm)	1 000	700
Cladding material	Al	Zr-2
Cladding diameter (mm)	22	8
Fuel element weight (kg)	6.2	0.25

Structure of China ADS VENUS-1 subcritical assembly

According to the above-mentioned design objectives and requirements, as well as the conceptual research on the reactor core physics for an accelerator-driven subcritical reactor, the core of VENUS-1 subcritical assembly is designed to be coupled core by a fast neutron spectrum zone and a thermal neutron spectrum zone. The fast neutron spectrum zone (including external neutron source) is located at the centre of the core and the thermal neutron spectrum zone is surrounding the fast neutron spectrum zone. The VENUS-1 is driven by an Am-Be neutron source or other steady neutron source (^{252}Cf, D-D reaction or D-T reaction) to research the effect of external neutron source with different energy on it, or is driven by a D-T pulsed neutron source through CPNG (CIAE Pulsed Neutron Generator) to research dynamic characteristic. There are no safety rod and control rods in the core, excepting the aluminium structure used for modelling sodium coolant in the fast neutron zone and the polyethylene structure used for modelling water coolant and moderator in the thermal neutron zone, thus satisfying the requirement for simple structure of the core. The maximum k_{eff} is designed less than 0.98, which is smaller than the shutdown margin of a PWR. The subcritical assembly is dependent on external neutron sources for its operation and can be shutdown by cutting off the deuterium ion current (for D-D or D-T reaction steady external neutron sources and D-T reaction pulsed neutron source) or by removing steady external neutron source (Am-Be isotope neutron source or ^{252}Cf spontaneous fission neutron source), thus exempting concerns in nuclear criticality safety for VENUS-1 subcritical assembly, unlike critical reactors depending on safety rods and control rods for operation.

After optimised calculation a suitable fuel arrangement is given. In the fast neutron spectrum zone of the VENUS-1 subcritical assembly, hexagonal aluminium structure with natural U fuel rods are arranged in an equilateral triangle and rod pitch is 25 mm. The fuel is loaded from the third to the tenth layer. The k_{eff} calculated by the MCNP code is approximately 0.5 and 0.3 with and without polyethylene reflector, respectively.

In the thermal neutron spectrum zone of the VENUS-1 subcritical assembly, LEU fuel rods with 3% enriched ^{235}U are inserted in the polyethylene with fuel rods arranged in pattern of equilateral triangle. An optimised calculation fuel rod pitch is 12 mm when thermal zone surrounding the fast zone exists, the calculated k_{eff} value can be larger than 1.0. (k_{eff} value is measured by experiment to satisfy requirement in the 0.90-0.98).

The final dimensions of the VENUS-1 subcritical assembly have been determined: 1 600 mm in diameter, 1 800 mm in length and 1 000 mm from the centre of the core to the floor surface. The subcritical assembly is laid on a chassis, which can be moved vertically or horizontally when coupled to the CIAE Pulsed Neutron Generator. The following describes the main parameters of the assembly:

1) *Neutron source zone*. The neutron source is sealed in an aluminium tube of 50 mm diameter, which is vacuumed and inserted in the centre hole of the subcritical assembly core. The centre hole is formed by remove 7 fuel elements from the fast neutron zone. As the neutron source aluminium tube occupies only half-length of the hole, 7 fuel rods of half-length can be loaded into the other section of the hole. Having finished experiments, the aluminium tube can be pulled out of the subcritical system.

2) *Fast neutron zone*. To form the fast neutron zone, natural uranium fuel rods are inserted in a hexagonal aluminium structure with rod pitch of 25 mm and in an equilateral triangle arrangement. The fuel rods totalling 264 of full length are loaded in 3 to 10 layers, with the 7 fuel rods of half-length inserted in the centre hole in 1 to 2 layers. The amounts of natural uranium fuel rods in each layer are listed in Table 2. Figure 1 shows the arrangement for neutron source zone and fast neutron zone.

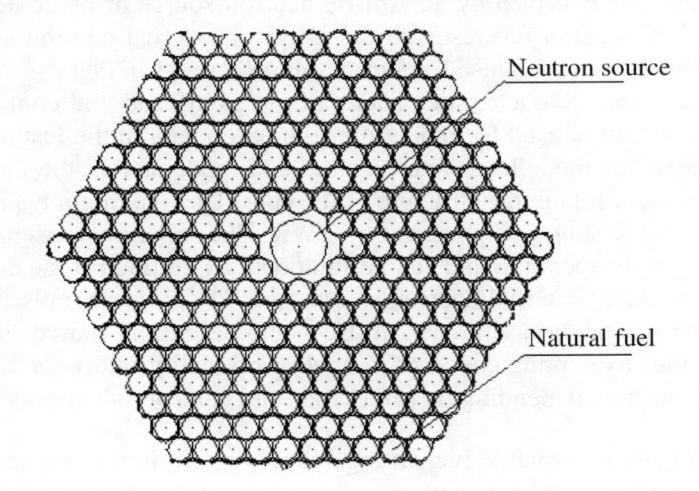

Table 2. Natural uranium fuel rods in each layer

Layer	1	2	3	4	5	6	7	8	9	10
Fuel element	1	6	12	18	24	30	36	42	48	54

Figure 1. Arrangement for neutron source zone and fast neutron zone

3) *Thermal neutron zone*. Thermal neutron zone is formed by UO$_2$ fuel rods (with 3% enriched ^{235}U) inserting in polyethylene. With rod pitch of 12 mm and in an equilateral triangle arrangement, a total of 2 268 fuel rods can be arranged in 15 layers. The final fuel loading of the thermal zone is determined through experiments. The amounts of fuel rods in each layer are listed in Table 3. As the core will take a quasi-cylindrical form, the amounts of fuel rods inserted in the outer three layers are less than those in the inner 11 layers of this zone.

Table 3. [3% UO$_2$] fuel rods in each layer

Layer no.	1	2	3	4	5	6	7	8	9	10	11	12	13	14	15
Fuel elements	126	132	138	144	150	156	162	168	174	180	186	174	156	126	96

4) *The reflector zone is formed by polyethylene in approximate 220 mm thickness*. Its side surface is cylindrical. The final thickness of the reflector is decided after the fuel loading of the thermal neutron zone.

5) *Shield zone and SS shell*. The shield is a cylinder made of polyethylene with boron in thickness of 200 mm. The outer shell of the assembly is a 10-mm thick stainless steel.

At present, neutrons in the VENUS-1 assembly are two-way coupled, i.e. if aluminium plate is filled in the gap between the fast zone and thermal zone, thermal and fast neutrons can travel to and from these two zones. If Cd or B absorber is inserted in the gap, fast neutrons can travel to and from these two zones, but thermal neutrons moderated by polyethylene in thermal zone cannot travel to fast zone, which means neutrons in the assembly are one-way coupled, hence characteristics of a reactor core with one-way coupled neutrons can be studied.

Figure 3 is longitudinal section arrangement of the VENUS-1 assembly and Figure 4 is a cross-section arrangement of the VENUS-1 assembly.

Figure 3. Longitudinal section arrangement of VENUS-1 assembly

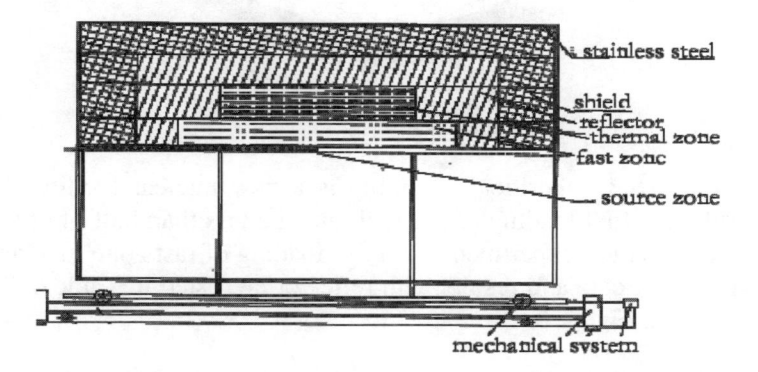

Figure 4. Cross-section arrangement of VENUS-1 assembly

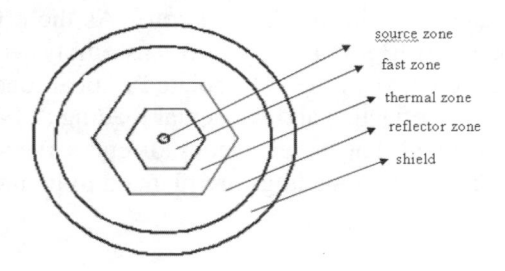

Experiment results

To determine the final loading of the Chinese ADS VENUS-1 subcritical assembly, the experiment of source multiplication method (i.e. inverse multiplication measurement) was used; it was undertaken in 2005. An Am-Be steady neutron source was used. During the experiment, detectors were placed at the outer of polyethylene reflector and in the outer layers of the fast zone, respectively, as is shown in Figure 5. The detectors are one 3He counter with dimensions of $\Phi 10 \times 150$ mm as 1# detector, two BF3 counter of $\Phi 65 \times 780$ mm as 2# and 3# detectors and two gamma-compensated ionisation chambers ($\Phi 50 \times 560$ mm). The 3He and BF3 proportional counters were used for inverse multiplication measurement (critical approach). Gamma-compensated ionisation chambers were used for power monitoring.

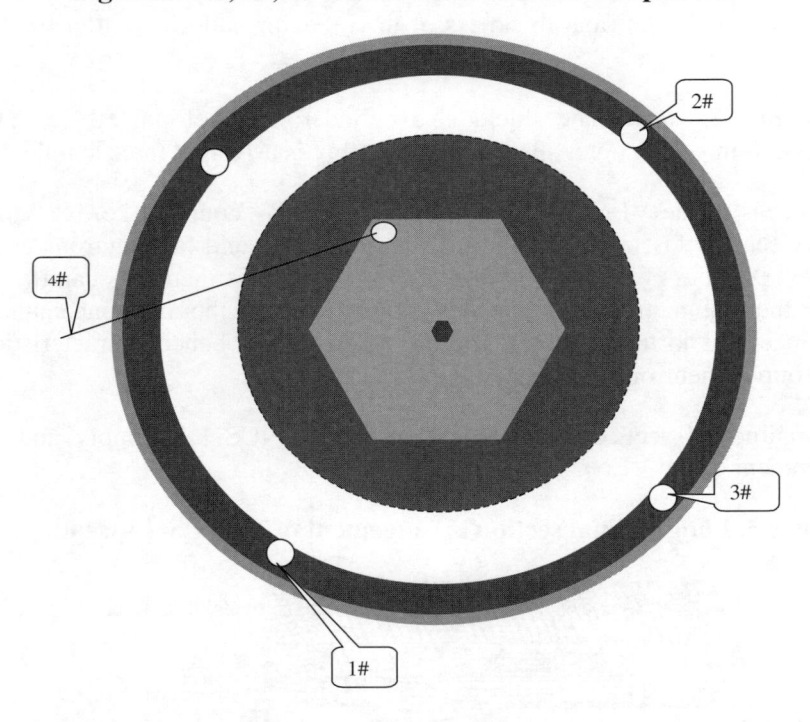

Figure 5. 1#, 2#, 3# and 4# neutron detectors position

The Chinese ADS VENUS-1 subcritical assembly is a new nuclear facility. According to the zero-power experiment rule, the first loading of the fuel must be less than half of critical loading from calculation. So we start extrapolation experiment from full loading of fast zone fuel, and then add fuels in thermal zone. The experiment data and results with reflector are listed in Table 4. The experimental loading of each step is shown in Figure 6.

The final extrapolated critical fuel loading of the VENUS-1 assembly is shown in Table 5. The determined final loading of fuel rods is 264 rods plus 7 half-rods of natural uranium fuel in the fast zone and 2 046 rods of 3% enriched ^{235}U in the thermal zone. As the efficiency of a fuel rod in the outmost layer of the thermal zone is 0.12 mk, the k_{eff} of the assembly with and without end reflector was obtained through the difference between the extrapolated critical loading (2 312 or 2 333 rods in the thermal zone with or without end reflector) and the actual loading (2 046 rods in the thermal zone), which was 0.9681 or 0.9656. It seemed that the efficiency of end reflector (~0.0025) was somewhat small. Accurate k_{eff} values of the above two loadings are planned to be measured by other methods in the future.

Calculations

Calculations have been performed using the Monte Carlo simulation method. The same loadings as extrapolation experiment are used for calculation. The average worth of each loading of fuel is calculated. The same means as with the extrapolation method are used to get critical loading of thermal fuel rods. The calculation results are shown in Table 6.

Discussion and conclusion

From the results of experiment and calculation, the worth of fuel rod decreases when the loading is adding. The efficiency of a fuel rod is 0.37 in inner side of thermal zone and is 0.12 mk in outmost of thermal zone. The critical loading of facility, affected by the efficiency of fuel rod, is also changed during the extrapolation experiment. There are some error between the experiment and simulation. More study is needed to analyse the reason of these errors.

The extrapolation method can be used for approaching the reactor critical. But it is difficult to use for determining the final loading of subcritical assembly. The loading of subcritical reactor is affected by efficiency of fuel rods, efficiency of detector, external neutron source.

Accurate k_{eff} value of the loading of facility is being planned to study for measurement by other methods in the future.

Table 4. Results of extrapolated experiment with reflector

Thermal fuel loading	Source position	Centre	17.5 cm	35 cm	Average critical loading	Critical level k_{eff}
	Detector	Extrapolation result (rods)				
540	1	929.3	1 041	886	1 005±148	0.8247
	2	905.6	957.6	1 349		
	3		964	1 008		
738	1	1 163	1 181	1 250	1 272±110	0.8190
	2	1 228	1 230	1 454		
	3	1 167	1 364	1 413		
948	1	1 426	1 430	1 430	1 413±50	0.8771
	2	1 390	1 391	1 311		
	3	1 494	1 442	1 405		
1 176	1	1 580	1 590	1 674	1 636±70	0.8992
	2	1 661	1 715	1 721		
	3	1 508	1 603	1 670		
1 392	1	1 845	1 872	1 794	1 859±93	0.9147
	2	1 774	1 833	1 859		
	3	1 810	1 852	2 091		
1 578	1	1 984	1 987	2 115	1 995±49	0.9358
	2	1 977	1 944	2 005		
	3	1 968	1 969	1 918		
1 746	1	2 100	2 111	2 078	2 109±27	0.9529
	2	2 117	2 136	2 148		
	3	2 061	2 112	2 115		
1 890	1	2 205	2 245	2 277	2 251±38	0.9570
	2	2 232	2 287	2 313		
	3	2 210	2 219	2 273		
2 046	1	2 304	2 314	2 319	2 312±21	0.9681
	2	2 298	2 324	2 311		
	3	2 266	2 338	2 332		

Figure 6. Fuel loading of experiment

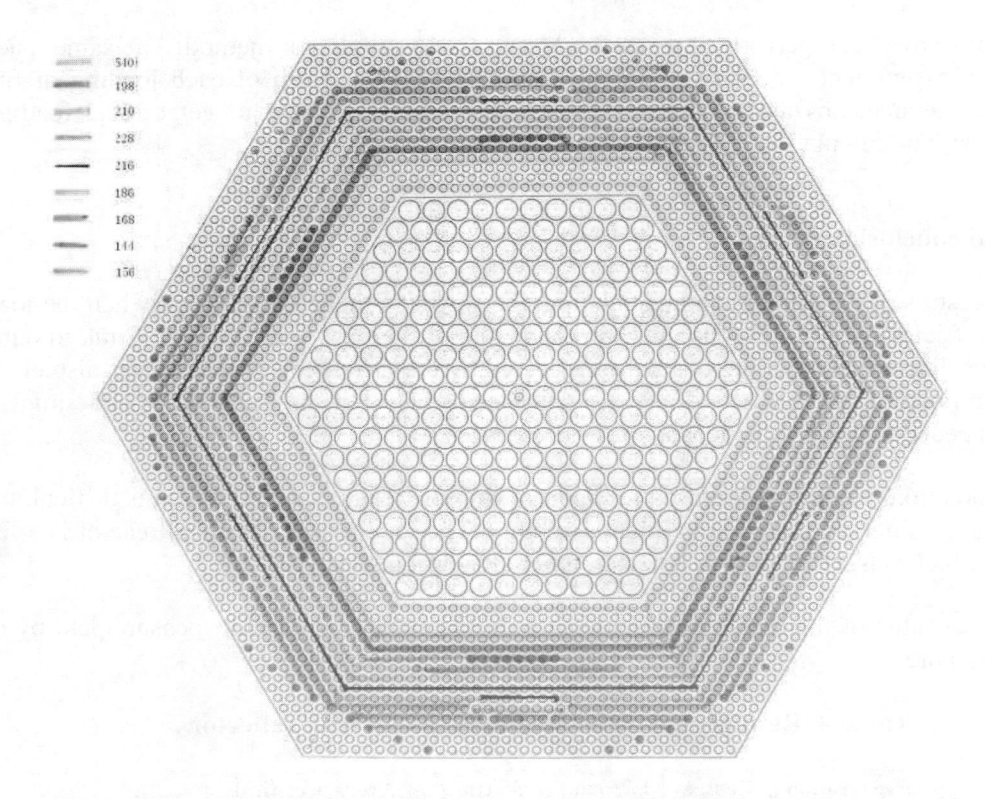

Table 5. Extrapolated critical fuel rod quantity

Source location (mm)			0	175	350
Final loading: 267.5$_{Nature}$ + 2 046$_{Low}$	Condition	Detector no.	Extrapolation result (rods)		
	Without reflector	1#	2 304	2 330	2 353
		2#	2 308	2 340	2 335
		3#	2 329	2 366	2 334
	Average		**2 333±19**		
	With reflector	1#	2 304	2 314	2 319
		2#	2 298	2 325	2 311
		3#	2 266	2 338	2 333
	Average		**2 312±21**		

Table 6. The k_{eff} calculation result for the experiment result

Fuel loading	Fuel adding	k_{eff}	Fuel rod worth (mk)	Critical adding	Critical loading	Critical level
267.5Nu		0.457221±0.0010				
267.5$_{Nature}$ + 540$_{Low}$	540	0.660819±0.0011	0.3770	899.6	1439.6	0.8247
267.5$_{Nature}$ + 738$_{Low}$	198	0.727929±0.0010	0.3389	802.7	1540.7	0.8190
267.5$_{Nature}$ + 948$_{Low}$	210	0.783438±0.0009	0.2643	819.3	1767.3	0.8771
267.5$_{Nature}$ + 1176$_{Low}$	228	0.833387±0.0009	0.2191	760.5	1936.5	0.8992
267.5$_{Nature}$ + 1392$_{Low}$	216	0.872844±0.0008	0.1827	696.0	2088.0	0.9147
267.5$_{Nature}$ + 1578$_{Low}$	186	0.901488±0.0008	0.1540	639.7	2217.7	0.9358
267.5$_{Nature}$ + 1746$_{Low}$	168	0.923257±0.0008	0.1296	592.3	2338.3	0.9529
267.5$_{Nature}$ + 1890$_{Low}$	144	0.940375±0.0007	0.1189	501.6	2391.6	0.9570
267.5$_{Nature}$ + 2046$_{Low}$	156	0.959429±0.0010	0.1221	332.2	2378.2	0.9681

SESSION V

ADS Experiments and Test Facilities

Chairs: H. Aït Abderrahim, M. Tanigaki

THE GUINEVERE PROJECT AT THE VENUS FACILITY

**P. Baeten, H. Aït Abderrahim, G. Vittiglio,
F. Vermeersch, G. Bergmans, B. Verboomen, D. Maes**
SCK•CEN, Belgium

Abstract

The GUINEVERE project is a European project in the framework of FP6 IP-EUROTRANS. The IP-EUROTRANS project aims at addressing the main issues for ADS development in the framework of partitioning and transmutation for nuclear waste volume and radiotoxicity reduction. The GUINEVERE project is carried out in the context of Domain 2 of IP-EUROTRANS, ECATS, devoted to specific experiments for the coupling of an accelerator, a target and a subcritical core. These experiments should provide an answer to the questions of on-line reactivity monitoring, subcriticality determination and operational procedures (loading, start-up, shutdown, etc.) in an ADS by 2009-2010. The GUINEVERE project will make use of the VENUS reactor, serving as a lead fast critical facility, coupled to a continuous beam accelerator. In order to achieve this goal, the VENUS facility has to be adapted and a modified GENEPI-C accelerator has to be designed and constructed. During the years 2007 and 2008, the VENUS facility will be modified in order to allow the experimental programme to start in 2009. The paper describes the main achievements with regard to the modifications for the VENUS facility.

Introduction

The GUINEVERE project is an international project in the framework of FP6 IP-EUROTRANS. The IP-EUROTRANS project aims at addressing the main issues for ADS development and consists of five domains. In the first domain, the design of a demonstration facility for the accelerator-driven system concept, XT-ADS, and a prototype industrial transmuter EFIT is carried out. The XT-ADS concept is strongly based on the current MYRRHA design and SCK•CEN has offered its site to the European forum for hosting this future fast experimental facility at its Mol site. The proposed GUINEVERE project is carried out in the framework of Domain 2, ECATS, devoted to specific experiments for coupling an accelerator, target and subcritical core. These experiments should provide an answer by 2009 to the questions of on-line reactivity monitoring, subcriticality determination and operational procedures (loading, start-up, shutdown, etc.) in an ADS.

Reanalysing the outcome of MUSE, two points were left open for significant improvement. To validate the methodology for reactivity monitoring, a *continuous beam* is needed, which was not present in the MUSE project. In the definition of the MUSE project, from the beginning a strong request was made for a *lead core* in order to have representative conditions of a lead-cooled ADS. This request was only partly satisfied by the creation of a lead central part in the last configuration of the MUSE programme. Due to programme limitations, the investigation of this core was strongly reduced.

Therefore, there is need for a lead fast critical facility connected to a continuous beam accelerator. Since such a programme/installation is not present at the European nor at the international level, SCK•CEN has proposed to use a modified VENUS critical facility located at its Mol site and to couple it to a modified GENEPI (used in MUSE) working in current mode: the GUINEVERE project (*G*enerator of *U*ninterrupted *I*ntense *NE*utrons at the Lead *VE*nus *RE*actor).

Although this will be a European project, responsibilities and tasks are clearly divided among the different partners. SCK•CEN will of course be responsible for all the modifications at the VENUS facility to transform the water-moderated facility into a fast lead core facility VENUS-F. CNRS will charge itself with the modifications to be applied to GENEPI to allow it to work in continuous mode: GENEPI-C. CEA will provide the fuel and lead rodlets needed for fuel assemblies and will be involved in the core characterisation. FZK will assist SCK•CEN in the safety evaluation of the new installation.

Description of the VENUS reactor and necessary modifications

The VENUS reactor was built in 1963-1964 and is a water-moderated reactor (zero-power critical facility). The fissile fuel is contained in fuel pins which are manually loaded into the reactor. The VENUS reactor is made critical by increasing the water level in the reactor. To control the reactor, one uses control bars which displace water and thereby change the water level. The reactor is shutdown by a fast water dump system. The water can then in case of an emergency stop (scram) be evacuated in a very short time by opening safety valves to so-called dump tanks.

The execution of this project will consist of two types of modifications at the SCK•CEN site. First, the modifications which are connected to the installation of the new GENEPI-C accelerator, working in continuous and pulsed mode, at the VENUS critical facility and its coupling to the core. The second type of modifications is linked with the adaptation of the VENUS critical facility to host a fast lead core, further on referred to as VENUS-F.

After consultation with the European partners, it was concluded that the vertical penetration beam line in the core would represent a significant added value to the project. To implement the vertical penetration option, the accelerator has to be put on top of the VENUS bunker in a technical room to be constructed. This means that civil engineering works are necessary to build a new technical area above the VENUS bunker.

To modify the water-moderated thermal reactor in a fast lead reactor, two main items were identified:

- A shutdown system based on shutdown rods (as in the first years of the VENUS facility) will have to be installed.

- Construction of fuel assemblies with lead blocks and uranium fuel for the core and large lead blocks for the reflector. This will represent a total weight of about 30 t in the vessel.

For the execution of this programme, a detailed planning exists. First of all, fuel from CEA will be transported in 2007-2008. After finishing the fuel assembly design around April 2007, the different parts will be ordered and fuel assembly construction will be performed by summer 2008. Meanwhile, construction works for the accelerator room will begin in 2008 and will be finished by summer 2008. The second half of 2008, the different newly designed internal parts of the reactor core will be installed. In January 2009, the commissioning of the installation will start. The commissioning and licensing procedure will be finalised by summer 2009, such that the envisaged experimental programme can be executed onwards from September 2009.

The GENEPI accelerator and necessary modifications

The GENEPI-1 accelerator was designed and built by the Accelerator Department of the *Laboratoire de Physique Subatomique et de Cosmologie* (CNRS/IN2P3/LPSC) of Grenoble, France, to achieve neutronic experiments in the MASURCA reactor at CEA Cadarache in the framework of 5FP MUSE. It was designed to generate neutron pulses with a time width of the order of magnitude of the mean neutron lifetime in a fast fissile medium, i.e. 1 µs. Moreover, to have a measurable effect, these pulses had to be as intense as possible. To reach these objectives a 250-kV accelerator coupled to a pulsed deuteron source and followed by a transport line bringing the beam onto a deuterium or tritium target was built, producing a pulsed neutron source thanks to $D(d,n)^3He$ or $T(d,n)^4He$ reactions.

Figure 1. The GENEPI-1 accelerator

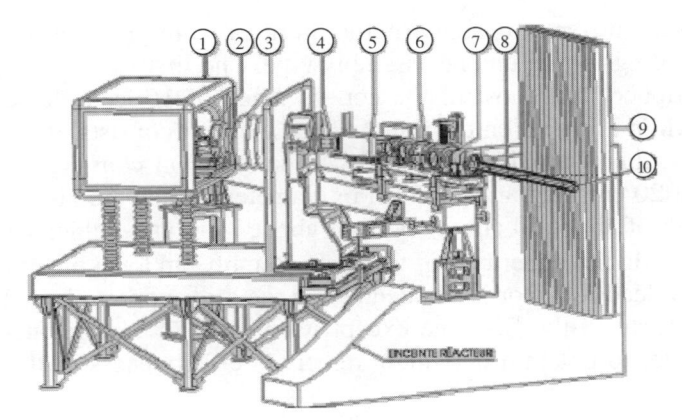

1) High voltage head
2) Duoplasmatron
3) Accelerator tube
4) Quad Q1
5) Magnet
6) Quad Q2
7) Quad Q3
8) Quad Q4 + T2 part
9) MASURCA tube
10) Target

The deuteron source is a "duoplasmatron" located on a 250-kV platform. Deuteron pulses are produced by a pulsed electron discharge which ionises the deuterium gas with a maximum repetition rate of 5 kHz. The high beam intensity at the source exit (around 80 mA in the pulse, with no compensation by the residual gas ionisation, energy 60 keV) requires great care in terms of focusing. The main acceleration is produced by an electrostatic tube to reach 250 keV at maximum. After acceleration ions cross a magnetic spectrometer chamber (45° deflection magnet) which allows the selection of charged particles and the elimination of parasitic ions produced in the source (essentially D_2^+ ions, twice as slow as D^+ ions) which may damage the pulse time structure. Transportation of the beam towards the target (of the order of 5 meters here) located at the end of a thimble is performed thanks to four focalisation lenses (quadrupoles) before the thimble, and six inside the thimble itself. The target ends the thimble: it is in copper, covered on its internal face by a layer of titanium loaded by deuterium or tritium. The peak current in a pulse is of the order of 40 mA, leading to a mean neutron production of about 10^6 neutrons per pulse (with a titanium-tritium target).

Based on the GENEPI 1-2 experience, a new GENEPI-C accelerator operating both in continuous and pulsed mode will be designed to be coupled to VENUS-F. In pulsed mode it should work with characteristics as close as possible of that of the GENEPI 1-2. In continuous mode, deuteron current to be reached should be in the 100 µA-1 mA range. It is recalled that a 1 µs peak of 40 mA intensity at 4 kHz is equivalent to a continuous beam of 160 µA, bringing the same quantity of source neutrons per second (around 10^9-10^{10} n/s), conditions of the MUSE experiment. To achieve the "beam trips" part of the experimental programme, prompt beam interruptions [transition time less than 1 µs (?)] should be performed, with a repetition rate of a fraction of Hz, with a duration bounded by a few hundreds of microseconds and a few tens of milliseconds.

These necessary modifications will be carried out by CNRS at CNRS laboratories in Grenoble where the GENEPI-C will be completely assembled and tested. In the beginning of 2009, the GENEPI-C accelerator will be dismounted at CNRS, transported to SCK•CEN and mounted again at SCK•CEN. The commissioning and licensing of the coupled GENEPI-C and VENUS-F reactor is scheduled to be finalised by summer 2009.

Evaluation of a critical fast lead core

The GUINEVERE project foresees the working of the fast lead VENUS both in a critical state (without accelerator) and in a subcritical state (with the GENEPI accelerator). Since for the critical state the largest number of fuel assemblies will be required, first an estimation of a critical fast lead configuration was made.

Based on a compromise between different design criteria, a critical configuration was proposed making use of two types of fuel loading for the unique assembly type. The first fuel loading assembly type which will constitute the periphery and most of the core is made of 16 lead elements, 9 fuel elements and 4 lead plates as shown in Figure 2 on the right. A lead element consists of 6 lead rodlets (of about 10 cm long and 1.27 cm thick) provided by CEA. A fuel element consists of three 30% enriched uranium rodlets (of about 20 cm long and 1.27 cm in diameter) which are also provided by CEA. The lead plates have a length of about 60 cm, a width of about 7 cm and a thickness of 4 mm and will be fabricated by SCK•CEN. In the upper part of the fuel assembly, an axial upper reflector of about 40 cm is inserted. The bottom lead reflector is situated below the grid support plate. The second fuel loading assembly type is identical to the first one except it contains four more fuel elements. The second type is used at the centre to obtain a maximum reactivity effectiveness of the additional fuel elements.

The core is thus enclosed in a square of 11 × 11 (88.66 cm × 88.66 cm) with an equivalent diameter of 100.04 cm. The bottom and top reflector have a thickness of 40 cm. For the radial lead reflector the distance between the fuel boundary and the core (with a diameter of 160 cm) is about

Figure 2. Critical configuration with a central hole for vertical beam penetration (left) and a detailed layout of the fuel assembly (right)

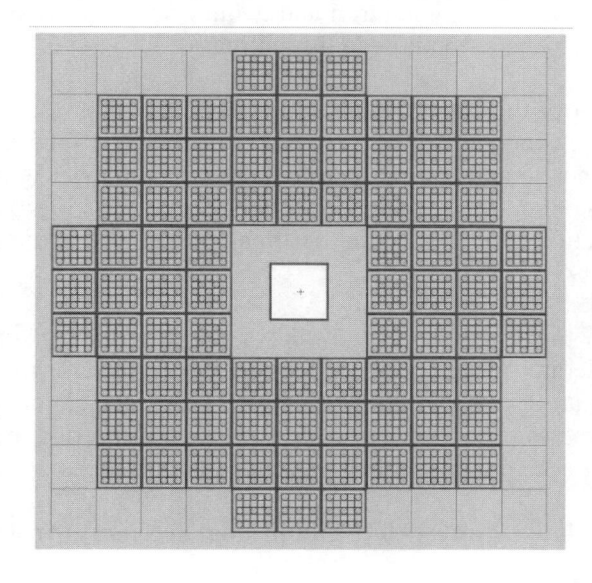

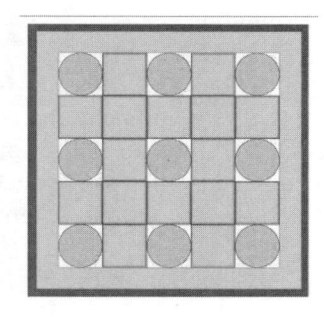

30 cm. The effective multiplication factor of this core without safety rods is $k_{eff} = 1.0071 \pm 0.0005$ and is thus slightly supercritical. Possible bias due to inaccurate libraries is evaluated to be about 1 500 pcm. Further calculations will be made to obtain the optimal configuration with the available fuel mass. To obtain criticality the number of fuel assemblies can be adjusted and the remaining reactivity difference will be compensated by control rod position.

The core will contain dedicated positions (at the interface core-reflector) for safety rods and control rods (not yet defined in the core). The subcritical configurations will be derived from this critical configuration by removing peripheral fuel assemblies (or insertion of control rods).

Figure 3. Vertical cross-section of the fast lead core with the core region and axial and radial lead reflectors

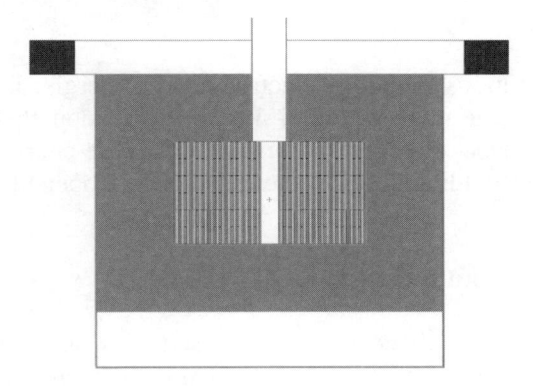

Description of the accelerator room

To allow a vertical penetration of the beam line into core, a new room has to be constructed on top of the VENUS hall. This "accelerator room" will have a direct penetration through the ceiling of the bunker enabling the accelerator tube to enter the reactor vertically. Shielding was preferred to be locally installed around the area of the bending instead of having a heavier shielding for the walls of the accelerator room to minimise the weight of the entire structure. The power supply and ion source were located outside the local shielding, since the primary neutron and gamma source is linked to the streaming of neutrons and gammas through the penetration hole in the ceiling.

The current shielding material of the roof of the current VENUS bunker consists of 12-cm barite concrete and 24-cm paraffin. The 12-cm barite concrete will be kept, but the paraffin will be removed. Instead an additional layer of 50-cm barite concrete is proposed together with a 25-cm layer of ordinary concrete. The shielding walls in the accelerator room are proposed to have a thickness of 40 cm and of barite concrete. Detailed shielding calculations will determine the final choice of thicknesses and materials for the shielding.

A stairway provides the entrance from the lower level and allows the personnel to enter the accelerator room. Large equipment and accelerator pieces can be lifted into the accelerator room from the floor of the VENUS reactor hall through an opening in the work floor of the accelerator room by means of a new tackle installed in the accelerator room. This new tackle allows transporting the pieces towards the accelerator through-put where the accelerator will be assembled and installed.

Figure 4. Cross-section of the VENUS hall layout with the additional accelerator room

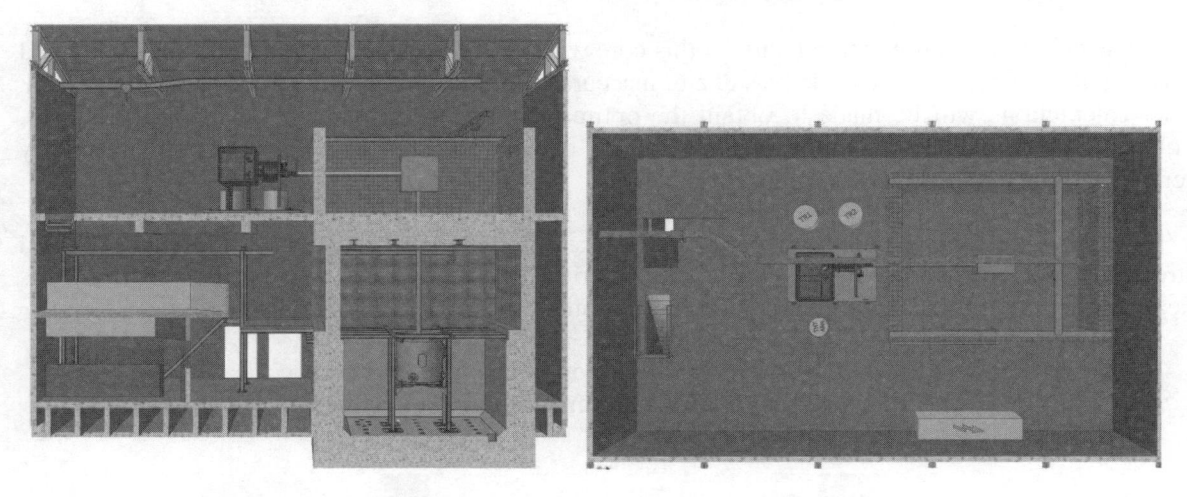

When the reactor is in operation, meaning "electromagnets energised", access to the accelerator room will be prohibited by physical and procedural means. When nonetheless such an unauthorised access is gained, the reactor will be shutdown automatically by de-energising the safety rods as is presently also the case when the shielding doors of the casement are opened.

Description of the core internals and fuel assembly

The internals of the vessel will mainly consist of fuel assemblies surrounded by a radial and bottom lead reflector. In Figure 5, a cut of the core is shown visualising different fuel assemblies supported by a stainless steel grid. This stainless steel grid rests on the vessel bottom by means of a dedicated support structure containing the bottom lead reflector. The central position is left open to allow the insertion of the accelerator beam tube. Around this central position there a lead buffer will be installed (at the place of the eight surrounding fuel assembly positions) in case of subcritical configurations to degrade the spectrum of the 14 MeV source neutrons.

Figure 5. Schematic view of the core internals (left) and the fuel assembly (right)

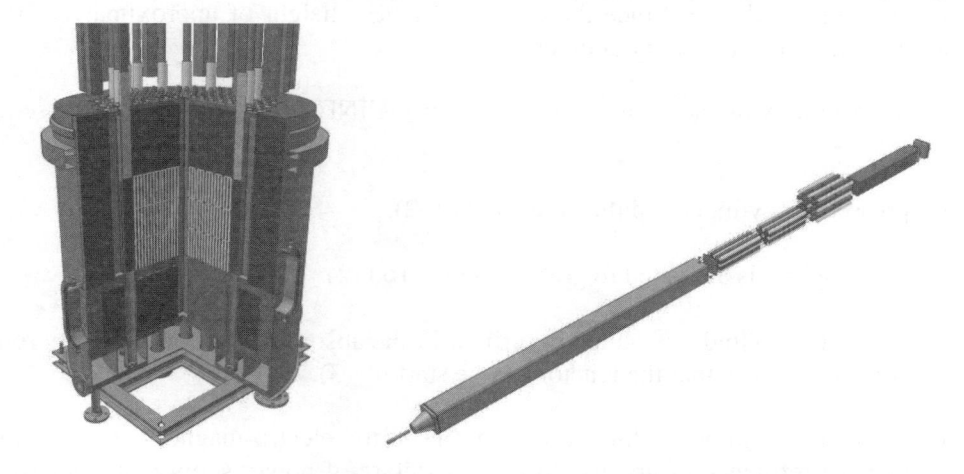

At the moment, eight positions for safety rods are foreseen at the interface of the core and radial lead reflector. Below the core, penetrations are foreseen for the four ionisation chambers which are currently used in the VENUS reactor and which will be reinstalled in the lead VENUS core beneath the core at approximately the same position. Also, the two BF_3-detectors for the start-up chains currently present in the VENUS reactor will be reinstalled at their positions in the radial reflector. A tube for the insertion of the Am-Be start-up source will also be foreseen.

Figure 5 shows a more detailed picture of a fuel assembly consisting of an upper part and lower part. The upper part simulates the upper lead reflector, whereas the lower part contains the fuel and lead rodlets in a square grid structure as outlined earlier. At the bottom, a conic end-plug is foreseen to allow an easy loading and fixation of the fuel assembly in the stainless steel grid support plate. At the upper part, an eye is foreseen for lifting the fuel assembly and positioning it into the grid support plate.

Description of the shutdown system based on safety rods

The purpose of the shutdown system is to provide sufficient anti-reactivity insertion in a time which is short enough to stop the chain reaction without introducing any damage to the installation. Moreover, the system should be intrinsically safe. Therefore, we have chosen to use the standard philosophy of shutdown rods which fall in the core under the influence of gravity upon receiving the signal for de-energising of the electro-magnets.

Such a system was already installed at the VENUS facility during the first years of its operation. Afterwards, the shutdown system was adapted and modified into a water dump shutdown system based on the opening of safety valves.

The presently proposed concept is based on a limited number of safety rod mechanisms (e.g. 8) for safety rods. The mechanisms will be distributed symmetrically on the periphery of the core. The safety rods will be placed in the reflector very close to the core in order to maximise its reactivity effect and avoid core perturbation. The safety rods will drop in a guide to assure that the free fall is mechanically guaranteed under a large set of (incidental) conditions.

The safety rods consist of an absorbing element followed by a lead follower. In case the safety rod is up, the lead follower is at the height of the core thereby eliminating reflector perturbation. In this way, the anti-reactivity insertion in case of rod drop will also be enhanced: removal of reflector and replacement by absorber. Of course, this means that at the position of the safety rod, we will have a hole in the lower lateral reflector. Since the core will have a height of approximately 60 cm, the safety rods will fall over a distance of about 60 cm.

To "start" the reactor with the shutdown system in GUINEVERE, the following sequence is foreseen (see Figure 6):

- The compressed air cylinder is pulled out (down) (2).

- The magnet coupling is activated to grab the safety rod (2).

- The compressed air cylinder is pulled in (up) to lift the absorbing part of the safety rod out of the reactor in such a way that the reactor can be start up (3).

In case the reactor needs to be shutdown, the current to the electro-magnet is cut and the safety rod falls down under the influence of gravity. In case of failure of power supply, the electro-magnets will be de-energised and the rods will drop automatically.

Figure 6. Working principle of the envisaged shutdown system for GUINEVERE

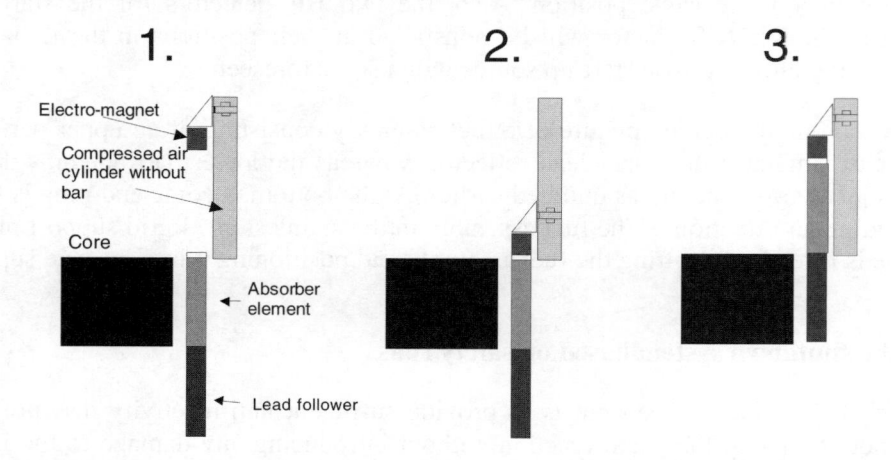

The dimensions of the safety rods were chosen to fit in the space of one fuel assembly in order to have maximum flexibility. The safety rods are 7 cm in diameter and consist of an active part of 60 cm made of B_4C with a density of 2.4 g/cc. The B_4C active part is followed by a lead follower of 60 cm.

The reactivity effect of the introduction of the eight identical safety rods by replacing lead with B_4C is of the order of 5 000 pcm. The reactivity effect of one safety rod therefore yields 625 pcm.

The reactivity weight of the safety system can be compared with the reactivity weight conditions of the safety rod system in the MASURCA facility at CEA Cadarache. The MASURCA reactor is a fast spectrum research reactor with a maximum power of 5 kW for the study of fast reactor systems: SFR and GFR. During the 5FP MUSE programme MASURCA also consisted of a significant central part of lead. In MASURCA, the minimum required reactivity weight of the safety rods is 2 000 pcm or -10$. We notice that our reactivity weight introduction of 5 000 pcm (or -7.7 $) largely satisfies the minimum criterion of 2 000 pcm. We can therefore preliminarily conclude that the reactivity weight of the safety rods is sufficient. The definitive analysis of the efficiency of the safety rods with regard to reactivity weight and speed of insertion will be made in view of the different possible positive reactivity insertion.

Evaluation of cooling requirements

The aim was to assess the temperature profile in the GUINEVERE lead core and reflector, supposing the system is only cooled by natural convection from the surrounding air in the room. In case the maximum lead temperature exceeds the lead melting temperature (327.5°C), the core needs additional cooling by forced convection. The analytical model used for this analysis is based on a similar configuration as described earlier, but did not reflect all details. However, it allowed obtaining a good temperature profile estimate. The analytical radial model is divided into several concentric circles, indicating the thermal conductivity discontinuities. A particular difficulty in this configuration arises from the modular structure of the fuel assemblies. Since each assembly consists of 16 lead rodlets and 4 lead plates, they might or might not be separated by (air) gaps, which introduce thermal resistances in the heat flux path. We could neglect these air gaps and admit that the heat conductivity equals that of homogeneous lead (k_{Pb} = 35.3 W/m/°C), but it would not be conservative. Since the effective heat conductivity $k_{Pb\text{-}eff}$ in radial direction also depends on the air gaps, it is more realistic (and conservative) to use a so-called effective conductivity defined as:

$$k_{Pb-eff} = \frac{2t_b + \sqrt{n_p}\, z_b}{\dfrac{2t_b + \sqrt{n_p}\, z_b}{k_{Pb}} + \dfrac{n_i \delta_a}{k_l}}$$

with t_b being the thickness of lead plates (6.35 mm), n_p the number of positions in one assembly (25), z_b the thickness of lead blocks (12.7 mm), n_i the number of intra-assembly air gaps (8) and δ_i the air gap thickness (0.3 mm).

The values of the number of intra-assembly air gaps and air gap thickness are chosen conservatively.

Figure 7 shows the radial temperature profile for a 100-W core power with an effective heat conductivity for the fuel assembly of $k_{Pb.eff}$ = 1.04 W/m/°C which corresponds to 76.2 mm lead in combination with 8 × 0.3 mm air gaps.

Figure 7. Radial temperature profile for a core of 100 W with $k_{Pb\text{-}eff}$ = 1.04W/m/°C

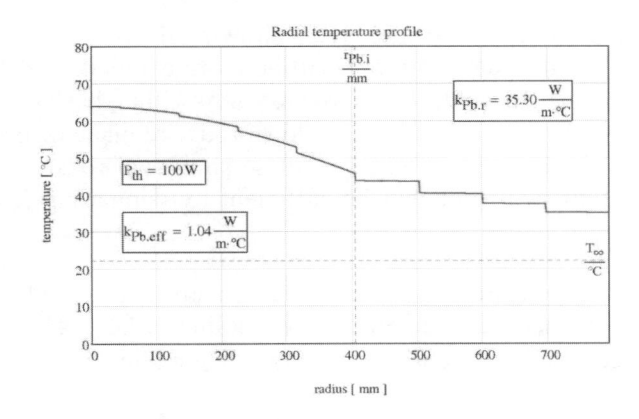

The lead only reaches a temperature of 65°C and hence lead melting is not present.

It must be noticed that the effectiveness of the vessel convective cooling considerably affects the maximum lead temperature. By enhancing the convective vessel cooling (e.g. by forced cooling instead of natural convection), the maximum core temperature might decrease by ten degrees for

100-W core power. Likewise, the temperature jumps in the reflector, caused by the radial gaps between the reflector modules, have a big impact on the core temperature. It is obvious that "pie-shaped" mono-blocks without any radial gap result in lower core temperatures.

Seeing the impact of $k_{Pb\text{-}eff}$ on the results, the value for $k_{Pb.eff}$ was investigated experimentally. The experiments pointed out that indeed a realistic value for the effective heat conductivity of a sandwich of lead plates is around 1-4 W/m/°C.

Based on the current assumptions, it can be concluded that the proposed configuration with a power level of 100 W with natural air cooling always allows to adequately evacuate the heat from the core without reaching the lead melting temperature (still a safety margin of 250°C).

Conclusions

The GUINEVERE project is an international project in the framework of FP6 IP-EUROTRANS. IP-EUROTRANS aims at defining the design of a demonstration facility for the accelerator-driven system concept, XT-ADS, and solving the main R&D issues related to the operation of such a facility by 2009. The determination and on-line monitoring of the subcriticality level is one of the major R&D issues to be addressed in this perspective. To investigate this topic a fast lead reactor coupled to a flexible neutron generator is needed.

The GUINEVERE project aims to fill this gap by coupling a modified GENEPI accelerator (CNRS design) to a modified VENUS reactor. The main modifications are the coupling of an accelerator and the adaptation of the VENUS facility to become a fast lead reactor. To couple the accelerator, a new room will be constructed on top of the VENUS reactor hall enabling a vertical penetration of the accelerator beam. For the adaptation of the VENUS facility into a fast lead reactor, a new shutdown system has to be installed, fuel assemblies and lead reflector pieces constructed.

A critical configuration based on fuel assemblies surrounded by an axial and radial lead reflector was calculated. The fuel assemblies contain lead plates, lead rodlets and fuel rodlets. The fuel rodlets are made of 30% enriched uranium metal. The configuration contains a central hole for the penetration of the beam tube which is surrounded by a limited lead zone to degrade the spectrum of the 14-MeV neutrons.

Drawings have already been made of the core layout showing the fuel assemblies, axial and radial reflectors and grid support structure. Later on these will be more detailed. A detailed drawing of the fuel assembly has already been made. The new shutdown system will be based on a limited set (e.g. 8) of shutdown rods at the periphery of the core. The shutdown rods are made of B_4C absorbing material followed by a lead-follower. The reactivity weight of these shutdown rods was calculated as about 5 000 pcm. This reactivity weight will probably be sufficient by comparing it to the criteria in the MASURCA installation.

Also, an assessment was made of the coolant requirement for a 100-W core. Based on a simplified model, it can be concluded that no forced air cooling is necessary and that a margin of 250°C is still present before lead melting occurs in the hottest assemblies.

Acknowledgements

This work was largely performed within the framework of the FP6 IP-EUROTRANS contract sponsored by the European Commission.

RESEARCH AND DEVELOPMENT PROGRAMME ON ADS IN JAEA

Hiroyuki Oigawa, Kenji Nishihara, Toshinobu Sasa, Kazufumi Tsujimoto, Takanori Sugawara, Kohei Iwanaga, Kenji Kikuchi, Yuji Kurata, Hayanori Takei, Shigeru Saito, Dai Hamaguchi, Hironari Obayashi, Yujiro Tazawa, Masao Tezuka, and Nobuo Ouchi
Japan Atomic Energy Research Institute, Japan

Abstract

JAEA has been promoting research and development (R&D) on accelerator-driven subcritical systems (ADS) as a dedicated system for the transmutation of long-lived radioactive nuclides. The ADS proposed by JAEA is a lead-bismuth eutectic (LBE) cooled, tank-type subcritical reactor with a thermal power of 800 MWth driven by a 30-MW superconducting linac. The R&D activities for ADS can be divided into two categories: one is design study and basic technical development for a future large-scale ADS, and the other is an experimental programme at the Transmutation Experimental Facility (TEF) under the J-PARC (Japan Proton Accelerator Research Complex) project. As for the design study of the future ADS, reduction of the maximum temperature of fuel claddings and verification of the feasibility of the beam window are under way. Regarding the TEF programme, although its construction has not yet been approved, design study including an experimental device to deal with minor actinide fuels is being conducted. To facilitate the research and development on ADS, an international common roadmap is necessary. The TEF programme can play an important role in such an international context as an experimental platform to conduct basic and flexible experiments.

Introduction

The Japan Atomic Energy Agency (JAEA) was established in October 2005 through the fusion of the Japan Atomic Energy Research Institute (JAERI) and the Japan Nuclear Cycle Development Institute (JNC). The JAEA is continuing research and development (R&D) on partitioning and transmutation (P&T) technology to reduce the burden for the management of high-level radioactive wastes (HLW) [1]. The accelerator-driven system (ADS) dedicated to the transmutation of long-lived nuclides such as minor actinides (MA) and long-lived fission products (LLFP) is mainly studied by the Nuclear Science and Engineering Directorate of JAEA through collaboration with the J-PARC (Japan Proton Accelerator Research Complex) Centre and the Advanced Nuclear System Research and Development Directorate which is in charge of the R&D for fast breeder reactors and related fuel cycles.

The R&D activities on ADS in JAEA are divided into two categories: one is design study and basic technical development for a future large-scale ADS, and the other is the planning of an experimental programme at the Transmutation Experimental Facility (TEF) under the J-PARC project. The present status and future perspective of these activities are described in this report. In addition, the importance of an international collaboration to share a common roadmap toward the realisation of transmutation technology using ADS is discussed.

Design study of future ADS

The ADS proposed by JAEA is an 800-MWth, lead-bismuth eutectic (LBE) cooled, tank-type subcritical reactor driven by a superconducting proton linac [2-4]. The basic design parameters are summarised in Table 1, and the conceptual view of the core region is shown in Figure 1. The engineering feasibility of such a system is being discussed mainly from three aspects: (1) the maximum temperature of the fuel claddings; (2) the structural feasibility of the beam window; (3) the tolerability against frequent beam trips. The latter issue is discussed in another report in the workshop [5].

Table 1. Basic design parameters of 800 MWth, LBE-cooled ADS

Parameters	Specification
Thermal power	800 MW
Electric power	270 MW
Coolant	LBE
Spallation target	LBE (window-type)
Accelerator	Superconducting linac
Proton beam energy	1.5 GeV
Active core diameter	234 cm
Active core height	100 cm
Fuel	(MA, Pu)N
Inert matrix	ZrN
Effective multiplication factor (k_{eff})	Max.: 0.970
Cycle length	600 EFPD
Transmutation rate	500 kgMA/cycle

Figure 1. Conceptual view of the core region of 800-MWth LBE-cooled ADS [4]

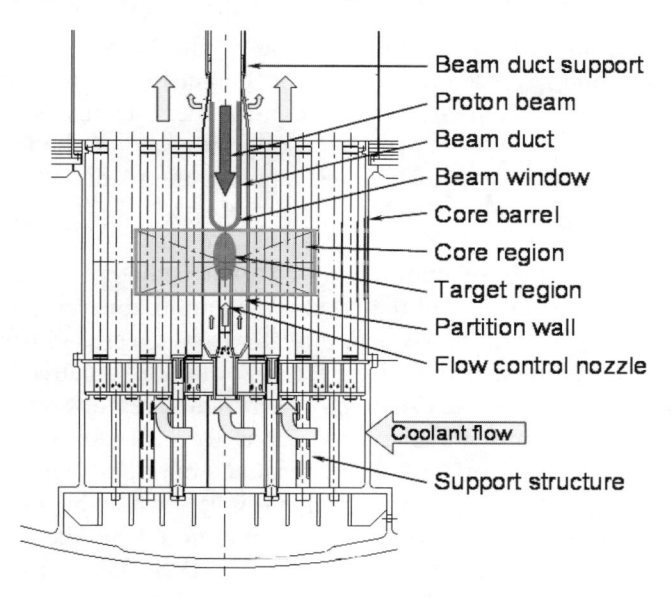

Reduction of maximum temperature of fuel claddings

In general, the core design of ADS is confronted with a large power peaking near the spallation target because of the subcriticality. This difficulty will be largely mitigated if the effective multiplication factor (k_{eff}) can be kept at relatively high value (e.g. 0.97).

Figure 2 shows the burn-up swing of k_{eff} and the radial temperature distributions of a typical ADS (single fuel zone) without active control of the k_{eff} and the power flattening, where inert matrix content (66.5 vol.% of fuel) is fixed throughout the fuel cycles and plutonium content (36.2 vol.% of heavy metal) at the first cycle is chosen to adjust the k_{eff} value to 0.97 at the beginning of cycle (BOC) of the equilibrium cycle without additional loading of plutonium into recycled fuel [6]. The maximum temperature of about 600°C at the surface of the fuel cladding is observed at the end of cycle (EOC) of the second fuel cycle where the minimum k_{eff} value of 0.935 is observed. As is well known, the corrosion of steel materials by LBE will be serious at this temperature range.

Figure 2. Core characteristics of a single-zone reference core [6]

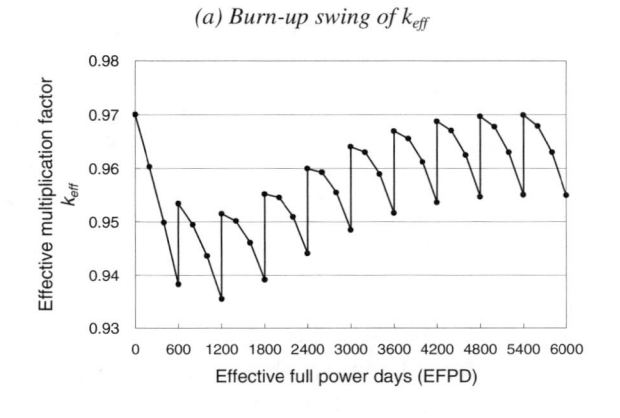

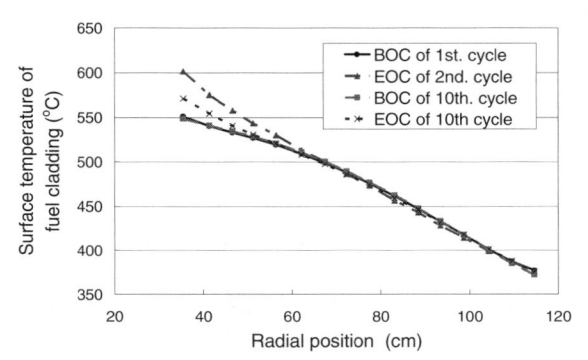

(a) Burn-up swing of k_{eff}

(b) Radial temperature distribution of fuel claddings

The design modification is, therefore, being investigated considering various methods: (1) control of burn-up swing of k_{eff} by changing fuel compositions for individual fuel cycle; (2) multi-zone configuration of the core in terms of fuel compositions and/or pin diameters; (3) optimisation of the vertical position of the beam window; (4) adoption of short-height fuel assemblies near the spallation target; (5) adoption of control rods or other reactivity control mechanisms. The optimisation of the core design should be done by coupling these methods taking into account the feasibility of fuel cycle processes including quality assurance in fabrication and reprocessing plants, the cost and the safety of the whole system, and so on.

One possible example obtained by coupling the first two methods, i.e. the adjustment of fuel compositions for the individual fuel cycle and the adoption of multi-zone configuration, is described below. At first, the k_{eff} values of BOC of the second and later cycles were adjusted to 0.97 by putting less plutonium at the first cycle and by adjusting the fractions of the inert matrix (ZrN) at BOC of each cycle. Then the core configuration was divided into four radial regions to flattening the power distribution. Table 2 shows the optimised fuel compositions for each zone and cycle. The volume fraction of ZrN varies over the range 50-78%. The resultant burn-up swing and the temperature distribution are shown in Figure 3. The minimum keff value is 0.958, while that of Figure 2(a) was 0.935. This improvement contributes to the reduction of the beam current to keep the reactor power constant. The maximum beam power was reduced form 34 MW to 27 MW. The maximum temperature was remarkably reduced to 493°C which is about 100°C lower than that in Figure 2(b).

Table 2. Fuel compositions of an improved four-zone core [6]

Parameters	Zone 1	Zone 2	Zone 3	Zone 4
Inner/outer radius (cm)	32.49/53.53	53.53/74.70	74.70/95.91	95.91/117.15
Initial Pu content (vol.%)	30.1	30.1	30.1	30.1
ZrN content (vol.%)				
1st cycle	71.1	64.2	59.3	50.8
2nd cycle	75.7	68.3	63.1	54.1
3rd cycle	75.8	68.5	63.2	54.2
4th and later cycles	77.8	70.2	64.9	55.6

Figure 3. Core performance of an improved four-zone core [6]

(a) Burn-up swing of k_{eff}   *(b) Radial temperature distribution of fuel claddings*

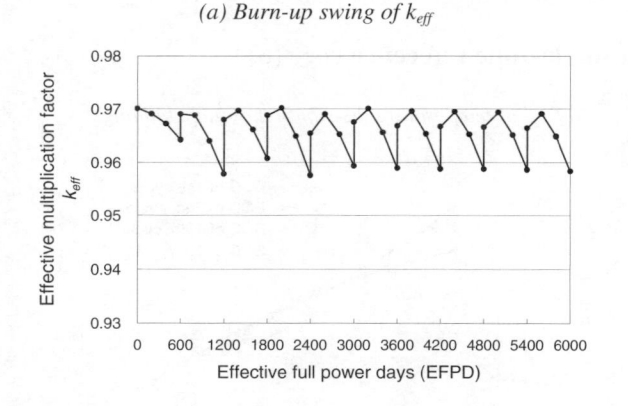

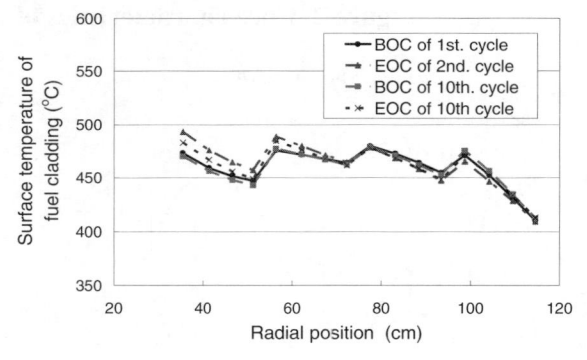

The above-mentioned discussion is based on the nominal calculation results. To design realistic ADS, several safety factors arising from the uncertainties of the calculation should be taken into consideration. In such uncertainties, the accuracy of the nuclear data must be carefully considered because the burn-up swing of k_{eff} is the result of a delicate balance between capture and fission reactions of MA and Pu isotopes. Although the nuclear data of these nuclides are being measured and evaluated extensively, it is considered necessary to conduct the integral verification of the nuclear data and the analysis method for the reactor physics parameters based on critical/subcritical experiments.

Feasibility of beam window

The beam window is unquestionably the most critical part of ADS, unless a windowless concept is adopted. Its feasibility should be discussed in terms of structural strength, corrosion and erosion, temperature, irradiation damage and so on. Our reference design of the beam window is shown in Figure 4 [7,8]. The beam window will be exchanged periodically, preferably every two years.

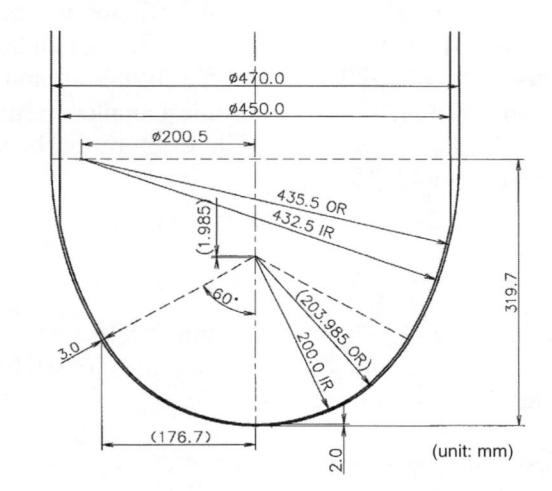

Figure 4. Reference beam window design [7,8]

The inner diameter of the beam duct is 45 cm, and the thickness of the window bottom is 2 mm, which is determined by considering the balance between the thermal stress caused by the heat generation inside the beam window and the structural strength against the LBE pressure (about 0.8 MPa). To assure the robustness against the buckling failure, the curvature of the window bottom is formed with a partial spherical shell whose inner radius is 20 cm.

The candidate material for the beam window is assumed as Mod.9Cr-1Mo steel. The irradiation damage in the beam window was estimated as 4.5-9.0 DPA/year by protons and 55-98 DPA/year in total depending upon the burn-up swing of k_{eff} and the beam current [9]. The helium gas production in the materials was estimated as 950-1 900 appm/year. In these calculations, the Gaussian distribution was assumed for the proton beam profile. The irradiation damage and the corrosion by LBE are major uncertain factors to determine the lifetime of the beam window. In JAEA, therefore, various experimental activities are being continued to investigate the feasibility of the beam window under the realistic operation conditions of the ADS.

Technical development for ADS

Various technical developments for ADS are under way in JAEA in the field of the LBE and related materials including spallation materials.

The static corrosion test and the loop corrosion test are two major activities for the material compatibility in LBE [10,11]. Recently, steel alloys modified to enhance the corrosion resistance by adding silicon were made and are being tested in the static device. On the other hand, in the loop test, the transfer of material compositions caused by the temperature heterogeneity and the differences of solubility is investigated by observing the deposits on the inner surface of the loop piping.

As for the irradiation damage of the structural materials by protons and neutrons, JAEA is participating in the irradiation programme for the spallation target materials using the SINQ facility at the Paul Scherrer Institute in Switzerland. Small pieces of samples irradiated by 580-MeV protons were transported to JAEA, and the tensile test and the fatigue test are being conducted at the hot cell facility. The results for the first phase of the programme for the specimens irradiated up to about 10 DPA have already been reported [12,13]. The tests for the second phase are under way for the austenitic steel irradiated up to about 20 DPA.

With regard to the test of the cooling performance of LBE for the beam window of ADS, a thermal-hydraulic loop test was also performed. The test was carried out under conditions in which the flow rate of LBE was changed as 100, 200, 300, 400 and 500 litre/min, and the inlet temperature of LBE was changed as 330, 380 and 430°C [3]. The Ultrasonic Doppler Method (UDM) is also being developed to measure the local LBE flow speed [14]. The evaporation behaviour of polonium and other elements from LBE was also measured [15].

J-PARC project

JAEA is conducting a multi-purpose high-intensity proton accelerator programme called J-PARC collaborating with the High Energy Accelerator Research Organization (KEK) [16]. The conceptual view of the accelerators and the facilities is shown in Figure 5.

The proton accelerators consist of a linac and two synchrotrons (3 GeV and 50 GeV). The proton energy of the linac is 400 MeV in Phase I of the project and will be upgraded to 600 MeV in Phase II. The experimental facilities consist of the Materials and Life Science Experimental Facility, the Hadron Beam Facility and the Neutrino Facility in Phase I, and the Transmutation Experimental Facility (TEF) in Phase II.

Figure 5. Conceptual view of J-PARC accelerators and facilities [17]

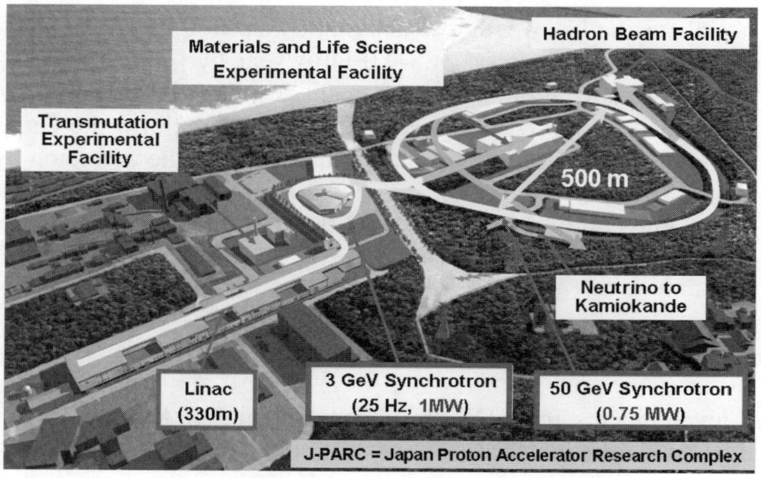

Present status of construction and commissioning of linac

The construction and installation of the linac part were completed up to 181 MeV, which is the incident energy for the 3-GeV synchrotron at the beginning of the Phase I operation. The linac energy is scheduled to be upgraded to 400 MeV which is the design energy for Phase I. The linac part is now in the commissioning phase. On 24 January 2007, the linac successfully achieved the planned energy of 181 MeV. Figure 6 shows the photos of linac part. It consists of an ion source, radio frequency quadrapole (RFQ) linac, drift tube linac (DTL), and separated-type DTL (SDTL) up to 181 MeV. The annular coupled structure (ACS) linac is being prepared for the energy range from 181 MeV to 400 MeV. The beam current of linac will be 0.33 mA in average, and operated in pulsed mode with 25 Hz in the final stage of Phase I. It will then be upgraded to 50 Hz in Phase II, where half of the pulses will be supplied to TEF.

Figure 6. Installation of J-PARC linac (as of December 2006)

(a) Low-energy part (RFQ and DTL)   *(b) Medium-energy part (SDTL up to 181 MeV)*

Transmutation Experimental Facility

The original design of TEF consists of two buildings: the Transmutation Physics Experimental Facility (TEF-P) and the ADS Target Test Facility (TEF-T) [18]. TEF-P is a zero-power critical facility wherein a low-power proton beam is available to research the reactor physics and the controllability of the ADS. TEF-T is a material irradiation facility which can accept a maximum 200 kW-600 MeV proton beam into the spallation target of LBE to conduct the material irradiation and target engineering tests.

Step-wise construction of TEF

Although the original concept of TEF is to construct both TEF-P and TEF-T simultaneously at the adjacent areas, the separation of TEF into two parts is now being discussed in order to start basic experiments as early as possible, considering the shortage of construction budget. The basic idea is to separate TEF into:

- Part 1: TEF-P (with 400-MeV proton beam) and a beam dump.

- Part 2: TEF-T and a 600-MeV superconducting linac.

In the first part (Part 1), about 30-kW (maximum) proton beam with 400 MeV will be available. Because the power of proton beam is too large to introduce it directly into TEF-P, a beam dump to accept the full power proton beam must be installed. A laser charge exchange device will be used to extract a 400-MeV, 10-W beam which is transported to TEF-P. The beam dump as well as TEF-P can be used for various experimental purposes. The conceptual view for this Part 1 facility is shown in Figure 7.

Figure 7. Conceptual view of Part 1 of Transmutation Experimental Facility (TEF)

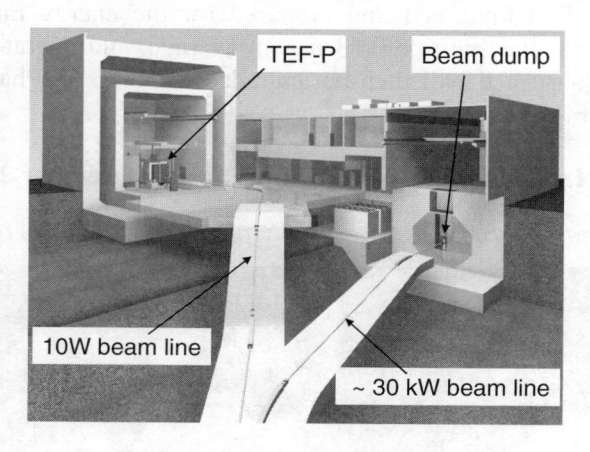

In the second part (Part 2), a 600-MeV, 200-kW (maximum) proton beam accelerated by a superconducting linac will be introduced into TEF-T. A laser charge exchange device will extract a 600-MeV, 10-W beam for TEF-P. The proposed layouts of the facilities in Parts 1 and 2 are shown in Figure 8.

Figure 8. Concept of step-wise construction of Transmutation Experimental Facility (TEF)

(a) Part 1 (TEF-P and a beam dump)   *(b) Part 2 (TEF-T and a superconducting linac)*

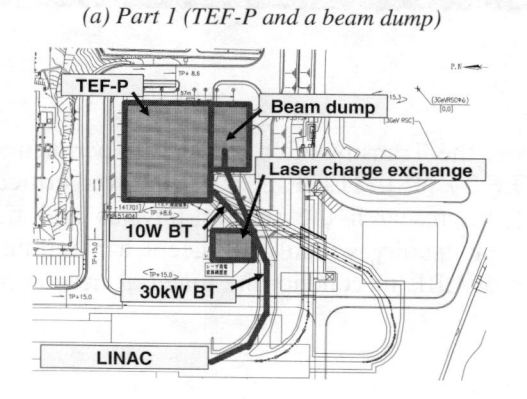

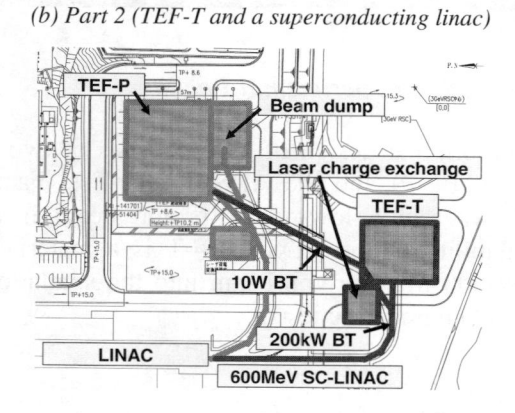

Specification of TEF-P

TEF-P is designed as the horizontal table-split type critical facility with rectangular lattice matrix, referring to Fast Critical Assembly (FCA) in JAEA/Tokai. Although the regular fuels of TEF will be plate-type, a partial mock-up region using pin-type fuels will also be possible. Figure 9 shows a schematic view of the partial loading of pin-type minor actinide (MA) fuels around spallation target. The central rectangular region (28 cm × 28 cm) will be replaced with hexagonal subassemblies. In order to measure the physics parameters of the transmutation system, the pin-type fuel containing MA is

Figure 9. Conceptual view of assembly of TEF-P and Pb block for pin fuel loading

(a) Conceptual view of assembly   *(b) Pb block for pin fuel loading*

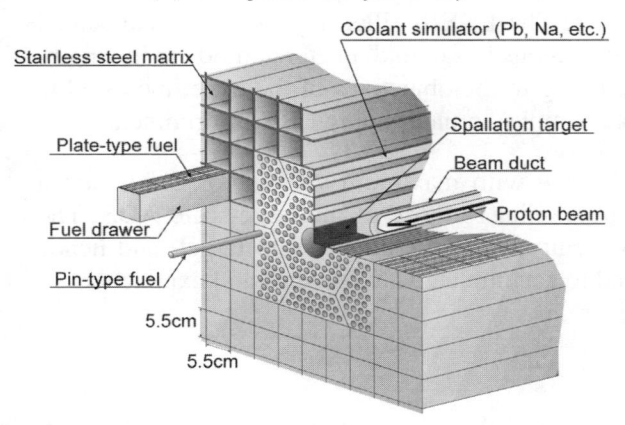

considered indispensable. To manage the decay heat and the radiations of MA fuels, TEF-P will equip an air-cooling system and a remote handling system. The power level of TEF-P is usually of the order of 10-100 W from the viewpoints of accessibility to the core and the maximum thermal power is temporarily fixed as 500 W. TEF-P is designed to contribute not only to the ADS development but also to the fast reactor development by the experiments in critical state.

In the proton beam experiment, the effective multiplication factor, k_{eff}, of the critical assembly will be kept less than 0.98. The proton beam is introduced horizontally from the centre of the fixed half assembly. For spallation target, solid materials such as lead and tungsten will be used. In TEF-P, it is also possible to perform the spallation target neutronics experiments without nuclear fuels.

Low current proton beam is extracted by a laser charge exchange technique from high-intensity beam line of 30 kW (0.075 mA, 400 MeV) in Part 1 and 200 kW (0.33 mA, 600 MeV) in Part 2, most beam is introduced into the beam dump or TEF-T. A pulse width of the proton beam delivered to TEF-P will be varied from 1 ns to 0.5 ms in accordance with the requirements of the experiments. The proton beam intensity can be controlled by a collimator device which will be installed in beam control area of TEF-P. This area can be used for various experiments that require the low power proton beam or very short pulse beam such as the time-of-flight measurement.

Experiments using MA fuels

To make meaningful experiments using MA fuels, a certain amount of MA fuels are necessary. For example, if a central test region of 28 cm × 28 cm × 90 cm (about 70 litres) is loaded with MA fuel, about 10 kg of MA is necessary to simulate 5% MA-added MOX fuels for fast reactors and about 40 kg of MA for ADS dedicated fuels.

The representativity and the effectiveness of the experiment are being discussed by coupling the covariance of nuclear data and the sensitivity analysis [19]. At the present time, however, this estimation does not correctly represent the usefulness of the experimental data because the current error data of MA nuclides are not fully provided and some data seem to have too small errors. It is, therefore, necessary to improve the error data as well as the nuclear data so as to discuss the necessity of the integral validation of the MA nuclear data.

Specification of TEF-T

TEF-T mainly consists of an LBE spallation target, an LBE primary loop, a helium gas secondary loop and an access cell to handle irradiation test pieces. The LBE is filled into a cylindrical sealed double-layer tube target. An effective size of the target is about 15 cm in diameter and 60 cm in length. Several kinds of target are planned and designed according to the objectives of the experiments. One of the target vessels is designed to irradiate ten or more samples in the flowing LBE environment.

A primary LBE loop is designed to allow LBE flow with maximum velocity of 2 m/s and a maximum temperature of 723 K. These conditions are similar to those of the large-scale ADS. The outstanding point of TEF-T is that it is a dedicated facility for the R&D of ADS target, and hence experiments under various conditions can be conducted to accumulate valuable data and experience.

Specification of beam dump

The low-power proton beam should be extracted from kW-order proton beam, most of which must be absorbed by a beam dump located near TEF-P. The beam power will be set at about 30 kW considering the stable operation of linac. The beam dump is made from copper disks cooled by water. The thickness of the disks was optimised to equalise the heat deposition of the proton beam. The dump unit is about 1 m in diameter and 2 m in length. It will be possible to provide a small area for material and/or sample irradiation in front of the beam dump.

Preliminary letters of intent

Although the TEF programme is not yet funded officially, the Project Team called for preliminary letters of intent (LOI) for experiments at the TEF in the year 2006. The purposes of preliminary LOI are: (1) to know which groups have an interest in this activity and what contributions from them can be expected; (2) to reflect upon new ideas and proposals on the specifications and the layout of TEF including the beam dump; (3) to establish an appropriate collaboration scheme between J-PARC and the anticipated outside users.

Figure 10 shows the results of preliminary LOI. Thirty-seven proposals were received in total. Experiments for both ADS and MA-loaded FBR were mainly proposed. In other fields, proposals were for nuclear data measurement, high-energy physics, LBE spallation target technology, and miscellaneous researches using protons and neutrons at the beam dump. Although the detailed discussions for the proposals have not yet been started, it is clear that TEF can serve as a basic experimental platform for nuclear science, engineering and applications. The project is still open to accept other proposals.

Importance of international roadmap

ADS and transmutation technology are increasingly important for the sustainable utilisation of nuclear energy all over the world. The technical challenges for ADS, however, spread over a wide range of scientific and engineering fields. It is therefore strongly desirable to share the experimental efforts in a systematic way by many countries. The MEGAPIE project is a good precursor for international collaboration in this field.

An intermediate goal before the realisation of transmutation by the ADS must be an experimental ADS. European counties are implementing intense R&D for the XT-ADS project, which is the experimental ADS with several-ten megawatt of thermal power. It is thus desirable to extend the XT-ADS to a global programme.

Figure 10. Results of preliminary LOI

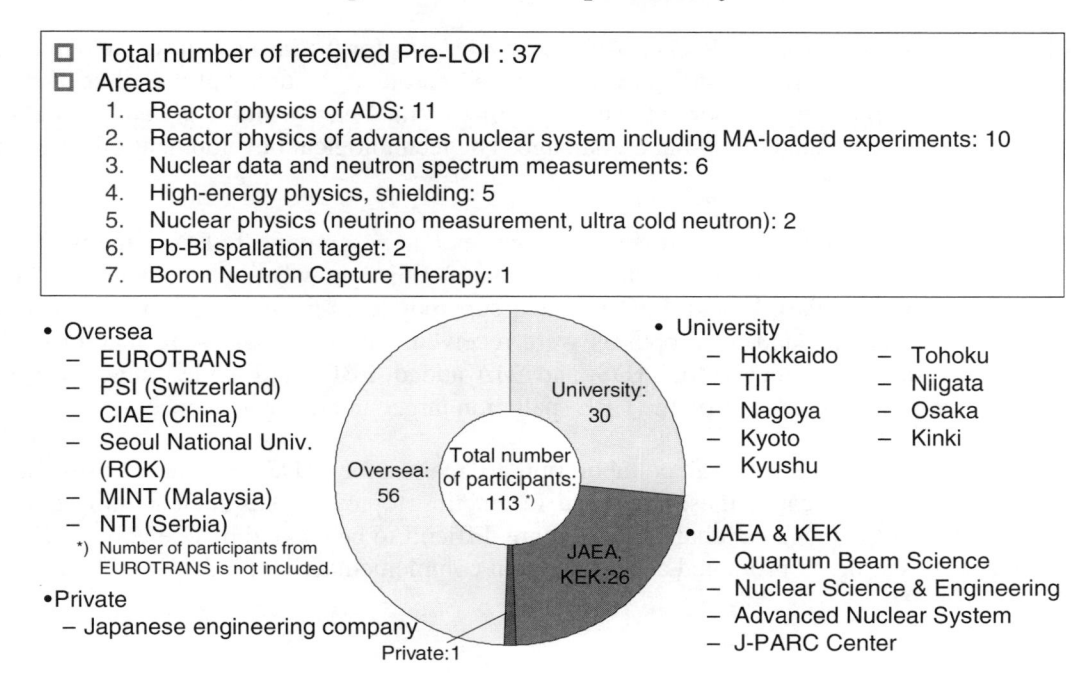

- Total number of received Pre-LOI : 37
- Areas
    1. Reactor physics of ADS: 11
    2. Reactor physics of advances nuclear system including MA-loaded experiments: 10
    3. Nuclear data and neutron spectrum measurements: 6
    4. High-energy physics, shielding: 5
    5. Nuclear physics (neutrino measurement, ultra cold neutron): 2
    6. Pb-Bi spallation target: 2
    7. Boron Neutron Capture Therapy: 1

- Oversea
    - EUROTRANS
    - PSI (Switzerland)
    - CIAE (China)
    - Seoul National Univ. (ROK)
    - MINT (Malaysia)
    - NTI (Serbia)

    *) Number of participants from EUROTRANS is not included.

- Private
    - Japanese engineering company

Total number of participants: 113 *)
University: 30
Oversea: 56
JAEA, KEK: 26
Private: 1

- University
    - Hokkaido
    - TIT
    - Nagoya
    - Kyoto
    - Kyushu
    - Tohoku
    - Niigata
    - Osaka
    - Kinki

- JAEA & KEK
    - Quantum Beam Science
    - Nuclear Science & Engineering
    - Advanced Nuclear System
    - J-PARC Center

Before proceeding to such a demonstrative stage, establishment of a technical base to deal with MA in nuclear energy system and to couple a proton accelerator with a fast-spectrum reactor is extremely important for the purpose of reliable design of the system, safety assessment and discipline of young scientists and engineers. From this viewpoint, TEF under the J-PARC project is expected to play important roles.

Figure 11 shows a draft of international roadmap for the realisation of ADS and transmutation technology. The TEF and XT-ADS will be complementary driving forces toward realising the challenging technology.

Figure 11. International roadmap toward realisation of transmutation technology by ADS

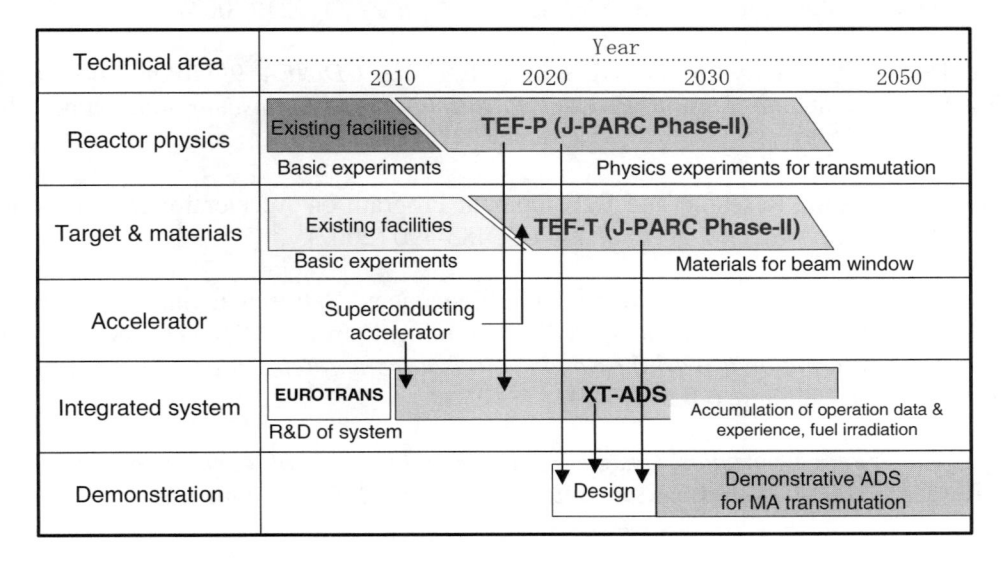

Technical area	Year			
	2010	2020	2030	2050
Reactor physics	Existing facilities / Basic experiments	TEF-P (J-PARC Phase-II)	Physics experiments for transmutation	
Target & materials	Existing facilities / Basic experiments	TEF-T (J-PARC Phase-II)	Materials for beam window	
Accelerator		Superconducting accelerator		
Integrated system	EUROTRANS / R&D of system		XT-ADS	Accumulation of operation data & experience, fuel irradiation
Demonstration			Design	Demonstrative ADS for MA transmutation

Conclusion

JAEA has been promoting various R&D activities on ADS. As for the design study, the reduction of the maximum temperature of fuel claddings was discussed and it was found that the adjustment of k_{eff} values for each burn-up cycle is important. The feasibility of the beam window is being discussed taking account of the corrosion and the radiation damage. The technological development on LBE and related materials is also under way.

Under the framework of the J-PARC project, the experimental programme TEF is being proposed as the Phase II programme of the project. The step-wise construction is now being considered for TEF; TEF-P and a beam dump as Part 1, and TEF-T and a superconducting linac as Part 2. Recently, preliminary LOI was called for and 37 proposals were received. The proposals were mainly in the fields of reactor physics experiments for ADS and MA-added FBR, and wide range of basic experiments were also proposed such as on the LBE spallation target and nuclear science.

The importance of the international collaboration to realise the ADS and the transmutation technology was emphasised because these are quite fascinating topics for sustainable utilisation of nuclear energy in spite of the technical challenges that are difficult to be tackled by one country alone. The TEF and XT-ADS should be promoted as international complementary and indispensable projects.

REFERENCES

[1] Minato, K., *et al.*, "Research and Development Activities on Partitioning and Transmutation of Radioactive Nuclides in Japan", *OECD/NEA 9th Information Exchange Meeting on Actinide and Fission Product Partitioning and Transmutation*, Nîmes, France, 25-29 September 2006.

[2] Tsujimoto, K., *et al.*, "Neutronics Design for Lead-bismuth Cooled Accelerator-driven System for Transmutation of Minor Actinide", *J. Nucl. Sci. Technol.*, 41, 21 (2004).

[3] Oigawa, H., *et al.*, "Activities on ADS at JAEA", *OECD/NEA 9th Information Exchange Meeting on Actinide and Fission Product Partitioning and Transmutation*, Nîmes, France, 25-29 September 2006.

[4] Tsujimoto, K., *et al.*, "Research and Development Program on Accelerator Driven Subcritical System in JAEA", *J. Nucl. Sci. Technol.*, 44 (3), 483-490 (2007).

[5] Takei, H., *et al.*, "Comparison of Beam Trip Frequencies Between Estimation from Current Experimental Data of Accelerators and Requirement from ADS Transient Analyses", *OECD/NEA 5th International Workshop on the Utilisation and Reliability of High-power Proton Accelerators*, Mol, Belgium, 6-9 May 2007.

[6] Iwanaga, K., *et al.*, *Neutronics Design for Power Flattening of Accelerator-driven System*, JAEA-Research 2007-025, Japan Atomic Energy Agency (2007) [in Japanese].

[7] Oigawa, H., *et al.*, "Design Study Around Beam Window of ADS", *OECD/NEA 4th International Workshop on the Utilisation and Reliability of High-power Proton Accelerators*, Daejeon, Republic of Korea, 16-19 May 2004.

[8] Saito, S., *et al.*, "Design Optimization of ADS Plant Proposed by JAERI", *Nuclear Instruments and Methods in Physics Research A*, 562, 646-649 (2006).

[9] Nishihara, K., K. Kikuchi, "Irradiation Damage to the Beam Window in the 800 MWth Accelerator-driven System", *8th International Workshop on Spallation Material Technology*, Taos, USA, 16-20 October 2006.

[10] Kurata, Y., *et al.*, "Comparison of the Corrosion Behavior of Austenitic and Ferritic/Martensitic Steels Exposed to Static Liquid Pb-Bi at 450 and 550°C", *J. Nucl. Mater.*, 343, 333-340 (2005).

[11] Kikuchi, K., *et al.*, "Lead-bismuth Eutectic Compatibility with Materials in the Concept of Spallation Target for ADS", *JSME International Journal*, Series B, 47 (2), 332-338 (2004).

[12] Saito, S., *et al.*, "Bend-fatigue Properties of 590 MeV Proton Irradiated JPCA and 316F SS", *J. Nucl. Mater.*, 329-333, 1093-1097 (2004).

[13] Saito, S., *et al.*, "Tensile Properties of Austenitic Stainless Steels Irradiated at SINQ Target 3", *J. Nucl. Mater.*, 343, 253-261 (2005).

[14] Kikuchi, K., *et al.*, "Measurement of LBE Flow Velocity Profile by UDVP", *J. Nucl. Mater.*, 356, 273-279 (2006).

[15] Ohno, S., *et al.*, "Equilibrium Evaporation Behavior of Polonium and its Homologue Tellurium in Liquid Lead-bismuth Eutectic", *J. Nucl. Sci. Technol.*, 43 (11), 1359-1369 (2006).

[16] The Joint Project Team of JAERI and KEK, *The Joint Project for High-intensity Proton Accelerators*, JAERI-Tech 99-056 (KEK Report 99-4, JHF-99-3) (1999).

[17] http://j-parc.jp/index-e.html

[18] Oigawa, H., *et al.*, "Concept of Transmutation Experimental Facility", *OECD/NEA 4th International Workshop on the Utilisation and Reliability of High-power Proton Accelerators*, Daejeon, Republic of Korea, 16-19 May 2004.

[19] Sugawara, T., *et al.*, Design of MA-loaded Core Experiments Using J-PARC, private communication (2007).

THE GENEPI NEUTRON SOURCES AT GRENOBLE: PROSPECTIVES FOR THE GUINEVERE PROGRAMME

J.M. De Conto, S. Albrand, M. Baylac, J.L. Belmont, A. Billebaud,
J. Bouvier, A. Fontenille, E. Froidefond, M. Fruneau, A. Garrigue,
M. Guisset, D. Marchand, R. Micoud, J.C. Ravel, C. Vescovi

Laboratoire de Physique Subatomique et de Cosmologie
Université Joseph Fourier Grenoble 1, CNRS/IN2P3, Institut National Polytechnique de Grenoble
53 avenue des Martyrs, F-38026 Grenoble Cedex, France

Abstract

In 2001, the first coupling between the subcritical MASURCA reactor (Cadarache) and the GENEPI-1 accelerator was achieved. GENEPI is a deuteron, high-intensity, electrostatic accelerator (250 keV, 40 mA peak intensity, 1 µs pulse, 5 kHz repetition rate) producing neutrons on a tritium target located in the reactor core. A second machine was built at LPSC-Grenoble in 2003 for neutron studies dedicated to innovative options for nuclear energy. A general and technical presentation of the GENEPI machines is given, and is placed in the technical prospective of the future GUINEVERE programme, where a new accelerator (GENEPI-3C) will be set up. This accelerator will have the characteristics of the previous machines, with new specifications: continuous beam (low intensity), possibilities of micro beam interruptions ("beam trips") for reactivity ADS monitoring. This work has been undertaken within a collaboration since the beginning of 2007 with other laboratories: LPC Caen, IPN Orsay and IPHC Strasbourg.

Introduction

The GENEPI (*GEnérateur de NEutrons Pulsés Intenses*) machines are a family of electrostatic neutron generators dedicated to accelerator-driven systems (ADS) studies. They are motivated by the necessity to extend the validation of calculation and measurement methods already made for critical fast reactors to ADS and/or to new reactor concepts based on new fuels. The first GENEPI machine was designed and built by the CNRS/IN2P3/LPSC (formerly ISN) laboratory. It was a specific pulsed 250-keV deuteron accelerator (1 µs or below, sharp edges pulses, no tails, 40-50 mA peak intensity). Neutrons (2.5 or 14 MeV) were produced by the fusion (D+D) or (D+T) reactions of the incident deuterons hitting a deuterium or tritium target located in the reactor core. The GENEPI-1 machine was designed and built from 1996 to 1999, and the final implantation at the Cadarache MASURCA reactor was completed in June 2000, and then dedicated to the MUSE4 experiments.

GENEPI-1 was partially dismantled in 2005. In parallel, GENEPI-2 was built in Grenoble, where it is still under operation. It remains quite similar to GENEPI-1, and is part of the PEREN facility, dedicated to the study of nuclear cross-sections for Generation IV reactor concepts.

The GENEPI-3C ("continuous") machine will be coupled to the VENUS reactor in Mol (Belgium) for the GUINEVERE programme planned for 2009. The main differences with the two first machines will be the topology (vertical coupling instead of horizontal) and the additional possibility to have a continuous beam with small interruptions inside ("beam trips"), at the millisecond level.

GENEPI-1

As already mentioned, the main objective of GENEPI-1 was the production of high-intensity, sharp-edged 14-MeV neutron pulses. Despite existing neutron generators commercially available, a new design was needed to generate short enough and, mainly, sharp-edged pulses: for example, the commercial generators deliver long, tailed pulses (around 1 ms), which are not compatible with the time resolution needed for fast reactor experiments.

Table 1. Summary of GENEPI-1 features

Peak current (deuterons)	~50 Ma (@10Hz-5kHz)
Mean current	100 µA at 4.7 kHz
Beam energy	140-240 keV
Pulse deuteron length	0.5-0.7 µs (mid height)
Neutron energy	2.5/14 MeV
Spot diameter	20-25 mm
Target	Deuterium or tritium/titanium
Neutron production (peak)	~5 10^6 n/pulse
Neutron production (average)	~25 10^9 n/s at 5 KHz
Reproducibility	1% from pulse to pulse

The pulsed ion source

The main feature of GENEPI is its deuteron source. Due to the very short pulses required, a duoplasmatron source has been chosen (rather than an ECR source, for example). The source (Figure 1) consists mainly of:

- an impregnated cathode (filament), to provide an electron arc;

- a first triggering electrode of conic shape, the "cone", with a small aperture on the beam axis;

- an anode;

- external coils, which provide an axial magnetic field, for plasma confinement.

Figure 1. Duoplasmatron

Under normal operation, a very small plasma is created between the cone and the cathode, where the deuterium gas is ionised before extraction. Deuterons are produced, as well as molecular ions D_2^+, the ratio depending on the plasma density. Typically, for GENEPI-1, the ratio is about 25% molecular ions, which have to be removed from the D^+ beam by a magnetic dipole located downstream.

In pulsed mode, the duoplasmatron is operated like a thyratron:

- The cone is used as a trigger (positive short pulse) to initiate the electron discharge.

- Once started, the electron beam is completely dominated by space charge and cannot be controlled by electrodes. Consequently, the only way to get a short pulse is to polarise the anode by a LC delay line, whose discharge has a square shape, and the appropriate duration.

- After the delay line is discharged (~1 µs), the electron beam stops.

Great care has been taken for the filament (Figure 2). First, it is prepared under a hydrogen atmosphere during several hours at 80-100 A for conditioning. Second, the filament is mounted inside the source and powered permanently under the nominal intensity to avoid any thermal shock, even when the accelerator is shutdown.

Figure 2. Filament (diameter ~4 cm)

Focusing and high-intensity aspects

After the extraction hole, the beam (around 80 mA peak) is focused by a set of electrostatic lenses of conical shape before entering inside the accelerating tube. A difficult challenge of GENEPI is the handling of high intensities. In the first part of the machine, the current is about 80 mA, because it also includes the molecular ions. The neutralisation by residual gas has a time constant of about 1 ms, and cannot occur with the GENEPI time structure, so all space charge has to be managed. So, the choice of the accelerating tube has been made with respect to two specific and mandatory characteristics:

- a sufficiently large aperture (60 mm in diameter);

- a strong gradient in order to have a maximum focusing effect at field entrance.

The acceptable losses in such a machine are very high, due to the very low power of the beam. These losses will produce some deuterium implantation along the machine, leading to D+D reactions and to 2.5 MeV neutron production, but this neutron noise will remain negligible.

Figure 3. General view of the duoplasmatron, the focusing electrodes and the accelerating tube

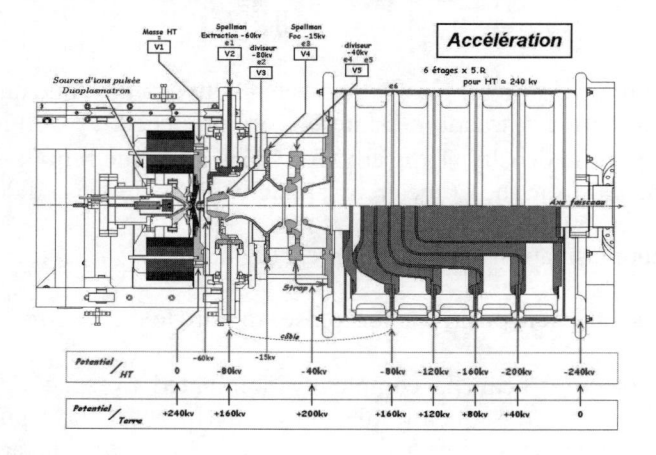

General structure

A view of GENEPI-1 is given in Figure 4. A high-voltage platform supporting the source and the accelerating tube is enclosed in a Faraday cage. A magnetic 45° dipole is used for removing the molecular ions from the beam. The focusing is done by several electrostatic quadrupoles over approximately 5 meters. The beam line is supported by a granite girder and can be translated by motion on air cushion. The target is located in the centre of the reactor core. The difficulty of handling high intensities is illustrated by the necessity of having permanent focusing. The GENEPI "rule of thumb" was to have a focusing element at least every 50 cms. So, a focusing electrostatic channel (the "thimble") goes through the reactor to the target. It is made of six electrostatic quadrupoles with planar electrodes. Detailed studies have shown that the non-linearities of these elements had no consequences on the beam quality. The dipole itself has 11.25° faces in order to have the same vergence in the two planes. The 45° shape of the machine has two additional advantages:

- The machine shape is well adapted to the confined MASURCA reactor room.

- It ensures an easy way to remove the target, by disconnecting the HV platform, and retracting it, by pulling back the beam line on top of the girder.

Figure 4. GENEPI-1 at Cadarache with HV platform (1), source (2), accelerating tube (3), electrostatic quadrupoles (4,6,7), dipole (5), beam line (8) in the reactor (9) and target (10)

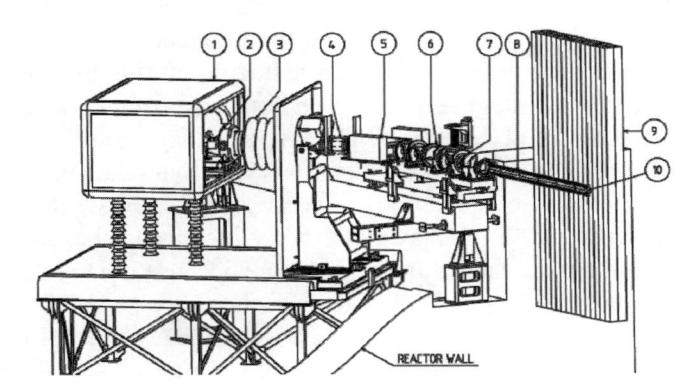

The machine was designed, built and commissioned in Grenoble, before its final installation in Cadarache. In particular, the whole calibration of the machine was done in Grenoble (profile, intensity, emittance measurements, beam position and steering calibration).

Target

The target is a 49-mm copper disc on which a thin layer of titanium is deposited over 30 mm (diameter). For Cadarache, about 1.2 mg of tritium or deuterium is then deposited (12Ci in case of tritium). The activity is defined by the local safety rules: in Grenoble, the load is limited to 0.9Ci. In operation, the target is air-cooled and kept below 100°C to avoid any desorption of tritium or deuterium.

Neutron production control

The beam intensity is controlled on the target via an amperemeter. The neutron production is monitored on-line by two backward silicon detectors located upstream the 45° magnet. The first detector detects α particles and protons produced backward during the fusion reaction and transmitted in the beam line while the second one, covered by an Al foil, collects only the protons. An absolute calibration is performed via Ni foils irradiations.

GENEPI-1/MASURCA: the first accelerator-driven system

The average deuteron intensity can be modulated via the modulation of the repetition rate. Typically, a factor 13 can be achieved in the intensity range by modulating from 100 Hz to 4 kHz. This modulation is done in two ways:

- very sharp edge falling times (a few microseconds) from I_{max} to I_{min};
- a plateau period, a low intensity;
- a progressive rise time from I_{min} to I_{max} (a few tens of seconds), longer than the reactor time constant.

By using this intensity modulation, the MUSE experiment demonstrated the real possibility of controlling the reactor power by the accelerator.

Operation feedback: reliability aspects

The GENEPI accelerator was an experimental device which was in constant evolution during all the MUSE4 experimental campaign. In spite of several component failures (hot filament, vacuum pump, electronic cards, etc.) operation of GENEPI was satisfactory during the considered period: more than 12 000 h of pumping and 130°C integrated charge on both types of targets TiD then TiT. In terms of reliability, the failure rate is about of 11%, the MTBF is 8.12 days and the mean time to repair is 4.3 days (eight effective hours per day). These considerations lead to a relatively low value of GENEPI availability of 65% which could be explained by several reasons. First, the constringent nuclear environment contributes to at least 10% of unavailability by increasing the intervention complexity. Second, the ions production had never reached the steady state because it had to be started and stopped every working day. This discontinuous operating mode associated with quite frequent complete stops promoted a failure rise. Third, and despite the common LPSC and CEA teams' effort, a correct technical maintenance level has never been reached. This is mainly due to an underestimation of this item in financial and manpower terms.

The main component behaviour feedback is the following:

- Filament has been changed only once in four years leading to a lifetime of at least 5000 h.

- Hardware located in the GENEPI room has suffered from n and γ emission coming from the core-inducing microcomputers PC104 and electronic device failures.

- The pumping system has been entirely satisfactory with only one failure of a secondary turbo-molecular pump in July 2004.

- Control command software has given entire satisfaction permitting quick remote failure diagnostics by the LPSC team and the record of a consistent operating parameter database.

- Concerning the use of tritiated target, even if tritium is confined in a very stable titanium hydride, it has been released by the beam effect. So special detectors, gloves box and air flowed diving-suits used for target manipulation have been of a real need. Of the ten targets purchased, only four, two TiD and two TiD have been used during the programme.

The GENEPI neutron source was characterised for each target by irradiation experiments of nickel foils inserted in the MASURCA core, as close as possible (10 cm) to the target. These irradiations were performed with all the safety rods down and at a repetition rate of about 4 kHz in order to minimise the activation due to the core inherent source. After irradiation the activity of each Ni target was measured in the Low Activity Laboratory of the LPSC using a low background germanium detector. These activities, compared to MCNP simulations, allow obtaining the neutronic yield per pulse.

The neutron production rate of the deuterium target was found equal to $3\pm0.3 \times 10^4$ n/pulse (60 mA peak intensity on target) in May 2001, and, for the tritium target, equal to $3.3\pm0.3 \times 10^6$ n/pulse (40 mA peak intensity) in January 2003. Due to the experimental needs, the target change has always been performed before reaching its lifetime limit. Literature indicates that desorption due to the impact of the beam is approximately 50% for a total charge of 20 Ci/cm^2. In our case it a desorption of almost 30% in the first few days of use has been noticed, and then a constant yield decay. In addition, on-line neutron production monitoring should have been upgraded in terms of reliability and on-line recording. These modifications will allow improving the accuracy of the experimental measurement techniques tested in the MUSE4 programme.

From the safety point of view no incident concerning the reactor-accelerator coupling was related during the almost four operating years. All the specific safety equipments and rules, upgraded all along the programme, have done their duty.

Finally, it is worth noting that GENEPI was designed to allow operation transfer to "local" technical staff with no initial accelerator background.

Safety aspects

In order to obtain the agreement to use GENEPI coupled with MASURCA from French safety authorities (DSIN), extended safety studies were performed in several phases from 1997 (preliminary safety report) to November 2001 (first coupling). The main recommendations are the following:

- *Explosion.* Deuterium tank pressure < 50 bars.

- *Fire.* Protection of the most inflammable parts and use of specific cables.

- *Radioprotection.* Rules of access to GENEPI room and reactor room strengthened (interlocks to the high-voltage acceleration voltage).

- *Tritium risk.* Glove box for target manipulation and use of air flowed diving-suits for equipped target-holder handling.

- *Ventilation.* Tritium detector added on exhaust gases circuits, GENEPI target forced air cooling, no oil in the vacuum system.

- *Emergency* reactor shutdown process modified.

- *Seism.* Safety demonstrations mainly on the core aggression by GENEPI glove finger and an upgrade of the accelerator mechanic structure.

- *Electro-magnetic compatibility* between GENEPI high-voltage cage and the reactor control command.

- *Passive beam intensity limitation* devices (beam rise time constant > reactor safety time constant).

The GENEPI-2/PEREN facility

In the frame of studies dealing with future reactor concepts using new fuels supposed to produce less minor actinides, the knowledge of numerous nuclear cross-sections are required. Thus, LPSC has developed the PEREN facility. It is dedicated to capture cross-section profile measurements between 0.1 eV to 30 keV using a lead-slowing-down-time spectrometer coupled with a pulsed neutron generator, GENEPI-2, an evolution of the GENEPI concept.

GENEPI-2 is very similar to GENEPI-1, except for the target (D or T) location. A first location (the shortest), is devoted to light slow-down materials, it is surrounded either by a graphite barrel or an assembly of Teflon® plates. The second position of the target is at the centre of the previously used eight-lead-block assembly. This topology needs additional electrostatic quadrupoles.

Figure 5. The GENEPI-2/PEREN facility at LPSC. The short configuration is shown, with the target inside the graphite barrel. A second one is possible, with the target inside the lead block, using an optic free and longer thimble.

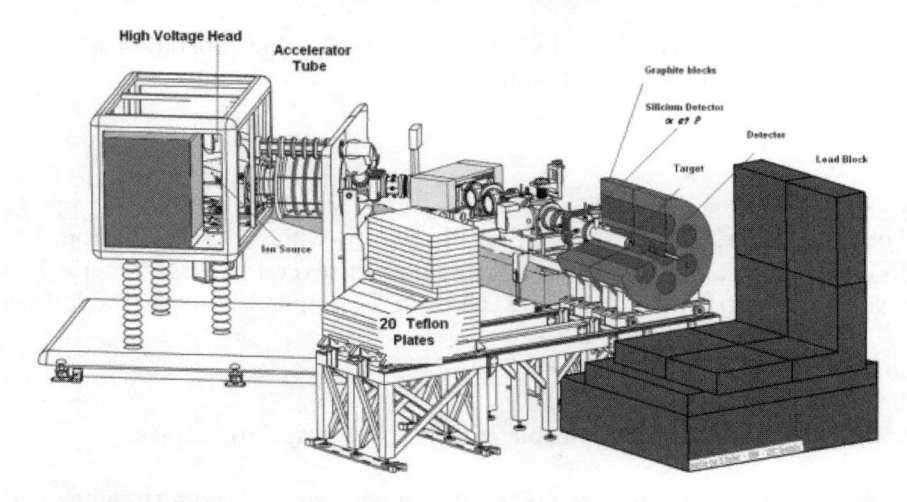

A minor modification has been made on the pulse duration, which has been decreased to 500 nsec by replacing the LC delay line by a coaxial line.

The future: The GENEPI-3C for the GUINEVERE programme at Mol

In 2006, the GUINEVERE programme was started. The objectives are to perform a low-power coupling experiment either with a pulsed beam but also with a continuous beam to study short beam interruptions. The aim of these interruptions is to demonstrate that it is possible to apply reactivity monitoring techniques investigated in MUSE without disturbing the CW operation mode of an ADS. It will be based on the GENEPI machine coupled to the modified VENUS reactor with two major differences:

- the possibility of having a continuous deuteron beam (200 µA minimum, 1 mA if possible) with some short programmable interruptions (a few ms);

- a vertical coupling.

Figure 6. Schematic and very preliminary layout of GENEPI-3C inside the VENUS reactor building (courtesy P. Baeten, SCK•CEN)

The source

The first question for GENEPI-3C is to know the possibility of running the same source either in the high-intensity pulsed mode or in a moderate-intensity continuous mode. The second question is to know the relative amount of molecular ions, which will be probably higher at low intensity. A bench test (Figure 7) has been built at LPSC and will be used at 40 kV in two stages:

- source test in pulsed and continuous mode and emittance measurements;

- beam analysis, by using a magnetic dipole downstream, to separate the molecular beam.

Figure 7. View of the test bench. Left: the 40-kV platform, right: the duoplasmatron.

In the first stage, the duoplasmatron has been run up to 1.2 mA deuteron total DC intensity without major difficulty. This corresponds to 48 W on the Faraday cup and 10-20 W on the cone. To prevent heating issues, we decided not to exceed these values in a first step. The total current is collected on a Faraday cup. The emittance measurement can be done by using this Faraday cup like a pepperpot, with holes on a vertical diameter only. A second, movable, Faraday cup with a 0.1 mm slit is located 50 mm downstream and gives the beam profile as a series of peaks (Figure 8). The RMS emittance is deduced by measuring the width of the peaks and their relative positions. The present (and preliminary) non-normalised value is about $20\pi\ 10^{-6}$ m.rad at 40 keV.

Figure 8. Side view of the test bench (left) and (right) beam profile on the second Faraday cup, used to measure the RMS emittance of the beam. Units are tenth of mm in horizontal and arbitrary in vertical.

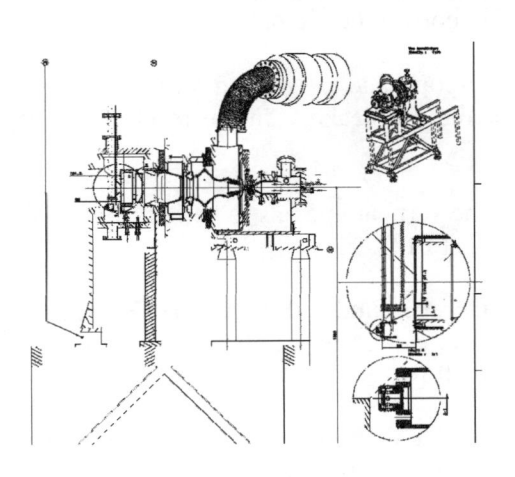

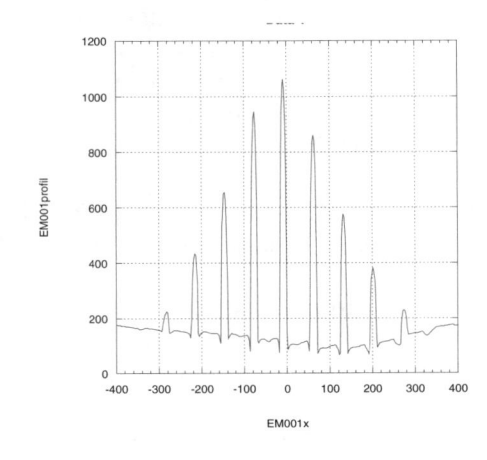

The source test must now enter the second phase (May 2007) as soon as possible. The first conclusion is that the same duoplasmatron can be used in the pulsed and continuous modes and that the main intensity goals can be met.

General structure

The general and preliminary structure is presented in Figure 9. The beam transport is made by electrostatic quadrupoles. A first quadruplet transports and focuses the beam at the 90° magnet entrance. A doublet is located at the dipole exit and there is no optical element in the width of the casemate ceiling. Two triplets are put above the reactor to focus the beam on the target. There is no optical element inside the reactor. Control rods limit the room available near the core. Pumps, steerers, diagnostics, are not represented on this figure.

Figure 9. General overview of the GENEPI-3C. The solid squares are electrostatic quadrupoles.

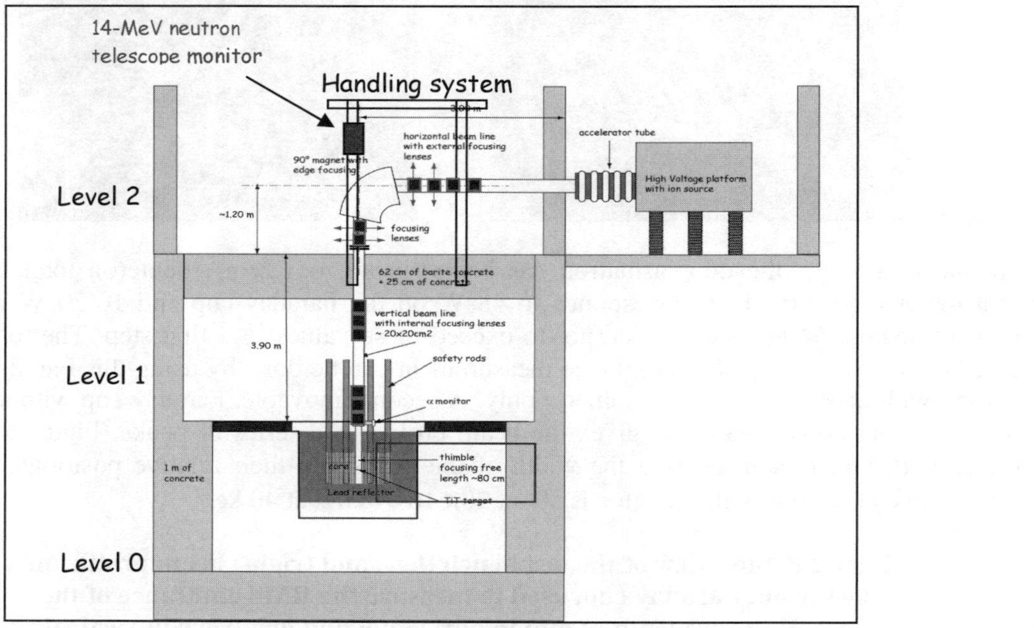

A lot of issues, taking into account all the reactor safety constraints, remain to be solved:

- The 90° dipole must accommodate a neutron telescope and presents a wide fringing field. Its cooling system requires special attention to avoid any leakage in the reactor. It is under design at the IPN Orsay.

- Target replacement is not possible at Level 1, and the vertical line has to be removed vertically for core loading. A handling system is under design at LPC Caen.

- The beam line from the reactor entrance to the target is under design at IPHC Strasbourg. As the beam power can now be 250 W in continuous mode (1 mA, 250 keV), versus 50 W for GENEPI-1, the cooling system has to be studied carefully.

- For the same reason (high beam power), the interceptive diagnostics must be reconsidered, as well as the dipole chamber, where the power deposited by the molecular ions will now not be negligible.

Following the GENEPI-1 experience, the accelerator is expected to be fully designed, assembled, tested and commissioned in Grenoble for the first quarter 2009. This first commissioning will be progressive and will require an easy access for beam calibration. It cannot be done on the VENUS site, in particular above the reactor. The first beam on the VENUS site is expected for mid-2009.

Acknowledgements

This work was partially supported by the 5th Framework Programme of the European Commission through the MUSE contract # FIKW-CT-2000-00063, and by the 6th FP through the EUROTRANS Integrated Project contract # FI6W-CT-2005-516520.

REFERENCES

[1] Belmont, J.L., J.M. De Conto, *GENEPI Accelerator – Design, Technology and Characteristics*, Internal report ISN00.77 CNRS/IN2P3-INPG-Joseph Fourier University (1998).

[2] Destouches, C., *et al.*, "The GENEPI Accelerator Operation Feedback at the MASURCA Reactor Facility", *Nuclear Instruments and Methods in Physics Research Section A*, Vol. 562, Issue 2, 23 June 2006.

REACTIVITY MEASUREMENTS IN SUBCRITICAL CORE: RACE-T EXPERIMENTAL ACTIVITIES

Mario Carta
ENEA, Via Anguillarese, 301, I-00060 Roma, Italy, carta@casaccia.enea.it

George R. Imel
ANL, 9700 S. Cass Avenue, Argonne, IL 60439, USA, imel@anl.gov

Christian C. Jammes
CEA, F-13108 Saint Paul-lez-Durance, France, christian.jammes@cea.fr

Stefano Monti
ENEA, Via Martiri di Monte Sole, 4, I-40129Bologna, Italy, monti@bologna.enea.it

Roberto Rosa
ENEA, Via Anguillarese, 301, I-00060 Roma, Italy, roberto.rosa@casaccia.enea.it

Abstract

The objective of the European Integrated Project EUROTRANS of the EURATOM 6th Framework Programme is to provide answers to the challenge of high-level nuclear waste transmutation in ADS. The EUROTRANS experimental activities have been joined into the ECATS domain, namely "Experiment on the Coupling of an Accelerator, a Spallation Target and a Subcritical Blanket."

The RACE-T experiment, formerly named TRADE, is part of ECATS. The experimental campaign presented in this paper was held in the period 2004-2006 in the TRIGA RC-1 reactor, operated by ENEA-Casaccia at the research centre near Rome, in order to propose experimental techniques for absolute reactivity calibration at either start-up or shutdown phases.

The RACE-T campaign includes fission rate measurements (performed with a special instrumented fuel element), investigation of different subcritical configurations (with D/T generator in the core centre), and development of special devoted instrumentation and acquisition systems.

The main outcomes of the experimental campaign, together with the specifications of an international IAEA benchmark focused on the subject, will be illustrated.

Introduction

The objective of the European Integrated Project EUROTRANS [1] of the EURATOM 6th Framework Programme is to provide answers to the challenge of high-level nuclear waste transmutation in ADS. The EUROTRANS experimental activities have been grouped into the ECATS domain, namely "Experiment on the Coupling of an Accelerator, a Spallation Target and a Subcritical Blanket".

The RACE-T experiment, formerly named TRADE [2], is part of ECATS. The European experimental teams performing the different experiments operated in the period 2004-2006 on different core configurations. In that period the reactor operation was exclusively in subcritical conditions, with the exception of the critical reference core assessment at the beginning of the campaign where the maximum power was 80 W. In this work we will focus on some selected RACE-T campaign outcomes:

- Characterisation of the critical phase performed by fission rate traverses.

- Evaluation of the applicability of various experimental techniques for assessing a subcritical level. Those techniques are based on:

  - The system response to a pulsed neutron source, in particular the area ratio method obtained by a D-T generator.

  - The system response to a source jerk (SJ), in particular the inverse kinetics (IK) method applied to a SJ obtained by a D-T generator (the difference between SJ results by D-T generator and ^{252}Cf sources are analysed in [3]).

  - The source multiplication technique.

By means of the methods mentioned above, reactivity estimates were performed at different core locations and for three different "clean" (control rods withdrawn) subcritical core configurations, namely SC0 (~ -500 pcm), SC2 (~ -2 500 pcm) and SC3 (~ -5 000 pcm).

Subcritical level measurements: theoretical outlines

The area ratio method

Pulsed neutron source (PNS) area methods are based on the analysis of the responses shown by a subcritical system to an external source pulse. A typical response of this kind is shown in Figure 1, where the contributions of prompt and delayed neutrons are evident.

Figure 1. Response shown by a subcritical system to an external source pulse

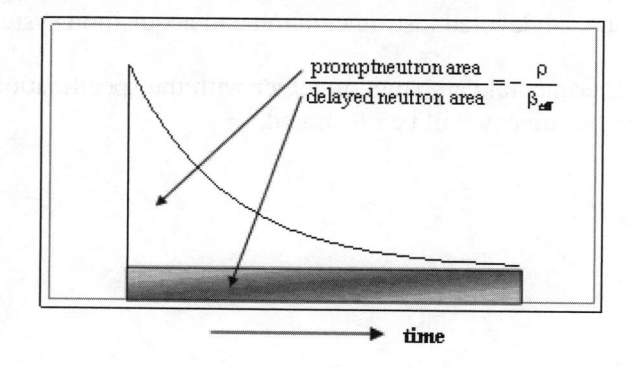

$$\frac{\text{prompt neutron area}}{\text{delayed neutron area}} = -\frac{\rho}{\beta_{\text{eff}}}$$

In particular, the fundamentals of the area ratio method [4] consist in the evaluation of the absolute level of reactivity (in dollars) by measuring, independently by the detector position, the ratio between the area under the prompt peak and the delayed one after the injection of a neutron pulse in the subcritical system:

$$-\rho_\$ = \frac{\text{Prompt neutron area}}{\text{Delayed neutron area}} = \frac{I_p}{I_t - I_p} \quad (1)$$

where the prompt area I_p is proportional to the detector response without delayed neutron precursors and the delayed area I_d is equal to the difference between the total area I_t and the prompt one. If the system response cannot be approximated by point kinetics, the reactivity value can apparently depend on the detector position $\mathbf{r}_D$, and we will obtain, in place of Eq. (1), relationships like:

$$-\rho_\$(\mathbf{r}_D) = \frac{\text{Local prompt neutron area}}{\text{Local delayed neutron area}} = \frac{I_p(\mathbf{r}_D)}{I_t(\mathbf{r}_D) - I_p(\mathbf{r}_D)} \quad (2)$$

In such cases, these spatial effects can be taken into account by calculating spatial correction factors obtained by:

- equivalent steady-state methods obtained by solving time-independent inhomogeneous transport problems (see for example [5] for an application to the MUSE-4 case).

- explicit time-dependent analysis where time-dependent inhomogeneous transport problems are solved.

Equivalent steady state methods

Consider the neutron source represented by $Q(r,E,\Omega,t) = Q(r,E,\Omega,t)\delta_+(t)$ and the signal due to the prompt neutrons alone; the prompt neutron flux $\Phi_p(r,E,\Omega,t)$ will obey to the prompt time-dependent ordinary transport equation, with the usual free-surface boundary conditions and the initial condition $\Phi_p(r,E,\Omega,0) = 0$. Integrating from $t = 0$ to $t = \infty$, and defining the time integrated prompt neutron flux as $\tilde{\Phi}_p(\mathbf{r},\mathbf{\Omega},E) = \int_0^\infty \Phi_p(\mathbf{r},\mathbf{\Omega},E,t)dt$, $\tilde{\Phi}_p$ will satisfy the inhomogeneous (with external source) time-independent prompt transport equation with the initial condition and the condition $\lim_{t\to\infty}\Phi_p = 0$ (because the system is subcritical):

$$\mathbf{\Omega}\cdot\nabla\tilde{\Phi}_p + \Sigma_t\tilde{\Phi}_p = \langle\Sigma_{ins}\tilde{\Phi}_p\rangle + \left(\chi - \sum_i \beta_i\chi_d^i\right)\langle\nu\Sigma_f\tilde{\Phi}_p\rangle + Q(\mathbf{r},E,\mathbf{\Omega})$$

where β_i and χ_d^i are, respectively, the delayed neutron fraction and the delayed energy spectrum for each precursor group i, and χ is the total (prompt + delayed) spectrum. Σ_{ins} is the in-scattering cross-section and, as usual, <> denotes integration. Analogously, the total time-integrated flux $\tilde{\Phi}(\mathbf{r},\mathbf{\Omega},E)$ can be defined by integrating the transport equation over time; $\tilde{\Phi}$ will then satisfy the inhomogeneous ordinary time-independent transport equation:

$$\mathbf{\Omega}\cdot\nabla\tilde{\Phi} + \Sigma_t\tilde{\Phi} = \langle\Sigma_{ins}\tilde{\Phi}\rangle + \chi\langle\nu\Sigma_f\tilde{\Phi}\rangle + Q(\mathbf{r},E,\mathbf{\Omega})$$

Therefore, the reactivity level of the system given by Eq. (2) can be evaluated by means of a steady-state Monte Carlo or deterministic multi-group methods. In particular, being σD the detector cross-section, we can write:

$$\frac{I_p(r_D)}{I_d(r_D)} = \frac{\iiint \sigma_D(r - r_D, E)\tilde{\Phi}_p \, dV \, dE \, d\Omega}{\iiint \sigma_D(r - r_D, E)(\tilde{\Phi} - \tilde{\Phi}_p) \, dV \, dE \, d\Omega}$$

Explicit time-dependent analysis

Explicit time-dependent analysis by numerical solution of the time-dependent inhomogeneous transport problems is the direct way to evaluate spatial correction factors for the area ratio method. In any case, for these methods we also will probably have to devise some stratagems to evaluate, *a priori*, the delayed background shown in Figure 1, without the need to run (useless?) multi-pulse CPU time-consuming cases in order to obtain a sufficient increase of the delayed background. Of course, it is of great interest to have explicit time-dependent calculations in order to compare these results with those obtained by equivalent steady-state methods.

Inverse kinetics applied to a source jerk

The inverse kinetics method is a widely known method used to analyse the transient behaviour of reactor systems. It is based on the inversion of the points kinetics equations which gives the reactivity versus time:

$$\rho_\$ = 1 - \frac{\Lambda}{\beta_{eff}} \cdot \frac{S_d(t) + S_{eff} - \frac{dn}{dt}}{n(t)}$$

with the delayed neutron source equal to $S_d(t) = \Sigma_i \lambda_i C_i(t)$. The delayed neutron precursors group concentrations $C_i(t)$ are calculated using the measured flux $n(t)$ and the effective source S_{eff} is adjusted so that the inferred reactivity is steady after the source jerk. Finally, the reactivity is obtained by averaging the reactivity on an appropriate time range after the source jerk. This method will be referred in the following as SJ-Gen technique.

The source multiplication technique

MSA approximation

Let us consider a reference subcritical state characterised by a negative reactivity level ρ_0 experimentally obtained by an independent method like a rod-drop measurement, and a generic subcritical core state i characterised by the reactivity ρ_i, assuming both states driven by the same external neutron source. If we measure for both states some counting rates T in a given core position **r**, the experimental reactivity for the state i will be given by:

$$\rho_i^{exp} = \rho_0^{exp} \cdot \frac{T_0^{exp}(\mathbf{r})}{T_i^{exp}(\mathbf{r})} \equiv \rho_i^{MSA} \qquad (3)$$

This is the so-called MSA approximation (from the French *Multiplication Source Approchée*), which presumes that the ratio $T_0(r)/T_1(r)$ does not depend on the position r.

MSM correction

If the above-listed MSA hypotheses break, (calculated) correction factors are needed. We can write for the reference state:

$$\rho_0 \cdot T_0(\mathbf{r}) = - <\phi_0^* S> \cdot \frac{T_0(\mathbf{r})}{<\phi_0^* F_0 \psi_0>} \equiv -S_{eff,0} \cdot \varepsilon_0(\mathbf{r}) \tag{4}$$

were $S_{eff,0}$ and $\varepsilon_0(r)$ denote, respectively, the effective source and the detector efficiency (at position r) for the state 0, and ϕ_0^* is the associated adjoint homogeneous flux solution. Analogously we can write for the generic subcritical state i:

$$\rho_i \cdot T_i(\mathbf{r}) = - <\phi_i^* S> \cdot \frac{T_i(\mathbf{r})}{<\phi_i^* F_i \psi_i>} \equiv -S_{eff,i} \cdot \varepsilon_i(\mathbf{r}) \tag{5}$$

From (4) and (5) we obtain:

$$\frac{\rho_i \cdot T_i(\mathbf{r})}{\rho_0 \cdot T_0(\mathbf{r})} = \frac{S_{eff,i} \cdot \varepsilon_i(\mathbf{r})}{S_{eff,0} \cdot \varepsilon_0(\mathbf{r})} \tag{6}$$

By comparison with Eq. (3) we see that the MSA approximation is equivalent to assume that effective source and detector efficiency remain unchanged when passing from state 0 to state i. Eq. (6) must be evaluated by calculation, and we have to impose that:

$$\frac{\rho_i^{exp} \cdot T_i^{exp}(\mathbf{r})}{\rho_0^{exp} \cdot T_0^{exp}(\mathbf{r})} = \frac{\rho_i^{cal} \cdot T_i^{cal}(\mathbf{r})}{\rho_0^{cal} \cdot T_0^{cal}(\mathbf{r})}$$

Thus, we have to replace the MSA formulation (3) with the modified formulation:

$$\rho_i^{exp} = \left(\rho_0^{exp} \cdot \frac{T_0^{exp}(\mathbf{r})}{T_i^{exp}(\mathbf{r})}\right) \cdot \left(\frac{\rho_i^{cal} \cdot T_i^{cal}(\mathbf{r})}{\rho_0^{cal} \cdot T_0^{cal}(\mathbf{r})}\right) \equiv \rho_i^{MSA} \cdot MSM_{i \to 0}^{cal}(\mathbf{r})$$

The correction factors $MSM_{i \to 0}^{cal}(\mathbf{r})$, depending on the position, have to be evaluated by calculation. This is the so-called MSM method (Modified Source Multiplication) which takes into account spatial and energetic effects, due to type and position of the detector.

Experimental set-up

Description of the TRIGA reactor

The RC-1 TRIGA reactor, located at the ENEA Casaccia research centre nearby Rome, is a 1-MW Mark II reactor built in 1960. It is a lightwater reactor cooled by natural convection, with a 284-mm thick annular graphite reflector. The core has a 565-mm diameter cylindrical configuration with 127 locations that are arranged in seven concentric rings, namely A, B, C, D, E, F and G (see

Figures 2 and 6). Each location can be filled with either a fuel element or some other component like graphite rods, a neutron source or a measurement channel. The stainless-steel clad fuel material with 37.3-mm diameter and 381-mm height is a homogeneous mixture of uranium and zirconium hydride terminated at both the bottom and top by 87.5-mm high cylindrical graphite slugs that act as axial reflectors. There are four boron-carbide control rods: two fuel-follower shim rods (SH1 and SH2), one fuel-follower safety rod (SEC) and one regulating rod (REG), not fuel followed.

The fuel elements consist of a stainless-steel clad (AISI-304, 0.05-cm thick, 7.5 g/cm^3 density) characterised by an external diameter of 3.73 cm and a total height of 72 cm, end cap included (Figure 2). The fuel is a cylinder (38.11 cm high, 3.63 cm external diameter, 5.9 g/cm^3 theoretical density) of a ternary alloy uranium-zirconium-hydrogen (H-to-Zr atom ratio is 1.7 to 1; the uranium, enriched to 20% in ^{235}U, makes up 8.5% of the mixture by weight; the total uranium content of a rod, on the average, is 197 g, of which 39 g is fissile) with a metallic zirconium rod inside (38.11 cm high, 0.5 cm in diameter, 6.49 g/cm^3 of density). There are two graphite cylinders (8.7 cm high, 3.63 cm in diameter, 2.25 g/cm^3 of density) at the top and bottom of the fuel rod. Externally two end-fittings are present in order to allow the remote movements and the correct locking to the grid.

Figure 2. TRIGA reactor overview

The three control rods are located above the REG rod, which is bottom centre in (b)

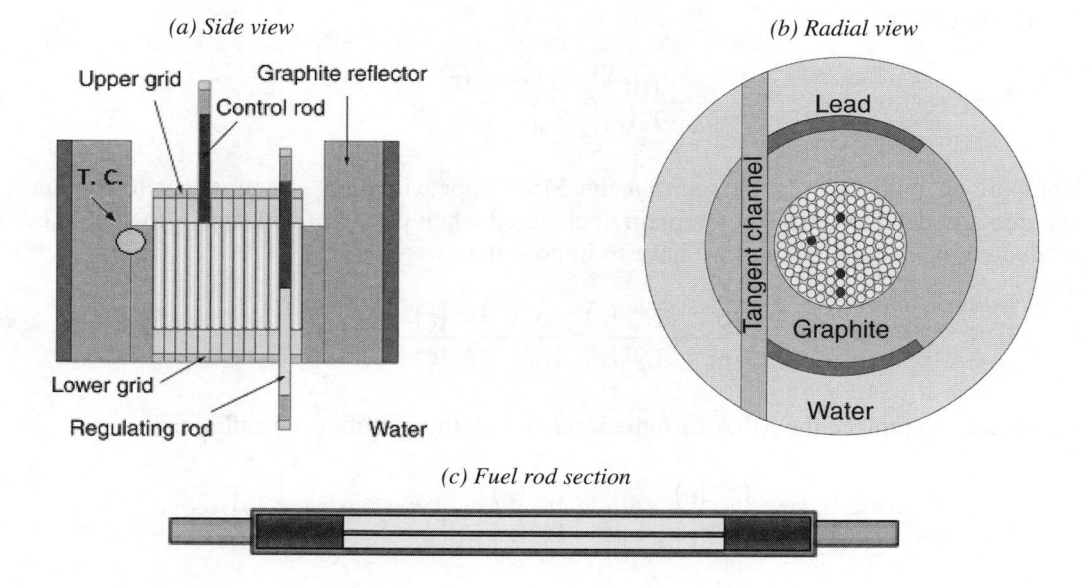

(a) Side view   *(b) Radial view*

(c) Fuel rod section

Neutron sources

For the subcritical configurations, the reactor was coupled with the following neutron sources:

- A pulsed deuterium-tritium neutron generator, accelerating deuterium ions onto a tritium target, and producing 14.1-MeV neutron bursts with strength 10^8 neutrons/s at maximal frequency. The frequency range spanned from 1 to 150 Hz. The pulse duration was less than 1 μs. The neutron generator was located at the core centre A01.

- A ^{252}Cf source, with a strength of 0.4 Ci, was used to perform source multiplication experiments using a fast rabbit (FR) location in the B02 position in ring B (Figure 6).

Instrumentation

For the characterisation of the critical phase by fission rate traverses, a special fuel pin was used. This fuel is hollowed in the centre, and enables the fission chamber to be inserted inside the fuel. The hole has a diameter of 5 mm, allowing the insertion of the ∅1.5 mm fission chambers. The cross-section of this special pin is shown in Figure 3.

Figure 3. Cross-section of the special fuel pin

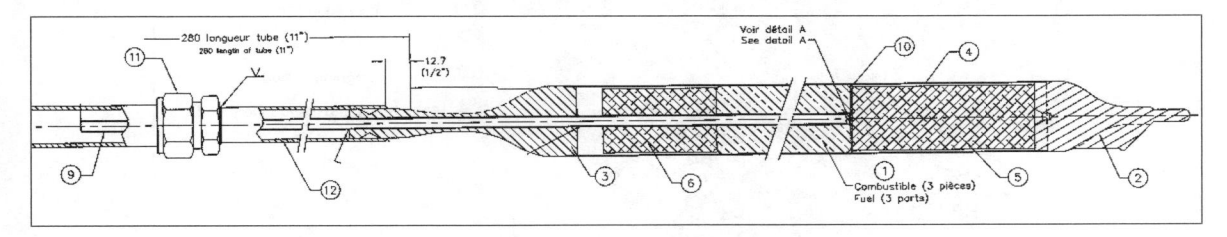

The position of the fission chamber in the special fuel pin corresponds to the mid-plane of the core. A tight tube is fixed at the end of fuel and enables the connexion cable to leave the pool of the reactor. Before each measurement, one withdraws a fuel pin and replaces it by the special fuel pin.

To evaluate the applicability of various experimental techniques for assessing a subcritical level, the instrumentation consisted of fission chambers (labelled from A to D in Figure 6), current-sensitive amplifiers and the X-MODE data acquisition system [6]. The fission chambers were placed within the core region close to the reflector (see Figure 6). The main purpose of X-MODE is to integrate in a single system all the features needed for reactor measurements. The main asset of X-MODE is an accurate time stamping capability that offers many methods of investigating acquired data.

Fission rate measurements

Radial traverses enable one to analyse the shape of the fission rate in different energy ranges, and to look at possible asymmetry effects. Measurements were carried out for the reference critical configuration shown in Figure 4, along the main G13-G31 diagonal (Figure 4), in the positions G13, F11, E09, D07, C05, C11, D16, E21, F26 and G31. The measurements were performed in critical condition with a reactor power in the range 10÷80 W.

In order to cover the broadest range in energy, various types of deposits were used:

- two ^{235}U fission chambers (n°1847 and n°331) for thermal fission rates;

- one ^{237}Np fission chamber (n°1523) for intermediate spectrum range;

- one ^{238}U fission chamber (n°861) for the fast energy range.

The results relative to the G13-G31 diagonal are plotted in Figure 5 for each fission chamber.

The effect of the asymmetry shown in Figure 5 is due to the presence of the tangential beam tube of important diameter situated near the core mid-plane, in the reflector (cf. Figure 2). This tube removes an important part of the water and induces a very important local void effect.

Figure 4. Reference critical core configuration for the radial fission rates measurement

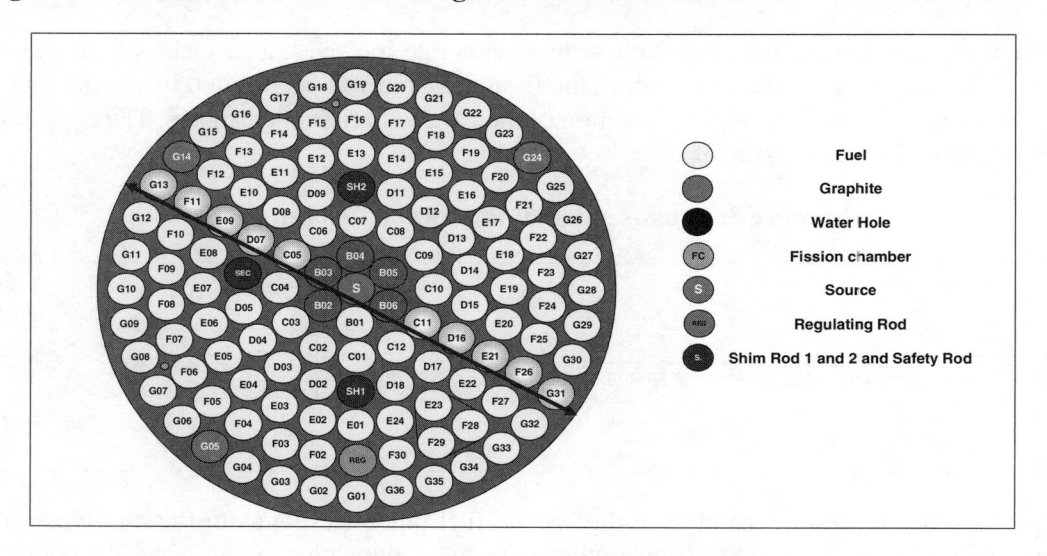

Figure 5. Normalised count rates on G13-G31 diagonal

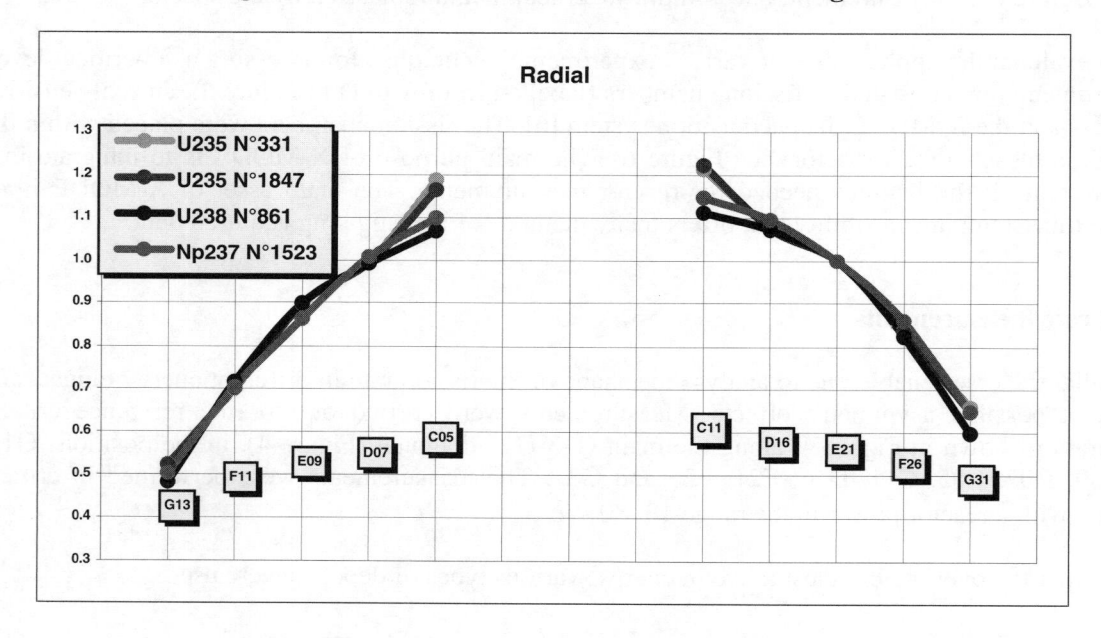

Subcritical level measurements

Core configurations

The investigated core configurations, all characterised by main control rods withdrawn, are as follows: one reference critical configuration (REF), the REF configuration with the regulation rod down (representing the reference subcritical configuration for the source multiplication method) and three subcritical configurations, namely SC0 (~ -500 pcm), SC2 (~ -2 500 pcm) and SC3 (~ -5 000 pcm), as shown in Figure 6. The fission chambers, the fast rabbit (FR) pipe and the neutron generator were always placed in-core during the experimental campaign. It should be mentioned that the REF core configuration was critical with the regulating rod (REG) 51% inserted (cf. Figures 2 and 6).

Figure 6. Core configurations for the subcritical level analyses

One removes three fuel elements in C-ring to shift from one subcritical configuration to another. The fast rabbit (FR) pipe is in B02 and the neutron generator (DT) in A0.

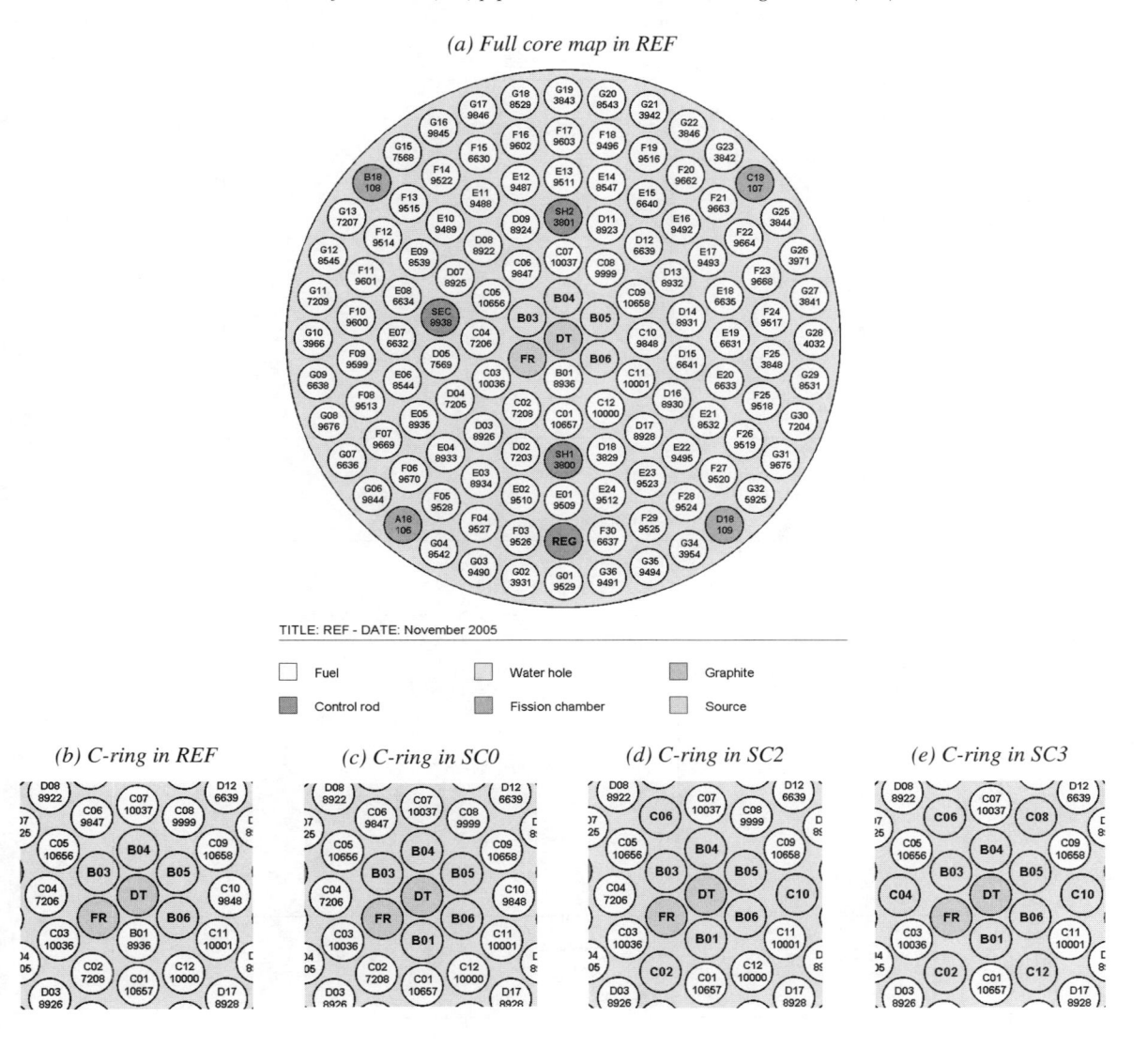

(a) Full core map in REF

TITLE: REF - DATE: November 2005

☐ Fuel   ☐ Water hole   ☐ Graphite
■ Control rod   ■ Fission chamber   ■ Source

(b) C-ring in REF   *(c) C-ring in SC0*   *(d) C-ring in SC2*   *(e) C-ring in SC3*

Experimental results and data analysis

The following methods have been into account for the intercomparison: the area ratio technique (PNS-Area), the inverse kinetics-source jerk technique based on the transient caused by the neutron generator shutdown (SJ-Gen) and the approximated source multiplication technique (MSA).

All the reactivity estimates for the three subcritical configurations are displayed in Figure 7. For SC0, the uncertainties are about 1% for the PNS-Area technique, 3.1% for the MSA technique, 4.0-4.4% for the SJ-Gen technique. For SC2, the uncertainties are about 0.4% for the PNS-Area technique, 3.1% for the MSA technique, 6-7% for the SJ-Gen technique. For SC3, the uncertainties are about 0.5% for the PNS-Area technique, 3.1% for the MSA technique, 5-6.5% for the SJ-Gen technique.

Figure 7. Comparison between PNS, SJ-Gen and MSA techniques with all rods up

The error bars corresponds to a confidence level of 95% (1$~700 pcm)

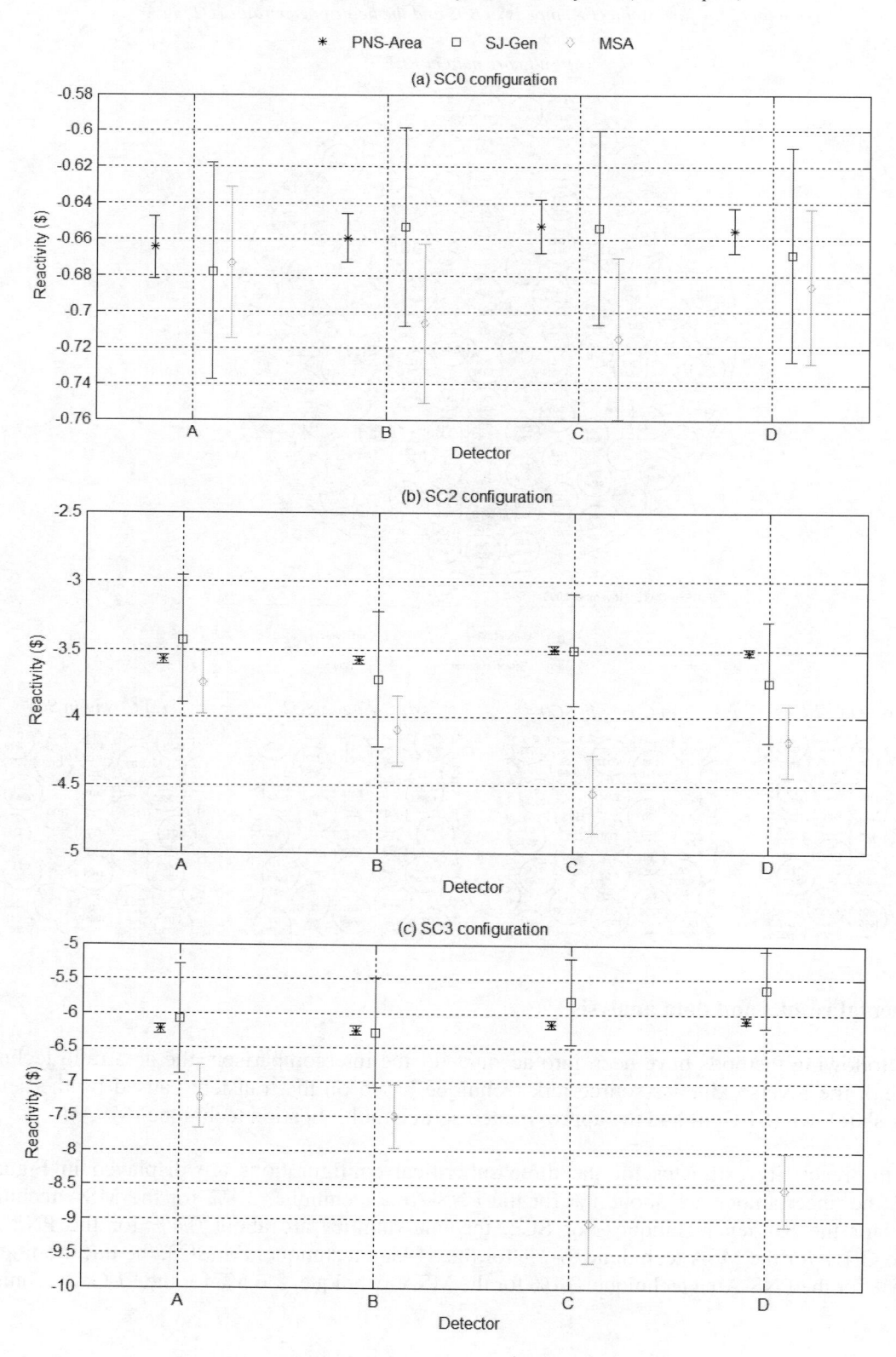

First, it appears that the PNS-Area and SJ-Gen reactivity estimates are always equal with a confidence level of 95%. Second, the MSA technique is clearly the most detector-location-dependent. The discrepancies from the PNS-Area estimates are about 1-5% for SC0, 5-19% for SC2 and 16-40% for SC3. Conversely, the PNS-Area technique is the least detector-location-dependent with a spread of 1.22% at most for SC3.

IAEA benchmark

Generalities

The results obtained from the flux measurements and from the subcritical level determination by different methods indicate, in the case of RACE-T experiments, the presence of the following effects:

1) asymmetry in the radial flux profile.

2) coherence between subcritical levels obtained by area ratio and SJ-Gen methods.

3) discrepancies between subcritical levels obtained by MSA and area ratio/SJ-Gen methods.

In particular, it cannot be excluded a priori the existence of a correlation between points 1) and 3). In this frame, an explanatory role may be played by the evaluation of the correction factors to be applied to the measurements (especially MSA results). A calculation benchmark focused on the evaluation of the correction factors for the RACE-T case not only can help to clarify the above-mentioned discrepancies but can also tackle the issue of the calculation methodologies to evaluate spatial/energy correction factors to be applied to subcritical level measurements in ADS.

Recently IAEA has endorsed the outcomes illustrated in this work from the RACE-T experimental campaign, and a computational benchmark has been launched in the framework of the Co-ordinated Research Project "Analytical and Experimental Benchmark Analyses of Accelerator-driven Systems (ADS)" co-ordinated by IAEA [7]. The benchmark, named "Pre-TRADE Experimental Benchmark", will be focused on the evaluation, via computation, of the correction factors to be applied to the PNS area ratio and MSA results for the selected reactivity estimates to take into account the role of the spatial/energy effects on the rough experimental data.

The TRIGA-RC 1 reactor burn-up issue

In this frame, a "basic" issue can be raised for what concerns the degree of knowledge of the fuel burn-up status of the RC-1 TRIGA reactor, although it is not expected, *a priori*, that a detailed description of the fuel composition will provide a significant impact on the corrections factors to be calculated for the above-mentioned subcritical configurations (SC0÷SC3), because several fresh pins are loaded into the core for such configurations.

As expected, in the case of the RC-1 TRIGA reactor, built in the sixties, the degree of the present knowledge of the fuel burn-up status is obviously low. In any case, seizing the opportunity given by the IAEA benchmark, a relatively more accurate approach to the burn-up issue is presently being undertaken, although in approximated ways taking into account the objective difficulties inherent to such an issue.

The old estimation for the reference critical configuration for the "Pre-TRADE Experimental Benchmark" is shown in Figure 8.

Figure 8. Old burn-up evaluation for the REF critical configuration

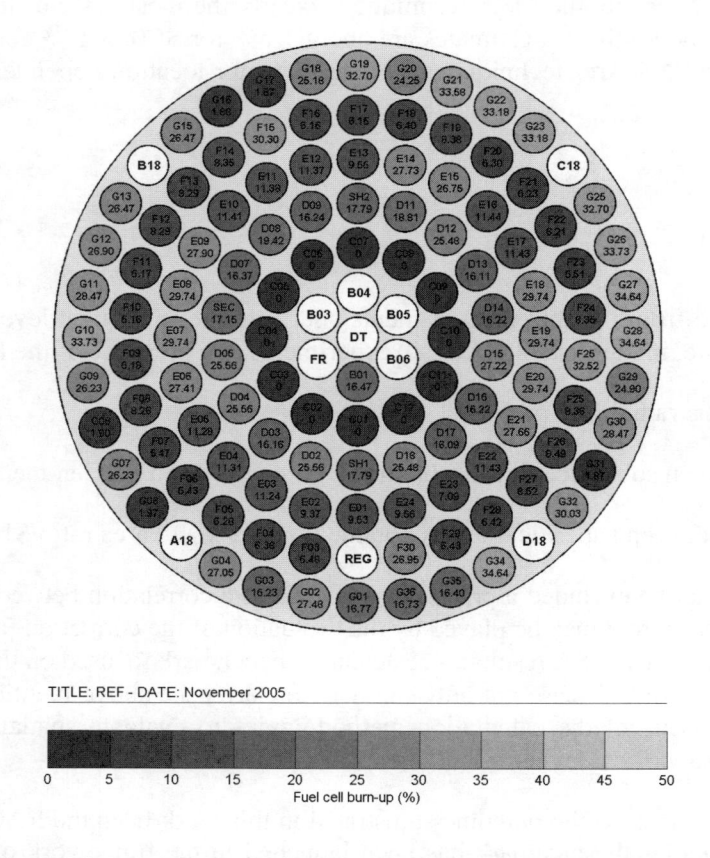

It may be seen from Figure 8 that, for this particular configuration, several fresh fuel elements have been loaded. To (slightly) improve the present knowledge of the burn-up status of the TRIGA RC-1 reactor, a new coherent burn-up factors data set has been prepared [8], which has been applied to all the core configurations loaded in the TRIGA RC-1 reactor since reactor start-up. Table 1 shows a synthesis of the results relative to the REF critical configuration shown in Figure 8, and it can be seen that the burn-up variation is around 3.5%.

Table 1. Burn-up evaluations intercomparison for the REF critical configuration

Evaluation	^{235}U consumption (g)	Burn-up (%)
New	554.25	12.73
Old	703.47	16.15
Δ	26.92%	3.43

Improvements with respect to this provisional new evaluation are presently ongoing by means of the ERANOS codes system [9], which takes into account the fission product build-up, together with the fuel consumption, in function of the actual energy produced by each fuel element during its life. The burn-up factors for the whole core and for each historical configuration of the TRIGA RC-1 reactor, together with the energy produced by each element, are obtained by the TRIGLAV code [10]. This new evaluation will provide the final compositions for the fuel elements to be used in the benchmark.

Conclusions

To have at one's disposal a reference critical state, contrary to the foreseen situations in ADS, allows the intercomparison of dynamic methods to measure the subcritical level (PNS area ratio, source jerk) with the static method MSA, which is usually taken as a reference but requires for large MSM spatial correction factors at deep subcriticalities.

Under this point of view, the RACE-T experiments have allowed the testing of different techniques to measure the subcritical level in ADS. Coherently with the outcomes from MUSE, the PNS area ratio method seems to be the most stable for what concerns the spatial effects, even if such stability has to be supported by theoretical and numerical confirmations.

As concerns the inverse kinetics/source jerk technique, the measurements provided reactivity estimates always in excellent agreement with those obtained by the area ratio technique, although a discrepancy between PNS area ratio and inverse kinetics/source jerk results can clearly be observed when increasing the subcriticality level.

The above-mentioned theoretical and numerical confirmations will be obtained, hopefully, thanks to a computational benchmark, recently endorsed by IAEA, focused on the evaluation of the correction factors to be applied to the PNS area ratio and MSA results described in this work to take into account the role of the spatial/energy effects on the rough experimental data. In order to prepare the benchmark, a new evaluation of the TRIGA RC-1 fuel burn-up level is presently ongoing.

Acknowledgements

The authors appreciate the efforts and support of all the scientists and institutions involved in EUROTRANS and the presented work, as well as the financial support of the European Commission through the contract FI6W-CT-2004-516520.

REFERENCES

[1] Knebel, J.U., et al., "IP EUROTRANS: A European Research Programme for the Transmutation of High-level Nuclear Waste in an Accelerator-driven System", *8th Information Exchange Meeting on Actinide and Fission Product Partitioning & Transmutation*, Las Vegas, Nevada, USA (2004).

[2] Imel, G., et al., "The TRADE Experiment and Progress", *GLOBAL 2003*, New Orleans, Louisiana, USA (2003).

[3] Jammes, C., et al., "Absolute Reactivity Calibration of Accelerator-driven Systems after RACE-T Experiments", *PHYSOR 2006*, Vancouver, BC, Canada (2006).

[4] Sjöstrand, N.J., *Arckiv. Fis.* 11, 233 (1956), see also in G.I. Bell and S. Glasstone, "Nuclear Reactor Theory", pp.546-549, Van Nostrand Reinhold Company (1970).

[5] Carta, M., *et al.*, "Reactivity Assessment and Spatial Time-effects from the MUSE Kinetics Experiments", *PHYSOR 2004*, Chicago, IL USA (2004).

[6] Geslot, B., *et al.*, "Multimode Acquisition System Dedicated to Experimental Neutronic Physics", *IMTC 2005*, Ottawa, Canada (2005).

[7] http://www.iaea.org.

[8] http://www.triga.enea.it/TRIGA/Eng/IAEA_Benchmark.htm.

[9] Rimpault, G., *et al.*, "The ERANOS Code and Data System for Fast Reactor Neutronic Analyses", *PHYSOR 2002*, Seoul, Korea (2002).

[10] Persic, A., M. Ravnik, S. Slavic, T. Zagar, *TRIGLAV, A Program Package for Research Reactor Calculations*, IJS-DP-7862, Version 1 (28 March 2000).

THE MATERIALS TEST STATION: A FAST SPECTRUM IRRADIATION FACILITY

Eric J. Pitcher
Los Alamos National Laboratory

Abstract

The US Department of Energy is developing technologies needed to reduce the quantity of high-level nuclear waste bound for deep geologic disposal. Central to this mission is the development of high burn-up fuel with significant inclusion of plutonium and minor actinides. Different fuel forms (e.g. nitrides, oxides and metal matrix) and composition are under study. The success of these cannot be judged until they have been irradiated and tested in a prototypic fast neutron spectrum environment. In 2005, the US Congress authorised funding for the design of the Materials Test Station (MTS) to perform candidate fuels and materials irradiations in a neutron spectrum similar to a fast reactor spectrum. The MTS uses a 1-MW proton beam to generate neutrons through spallation reactions. The peak neutron flux in the irradiation region approaches that of the world's most powerful fast-spectrum research reactors. The estimated cost for building the MTS is $79M, including 25% contingency. Once approved for construction, the MTS can be operational in four years.

Introduction

Nuclear power is predicted to play a crucial role in the coming decades as a means to reduce global emissions of carbon dioxide. Proliferation concerns and the disposition of high-level nuclear wastes are seen as impediments to a broad expansion in the worldwide use of nuclear power. These concerns can be greatly alleviated by transitioning to a closed nuclear fuel cycle wherein spent nuclear fuel is recycled and used to power a new generation of fast reactors. New fuel forms containing neptunium, americium and possibly curium are needed to close the fuel cycle. The economics of a closed nuclear fuel cycle are closely tied to the time that the new fuels and associated cladding are able to reside in the harsh radiation environment of a fast reactor. Therefore a primary goal in the development of reactor fuel for the next generation of fast reactors is the creation of radiation-resistant fuel and cladding that can tolerate very high neutron fluence (4×10^{23} n·cm^{-2} or greater).

The development of new forms of nuclear fuel requires testing in prototypic neutron spectra. Within the United States of America, fast reactor fuel development is hampered by the lack of a domestic fast-spectrum irradiation facility. The Materials Test Station (MTS), currently under development and slated to be built in the Area A experimental hall at the Los Alamos Neutron Science Center (LANSCE) within Los Alamos National Laboratory (LANL), is well suited for providing the intense radiation environment needed to test the fast reactor fuels being developed for a closed nuclear fuel cycle. As a fast-neutron spectrum irradiation facility, the MTS will be unique in that the vast majority of neutrons by which the fuel is irradiated are not created through low-energy fission reactions, but rather by spallation. The spallation reactions are driven by a powerful 800-MeV, 1-MW beam of protons delivered by the LANSCE linear accelerator.

Design overview

The MTS project is currently in the pre-conceptual design phase. The design requirements include:

- a neutron energy spectrum in the fuel irradiation region similar to that of a fast reactor;

- a peak fast (E > 0.1 MeV) neutron flux in the fuel irradiation region of at least 1×10^{15} n·cm^{-2}·s^{-1};

- an irradiation volume sufficient to irradiate at least forty 0.5-cm high fuel pellets in a fast neutron flux of 1×10^{15} n·cm^{-2}·s^{-1} or greater;

- an overall facility availability that provides at least 1.5×10^{22} n·cm^{-2} fast neutron fluence on pellets in the peak flux position;

- an ability to maintain the fuel clad temperature at up to 550°C during irradiation.

The current MTS design meets or exceeds all of these design requirements.

The target assembly forms the heart of the MTS. As depicted in Figure 1, it consists of two spallation target sections separated by a fuel irradiation region. The active fuel height of the fuel rodlets in the fuel irradiation region is 12 cm. In order to provide near-uniform neutron flux over typical pellet dimensions of 5 mm diameter by 5 mm height, alternate macropulses of the LANSCE proton beam are directed onto the two spallation target sections. This scheme generates a time-averaged flux in the fuel irradiation region with a low gradient in all three transverse dimensions. The beam pulse structure for nominal MTS operation will consist of 750-μs long macropulses delivered at a

Figure 1. Basic configuration of the spallation target and fuel irradiation region

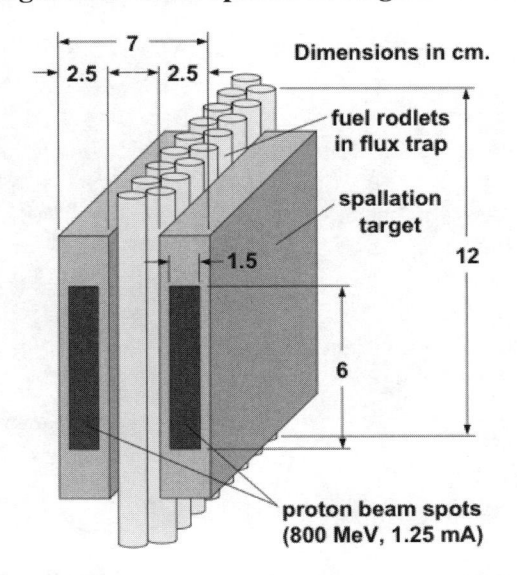

100-Hz repetition rate. A beam raster system composed of a series of ten magnets (up to 60 kHz raster frequency) delivers a beam spot on each target section of nearly uniform 70-μA proton current density with dimensions 15 mm wide by 60 mm high. The LANSCE accelerator is slated to undergo a major refurbishment within the next five years that will assure stable and reliable beam delivery to the MTS for a decade or more. Once completed, the MTS is expected to receive at least 4 400 hours of beam at a nominal current of 1.25 mA.

A horizontal cut through the target assembly is shown in Figure 2. The spallation target modules are comprised of a series of ever-thicker tungsten plates separated by 1-mm coolant channels. The coolant in both the spallation target modules and the fuel module is Pb-Bi eutectic, which has a melting temperature of 125°C. The primary motivation for using Pb-Bi to cool the fuel module is that its selection allows irradiation temperature of the fuel clad up to 550°C. At temperatures over ~400°C, active control of oxygen concentration in the Pb-Bi must be employed. Los Alamos National Laboratory has developed oxygen control expertise through the operation of the Delta Loop [2], and this experience will be applied to the design and operation of the Pb-Bi loops used in the MTS. Lead-bismuth was recently demonstrated in a spallation target application in the successful MEGAPIE experiment [1]. Two primary Pb-Bi loops are employed, one cooling the target modules and the other cooling the fuel module and backstop. The fuel module and spallation target modules share common sidewalls that allow active removal of decay heat from neighbouring modules should either of the primary Pb-Bi loops fail during operation.

Neutronic performance

The spatial distribution of the proton flux at the horizontal mid-plane of the target is depicted in Figure 3(a). The incident beams are evident at the left of the figure. As the protons interact with tungsten atoms, the beams attenuate and spread. The protons are reasonably well confined to the spallation target modules, although some protons do enter the irradiation regions. At the peak neutron flux position in the fuel module, the proton flux is about 1% of the neutron flux. The spatial distribution of the neutron flux is shown in Figure 3(b). The neutron flux peaks in the spallation targets where the neutrons are born. The peak flux in the fuel module is 1.6×10^{15} n·cm^{-2}·s^{-1} and is nearly constant over the dimensions of a typical fast reactor fuel pellet.

Figure 2. Horizontal cut through the target assembly

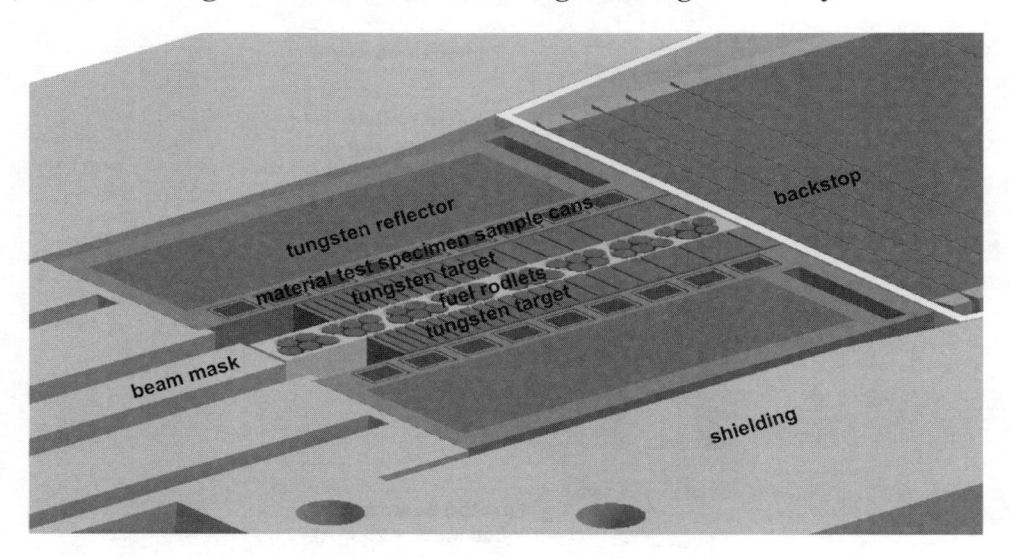

Figure 3. (a) Proton and (b) neutron flux spatial distributions

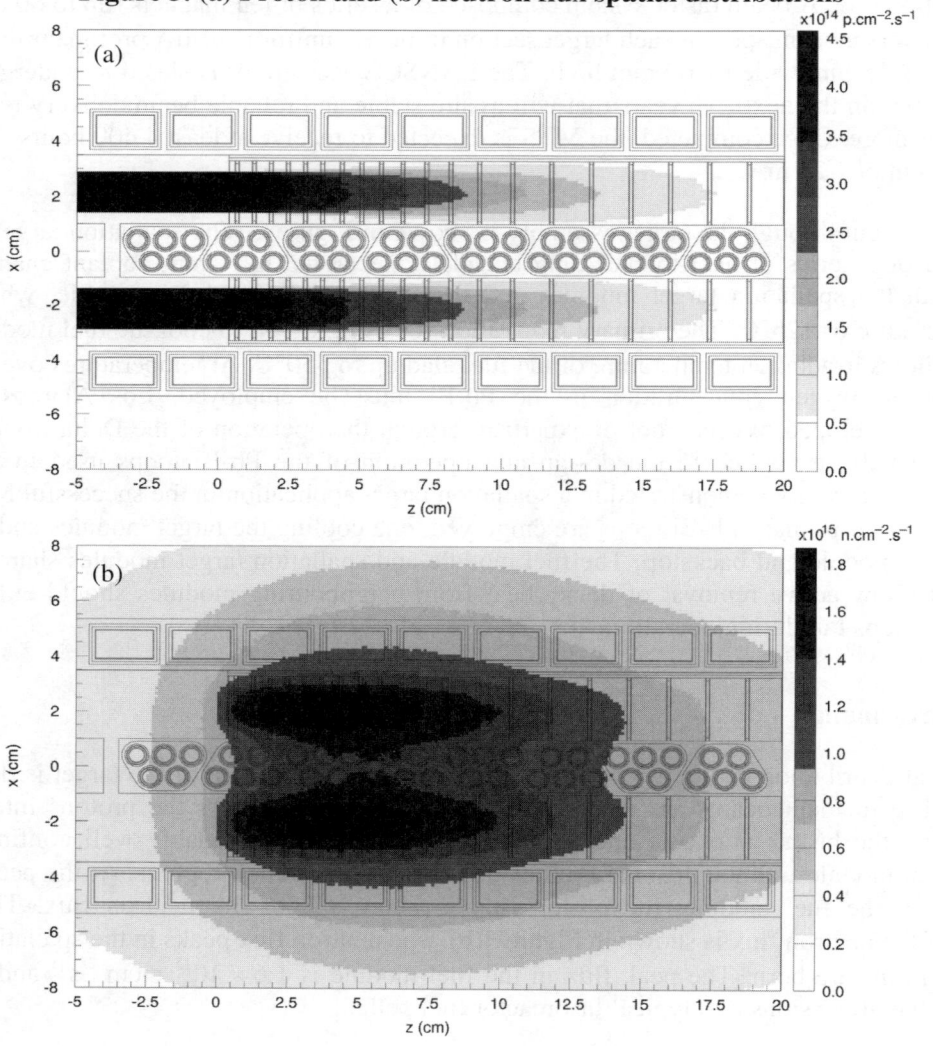

The energy spectrum at the peak flux position in the fuel module is shown in Figure 4. Also shown for comparison is the spectrum of a sodium-cooled fast reactor. The two spectra are quite similar below 1 MeV, but the MTS exhibits a high-energy tail above 10 MeV that is not prototypic of fast reactors. These neutrons arise from the intranuclear cascade phase of spallation reactions, and have energies up to the incident proton beam energy. While studies indicate that this high-energy tail should not impact fuel performance, higher helium production in steel cladding will occur. At the peak irradiation position, the helium production rate in iron is 60 times greater than that typically seen in a fast reactor at equal flux. This higher helium production rate may give rise to creep rupture or embrittlement at concentrations of 600 appm, which corresponds to about 12% burn-up for typical fuels with low fertile content.

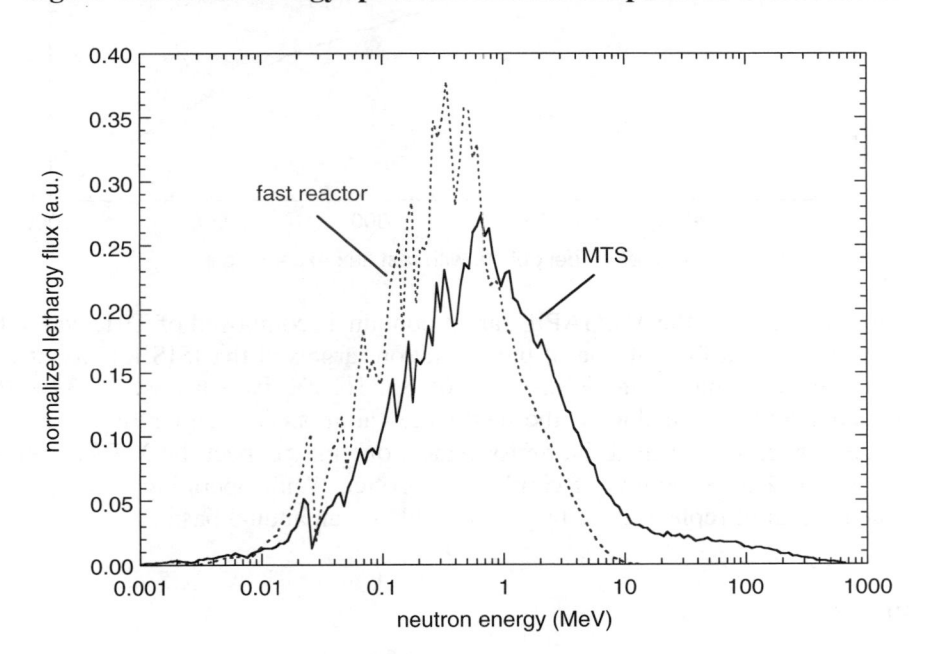

Figure 4. Neutron energy spectrum of MTS compared to a fast reactor

The MTS can accommodate up to forty 40-cm-long rodlets. This length allows an active fuel length of 12 cm and a 24-cm-high gas plenum region above the fuel. The total length of fuel that can be loaded into the MTS is 4.8 m. Figure 5 shows a histogram of the number of 0.5-cm-high fuel pellets as a function of the fast neutron flux. Over 90 cm of fuel experiences a fast flux exceeding 1×10^{15} n·cm^{-2}·s^{-1}. Heat loads for the fuel module (and backstop) and target modules are calculated to be 290 kW and 520 kW, respectively. As the LANSCE accelerator has historically experienced numerous beam trips per day, the heat removal system must be designed to accommodate these trips in a manner that does not subject the fuel rodlets and module components to unacceptable thermal cycling. To meet this design criterion, the current concept calls for shell-and-tube heat exchangers that can vary heat exchange surface area in a matter of seconds. Between the primary flowing Pb-Bi and the secondary water coolant is an intermediate volume of static Pb-Bi whose elevation in the heat exchanger can be controlled by changing the pressure of a cover gas in the static Pb-Bi reservoir. The performance of this innovative heat exchanger is sufficiently important that testing of a scaled version of the heat exchanger will be conducted on the Delta Loop [2] in the near future.

The life-limiting component of the integrated target assembly will be the target front face, which is subjected to a proton current density within the beam spots of 70 µA/cm^2. Assuming 4 400 hours of operation at full beam power per calendar year, the proton fluence will be 6.9×10^{21} p·cm^{-2} on the

Figure 5. Histogram showing number of fuel pellets as a function of fast neutron flux

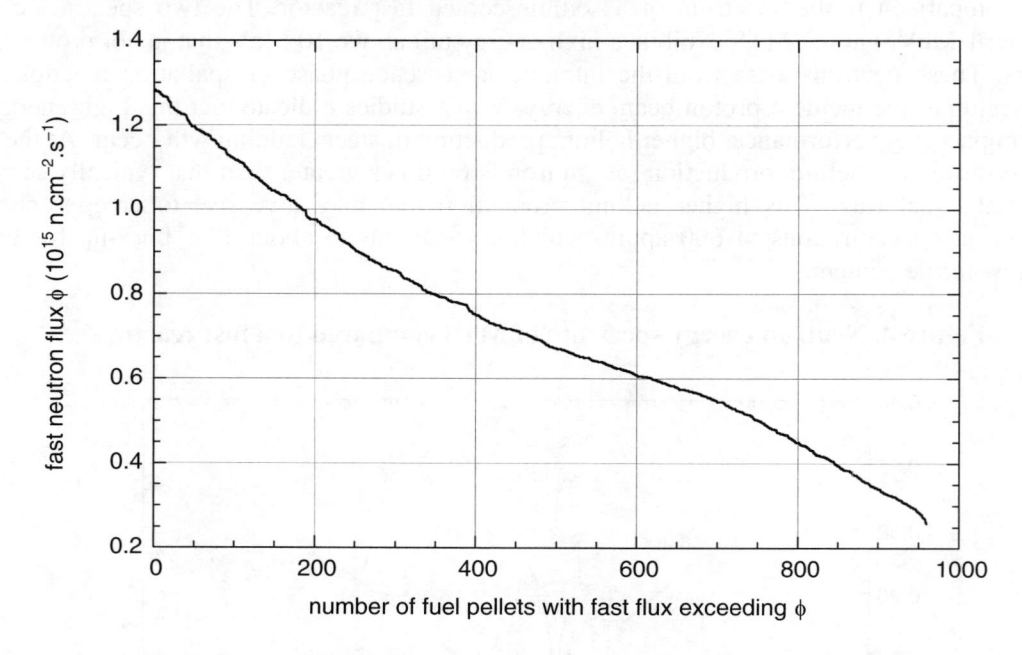

target front face. For comparison, the MEGAPIE target container, composed of T91, was subjected to 1.9×10^{21} p·cm^{-2}, while the front face of one of the spallation targets at the ISIS facility at Rutherford Appleton Laboratory in the United Kingdom, made of SS316, received a dose of 3.0×10^{21} p·cm^{-2} before being removed from service due to thermocouple failures. As neither of these targets was removed from service for reasons of material performance, one may expect the MTS target to survive to at least these doses, or 2 to 3 months of continuous service. With operating experience, it is not unreasonable to reach a goal of replacing the target assembly on an annual basis.

Safety and permitting

The design process places paramount importance on the safe operation of the MTS. The "safety by design" philosophy has the goal of precluding the possibility of accidents leading to a release of radioactivity to the environment, or to mitigate the consequences of an accident should one occur. Defence in depth is employed, as well as reliance on passive safety features where viable. The overarching safety goal is to preclude temperature excursions in the tungsten target material or fuel rodlets that might result in melting or vaporisation, as these components contain the majority of the radioactive inventory in the target assembly. Those components that are critical for safe operation are deemed credited controls, and meet stringent quality specifications in their manufacture and will undergo regular testing during operation. Fewer than 5% of neutrons generated in the MTS are born from fissions induced by neutrons, which means the target assembly is deeply subcritical. The fissile inventory contained in the 40 fuel rodlets will be less than 1 kg. As such, the MTS is expected to operate under the Department of Energy's Accelerator Safety Order 420.2B [3], and not as a nuclear facility.

Summary

In nearly every respect (e.g. fuel burn-up to clad dpa ratio, radial dependence of the fission rate within the fuel pellet, cladding temperature), the MTS irradiation environment is prototypic of fast reactor conditions. The significant differences between fuel irradiations in MTS and a fast reactor are the higher helium production rate in the fuel cladding and the frequent thermal cycling due to beam trips. Thermal cycling may be an issue for oxide fuel irradiations, but are not thought to be a concern with respect to cladding or metal fuel performance. Investigations of oxide fuel performance in pulsed reactor environments can provide some insights on this issue. The higher helium production in the cladding may be a concern beyond about 600 appm, although the high He concentrations on the clad inner surface resulting from alpha decay of the americium-bearing fuels that are expected to be developed under GNEP dwarfs the He production that comes about from (n,α) reactions in the steel alloys.

The MTS flux magnitude, irradiation volume, and facility availability are sufficient to support a vigorous fast reactor fuels and materials development programme in support of GNEP. The MTS can provide a near-term domestic fast-spectrum irradiation capability for rodlet-scale testing, which is required prior to lead test assembly irradiations in a fast reactor. Tests on fuels and materials performed in the MTS will provide essential and timely information that will meet the needs of the GNEP programme.

REFERENCES

[1] Wagner, W., *et al.*, "Operational Experience with MEGAPIE, the First Liquid Metal Target Driven by a Megawatt Class Proton Beam," to be published in the *Proceedings of the 8th International Topical Meeting on Nuclear Applications and Utilisation of Accelerators* (2007).

[2] Tcharnotskaia, V., *et al.*, "Liquid Lead-bismuth Materials Test Loop", *Proceedings of ADTTA/ACCAPP'01* (2001).

[3] See http://www.directives.doe.gov/pdfs/doe/doetext/neword/420/o4202b.pdf.

BENCHMARK EXPERIMENTS OF ACCELERATOR-DRIVEN SYSTEMS (ADS) IN KYOTO UNIVERSITY CRITICAL ASSEMBLY (KUCA)

Cheol Ho Pyeon, Tsuyoshi Misawa, Hironobu Unesaki, Kaichiro Mishima, Seiji Shiroya

Nuclear Engineering Science Division, Research Reactor Institute, Kyoto University
Asashiro-nishi, Kumatori-cho, Sennan-gun, Osaka 590-0494, Japan

Abstract

The ADS benchmark problems in the KUCA were based on both the 14-MeV neutrons generated from pulsed neutron generator and 150-MeV protons generated from the FFAG accelerator. Among the benchmark problems, the valuable and important information on the neutron spectrum measurement of 14-MeV neutrons using the foil activation method was described. The experiments and the numerical analyses by MCNP-4C2 with ENDF/B-VI.2 revealed that the 27Al foil was experimentally confirmed to be one of suitable activation foils for obtaining the neutron spectrum information in the ADSR. For the high-energy protons generated from FFAG accelerator, the foil activation method was found to be a useful measuring technique for examining the neutron spectrum of the ADSR through the reaction rates obtained by MCNPX with LA150 and the unfolding analyses by SANDII with ENDF/B-VI.2.

Introduction

The Research Reactor Institute of Kyoto University (KURRI) is going ahead with a research project [1] on an accelerator-driven subcritical reactor (ADSR) using the fixed-field alternating gradient (FFAG) accelerator [2]. The FFAG accelerator, which is a synchrotron-type accelerator, was developed in the High Energy Accelerator Research Organization (KEK) of Japan. The goal of the research project is to demonstrate the basic feasibility of an ADSR as a next-generation neutron source multiplication system using the Kyoto University Critical Assembly (KUCA) coupled with a newly developed variable energy FFAG accelerator. In ADSR experiments starting in August 2007, high-energy neutrons generated by nuclear reactions with a 150-MeV proton beam in a tungsten target will be injected into a solid-moderated and -reflected core (A-core) in the thermal neutron field of the KUCA.

Prior to operate the FFAG accelerator, it is inevitable to evaluate neutronic characteristics for the ADSR of the KUCA and to establish measurement techniques for several neutronic parameters in the ADSR. For these purposes, a series of preliminary experiments in the ADSR with 14-MeV pulsed neutron generator [3] by D-T reactions at a Cockcroft-Walton type accelerator had been carried out at the KUCA A-core. In the experiments, several neutronic parameters had been measured: neutron multiplication [4], neutron decay constant [4], reaction rate distribution [5], neutron spectrum [6] and subcriticality [7]. The numerical analyses for the experiments had been executed by using Monte Carlo calculation code MCNP-4C2 [8] coupling with nuclear data libraries ENDF/B-VI.2 and JENDL-3.3. Through the analyses, very important and valuable results have been obtained for neutronic characteristics in the ADSR. After completing the examinations of the FFAG accelerator, consequent ADSR experiments are planned to start by using 150-MeV protons from the FFAG accelerator including those topics of the preliminary experiments by using the 14-MeV pulsed neutrons: γ-ray spectrum detection at the target and in the core region, power monitoring of the core during beam current change and moving control rods. Moreover, these experiments could be possibly carried out in several neutron spectrum and γ-ray fields using cores consisting of several kinds of fuel and reflectors; highly-enriched uranium, thorium fuel and natural uranium; polyethylene and graphite.

The ADS benchmark problems in the KUCA are based on both the 14-MeV neutrons generated from pulsed neutron generator and 150-MeV protons generated from the FFAG accelerator. Among the benchmark problems, the valuable and important information on neutron spectrum measurement using the foil activation method in the preliminary experiments was described in this paper. Moreover, the results in experimental and numerical analyses in the FFAG accelerator were presented. The FFAG accelerator, the KUCA A-core configuration and the ADS benchmark problems are presented; the results of experiments and analyses by MCNP-4C2 code with ENDF/B-VI.2, MCNPX [9] code with LA150 [10] and SANDII [11] code with ENDF/B-VI.2, and the summary of the study.

ADS in KUCA

FFAG accelerator

The conceptual image of the ADSR in the KURRI is shown in Figure 1. In this ADSR system, all of ion beta, booster and main accelerator is composed of the FFAG accelerators, and maximum power of the A-core and maximum neutron yield could be 100 W and 1×10^{10} neutrons/s, respectively, when 150-MeV protons generated from the FFAG accelerator are injected onto the tungsten target. The main characteristics of the FFAG accelerator are indicated in Table 1. Maximum beam current of the FFAG accelerator at the target is 1 μA and average one is 1 nA.

Figure 1. Conceptual image of ADSR with FFAG accelerators in KURRI

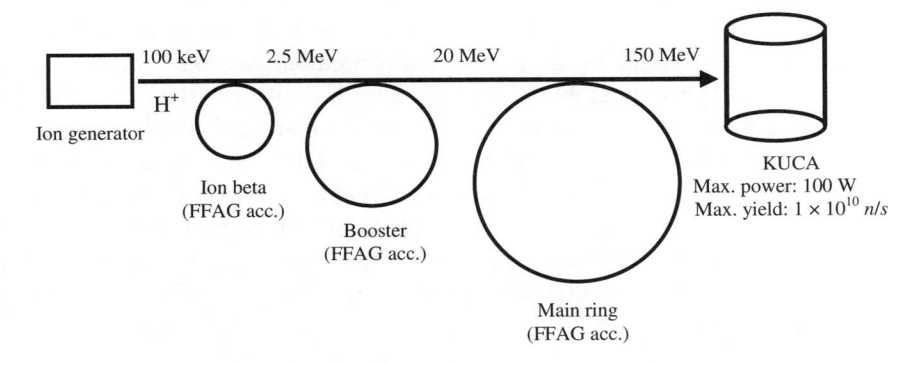

Table 1. Main characteristics of FFAG accelerator

Number of sectors	12
Proton energy	2.5-150 MeV
Repetition rate	120 Hz
Pulsed width	60 ns
Beam current	1 µA (max.), 1 nA (ave.)
Rf frequency	1.5-4.6MHz
Field index	7.5
Closed orbit radius	4.4-5.3 m

KUCA A-core configuration

The KUCA comprises solid-moderated and -reflected type-A and -B cores, and a water-moderated and -reflected type-C core. In the series of preliminary experiments, the solid-moderated and -reflected type-A core was combined with a Cockcroft-Walton type pulsed neutron generator installed at the KUCA. The materials used in critical assemblies were always in the form of rectangular parallelepipe, normally 2″ sq. with thickness ranging between 1/16″ and 2″. The upper and lower parts of the fuel region were polyethylene reflector layers of more than 500 mm long, as shown in Figure 2. The fuel rod, a highly-enriched uranium-aluminium (U-Al) alloy, consisted of 36 cells of polyethylene plates 1/8″ and 1/4″ thick, and a U-Al plate 1/16″ thick and 2″ sq. The functional height of the core was approximately 400 mm.

Figure 2. Full sideways view of configuration of fuel rod of A3/8"P36EU(3) in KUCA A-core

EU – enriched uranium, Al – aluminium, P – polyethylene

The present configuration of the ADSR at KUCA is a slightly different from several ADS, because a target is located outside the core. In the experiments, therefore, neutron guide composed of several shielding materials and beam duct (void) were installed in the polyethylene reflector region, as shown in Figure 3. The main purpose of installing the neutron guide was to direct the highest number possible of the high-energy neutrons generated at the target to the centre of the core. For shielding the high-energy and thermal neutrons, collimator comprises several materials inserted into the core, iron (Fe) for shielding the high-energy neutrons generated in the target region by inelastic scattering reactions; the polyethylene containing 10% boron [polyethylene + boron (10%)] for shielding the thermal neutrons, moderated by absorption reactions, in the reflector region; beam duct for directing collimated high-energy neutrons, by streaming effect, to the core region. Note that a new target of the ADSR with the FFAG accelerator will be also placed outside the A-core, and that the neuron guide be installed in target region to the centre of the core.

Figure 3. Top view of configuration of A-core with neutron guide conducted in neutron spectrum experiments

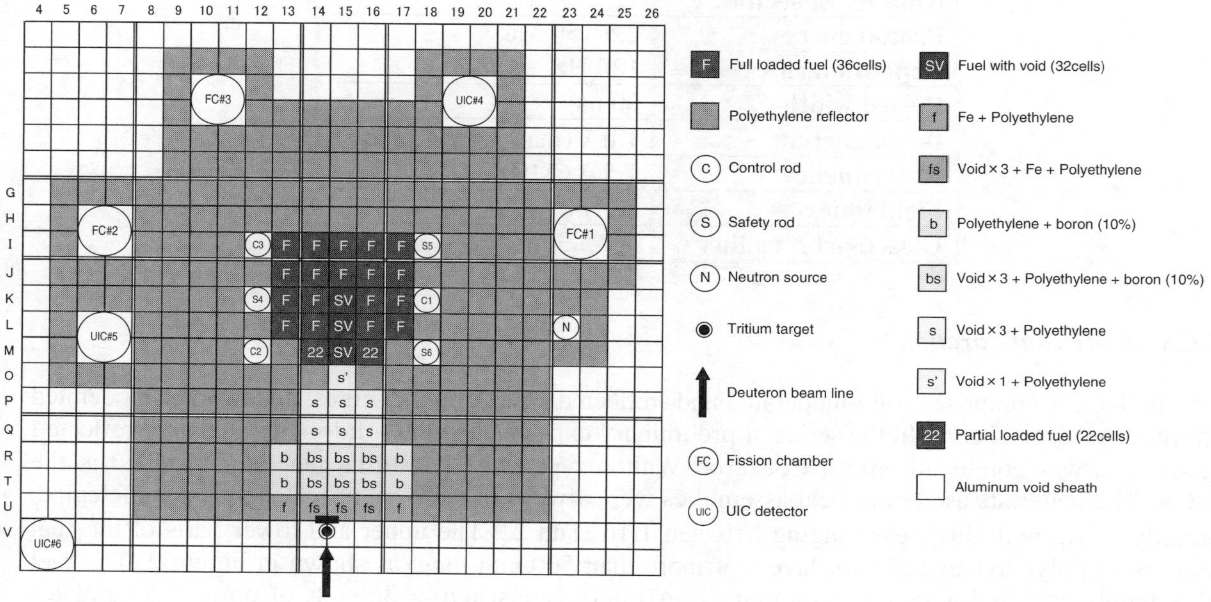

ADS benchmark problems

The KUCA is equipped with these following cores coupling several kinds of fuel and reflectors:

- polyethylene moderated and reflected core with highly enriched uranium fuel;

- graphite-moderated and -reflected core with thorium fuel;

- polyethylene + graphite-moderated and -reflected core with (thorium + natural uranium) fuel;

- neutronic decoupling core modelling large size core.

After introducing the FFAG accelerator, reactor physics experiments for the ADSR are planned at a next stage as follows:

- measurement of subcriticality by pulsed neutron method, neutron noise methods and source multiplication method;

- measurement of reaction rate distribution in the core by the foil activation method and optical fibre detection system;

- measurement of neutron spectrum by the foil activation method and organic scintillator;

- evaluation of neutron multiplication characteristics ($M = S/1 - k_{eff}$);

- γ-ray spectrum detection at the target and in the core region;

- power monitoring of the core in case of beam current change or moving control rods;

- optimisation of the neutron guide installed in the A-core.

Both new experiments and numerical simulations could be conducted for the high-energy neutrons obtained by 150-MeV protons generated from the FFAG accelerator, on the basis of the important and valuable information obtained in the preliminary experiments using 14-MeV pulsed neutrons. At the next stage, several benchmark experiments could be opened not only in Japan but also in other countries in the future. On the other hand, these benchmark problems are positioned as basic research for the ADS development and nuclear transmutation technology in co-operation with international collaboration research programme [12] published by the J-PARC project of JAEA in February 2006.

Neutron spectrum experiments

14-MeV pulsed neutrons

The neutron spectrum experiments were carried out at the A-core in subcritical system and the neutron spectrum was measured by the foil activation method. In these experiments, activation foils of five materials (Fe, Al, In, Nb and Au) were irradiated with the aggregate of the activation foils at once, and, among these, Au foil was utilised as normalisation factor including the reactor power and neutron source generation in the core. These activation foils were chosen for covering range of threshold energy up to 14 MeV. The size of ^{56}Fe and ^{27}Al is $45 \times 45 \times 5$ mm^3 for each of them, that of ^{115}In $45 \times 45 \times 3$ mm^3, and that of ^{197}Au $1\phi \times 0.05$ mm^3. The foils were set at void region in streaming void (SV) of (15, K) position shown in Figure 3. The subcriticality of the core was adjusted by inserting a variety of combination of three control rods (C1, C2 and C3) and three safety rods (S4, S5 and S6) shown in Figure 3. After the irradiation, γ-ray emission of each activation foil was measured by the high-purity germanium (HPGe) detector.

The comparison between reaction rates by experiments and MCNP calculations with ENDF/B-VI.2 for the subcriticality $\rho = 1.23\%\Delta k/k$ is shown in Table 2. It was observed that the result of the calculation in ^{27}Al agreed with that of the experiment within the relative difference of 5% in C/E value, while ^{56}Fe, ^{93}Nb and ^{115}In were observed to be a large discrepancy in the results of the C/E value. And, the same tendency was also obtained in other subcriticalities (0.87 and 1.75%$\Delta k/k$). From these results, it was concluded that the large discrepancy of three foils (^{56}Fe, ^{93}Nb and ^{115}In) was

Table 2. Comparison of measured and calculated results of reaction rates obtained at position of (15, K) shown in Figure 4 for subcriticality $\rho = 1.23\%\Delta k/k$

Reaction	Threshold (MeV)	Measured reaction rate (1/s/cm^3)	C/E value
^{115}In$(n,n')^{115m}$In	0.32	$(3.15\pm0.02) \times 10^3$	2.22 ± 0.05
^{56}Fe$(n,p)^{56}$Mn	2.97	$(4.09\pm0.06) \times 10^2$	0.17 ± 0.01
^{27}Al$(n,\alpha)^{24}$Na	3.25	$(5.20\pm0.02) \times 10$	1.05 ± 0.03
^{93}Nb$(n,2n)^{92}$Nb	9.08	$(4.78\pm0.28) \times 10^2$	0.10 ± 0.01

attributed to the influence of the self-shielding by each activation foil and the neutron flux depression by interaction of mutual activation foil. From Table 2, it was considered that ^{27}Al is one of suitable activation foils for obtaining the neutron spectrum information on high-energy range up to 14 MeV at the present state. On the other hand, the neutron spectrum information by unfolding analyses using SANDII code with ENDF/B-VI.2 on the basis of the results of Table 2 was shown in Figure 4. It was observed that configuration of neutron spectrum obtained by the unfolding analyses was approximately good shown in Figure 4, compared with that of initial guess by MCNP code with ENDF/B-VI.2.

Figure 4. Comparison of results in neutron spectrum of unfolding by SANDII and initial guess by MCNP at (15, M) shown in Figure 4 for $\rho = 1.23\%\Delta k/k$

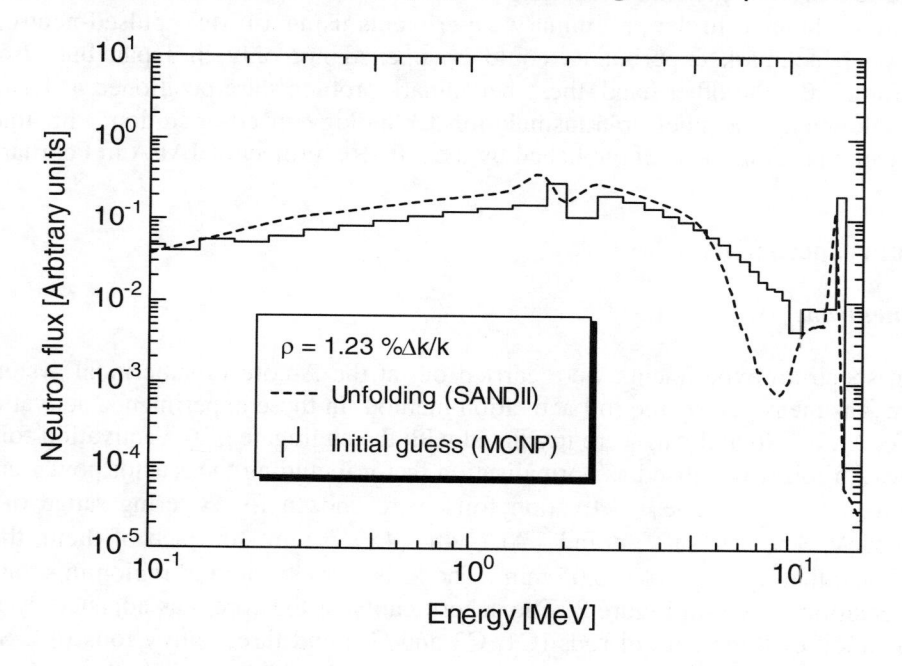

High-energy protons by FFAG accelerator

The FFAG accelerator in proof-of-principle (PoP) model had been constructed at the KEK on March 2005, and the neutron spectrum experiments were conducted in an irradiation hole shown in Figure 5(a) of the FFAG accelerator in PoP model, although high-energy proton beam was not concentrated adequately with a slender due to just commissioning. A specific flange shown in Figure 5(b) setting a tungsten target (^{184}W), ^{93}Nb and ^{209}Bi was placed in the irradiation hole. In these

Figure 5. Simple image of irradiation hole at the FFAG accelerator and flange setting activation foils

(a) Irradiation hole at the FFAG accelerator setting outside beam dump

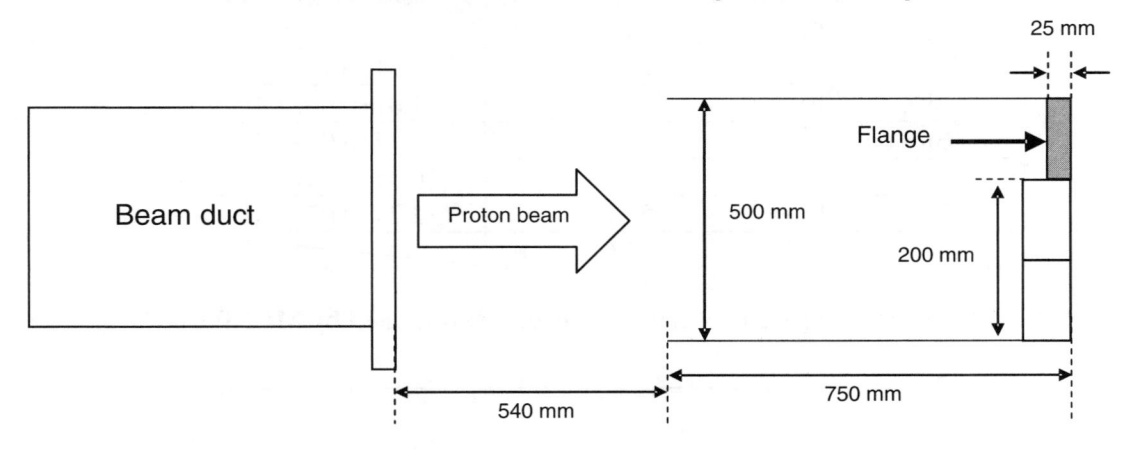

(b) Flange setting activation foils (W, Nb and Bi)

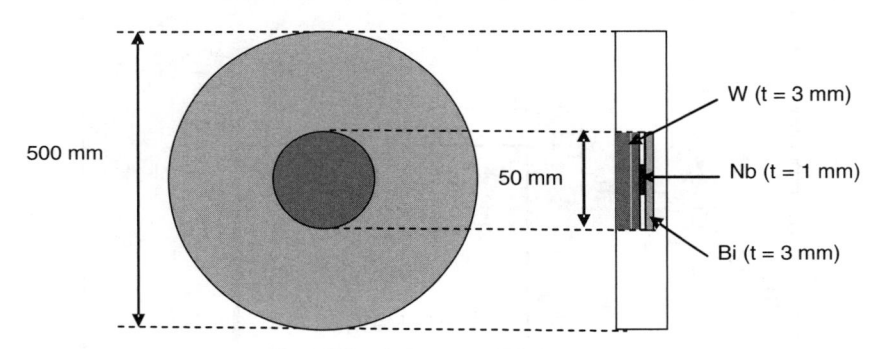

experiments, ^{209}Bi was utilised for measuring reactions rates, and ^{93}Nb as normalisation factor of neutron yield generated from the tungsten target. The size of ^{184}W and ^{209}Bi 50ϕ × 3 mm^3 for each of them, that of ^{93}Nb 20 × 20 × 1 mm^3. The beam injection onto the tungsten target was conducted with below 100 MeV of proton energy and 0.4 nA of beam intensity, and in about three hours of irradiation time. The reaction rates were obtained by nuclear reactions of ^{209}Bi$(n,xn)^{210-x}$Bi (x = 3 to 12) covering a wide range of threshold energy about 15 MeV to 90 MeV according to changing the value of x. In other words, the neutron spectrum information on this irradiation field can be easily acquired by irradiating the ^{209}Bi foil once, especially in high-energy region more than 15 MeV.

The results in measured reaction rates obtained by the proton beam injection are shown in Table 3. From the results in Table 3, it was considered that the high-energy neutrons from about 20 to 50 MeV were generated by the injection of the high-energy protons onto the tungsten target. As shown in Figure 6, the results of calculated neutron spectrum generated from the tungsten target were obtained by using MCNPX code with LA150. From these results, it was inferred that the high-energy protons in about 70 MeV were generated from the FFAG accelerator and injected onto the tungsten target. In these neutron spectrum experiments using the FFAG accelerator, it was concluded that neutrons up to 50 MeV can be obtained by about 70 MeV proton beam injection onto the tungsten target.

Table 3. Measured reaction rates obtained at irradiation hole of the FFAG accelerator

Reaction	Threshold (MeV)	Measured reaction rate (1/s/cm³)
^{209}Bi$(n,3n)^{207}$Bi	14.42	
^{209}Bi$(n,4n)^{206}$Bi	22.55	$(1.51\pm0.01) \times 10^5$
^{209}Bi$(n,5n)^{205}$Bi	29.62	$(1.01\pm0.03) \times 10^5$
^{209}Bi$(n,6n)^{204}$Bi	38.13	$(2.37\pm0.02) \times 10^4$
^{209}Bi$(n,7n)^{203}$Bi	45.37	$(6.35\pm0.16) \times 10^3$
^{209}Bi$(n,8n)^{202}$Bi	54.24	$(2.74\pm0.07) \times 10^2$
^{209}Bi$(n,9n)^{201}$Bi	61.69	

Figure 6. Calculated results of neutron spectrum obtained by MCNPX code

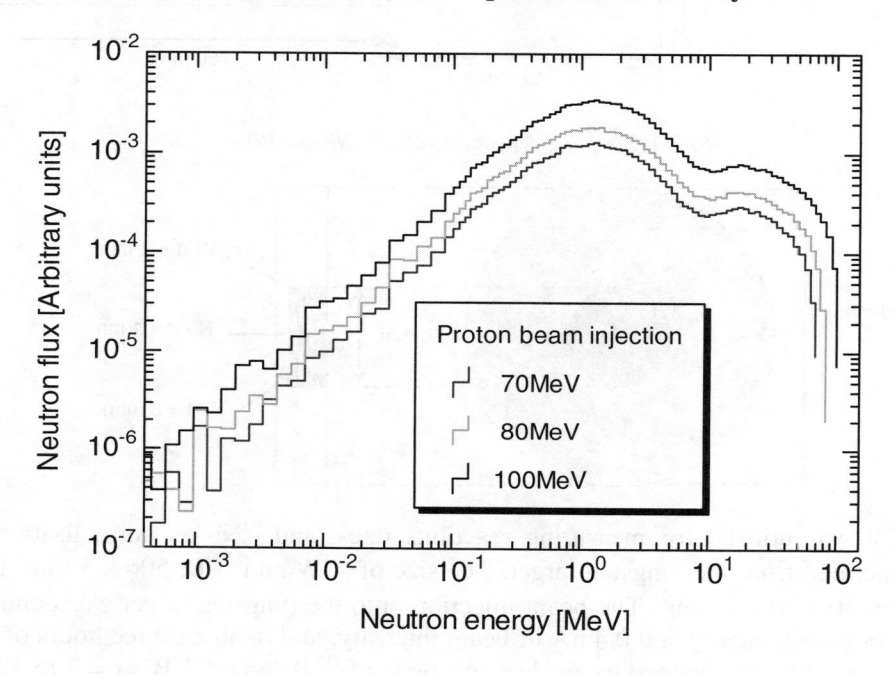

Summary

A series of preliminary experiments were carried out at KUCA A-core by using 14-MeV pulsed neutron generator to evaluate neutron spectrum. The experiments and analyses by MCNP-4C2 with ENDF/B-VI.2 revealed the following: the ^{27}Al foil was experimentally confirmed to be one of suitable activation foils for obtaining neutron spectrum information in the ADSR of the KUCA.

For the high-energy proton beam generated from FFAG accelerator, the foil activation method was found to be a useful measuring technique for examining the neutron spectrum through the reaction rates and the unfolding analyses.

In the future, it is expected that the basic research activity of ADS could be progressed through these ADS benchmark problems in the KUCA suggested in this paper, especially about neutronic characteristics in terms of reactor physics.

Acknowledgements

This work was supported financially by the Ministry of Education, Culture, Sports, Science and Technology of Japan within the task of "Research and Development for an Accelerator-driven Subcritical System Using an FFAG Accelerator."

The authors extend special thanks to all KUCA and KEK staff members for carrying out these experiments and to Mr. Hiroshi Shiga of Kyoto University for his contribution in executing the computational procedures.

REFERENCES

[1] Shiroya, S., H. Unesaki, Y. Kawase, H. Moriyama, M. Inoue, "Accelerator Driven Subcritical System as a Future Neutron Source in Kyoto University Research Reactor Institute (KURRI) – Basic Study on Neutron Multiplication in the Accelerator Driven Subcritical Reactor", *Prog. Nucl. Energy*, 37, 357 (2000).

[2] Mori, Y., "FFAG Proton Driver for Muon Source", *Nucl. Instrum. Methods A*, 451, 300 (2000).

[3] Ichihara, C., *et al.*, "Characteristics of KUCA Pulsed Neutron Generator", Kyoto University Research Reactor Institute Technical Report, KURRI-TR-240, (1983) [in Japanese].

[4] Shiroya, S., *et al.*, "Experimental Study on Accelerator Driven Subcritical Reactor Using the Kyoto University Critical Assembly (KUCA)", *Proc. Int. Conf. on New Frontiers of Nucl. Technol.: Reactor Physics, Safety and High-performance Computing (PHYSOR2002)*, Seoul, Korea, 7-10 October, 7C-01, American Nuclear Society (2002).

[5] Pyeon, C.H., Y. Hirano, T. Misawa, H. Unesaki, S. Shiroya, "Preliminary Study on ADSR by Using FFAG Accelerator in KUCA", *Proc. Topl. Mtg. (GLOBAL2003)*, New Orleans, Louisiana, 16-20 November, CD-ROM, pp. 2193-2200, American Nuclear Society (2003).

[6] Pyeon, C.H., T. Misawa, H. Unesaki, S. Shiroya, H. Tagei, K. Wada, T. Iwasaki, "Experimental Analyses for Accelerator Driven Subcritical Reactor in Kyoto University Critical Assembly by Using Foil Activation Method", *Proc. Int. Topl. Mtg. on Mathematics and Computation, Supercomputing, Reactor Physics and Nucl. Biological Applications (M&C2005)*, Avignon, France, 12-15 September, CD-ROM, American Nuclear Society (2005).

[7] Hervault, M., C.H. Pyeon, T. Misawa, H. Unesaki, S. Shiroya, "Monte Carlo Analysis of Subcriticality for Accelerator Driven Subcritical Reactor Mock Up in Kyoto University Critical Assembly", *Proc. Int. Topl. Mtg. on Mathematics and Computation, Supercomputing, Reactor Physics and Nucl. Biological Applications (M&C2005)*, Avignon, France, 12-15 September, CD-ROM, American Nuclear Society (2005).

[8] "MCNP – A General Monte Carlo N-particle Transport Code, Version 4C", J.F. Briesmeister (Ed.), LANL Report LA-13709-M, Los Alamos National Laboratory (2000).

[9] "MCNPX User's Manual Version 2.3.0", L.S. Waters (Ed.), LA-UR-02-2607, Los Alamos National Laboratory (2002).

[10] Chadwick, M.B., *et al.*, "LA150 Documentation of Cross Sections, Heating, and Damage", LA-UR-99-1222, Los Alamos National Laboratory (1999).

[11] McEloy, W.N., S. Berg, G. Gigas, "Neutron-flux Spectral Determination by Foil Activation", *Nucl. Sci. Eng.*, 27, 533 (1967).

[12] JAEA Homepage, http://j-parc.jp/documents/pdf/loi/Call_for_pre_LOI.pdf

WORKSHOP CONCLUSION

At the end of the conference, session co-chairs were invited to summarise the technical sessions and to form a panel for an open discussion with the HPPA5 attendees.

During this workshop it became clear that the accelerator community is designing, constructing, testing and operating more and higher-powered proton accelerators with high reliability properties. The Spallation Neutron Source (SNS, USA) shows us the way forward in order to improve beam reliability for an operating accelerator by adopting the fault-tolerant design approach proposed by the accelerator community working for ADS application. Major components are tested on a one-to-one scale.

Furthermore, the spallation target, either a stand-alone neutron source or the neutron source as interface between the proton beam and the subcritical core in the case of an ADS, received quite a bit of attention. MEGAPIE demonstrated that a lead-bismuth (Pb-Bi) technology-based spallation source can be built and operated under reliable conditions. A great deal of progress was also made for the windowless spallation source design. Especially, the WEbExpIr experiment performed by SCK•CEN, IBA (Belgium) and CNRS/SUBATECH (France) got appraisal from the community, as this was a cornerstone experiment for the coupling of a high-power electron beam simulating the high-power proton beam interaction with a lead-bismuth eutectic windowless concept at the free surface level.

In the session on subcritical system design both ADS systems, XT-ADS and EFIT, developed within the European Commission's FP6 EUROTRANS project were presented. Neutron code and neutron/thermal-hydraulics coupled code development for ADS systems also received attention.

Several existing and upcoming ADS experiments were discussed in the last session. Coupling experiments with a subcritical core are started and promising results are obtained: MUSE (France), TRADE (Italy), RACE (USA); and further coupling experiments are planned: YALINA (Belarus), KIPT (Ukraine), KART (Japan), VENUS-1 (China). Recently, within the context of FP6 EUROTRANS, it was decided to launch the GUINEVERE project at SCK•CEN, which studies the transformation of the VENUS zero power light water reactor into a solid lead fast spectrum subcritical core coupled with a GENEPI accelerator.

Moreover, it is clear that progress has been made within the ADS community, which may pave the way for more ambitious projects such TEF-P (Japan), TEF-T (Japan) and MYRRHA/XT-ADS (Belgium/EU).

The comment participants made the most was that these workshops contributed significantly to improving communication between the accelerator community and the nuclear reactor community. This, of course, is indispensable to the efficient development of accelerator driven systems.

In response to the worry expressed by the NEA Nuclear Science Committee and by the nuclear community in general about transferring know-how to the young generation and building up new competencies in the nuclear engineering and nuclear research fields, IAP Frankfurt and SCK•CEN took advantage of the experts and experienced professionals in the field of HPPA present at this workshop, to organise an International Training Course on Particle Accelerator Technology in conjunction with HPPA5. This training course is part of the FP6 EUROTRANS project. The students were allowed to follow the complete HPPA5 workshop and they all greatly appreciated this initiative.

LIST OF PARTICIPANTS

BELGIUM

AIT ABDERRAHIM, Hamid
Director, Institute Advanced
Nuclear Systems (ANS)
SCK•CEN
Boeretang 200
B-2400 Mol

Tel: +32(0)14 33 22 77
Fax: +32(0)14 32 15 29
Eml: haitabde@sckcen.be

AOUST, Thierry
ANS/Reactor Modelling & Safety
SCK•CEN
Boeretang 200
B-2400 Mol

Tel: +32(0)14 33 21 86
Fax: +32(0)14 32 15 29
Eml: taoust@sckcen.be

BAETEN, Peter
Deputy Director ANS
ANS/Research Reactors Operation
SCK•CEN
Boeretang 200
B-2400 Mol

Tel: +32(0)14 33 21 93
Fax: +32(0)14 32 15 29
Eml: pbaeten@sckcen.be

CUGNON, Joseph
Physics Department B5
University of Liège
B-4000 Sart Tilman Liège 1

Tel: +32(0)4 366 36 01
Fax: +32(0)4 366 36 72
Eml: cugnon@plasma.theo.phys.ulg.ac.be

D'HONDT, Pierre
Deputy General Manager
SCK•CEN
Boeretang 200
B-2400 Mol

Tel: +32(0)14 33 25 95
Fax: +32(0)14 31 89 36
Eml: pierre.dhondt@sckcen.be

DIERCKX Marc
ANS/Reactor Technology Research
SCK•CEN
Boeretang 200
B-2400 Mol

Tel: +32(0)14 33 22 24
Fax: +32(0)14 32 15 29
Eml: mdierckx@sckcen.be

GYSEMBERGH, Robert
Université de Liège
Allée du 6 Aout, 3
B-4000 Sart Tilman B12B

Tel: +32(0)4 366 22 07
Fax: +32(0)4 366 28 76
Eml: r.gysembergh@belgacom.net

HEUSDAINS, Simone
ANS/Reactor Modelling & Safety
SCK•CEN
Boeretang 200
B-2400 Mol

Tel: +32(0)14 33 21 92
Fax: +32(0)14 32 15 29
Eml: sheusdai@sckcen.be

HEYSE, Jan
ANS/Reactor Technology Research
SCK•CEN
Boeretang 200
B-2400 Mol

Tel: +32(0)14 33 21 60
Fax: +32(0)14 32 15 29
Eml: jan.heyse@sckcen.be

JONGEN, Yves
Chief Research Officer
IBA s.a.
Chemin du Cyclotron, 3
B-1348 Louvain-la-Neuve

Tel: +32(0)10 47 58 95 (or 854)
Fax: +32(0)10 47 58 10
Eml: yves.jongen@iba-group.com

ROSSEEL, Kris
ANS/Reactor Technology Research
SCK•CEN
Boeretang 200
B-2400 Mol

Tel: +32(0)14 33 21 97
Fax: +32(0)14 32 15 29
Eml: kris.rosseel@sckcen.be

SCHUURMANS, Paul
ANS/Reactor Technology Research
SCK•CEN
Boeretang 200
B-2400 Mol

Tel: +32(0)14 33 22 93
Fax: +32(0)14 32 15 29
Eml: Paul.Schuurmans@sckcen.be

SOBOLEV, Vitaly
NMS/Fuel Materials
SCK•CEN
Boeretang 200
B-2400 Mol

Tel: +32(0)14 33 22 67
Fax: +32(0)14 32 15 29
Eml: vsobolev@sckcen.be

VAN DEN EYNDE, Gert
ANS/Reactor Modeling & Safety
SCK•CEN
Boeretang 200
B-2400 Mol

Tel: +32(0)14 33 22 30
Fax: +32(0)14 32 15 29
Eml: gvdeynde@sckcen.be

VANDEPLASSCHE, Dirk
IBA, Chemin du Cyclotron 3
B-1348 Louvain-la-Neuve

Tel: +32(10)47 58 49
Fax: +32(10)47 58 47
Eml: dirk.vandeplassche@iba-group.com

VAN TICHELEN, Katrien
ANS/Reactor Technology Research
SCK•CEN
Boeretang 200
B-2400 Mol

Tel: +32(0)14 33 21 96
Fax: +32(0)14 32 15 29
Eml: kvtichel@sckcen.be

FRANCE

BIARROTTE, Jean-Luc
IPN Orsay
15 rue Georges Clémenceau
F-91406 Orsay Cedex

Tel: +33 (0)1 69 15 79 30
Fax:
Eml: biarrott@ipno.in2p3.fr

BILLEBAUD, Annick
Réacteurs Hybrides
Institut des Sciences Nucléaires
53 avenue des Martyrs
F-38026 Grenoble Cedex

Tel: +33 (0)4 76 28 40 57
Fax: +33 (0)4 76 28 40 04
Eml: billebaud@lpsc.in2p3.fr

DE CONTO, Jean-Marie
Institut de Sciences Nucléaires
CNRS/IN2P3
53 avenue des Martyrs
F-38026 Grenoble Cedex

Tel: +33 (0)4 76 28 40 98
Fax: +33 (0)4 76 28 41 43
Eml: deconto@lpsc.in2p3.fr

LUKOVAC, Lucija
PhD Student
Institut de Physique Nucléaire
15 rue Georges Clémenceau
F-91406 Orsay Cedex

Tel:
Fax:
Eml: lukovac@ipno.in2p3.fr

MUELLER, Alex
Department Head
R&D Accelerators/Exotic Beams
Institut de Physique Nucléaire
15 rue Georges Clémenceau, Bâtiment 106
F-91406 Orsay Cedex

Tel: +33 (0)1 69 15 62 40
Fax: +33 (0)1 69 15 77 35
Eml: mueller@ipno.in2p3.fr

PONTON, Aurélien
PhD Student
IPNO
15 rue Georges Clémenceau, Bâtiment 106
F-91406 Orsay Cedex

Tel:
Fax:
Eml: ponton@ipno.in2p3.fr

RICHARD, Pierre
CE Cadarache, Bâtiment 212
DER/SESI/LCSI
F-13108 Saint-Paul-lez-Durance

Tel: +33 (0)4 42 25 31 54
Fax: +33 (0)4 42 25 71 87
Eml: pierre.richard@cea.fr

RIDIKAS, Danas
CEA Saclay
DSM/DAPNIA/SPhN
F-91191 Gif-sur-Yvette Cedex

Tel: +33 (0)1 69 08 78 47
Fax: +33 (0)1 69 08 75 84
Eml: danas.ridikas@cea.fr

GERMANY

BUSCH, Marco
Institut für Angewandte Physik
Max-von-Laue-Strasse 1
D-60438 Frankfurt

Tel:
Fax:
Eml: Busch@iap.uni-frankfurt.de

DELAPERRIERE, Marc
Serviceability Engineer
Siemens
Hoffmanstrasse 26
D-91052 Erlangen

Tel: +49 91 31 84 34 08
Fax:
Eml: Marc.delaperriere@siemens.com

FISCHER, Philipp
Institut für Angewandte Physik
Max-von-Laue-Strasse 1
D-60438 Frankfurt am Main

Tel:
Fax:
Eml: P.Fischer@iap.uni-frankfurt.de

GRAMS, Harald
Siemens AG
P.O. Box 32 40
D-91050 Erlangen

Tel: +49 91 31 84 65 04
Fax:
Eml: harald.grams@siemens.com

KLEIN, Horst
Director of Institute
Institut für Angewandte Physik
Max-von-Laue-Strasse 1
D-60438 Frankfurt am Main

Tel: +49 69 79 84 74 16
Fax:
Eml: horst.klein@iap.uni-frankfurt.de

LIEBERMANN, Holger
Institut für Angewandte Physik
Max-von-Laue-Strasse 1
E-60438 Frankfurt am Main

Tel: +49 69 79 84 74 52
Fax:
Eml: Liebermann@iap.uni-frankfurt.de

PODLECH, Holger
Group Leader, Linac Development at IAP
IAP – University of Frankfurt
Max-von-Laue-Strasse 1
E-60438 Frankfurt am Main

Tel: +49 69 79 84 74 53
Fax: +49 69 79 84 74 07
Eml: h.podlech@iap.uni-frankfurt.de

STIEGLITZ, Robert
Institute for Nuclear and Energy
Technologies (IKET)
Herrmann von Helmholtz Platz 1
P.B. 3640
D-76344 Eggenstein-Leopoldshafen

Tel: +49 72 47 82 34 62
Fax: +49 72 47 82 63 21
Eml: robert.Stieglitz@iket.fzk.de

WIESNER, Christoph
Institut für Angewandte Physik
Max-von-Laue-Strasse 1
D-60438 Frankfurt am Main

Tel:
Fax:
Eml: wiesner@iap.uni-frankfurt.de

ZHANG, Chuan
Institut für Angewandte Physik
Max-von-Laue-Strasse 1
D-60438 Frankfurt am Main

Tel:
Fax:
Eml: zhang@iap.uni-frankfurt.de

ITALY

ARTIOLI, Carlo
ENEA, Energy Department
Nuclear Fission Branch
Via Martiri di Monte Sole 4
I-40129 Bologna

Tel: +39 05 16 09 84 36
Fax: +39 05 16 09 86 74
Eml: artioli@bologna.enea.it

BARBANOTTI, Serena
INFN Sezione di Milano LASA
Via Fratelli Cervi, 201
I-20090 Segrate MI

Tel: +39 02 50 31 95 65
Fax: +39 02 50 31 95 43
Eml: serena.barbanotti@mi.infn.it

CALABRETTA, Luciano
Senior Scientist
INFN-LNS
Via Enrico Pantano, 70
I-95129 Catania

Tel: +39 95 54 22 59
Fax: +39 95 54 23 00
Eml: calabretta@lns.infn.it

CARTA, Mario
ENEA C.R.E.
Casaccia via Anguillarese, 301
S. Maria de Galeria (Rome) 60

Tel: +39 06 30 48 31 83
Fax: +39 06 30 48 63 08
Eml: carta@casaccia.enea.it

MONTI, Stefano
ENEA
Responsible for Nuclear Fission Programme
Via Martiri di Monte Sole, 4
I-40129 Bologna

Tel: +39 051 609 84 62
Fax: +39 051 609 87 85
Eml: stefano.monti@bologna.enea.it

NAPOLITANO, Marco
Dipartimento de Scienze Fisiche
Univ. "Federico II" & INFN
Complesso Univ. Monte Sant'Angelo
Via Cintia
I-80126 Napoli

Tel: +39 081 67 61 29
Fax: +39 081 67 63 46
Eml: marco.napolitano@na.infn.it

PANZERI, Nicola
INFN Milano LASA
Via Fratelli Cervi, 201
I-20090 Segrate MI

Tel: +39 02 50 31 95 44
Fax: +39 02 50 31 95 43
Eml: nicola.panzeri@mi.infn.it

PIERINI, Paolo
INFN-Milano-LASA
Via Fratelli Cervi, 201
I-20090 Segrate MI

Tel: +39 02 239 25 60
Fax: +39 02 239 25 43
Eml: paolo.pierini@mi.infn.it

SAROTTO, Massimo
ENEA
Via Martiri di Monte Sole 4
I-40129 Bologna

Tel: +39 051 609 84 83
Fax: +39 051 609 87 85
Eml: massimo.sarotto@bologna.enea.it

VACCARO, Vittorio Giorgio
National Institute of Nuclear Physics
Complesso Universitario MSA
Via Cinthia ed. 6
I-80126 Napoli

Tel: +39 081 67 61 39
Fax: +39 081 67 62 55
Eml: vaccaro@na.infn.it

JAPAN

ABE, Kazuaki
Neutron Device Laboratory
Tohoku University
980-8579 Aoba 6-6-01-2
Aramaki, Aoba-ku
Sendai, 980-8579

Tel: +81 22 795 79 09
Fax: +81 22 795 79 09
Eml: kazuaki@neutron.qse.tohoku.ac.jp

CHISHIRO, Etsuji
J-PARC Center, JAEA
Tokai-mura, Naka-gun
Ibaraki-ken 319-1195

Tel: +81 29 282 60 07
Fax: +81 29 284 37 19
Eml: estuji.chishiro@j-parc.jp

OIGAWA, Hiroyuki
Group Leader, Nuclear Transmutation
Technology Group
Nuclear Science and Engineering Directorate
Japan Atomic Energy Agency
2-4, Shirane, Shirakata, Tokai-mura,
Ibaraki-ken, 319-1195

Tel: +81 29 282 69 35
Fax: +81 29 282 56 71
Eml: oigawa.hiroyuki@jaea.go.jp

PYEON, Cheol Ho
Kyoto University
Research Reactor Institute (KURRI)
Asashiro-nishi, Kumatori-cho, Sennan-gun
Osaka 590-0494

Tel: +81 724 51 23 56
Fax: +81 724 51 26 03
Eml: pyeon@kuca.rri.kyoto-u.ac.jp

SUZUKI, Motomu
Tohoku University
Aoba 6-6-01-2, Aramaki,Aoba-ku
Sendai, 980-8579

Tel: +81 22 795 79 09
Fax: +81 22 795 79 09
Eml: szk@neutron.qse.tohoku.ac.jp

TAKAHASHI, Minoru
Research Laboratory for Nuclear Reactors
Tokyo Institute of Technology
N1-18, 2-12-1, O-Okayama, Meguro-ku
Tokyo 152-8

Tel: +81 3 5734 29 57
Fax: +81 3 5734 29 59
Eml: mtakahas@nr.titech.ac.jp

TAKEI, Hayanori
Japan Atomic Energy Agency, Japan
2-4, Shirane, Shirakata, Tokai-mura
Ibaraki-ken, 319-1195

Tel: +81 29 282 69 48
Fax: +81 29 282 56 71
Eml: takei.hayanori@jaea.go.jp

TANIGAKI, Minoru
Research Reactor Institute
Kyoto University
2-1010 Asashironishi, Kumatoricho, Sennan
Osaka

Tel: +81 724 51 24 76
Fax: +81 724 51 26 20
Eml: tanigaki@rri.kyoto-u.ac.jp

PORTUGAL

PIRES, Rui
FEUCP
Estrada de Talaíde
P-2635-631 Rio de Mouro

Tel: +35 12 14 26 97 79
Fax:
Eml: pires@fe.ucp.pt

P.R. OF CHINA

SHI, Yongqian
China Institute of Atomic Energy
P.O.Box 275-45
102413, Beijing

Tel: +86 10 69 35 80 28
Fax: +86 10 69 35 80 28
Eml: shiyq@ciae.ac.cn

XIA, Pu
China Institute of Atomic Energy
P.O. Box 275-75
102413, Beijing

Tel: +86 10 69 35 78 92
Fax: +86 10 69 35 70 08
Eml: xiapu@ciae.ac.cn

REPUBLIC OF KOREA

CHO, Yong-Sub
Accelerator Development & Operation Div.
Proton Engineering Frontier Project
KAERI
P.O. Box 105, Yuseong
305-600 Daejeon

Tel: +82 42 868 29 75
Fax:
Eml: choys@kaeri.re.kr

CHOI, Byung-Ho
Nuclear Physico-Engineering
Team Director
KAERI
P.O. Box 105, Yusong
305-600 Daejeon

Tel: +82 42 868 89 04
Fax: +82 42 868 88 81
Eml: bhchoi@kaeri.re.kr

SWITZERLAND

GROESCHEL, Friederich
Head, Target Development Section
ODGA/C101
Paul Scherrer Institute
CH-5232 Villigen PSI
Tel: +41 56 310 21 96
Fax: +41 56 310 31 31
Eml: friedrich.groeschel@psi.ch

HERRERA-MARTINEZ, Adonai
CERN
AB Department, CERN
CH-1211 Geneva
Tel: +41 22 767 81 65
Fax: +41 22 766 95 28
Eml: adonai.herrera.martinez@cern.ch

NEUHAUSEN, Jörg
Institut für Radio- und Umweltchemie
OFLB/110
Paul Scherrer Institut
CH-5232 Villigen PSI
Tel: +41(0)56 310 24 07
Fax: +41(0)56 310 44 35
Eml: joerg.neuhausen@psi.ch

PODOFILLINI, Luca
Researcher
Paul Scherrer Institute
CH-5232 Villigen PSI
Tel: +41 56 310 53 56
Fax:
Eml: luca.podofillini@psi.ch

SCHMELZBACH, Pierre A.
Head of the Accelerator Division
Paul Scherrer Institute, WBGA/C22
CH-5232 Villigen PSI
Tel: +41 56 310 40 73
Fax: +41 56 310 33 83
Eml: Pierre.Schmelzbach@psi.ch

SCHUMANN, Dorothea
Paul Scherrer Institute
CH-5232 Villigen PSI
Tel: +41 56 310 40 04
Fax:
Eml: dorothea.schumann@psi.ch

SEIDEL, Mike
Department Head
Accelerator/Operation and Development
Paul Scherrer Institute
Building WBGA/C19
CH-5232 Villigen PSI
Tel: +41 56 310 33 78
Fax: +41 56 310 33 83
Eml: mike.seidel@psi.ch

THOMSEN, Knud
Project Director
Paul Scherrer Institute
CH-5232 Villigen PSI
Tel: +41 56 310 42 10
Fax: +41 56 310 31 31
Eml: knud.thomsen@psi.ch

UNITED STATES OF AMERICA

GALAMBOS, John
Accelerator Physics Group Leader, SNS
Oak Ridge National Laboratory
MS 6462, Oak Ridge, TN, 37830-6462
Tel: +1 (865) 576-5482
Fax:
Eml: jdg@ornl.gov

GOHAR, Yousry
Argonne National Laboratory
9700 South Cass Avenue
Argonne, Illinois 60439

Tel: +1 (630) 252-4816
Fax: +1 (630) 252-4007
Eml: gohar@anl.gov

PITCHER, Eric
Los Alamos National Laboratory
LANSCE-12, MS H805
Los Alamos, NM 87545

Tel: +1 (505) 665 0651
Fax: +1 (505) 665 2676
Eml: pitcher@lanl.gov

SHEFFIELD, Richard
Los Alamos National Laboratory
LANSCE-DO, MS H851
Los Alamos, New Mexico 87545

Tel: +1 (505) 667 1237
Fax: +1 (505) 667 7443
Eml: sheff@lanl.gov

INTERNATIONAL ORGANISATION

CHOI, Yong-Joon
OECD Nuclear Energy Agency
Le Seine St-Germain
12 boulevard des Iles
F-92130 Issy-les-Moulineaux

Tel: +33 (0)1 45 24 10 91
Fax: +33 (0)1 45 24 11 28
Eml: yongjoon.choi@oecd.org

OECD PUBLICATIONS, 2, rue André-Pascal, 75775 PARIS CEDEX 16
PRINTED IN FRANCE
(66 2008 01 1 P) ISBN 978-92-64-04478-4 – No. 56125 2008